Experimental Techniques in Physics and Materials Science

Principles and Methodologies

Experimental Techniques in Physics and Materials Science

Principles and Methodologies

R Srinivasan, retired
Indian Institute of Technology Madras, India

T G Ramesh, retired
CSIR – National Aerospace Laboratories, India

G Umesh, retired
National Institute of Technology Karnataka, India

C S Sundar, retired
Indira Gandhi Centre for Atomic Research, India

World Scientific

NEW JERSEY • LONDON • SINGAPORE • BEIJING • SHANGHAI • HONG KONG • TAIPEI • CHENNAI • TOKYO

Published by

World Scientific Publishing Co. Pte. Ltd.
5 Toh Tuck Link, Singapore 596224
USA office: 27 Warren Street, Suite 401-402, Hackensack, NJ 07601
UK office: 57 Shelton Street, Covent Garden, London WC2H 9HE

British Library Cataloguing-in-Publication Data
A catalogue record for this book is available from the British Library.

EXPERIMENTAL TECHNIQUES IN PHYSICS AND MATERIALS SCIENCE
Principles and Methodologies

ISBN 978-981-127-888-4 (hardcover)
ISBN 978-981-127-889-1 (ebook for institutions)
ISBN 978-981-127-890-7 (ebook for individuals)

For any available supplementary material, please visit
https://www.worldscientific.com/worldscibooks/10.1142/13485#t=suppl

Desk Editor: Rhaimie Wahap

Typeset by Stallion Press
Email: enquiries@stallionpress.com

PREFACE

In the past century, there has been tremendous progress in the techniques for the preparation of materials, their characterization, and the methods of measurement of various physical properties of these materials. At the same time, advances in vacuum techniques, low-temperature techniques, and high-pressure techniques have led to extensive studies on the changes in the properties of materials when subjected to such environmental conditions. Research in materials science has produced novel materials which have changed the lifestyle of humans drastically. It is no surprise that master's degree courses in materials science have been started in many universities globally.

The curriculum for the MSc Physics course in all universities in India is dominated by theoretical courses, such as mathematical techniques in physics, classical and quantum mechanics, statistical physics, and condensed matter physics. Although it is an accepted fact that experiments form the bedrock of science, it is ironic that new advances in experimental physics do not find a place in the curriculum. True, the students do some experiments in the laboratory, but the instructions given to them to do the experiments are rather cursory. The principle behind an experiment is not explained in sufficient detail. It is necessary to explain why a particular method of measurement is adapted, why the analysis of the data should be carried out in the specified way, and why a certain formula should be used in calculating the results. Most often, the student does the experiment mechanically and records the data following the instructions without any understanding of the aim of the experiment, the theory behind the experiment, or the limitations of the technique he or she is using.

In the past 20 years, two of the authors of this book, along with their colleagues from Goa University, have developed low-cost experiments at the BSc, MSc, and post-MSc levels in physics and materials science. All four authors, along with other resource persons, have conducted more than a hundred two-week refresher courses to train more than 2,500 teachers and student participants from across India in conducting these experiments. These courses have been sponsored by the Indian Academy of Sciences, the Indian National Science Academy, and the National Academy of Sciences, Allahabad. A detailed laboratory manual, covering 62 such experiments, has been published as an e-book by the Indian Academy of Sciences, Bengaluru. These experiments can be performed using the low-cost equipment manufactured by Ajay Sensors and Instruments, Bengaluru, under license from the Indian Academy of Sciences.

The COVID-19 pandemic in 2020–2022 resulted in a temporary suspension of the refresher courses in experimental physics. During this period of inactivity, the authors of this book decided to write brief chapters on the techniques for preparing solid-state materials in bulk and thin-film forms and on the characterization techniques, such as powder X-ray and neutron diffraction, ESCA, ellipsometry for thin films, electron microscopy and surface probe techniques, and positron annihilation as a tool to study defects in materials. The methods for measuring the elastic, thermal, electrical transport, dielectric, and magnetic properties are discussed. Spectroscopic techniques, such as NMR, EPR, IR, visible, UV, and Mossbauer spectroscopies, are discussed. A fairly detailed chapter on the study of phase transitions is included.

Each chapter is about 20–30 printed pages in length and is self-contained. Naturally, in such short chapters, one can only present the basic principles of each technique with some typical examples. Different chapters have been written by different authors. So, some differences in the style of presentation could not be avoided. There could be some overlap, but this has been reduced to a minimum.

The scope of the book is limited to meeting the requirements of students at the level of a master's degree in physics or materials science. At the master's level, the curriculum can be made more balanced by introducing at least a one-semester course in experimental techniques, choosing appropriate chapters from the book. The book should also be useful for students engaged in research in the domain of materials science and for teachers teaching courses in experimental physics.

R. Srinivasan T. G. Ramesh G. Umesh C. S. Sundar

ACKNOWLEDGEMENT

The authors are grateful to World Scientific Publishing Company, Singapore, Ms. Lakshmi Narayanan in particular, for agreeing to publish the book. We are thankful to Mr. Rhaimie Wahap and his Team for their cooperation in agreeing to our repeated requests for proof correction and bringing out the book in print.

CONTENTS

Preface v

Acknowledgement vii

Part I Techniques for Preparation of Materials 1

Chapter 1 Techniques for Preparation of Solid-State Materials and Nanoparticles 3

Chapter 2 Deposition of Thin Films 23

Part II Techniques for Materials Characterization 45

Chapter 3 X-Ray and Neutron Powder Diffraction 47

Chapter 4 Electron Spectroscopy for Chemical Analysis 71

Chapter 5 Ellipsometry for Thin-Film Analysis 93

Chapter 6 Electron Microscopy 123

Chapter 7 Surface Probe Techniques 145

Chapter 8 Positron Annihilation Spectroscopy as a Tool for the Study of Defects in Solids 161

Part III Techniques for Measurement of Physical Properties 193

Chapter 9 Elastic Properties 195

Part III.1 Thermal Properties **217**

Chapter 10 Specific Heat 219
Chapter 11 Thermal Expansion of Solids 235
Chapter 12 Thermal Conductivity and Diffusivity 253

Part III.2 Electrical Transport Properties **275**

Chapter 13 Electrical Conductivity of Metals and Semiconductors 277
Chapter 14 Seebeck Coefficient in Metals and Semiconductors 319
Chapter 15 Dielectric Properties 347
Chapter 16 Magnetic Properties 371

Part IV Spectroscopic Techniques **395**

Chapter 17 NMR and EPR Spectroscopy 397
Chapter 18 IR, Visible, and UV Spectroscopies 427
Chapter 19 Mossbauer Spectroscopy 447

Part V Phase Transition **471**

Chapter 20 Phase Transitions 473

Index 513

Part I

Techniques for Preparation of Materials

Chapter 1

TECHNIQUES FOR PREPARATION OF SOLID-STATE MATERIALS AND NANOPARTICLES

1. Introduction

In the 20th century, the creation of new materials revolutionized the life of the common man. This became possible due to innovations in techniques on the preparation of materials. It is not enough to just prepare a new material. We must characterize the material precisely to know its composition, structure, and physical properties. This knowledge is essential in order to make proper use of the material for any application. It was fortunate that this period saw the invention of new techniques for the synthesis and characterization of materials. It also saw a revolution in techniques for the fabrication of electronic devices and the development of computers, along with computational techniques. These advances enabled one to make fast and precise automated measurements of different properties of the material and, simultaneously, make intricate theoretical calculations to understand the material properties.

2. Preparation of Materials in Solid State

New compounds need to be prepared from available materials. One must take proper precautions to see that the compound prepared is in the correct stoichiometric ratio and does not contain impurity phases. A sound knowledge of solid-state chemistry is essential for the preparation of new materials. A good book in solid-state chemistry is the one by Anthony West (Ref. 1).

Materials can be prepared as bulk solid materials, thin films, or nanoparticles. In this chapter, we deal with the preparation of materials in bulk solid state and as nanoparticles. The following chapter deals with the preparation of materials as thin films.

3. Preparation in Bulk Form

3.1. *Solid-state reaction technique*

This is a technique often used for preparing materials. When an intimate mixture of two materials, A and B, is heated, the cations (or the anions) from one material diffuse into the other due to a concentration gradient. This process results in the production of a new material. For example, zirconium silicate can be prepared by heating a mixture of ZrO (A) and SiO_2 (B) in the correct stoichiometric proportions.

This diffusion process will depend on (1) the surface area of the grains of the materials, (2) the temperature of the mixture, and (3) the distance between adjacent grains across which diffusion takes place.

To illustrate how the surface area varies with grain size, we take the following example. In a powdered material of 1 cm^3 volume, comprising coarse grains of 10^{-2} cm size, the total number of grains is around 10^6, and the total surface area of all the grains is around 600 cm^2, assuming the grains to be cubical in shape. If the grain size is 10^{-4} cm, the number of grains will be 10^{12}, and the total surface area of the grains will be 6×10^4 cm^2. The surface area increases enormously as the grain size decreases. So, to facilitate diffusion, we take the starting materials in the form of very fine powder.

Diffusion occurs by random hopping across a potential barrier. The rate of diffusion will increase as the temperature is increased. For example, let us take a single crystal of MgO and a single crystal of Al_2O_3, put them in contact, and heat them. The Mg and Al ions will diffuse across the interface between the two crystals in opposite directions. A thin layer of $MgAl_2O_4$ will be formed in the interfacial region. Figure 1 (from Ref. 2) shows how the square of the thickness of the $MgAl_2O_4$ region varies with time at different temperatures.

From this figure, we note that the diffusion process is accelerated significantly at higher temperatures. Even so, to get a growth in thickness, x,

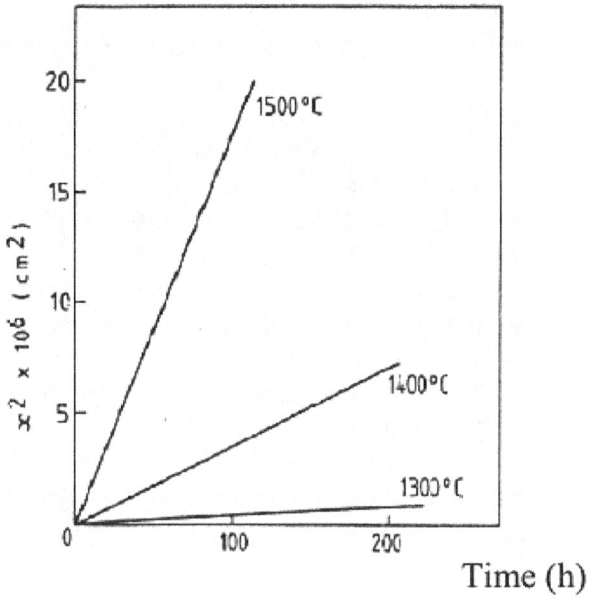

Figure 1. Growth in thickness, *x*, of $MgAl_2O_4$ region with time at different temperatures (extracted from Ref. 2).

of 30 μm of $MgAl_2O_4$ at a temperature of 1500°C, one will have to heat the materials for about 50 h. That indicates how slow diffusion is.

It is not enough to grind the starting materials into a fine powder and heat the mixture. The mixture will have to be pelletized to a high density to bring the reacting grains close together.

The rate of diffusion depends on the size of the diffusing ions and the host lattice into which the diffusion takes place. Therefore, it is necessary to choose the proper starting materials so that they have nearly the same crystal structure as the final product.

The material A or B should not melt, and the resulting compound formed by diffusion should not decompose at the temperature to which the pellet is heated. Usually, the temperature to which the mixture is heated is chosen to be about two-thirds the temperature of the lower melting points of the two materials, A and B.

Since diffusion is a slow process, it will take a long time for the two starting materials to be converted completely into the resultant compound. Often, one will have to regrind the material and reheat the mixture many times for the reaction to be complete.

Many ceramic oxides are prepared by this technique. The purity of the compound will depend on the purity of the starting materials and on the nature of the crucible used. An important requirement is that the material of the crucible must be stable at the temperature to which it is heated and it should not react with materials A and B or the compound. Commonly used crucible materials are silica up to 1400 K, alumina up to 2200 K, and zirconia up to 2300 K. Platinum crucibles can be used only up to 1800 K because the melting point of platinum is 2045 K.

The atmosphere in which the mixture is heated is important. When preparing oxides under oxidizing conditions, the atmosphere can be air or oxygen, and the temperature should be low. If oxides are prepared under reducing conditions, the atmosphere should be a mixture of hydrogen gas and an inert gas, such as argon.

One normally uses resistive heating in the furnace. The resistive materials may be wires or strips of molybdenum, Kanthal, or silicon carbide. A commercially available furnace is shown in Figure 2.

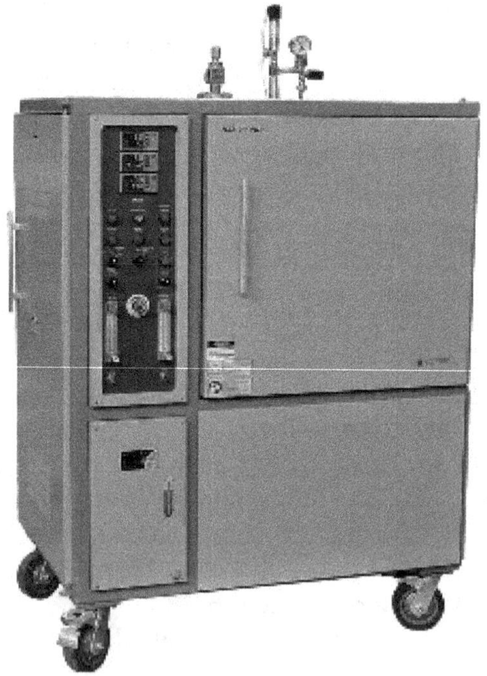

Figure 2. Furnace going up to 1700°C for material preparation (extracted from MTI Corporation - https://www.mtixtl.com).

This furnace has a double-layered stainless-steel structure designed for preparing materials under vacuum. The front door is water-cooled, and the shell is air-cooled. There is a built-in vacuum pump, and channel flow meters are mounted on the front panel to circulate different gases. The volume of the heating chamber is 8 liters. Thermal insulation is provided by alumina powder. The heating rate is controlled by a 30-segment PID controller. This furnace, manufactured by MTI Corporation, can operate up to a maximum temperature of 1700°C.

If one wants to reach a temperature of around 3000 K, one may heat the reactant mixture by an electric arc directed at the material. One may also use a focused carbon dioxide laser beam to reach a temperature of around 4000 K.

How do we know when the reaction is complete? After cooling the sample to room temperature, we regrind it and obtain its X-ray powder diffraction. If the reaction is not complete, the X-ray powder spectrum will have characteristic diffraction peaks of the starting materials along with the diffraction peaks of the compound, superposed on one another. With repeated grinding and heating, the intensity of the peaks of the starting material will get reduced progressively, and the intensity of the peaks of the compound will keep growing. When the reaction is complete, the peaks of the starting material will disappear, leaving only the diffraction peaks due to the compound. This will happen only if the starting materials are nonvolatile, and they are mixed in the correct stoichiometric ratio.

As an example of the solid-state reaction method, we consider the preparation of Sr_2CrTaO_6 (Ref. 2). The starting materials could be SrO (mp: 2703 K) or $SrCO_3$ (decomposes to SrO at 1643 K), Ta metal (mp: 3269 K) or Ta_2O_5 (mp: 2073 K), and Cr_2O_3 (mp: 2708 K). The convenient starting materials will be $SrCO_3$, Ta_2O_5, and Cr_2O_3. The chemical reaction is

$$4\,SrCO_3 + Ta_2O_5 + Cr_2O_3 \rightarrow 2\,Sr_2CrTaO_6 + 4\,CO_2$$

The molecular weights of Sr_2CrTaO_6, $SrCO_3$, Ta_2O_5, and Cr_2O_3 are 504.2, 147.6, 441.9, and 152.0 g, respectively. To prepare 0.01 mole of Sr_2CrTaO_6 (i.e., 5.04 g) in the correct stoichiometric proportion, we need to take 0.02 mole of $SrCO_3$ (i.e., 2.96 g), 0.005 mole of Ta_2O_5 (i.e., 2.21 g), and 0.005 mole of Cr_2O_3 (i.e., 0.76 g). We grind the mixture in a ball mill and pelletize it. The value of two-thirds of the melting point of Ta_2O_5 is 1382 K. One could prepare the material at a temperature of 1400 K; however, at such a low temperature, there will be a fair amount of unreacted Cr_2O_3. So, the mixture is heated to 1800 K.

It may happen that one of the reactants is volatile. This happens, for instance, in the preparation of sulfides. If the reactants are volatile, one may seal the mixture in a suitable tube after evacuating the tube to low pressure. The sealed ampule with the mixture may then be heated in a furnace.

If the materials react with air on heating, one may heat the materials in a rare gas atmosphere. The crucible is kept in a tube in which a rare gas (e.g., argon) flows at a continuous rate. The crucible is heated after the air in the tube is completely flushed out by the rare gas.

3.2. *Precursor method*

This method has the following advantages:

(a) It is a low-temperature method. Unlike the solid-state reaction method, the precursor is heated to a lower temperature for the reaction to be complete.
(b) It produces materials of uniform grain size. This is especially important when sub-micrometer-sized grains are to be produced. At sub-micrometer size, the properties of the material will vary with the size of the grain. A material with a uniform grain size will have physical properties with values defined in a narrow range.
(c) One can achieve a higher level of purity in the compound.

A precursor is an organic compound of the two metal ions in the correct proportion which, on heating, produces the desired compound in the correct stoichiometric ratio.

We briefly describe a precursor method to produce $BaTiO_3$, with grain sizes in the range of 10–100 nm. This method is described by Vinothini *et al.* (Ref. 3).

Titanium tetra-isopropoxide was dissolved in a solution of citric acid and ethylene glycol mixed in a molar ratio of 1:4. The amount of $BaCO_3$, which gives an atomic ratio of 1:1 of Ba to Ti, was dissolved in the solution. The solution was first heated at 90°C, with constant stirring until it became a clear and transparent yellow solution. The solution was then heated to 200°C and kept for 5 hours (h) at this temperature till it solidified to form a dark brown glassy resin. The resin was then charred at 400°C for 2 h to form a black solid mass. **This was the barium titanium citrate polyester resin.** This resin was ground into a fine powder and used as the precursor. This precursor was heated for 5 h at various temperatures ranging from 500°C to 900°C. The material thus obtained was analyzed using X-ray powder diffraction.

Using thermogravimetric analysis (TGA) and differential thermal analysis (DTA), it was concluded that the precursor started decomposing at 140°C and the decomposition was complete at 900°C, resulting in the formation of $BaTiO_3$. For further details, readers may consult the paper.

Oxalates or citrates of the metals can be used as precursors. Thus, ferrites are formed by the decomposition of the corresponding oxalates.

3.3. *Sol–gel method*

This is a convenient method to prepare certain oxides from alkoxides or carbonates.

An alkoxide RO^- is the conjugate base of the alcohol ROH, where the hydrogen atom in OH is removed. R is an alkyl group, such as CH_3CH_2. A metal atom M can attach to RO^- to form ROM, the metal alkoxide. For example, CH_3CH_2ONa is sodium ethoxide.

A sol is a colloidal suspension of particles in a liquid. When a sol is allowed to age or is heated, it forms a gel in which the concentration of the particles is higher. The gel contains the solvent in cages within a framework which is colloidal or polymeric. When this gel is heated to a high temperature, the oxide is obtained. This is a low-temperature preparation technique.

As an example, we describe how lithium niobate is prepared by this method (Ref. 4) in the following. The starting materials are lithium ethoxide ($LiOC_2H_5$) and niobium ethoxide ($Nb_2\,(OC_2H_5)_{10}$). Each of these ethoxides is dissolved in absolute ethanol. The addition of water leads to hydrolysis, giving hydroxy alkoxides as per the following reaction:

$$Nb_2(OC_2H_5)_{10} + 2H_2O \rightarrow 2Nb\,(OEt)_4(OH) + 2\,EtOH$$

"Et" stands for the ethyl group, $-CH_2-CH_3$. The two liquids are mixed in the correct ratio to yield a ratio of Li to Nb of 1:1. The gel is formed by the condensation of hydroxyl metal oxides to give metal–oxygen–metal bonds. Condensation is a chemical reaction in which two molecules combine to yield a single molecule and molecules of water. When this gel is heated, $LiNbO_3$ is formed. The remaining water is evaporated, and the ethanol groups are pyrolyzed.

The sol–gel method operates at a reduced temperature and takes less time to prepare the oxide.

Methods for preparing some other oxides using the sol–gel method are described in Ref. 4. Table 1 (from Ref. 2) lists some alkoxides used for processing ceramics by the precursor method.

Table 1. Some metal alkoxides for sol–gel processing of ceramics (from Ref. 2).

Single-cation alkoxides		Alkoxide
I A (1) group	Li, Na	$LiOCH_3$(s), $NaOCH_3$(s)
I B (11) group	Cu	$Cu(OCH_3)_2$(s)
II A (2) grp	Ca, Sr, Ba	$Ca(OCH_3)_2$(s), $Sr(OCH_3)_2$(s), $Ba(OCH_3)_2$(s)
II B (12) grp	Zn	$Zn(OCH_3)_2$(s)
III A (3) grp	B, Al, Ga	$B(OCH_3)_3$(l), Al(i-$OC_3H_7)_3$(s), $Ga(OC_2H_5)_3$(s)
III A (13) grp	Y	$Y(OC_4H_9)_3$
IV A (4) grp	Si, Ge	$Si(OC_2H_5)_4$(l), $Ge(OC_2H_5)_4$(l)
IV B (14) grp	Pb	$Pb(OC_4H_9)_4$(s)
V A (5) grp	P, Sb	$P(OCH_3)_3$(l), $Sb(OC_2H_5)$3(l)
V B (15) grp	V, Ta	$V(OC_2H_5)_3$(l), $Ta(OC_3H_7)_5$(l)
VI B (16) grp	W	$W(OC_2H_5)_6$(s)
Lanthanides	La, Nd	$La(OC_3H_7)_3$(s), $Nd(OC_3H_7)_3$(s)
Alkoxides with various alkoxyl groups		
	Si	$Si(OCH_3)_4$(l), $Si(OC_2H_5)_4$(l), Si(i-$OC_3H_7)_4$(l) Si(l-$OC_4H_9)_4$(l)
	Ti	$Ti(OCH_3)_4$(s), $Ti(OC_2H_5)_4$(l), Ti(i-$OC_3H_7)_4$(l) $Ti(OC_4H_9)_4$(l)
	Zr	$Zr(OCH_3)_4$(s), $Zr(OC_2H_5)_4$(s), $Zr(OC_3H_7)_4$(s) $Zr(OC_4H_9)_4$(s)
	Al	$Al(OCH_3)_3$(s), $Al(OC_2H_5)_3$(s), Al(i-$OC_3H_7)_3$(s $Al(OC_4H_9)_3$(s)

(*Continued*)

Table 1. *(Continued)*

Single-cation alkoxides		Alkoxide
Double-cation alkoxides	La–Al	La $[Al(i\text{-}OC_3H_7)_4]_3$
	Mg–Al	Mg $[Al(i\text{-}OC_3H_7)_4]_2$, Mg $[Al(s\text{-}OC_4H_9)_4]_2$
	Ni–Al	Ni $[Al(i\text{-}OC_3H_7)_4]_2$
	Zr–Al	$(C_3H_7O)_2Zr\ Al(OC_3H_7)_2$
	Ba–Zr	Ba $[Zr_2(OC_2H_5)_3]_2$

Figure 3. Propagation of ignition in a powdered mixture of Tl–C–Al system (extracted from Ref. 4).

3.4. *Combustion synthesis*

To prepare many refractory materials, such as borides, nitrides, and silicides, the combustion synthesis method is used. This method is used in situations where the reaction is highly exothermic (it generates heat) or even explosive. The reactants are powdered, mixed in the proper ratio, and pelletized. The mixture is ignited using a heating coil or a laser pulse. Once ignited, the heat released by the reaction enables the ignition to propagate as a wave through the material, producing the high temperature required for synthesis. It need not be emphasized that the material should lose less heat than what is generated by ignition. Otherwise, the ignition will be quenched. One may maintain a temperature of up to 3000°C due to the heat generated by the reaction. Figure 3 (Ref. 4) shows how the ignition propagates in a mixture of Tl–C–Al system that are pressed to form pellets.

Combustion synthesis is useful for the preparation of hydrides (for the storage of hydrogen), borides, and carbides (for abrasives and cutting tools), nitrides (high-strength heat-resistant ceramics), silicides (heating elements), and oxides.

3.5. *High-pressure synthesis*

Some materials are best prepared by applying high pressure. A high pressure can be achieved by superheating the solvent in an autoclave, such as in the hydrothermal process for synthesizing quartz. If one is using a reactive gas, the pressure of the gas may be increased. High pressure for synthesis may also be achieved by detonation or by ultrasonics.

CrO_2 is a magnetic material used in magnetic memory tapes. Of the several oxides of chromium, CrO_2 is the only ferromagnetic oxide. The pressure–temperature phase diagram of the oxides of chromium is shown in Figure 4. It is seen that one needs to maintain oxygen gas at a pressure of at least 50 bar in the reaction chamber to prepare chromium dioxide.

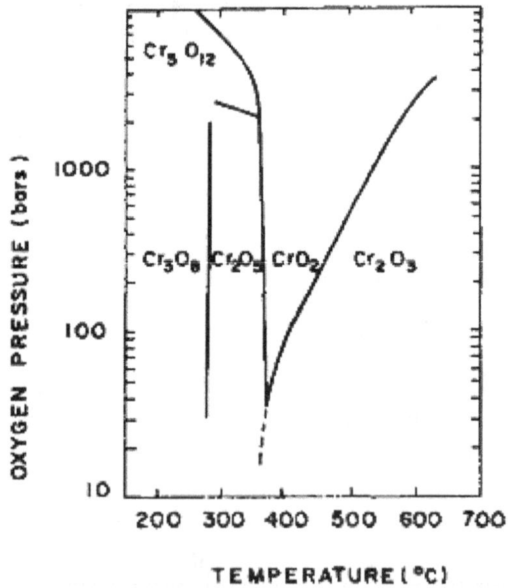

Figure 4. **Phase diagram of oxides of chromium (reprinted with permission from Ref. 5).**

After preparation, CrO_2 can be quenched to room temperature under pressure, and the pressure is reduced to ambient pressure. The oxide remains fairly stable.

Abdul Jaleel *et al.* (Ref. 5) used a single-step hydrothermal decomposition of Cr_2O_3 dissolved in water to produce CrO_2. Hydrothermal synthesis was achieved in autoclaves made of forged and heat-treated EN24 carbon steel. Water was added in the ratio Cr_2O_3: H_2O :: 1: 6. Lithium carbonate, 10% by weight of CrO_3, was used as a mineralizer. The mineralizer aids in the solubilization of the nutrient solid. Oxygen at a pressure of around 120 bar was filled in the autoclave. The temperature was raised to a value ranging from 325°C to 525°C. The final pressure at these temperatures ranged from 300 to 1200 bar, depending on the filled volume of the autoclave. The autoclave was maintained at this temperature and pressure for a time ranging from 30 min to 4 h. The autoclave was then cooled without releasing the internal pressure. The pressure was then released, and the final product was powdered, washed, and dried. The product was characterized for microstructure by electron microscopy, and its phase purity was determined by employing X-ray diffraction.

Normally, CrO_2 forms in the shape of platelets. The addition of Sb_2O_3 in the initial charge makes the CrO_2 particles needle-shaped, with a larger fraction having a smaller particle size. A particle of small size has a single magnetic domain, while a particle of large size has multiple domains. The improvement in magnetic properties with increasing additions of Sb_2O_3 is shown in Figure 5. Hysteresis loops become larger as the amount of Sb_2O_3 increases, showing an increase in remanent magnetization and coercivity.

Hydrothermal synthesis is also used to produce zeolites. These materials have large cavities in their structure and are good adsorbents.

4. Preparation of Nanomaterials

Nanomaterials have a particle size ranging from 1 to 100 nm. Because of their small size, the surface-to-volume ratio is very high. This affects the mechanical and catalytic properties of these particles. The structure of the quantized electronic energy levels for nanoparticles varies significantly with the actual size of the particles. Due to this dependence, the electrical, magnetic, and optical properties depend on the size of the particle. Table 2 (from Ref. 6) lists the different techniques used for producing nanoparticles.

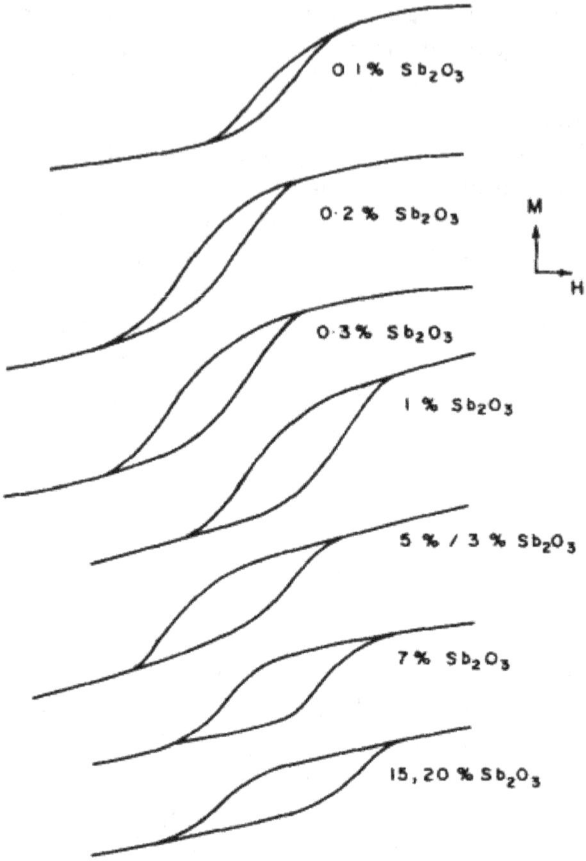

Figure 5. Improvement in magnetic properties of CrO_2 particles with addition of Sb_2O_3 (reprinted with permission from Ref. 5).

4.1. *Physical methods*

4.1.1. Flash spray pyrolysis

This process is described in Ref. 6 and illustrated in Figure 6. The precursor is in liquid form, mixed with an organic solvent. This is a single-stage scalable process. The flame is at a high temperature and has a large temperature gradient. The flame is a self-sustaining flame due to the large combustion enthalpy of the precursor. The process involves the following sequential steps: The fuel and precursor enter the bottom of the flame through a nozzle. Air also enters the flame through a tube surrounding

Table 2. Methods for synthesizing nanoparticles (from Ref. 6).

(a) Physical Methods:
High-energy ball milling
Flash spray pyrolysis
Laser pyrolysis
Pulsed vapor deposition
Inert gas condensation
Electrospraying
Melt mixing
(b) Chemical Methods:
Sol–gel synthesis
Microemulsion techniques
Hydrothermal synthesis
Polyol synthesis
Chemical vapor synthesis
Plasma-enhanced chemical vapor synthesis
(c) Biological Methods
Of these processes, thin-film and chemical vapor deposition techniques will be discussed in the following chapter. In this chapter, we deal only with flash spray pyrolysis, laser pyrolysis, high-energy ball milling, microemulsion, and polyol methods.

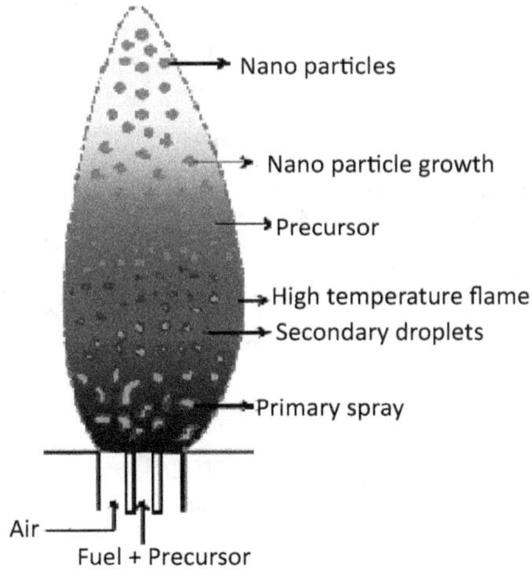

Figure 6. Schematic illustration of the spray pyrolysis technique (reprinted with permission from Ref. 6).

the nozzle. The precursor evaporates or decomposes to give off metallic vapors near the middle of the flame. Nucleation takes place due to supersaturation, and the particles grow by agglomeration and sintering through chemical bonding and physical interactions near the top of the flame. Single- and multicomponent nanoparticles have been prepared through this technique. For further details, the reader should consult Ref. 6.

4.1.2. Laser pyrolysis

Laser pyrolysis (Figure 7) may be employed to prepare nanoparticles of metal oxides, such as Al_2O_3 and Fe_2O_3, and high-temperature materials, such as silicon carbide and silicon nitride. The vapors carrying the reactants flow into the reaction chamber along with other gases. The CO_2 laser beam induces chemical reactions between the reactants. One of the vibration modes of the reactants must resonantly absorb the laser photons. Otherwise, gases, such as ammonia, can be added as inert photosensitizers to enable energy transfer from the laser beam to the reactants to facilitate the chemical reaction. The energy transfer process causes an increase in the temperature of the gas, often leading to the production of a flame. The condensable reaction product is collected. Because of the rapid nucleation followed by rapid quenching, one can get particles with a very narrow distribution in the particle size, in the high temperature zone. One problem is that cooling leads to the agglomeration of nanoparticles. So,

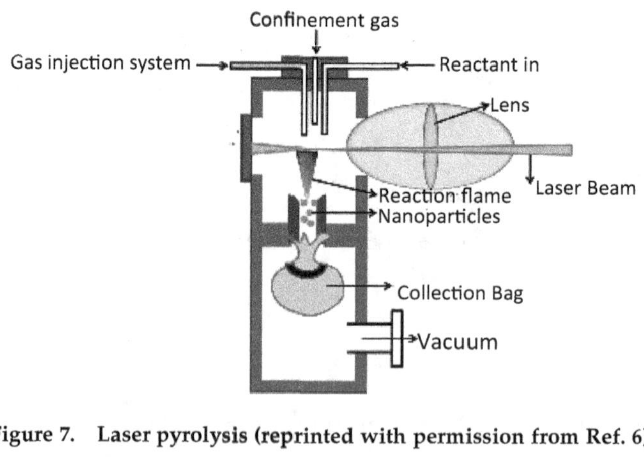

Figure 7. Laser pyrolysis (reprinted with permission from Ref. 6).

the final product must be further subjected to ball milling to reduce the particle size.

4.1.3. High-energy ball milling

In this technique, we take the material in the form of a coarse powder with a particle size of a few micrometers and grind the powder in a ball mill (using steel balls) at rotation speeds of a few thousand RPM. The kinetic energy of the steel balls is transferred to the grains, rupturing chemical bonds and breaking the grains into nanosized particles. There are different types of ball mills available: vibratory mills, planetary mills, tumbler ball mills, etc. The efficiency of energy transfer varies with the type of ball mill, the weight of the balls, and the speed of rotation. The grinding can be a dry or wet process. Adding a surfactant causes the particle to be coated with the surfactant molecules, preventing reaggregation. One can prepare particles ranging in size from 20 to 50 nm.

Chen *et al.* (cited in Ref. 6) used microwave-assisted high-energy ball milling at a temperature below 100°C to prepare cobalt ferrite nanoparticles of 20 nm size with a high saturation magnetization. No subsequent calcination was required.

4.2. *Chemical methods*

In contrast to the top-down method in high-energy ball milling, in chemical methods, nanoparticles are formed by a bottom-up process of assembling atoms or molecules one by one. One such method is the sol–gel method, which has been described earlier in this chapter.

4.2.1. Microemulsion technique

The second technique is the microemulsion method. A microemulsion is an optically transparent dispersed phase containing three components, one of which is a polar liquid, such as water, the second is a hydrocarbon, such as oil, and the third is a surfactant, which keeps the microemulsion drops separated. Water in oil is such a microemulsion.

In the single-emulsion process, the drops contain the precursor. Nucleation in the precursor is triggered by an energy source, such as a laser pulse. One may also prepare a micro-emulsion containing one of the reactants and add the other reactant to the microemulsion.

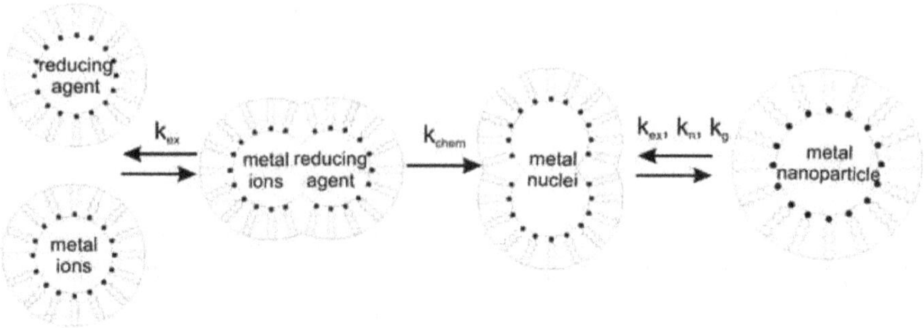

Figure 8. Two-microemulsion technique for producing metal nanoparticles (reprinted with permission from Ref. 7).

The two-emulsion process of synthesizing metal nanoparticles is shown in Figure 8 (Ref. 7). In this process, two water-in-oil microemulsions, one containing a metal salt and the other the reducing agent, are mixed together. Brownian motion results in the collision of micelles containing the metal ions and the reducing agent. The micelles fuse, and the metal ions are reduced to form the metal nanoparticles. Nanoparticles of gold of 2.5–10 nm size can be produced by this process.

The microemulsion technique has been successfully used to synthesize nanoparticles of metals, the sulfides and selenides of copper, cadmium, and lead, the carbonates of alkali earth metals such as calcium, barium, and strontium, the oxides of metals, such as titanium and zinc, and the nanoparticles of magnetic materials, such as ferrites.

4.2.2. Polyol process

A polyol is a compound containing multiple hydroxy groups. For example, ethylene glycol is $HOCH_2CH_2OH$.

The polyol process is a very commonly used technique for synthesizing nanoparticles containing metals. In this process, the metal precursor is suspended in a glycol solvent, and the solution is heated to the refluxing temperature. The polyol is used as a solvent, a reducing agent, and a ligand to prevent agglomeration of the nanoparticles. Convective heating takes a long time. Microwave heating shortens the time. This process is discussed at length in Ref. 8.

Polyethylene glycol is used as a solvent. It also serves as a reducing agent. Polyvinyl pyrrolidine (PVP) is often used as the protecting agent.

PVP facilitates reduction on certain facets of the crystal and prevents reduction on other facets. The morphology, size, and uniformity of metal nanoclusters are affected by the polymer chain length of PVP. The reactant concentration and reaction temperature also affect the morphology and size of the metal nanoparticles.

The synthesis of metal nanoparticles with a high degree of control over their size dispersion is possible through the polyol process. One can control the rate of reduction by using different polyols, such as ethylene glycol and 1,2-propylene glycol, to produce nanoparticles of different morphologies. Table 3 (from Ref. 8) gives the synthesis parameters and products of different polyol reactions.

Table 3. Synthesis parameters and products of different polyol reactions (from Ref. 8).

Material	Precursors	Conc. Range Used (mole/l)	Average Crystalline Size of Powder (nm)	Average Crystalline Size of Coating (nm)	Reaction Time (h)
Fe	Fe(II)acetate	0.01–0.20	20		2
Co	Co acetate	0.05–0.20	12.1	15 (K)	2
	Tetrahydrate			14 (P)	
	Co(II)Chloride		14	23 (T)	
	Hexahydrate				
Ni	Ni(II)acetate	0.02–0.20	20	9 (K)	1
	Tetrahydrate			30 (K)	1
				15 (P)	
Cu	Cu(II)acetate	0.02–0.025	10–80	12 (AIN)	2
	Tetrahydrate			43 (K)	
Ru	Ru(III)	0.021	5		1
	Chloride				
Rh	Rh(III)	0.01		9 (P)	1
	Chloride				
Pd	Pd(II)	0.02–0.15	10	18 (K)	1
	Chloride			22 (P)	
Ag	Ag nitrate	0.05–0.2	40	34 (T)	1
				43 (K)	
				50 (P)	

(Continued)

Table 3. (*Continued*)

Material	Precursors	Conc. Range Used (mole/l)	Average Crystalline Size of Powder (nm)	Average Crystalline Size of Coating (nm)	Reaction Time (h)
Sn	Sn(II)Oxide	0.01–0.03	36		2
Re	Re(III) chloride	0.02	1	14(P)	1
W	Tungstic acid	0.012–0.20	8	12(P)	3
	Na Tungstate	0.03	10	12(P)	3
Pt	Potassium Hexachloro Platinate	0.01–0.20	2	10 (K) 12 (T) 14 (GF)	1
Au	Au(III) Chloride	0.01–0.20		15 (AF)(SF)	
Fe–Cu	Fe(II)acetate	0.016–0.16	28	32(P)	2
	Cu(II)acetate Tetrahydrate	0.018–0.14	27–47		2
Co–Cu	Co(II)acetate tetrahydrate Cu(II)acetate Tetrahydrate	0.01–0.20	17–35		2
Ni–Cu	Ni(II)acetate Tetrahydrate	0.0321	8		1
	Cu(II)acetate Tetrahydrate	0.0963			

Note: K: Kapton; P: Pyrex; T: Teflon; G: graphite; A: alumina; S, sapphire; F: fiber.
The range of crystallite size is given when it is concentration dependent [205].

4.3. *Bio-assisted methods*

Bio-assisted methods provide low-cost, efficient, and environment friendly protocols for the production of nanoparticles. In these methods, one uses bio-organisms, such as bacteria, fungi, or plant materials, for the preparation of nanoparticles. For a discussion of these methods, please see Ref. 6.

5. Conclusion

In this chapter, some of the common methods of preparation of materials and nanoparticles are discussed. One cannot overemphasize the necessity of a good understanding of solid-state chemistry in the preparation of materials. Though the basic principles have been outlined above, material preparation is still an art, and one can gain experience in these techniques through repeated trials and improvisations.

References

1. West, A. (2014). *Solid State Chemistry and Its Applications*. Wiley.
2. PPT — Solid State Synthesis PowerPoint Presentation, free download — ID:791725. www.slideserve.com.
3. Vinothini, V., Singh, P., and Balasubramanian, M. (2006). *Ceram. Int.* **32**, 99–103.
4. Chapter 3: Preparative methods, www.unf.edu › chem4627ch3_solid_state.
5. Jaleel, V. A., and Kannan, T. S. (1983). *Bull. Mater. Sci.* **5**, 231.
6. Dhand, C., Dwivedi, N., Loh, X. J., Ying, A. Y. J., Verma, N. K., Beuermann, R. W., Lakshminarayanan, R., and Ramakrishna, S. (2015). Methods and strategies for synthesis of diverse nano-particles and their applications: A comprehensive overview. *RSC Adv.* (Accepted manuscript) **5**, 105003–105037.
7. Cele, T. (2020). Preparation of nanoparticles. In *Engineered Nanomaterials — In Health and Safety*. Intech Open, London, United Kingdom. Available https://www.intechopen.com/chapters/71103. DOI: 10.5772/intechopen.90771.
8. Benseba, F. (2013). Chapter 2 — Wet production methods. In *Interface Science and Technology*", vol. 19, p. 85.

Chapter 2

DEPOSITION OF THIN FILMS

1. Introduction

Since the end of World War II, industrial progress has been marked by the revolution caused by the invention of the semiconductor transistor in the early 1950s. Rapid developments following this led to the miniaturization of electronic devices and IC chips through VLSI technology. This technology has led to the invention of electronic systems which consume far less electrical power than vacuum-tube-based systems and are much smaller in size. Two good examples of this revolution are mobile phones and laptop computers, which we all use in our homes and offices. The development of thin-film deposition techniques played an important role in this revolution. In this chapter, we discuss the techniques for the deposition of thin films briefly.

Many materials can be prepared in the form of thin films. The following techniques have been employed to prepare thin films of any solid material:

(a) dip coating,
(b) spin coating,
(c) spray pyrolysis,
(d) chemical vapor deposition (CVD),
(e) thermal evaporation or physical vapor deposition sputtering.

In each case, the thin film is created on the surface of a suitable plate called the substrate. This chapter provides a brief introduction to these techniques.

2. Dip Coating

Reference 1 gives a brief description of the theory of dip coating. Figure 1 illustrates this process.

This is an inexpensive wet method for preparing a thin film using a solution at room temperature. In this method, a substrate is (1) immersed in a precursor solution containing the molecules of the material to be deposited, (2) kept dipped in the solution for the required time, and (3) then withdrawn. This leaves a thin layer of the solution on the surface of the substrate. The coated substrate is dried in open air, which leaves a thin layer of solid material on the substrate. If required, the deposited thin film can be further cured by keeping it in an oven at a suitable temperature.

The thickness of the film depends on the duration of immersion and the rate of withdrawal. For high rates of withdrawal from a highly viscous liquid, viscosity is the dominant force opposing the gravitational force, which acts to drain out the solution sticking to the substrate. Thus, the film thickness depends on the balance between the viscous and gravitational forces. When the withdrawal speed is low or the viscosity is low, one should take into consideration the surface tension forces along with

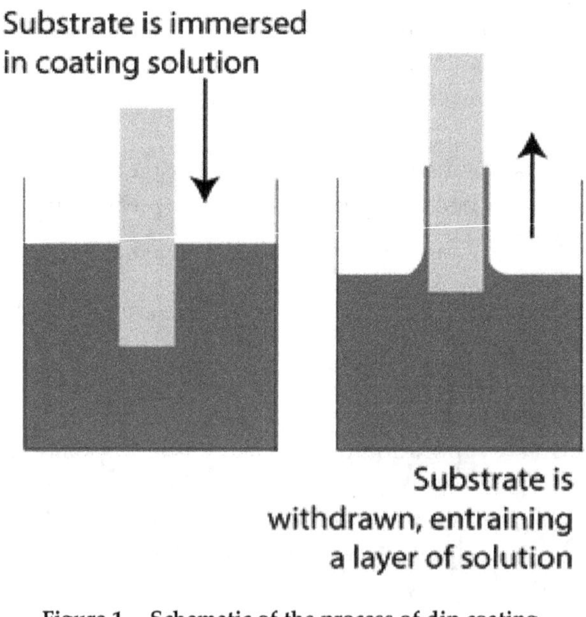

Figure 1. Schematic of the process of dip coating.

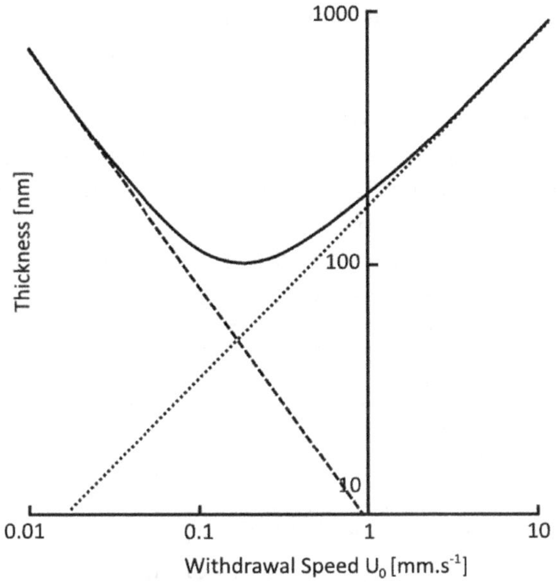

Figure 2. Variation of film thickness as a function of withdrawal speed (reprinted with permission from Ref. 1).

the viscosity. When the speed is very low, surface tension plays a dominant role in determining the thickness of the deposited film.

Figure 2 (Ref. 1) shows the variation in film thickness as a function of withdrawal speed.

When the withdrawal speed is high, the viscous force is high, which acts to prevent the drainage of the film, and gravity has less time to drain the film. So, the thickness of the film is high. When the withdrawal speed is low, surface tension provides a force, in addition to the viscous force, that opposes gravity, so the thickness of the film is again high.

Using this method, one can coat a single layer of the film or multiple layers of different materials, one by one. Commercially made dip-coating instruments are available for use in the laboratory and industry.

3. Spin Coating

A guide to spin coating is given in Ref. 2.

Spin coating is a technique commonly used to deposit films of molecules in a solvent onto a substrate. It produces films of uniform thickness. The thickness can vary from a few nanometers to a few micrometers. A schematic of the four stages of spin coating is shown in Figure 3 (from Ref. 2).

The dispenser contains the molecules dissolved in a solvent. In stage 1, one drop of the solution is placed on the substrate. In stage 2, the substrate with the drop on it is set to spin at a high speed. The centrifugal force generated by the rotation acts to spread out the drop to cover the entire surface of the substrate. In the process, most of the solution is thrown out of the substrate, leaving a thin film on the substrate. The thickness of the film is determined by the competition between the centrifugal force and the surface tension. Typically, the edge of the film is a little thicker due to beading. In stage 3, the solvent evaporates, while the substrate continues to spin, and a thin semi-solid film is formed on the substrate. In stage 4, the substrate, with the film on it, is taken out of the spin coater and subjected to further drying. At the end, we obtain a solid thin film deposited on the substrate.

The thickness of the film will depend on the speed of rotation of the substrate. Typically, the rotation speeds used vary from 600 to 6,000 RPM.

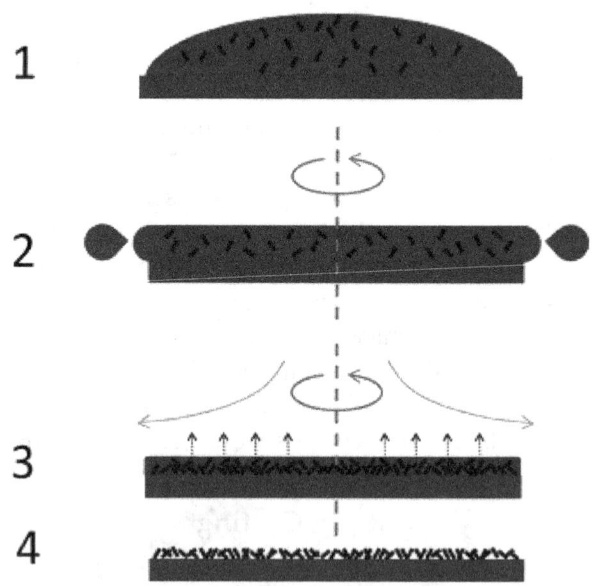

Figure 3. Schematic of the four stages of spin coating with a static dispenser (reprinted with permission from Ref. 2).

One may also go up to 12,000 RPM if required. Figure 4 (from Ref. 2) illustrates how the thickness varies with speed.

Apart from the rotation speed, the film thickness also depends on the nature of the material and the viscosity of the solvent.

The advantages of spin coating are as follows:

(a) It is a simple method.
(b) The film thickness is uniform over a large area.
(c) Due to the high speeds of rotation and the consequent air flow, the drying time is small.

The disadvantages are as follows:

(a) Only about 10% of the starting material is used in making the film. The rest of the material is flung off due to the high speed of rotation.
(b) It is a single-substrate process and thus has a relatively low throughput.
(c) For some applications in nanotechnology, fast drying may lead to lower performance.

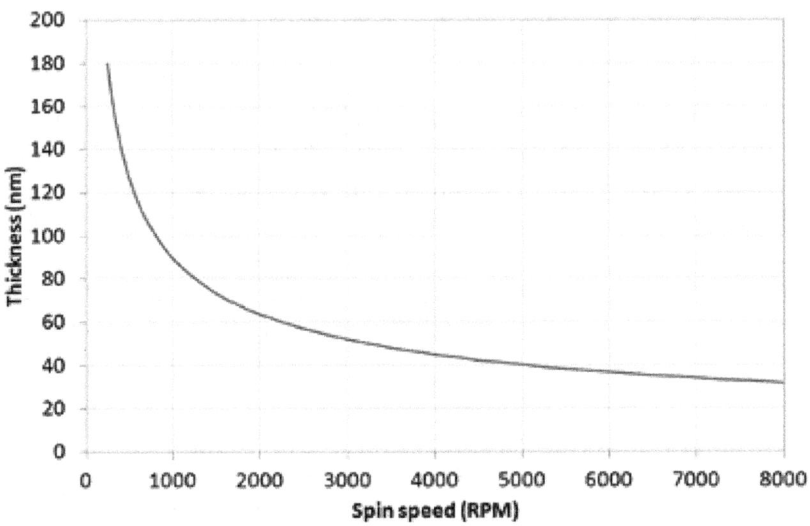

Figure 4. **Variation of thickness as a function of spin speed in RPM (reprinted with permission from Ref. 2).**

Spin coating is suitable for coating substrates with photoresists, insulators, organic semiconductors, synthetic metals, nanomaterials, metal and metal oxide precursors, transparent conductive oxides, polymers, and many other materials. For further details, readers may consult Ref. (2).

4. Chemical Vapor Deposition

This is a widely used method for preparing thin films of high purity copper for microelectronics, films of semiconductors doped with rare earth elements, such as GaAs doped with Nd, used in optical fibers, films of nitrides, such as TiN, for wear-resistant gold-like coatings, and sulfide films. The films are deposited inside vacuum chambers under a controlled atmosphere. Hence, the films are free from impurities.

The precursors chosen must be volatile. Figure 5 illustrates the process. The precursors are heated to form vapors. They are transported by a carrier gas into a region of high temperature in a tube. Chemical reaction takes place in the vapor phase at this high temperature. The product of the reaction is deposited on a solid substrate kept in the hot zone to form a thin film of the material.

The starting materials are volatile hydrides or halides of the metal. If an organometallic precursor is used, the process is called metal–organic chemical vapor deposition (MOCVD). As an example, precursors for the preparation of compounds containing Y, Ba, and Cu are the tetramethyl hepta-dionates (TMHDs) of these metals. The sublimation temperatures, T_s, of these TMHDs are low; for example, $Y(TMHD)_3$ has a sublimation temperature of 160°C. So, they can be vaporized easily.

Earlier, in Chapter 1, we had discussed the preparation of $LiNbO_3$ by the sol–gel method. The same compound can also be prepared by CVD. Lithium ethoxide is less volatile than niobium ethoxide. So, for the CVD method, one uses a β-diketonate of lithium. The lithium compound was

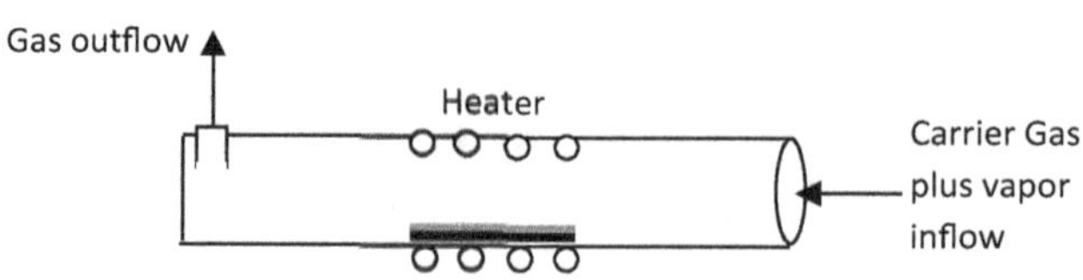

Figure 5. Chemical vapor deposition.

heated to 250°C and mixed with niobium penta-methoxide heated to 200°C in a stream of argon containing oxygen gas. Through this process, $LiNbO_3$ was successfully deposited on a substrate at 450°C in a reaction chamber.

CVD, assisted by plasma, produced by a high-frequency discharge in a gas in a reactor, is called plasma-assisted CVD (PCVD). The plasma can be produced by microwaves fed into the reactor through an antenna. As an example of microwave-assisted plasma CVD (MPCVD), we describe in the following a reactor for the commercial preparation of diamond thin films.

Diamond films have a high thermal conductivity (four times that of copper at room temperature). They are electrically insulating. They are highly wear resistant and are used in coating cutting tools. The films can be prepared at a temperature of 1000°C by the decomposition of CH_4, present as a small percentage in a mixture of CH_4 and hydrogen. A reactor for producing diamond films is described by Wang *et al.* (Ref. 4). A schematic diagram of the reactor is shown in Figure 6.

In this reactor, the substrate was a HPHT diamond of $4 \times 4 \times 1$ mm^3 in size. HPHT implies that the diamond was prepared by a process involving high pressure and high temperature. It was heated to 1000°C. Microwave power of 3 kW at a frequency of 2.4 GHz was used to produce the plasma. A mixture of high-purity (99.9999%) CH_4 in high-purity (99.9999%) hydrogen was used. The CH_4 fraction in the mixture ranged

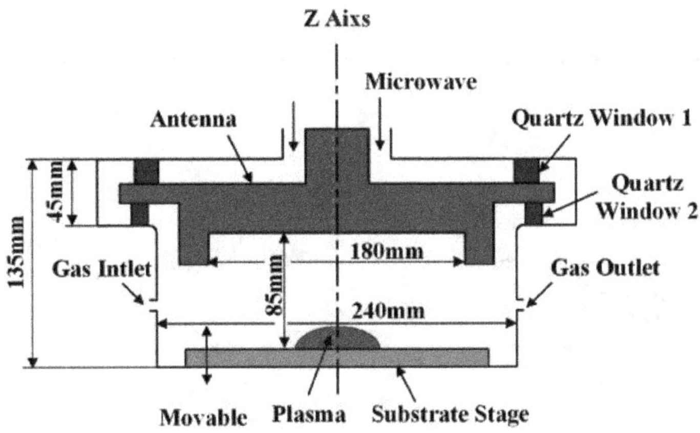

Figure 6. Schematic diagram of an MPCVD reactor for deposition of diamond films (reprinted with permission from Ref. 4).

from 1.5% to 3.5%. The pressure in the reactor was about 150 mbar. Diamond film grew at a rate of about 6 μm/h. Raman spectroscopy indicated a single narrow line at 1332 cm^{-1}, which showed that the film was very pure and free from any other carbonaceous content. Rocking curves of the X-ray diffraction peak of the (4,0,0) reflection of both the film and the substrate showed that the full width at half maximum was 43 and 28 s, respectively, indicating that the film was of high quality.

CVD can also be used to produce nanoparticles.

In Sections 5 and 6, we deal with thin-film deposition by evaporation and sputtering. Srinivasan, Ramesh, and Priolkar have written an article on this topic in *Manual on Experiments in Physics*, published by the Indian Academy of Sciences in Bangalore (Ref.7). Sections 5 and 6 are reproductions of this article with the permission of the Indian Academy of Sciences, Bangalore.

5. Deposition of Thin Films by Thermal Evaporation

For a detailed discussion of evaporation and sputtering techniques for depositing thin films, the books by Smith (Ref. 5) and George (Ref. 6) may be consulted.

Often, metals are deposited by evaporation. Figure 7 shows a schematic of two types of evaporation deposition systems.

A small strip of the metal to be deposited is put in a crucible, which is then placed on a resistive heater inside a vacuum chamber (Figure 7(a)).

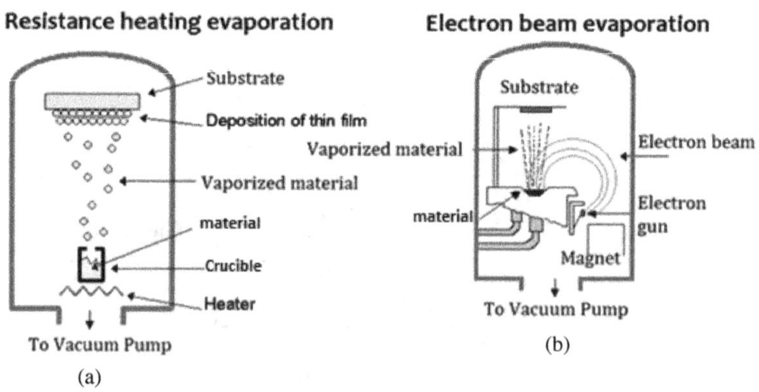

Figure 7. Deposition of thin film by evaporation: (a) thermal evaporation; (b) electron beam evaporation.

The vacuum chamber contains the crucible with the material and the substrate, on which the film is to be deposited. This chamber is evacuated to a vacuum of 10^{-6} mbar pressure or lower using a diffusion or turbomolecular pump, backed by a rotary pump. By driving a current through the heater, the temperature of the crucible is raised above the melting point of the metal. The metal melts, and the evaporating atoms of the metal freely effuse out of the crucible due to the vacuum inside the chamber. As the gas pressure is very low in the chamber, the mean free path of the atoms is large. Most of the atoms travel in a straight path, reach the substrate, and stick to it. The substrate can be heated if required. This method of depositing thin films has the advantage that the deposition rate is high.

Alternatively, a focused beam of electrons is directed toward the metal strip, resulting in the heating of the metal. This is especially useful for metals with a high melting point. The metal piece is heated rapidly, causing it to evaporate. This process is called electron beam evaporation, and it is illustrated in Figure 7(b). In this case, the heat deposited by the electrons in the crucible is high; therefore, there is a need to cool the crucible by circulating cold water through tubes wrapped on the outer surface of the crucible.

The rate of production of the vapor atoms, R, at a vapor pressure p and temperature T (in K) is given by the Langmuir formula:

$$R = 0.0583\,(M/T)^{1/2}p \tag{1}$$

Here, R is the rate of evaporation in g/cm^2/s and M is the molecular weight of the material in g. The rate of deposition of the atoms on the surface of the substrate is proportional to R and decreases as the distance of the substrate from the source (crucible) is increased.

The various types of heater coils and crucibles used to evaporate or sublimate the material are shown in Figure 8. For melting temperatures, support materials, and other details, Ref. 8 maybe consulted.

5.1. *Distribution of film thickness on substrate*

Let us consider a perpendicular line of length h from a certain point in the source to the substrate. The deposition rate of the atoms evaporating from this point in the source will be a maximum at the point P where the perpendicular meets the substrate. This is because the distance between the source point and P is the shortest distance between source and substrate. At any point on the surface of the substrate at a distance ℓ from P, the

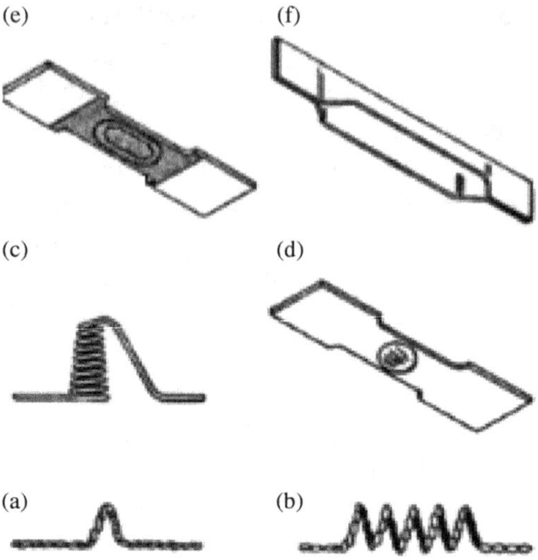

Figure 8. Different types of heater coils (a, b, c) and sample holders (d, e, f) used for thermal evaporation (reprinted with permission from Ref. 8).

deposition rate will be lower than that at P because there the atoms hit the substrate at oblique incidence. For an extended source of material, one will have to sum over the deposition rates from different points in the material.

For a given duration of deposition, the film thickness d will be maximum at one point on the substrate and will decrease as we move away from this point. The variation in the thickness d/d_0 (d_0 is the maximum thickness) is shown as a function of (ℓ/h) in Figure 9 (from Ref. 8). The graphs are drawn for different radii, r, of the source. h is the distance from the source to the point at which the deposition rate is maximum.

To get a film of uniform thickness, the source should have a large radius compared to the distance h between the source and the substrate. By rotating the source, one may improve the uniformity of film thickness. For very large area coatings, one may use multiple sources, distributed on a ring.

When depositing metallic alloys or intermetallic compounds, one should be aware that the different elements in the alloy or compound may deposit at different rates on the substrate. Thus, the stoichiometry (proportion of the various components in the alloy) of the deposited film may

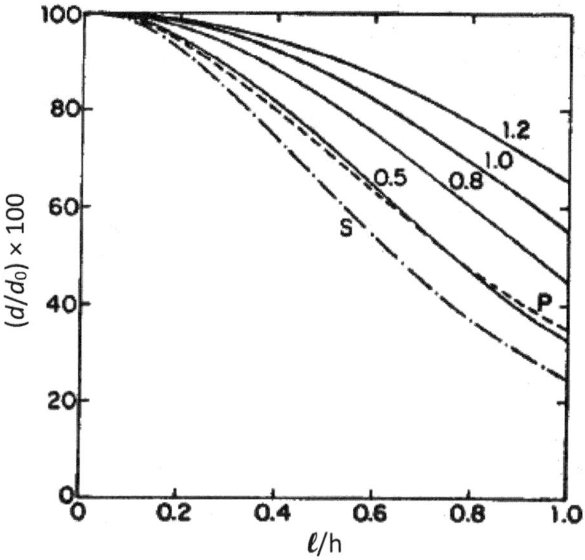

Figure 9. Deposited film thickness *d*, expressed as a fraction d/d_0 of maximum thickness d_0, as a function of ℓ/h for sources of different r/h values. Curve P is for a point source, and S is for a small area source (reprinted with permission from Ref.8).

differ from the stoichiometry of the source. Controlling the substrate temperature can improve the stoichiometry.

For evaporating oxides, one needs to achieve high temperatures of the order of 1500°C. At these temperatures, the oxides may dissociate. The oxygen released thereby may react with the material of the evaporation boat or with residual hydrocarbons in the vacuum chamber. This will affect the oxygen stoichiometry of the deposited film. To overcome this problem, one may employ laser beam evaporation or electron beam evaporation.

In pulsed laser deposition (Figure 10), pulses of nanosecond duration from a high- power excimer laser or Nd:YAG laser are focused on a rotating target through a transparent window in the vacuum chamber. Due to the high electric field of the laser light, high-energy electrons are produced, which oscillate in the electric field of the laser and collide with the atoms, transferring their energy to the solid material. The material gets heated locally and ejects a plume, which expands at a high velocity. The neutral molecules in the plume deposit on the substrate. The thickness

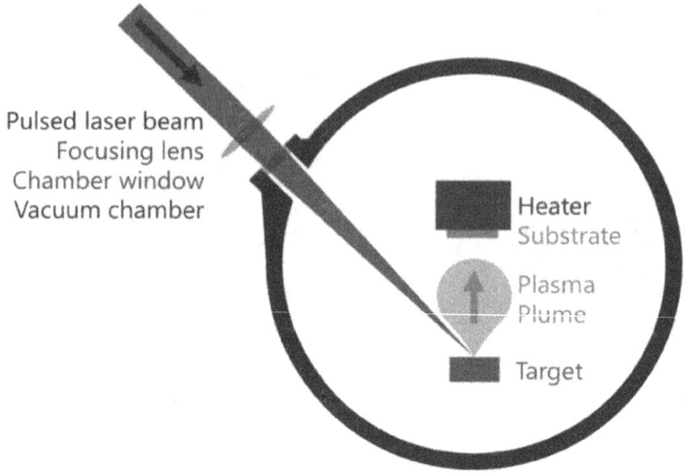

Figure 10. Schematic illustration of pulsed laser deposition.

of the deposited film is not uniform. The quality of the film depends on the pulse energy, the pulse repetition rate, and the temperature of the substrate. This method is best suited for depositing good-quality films of oxides and high-melting-point materials. The deposition rate is in the hundreds of angstroms per minute.

5.2. *Growth of film*

The atoms or molecules arrive individually on the substrate. After getting deposited, they migrate to nearby points on the substrate due to their kinetic energy. Some of them may leave the substrate through reevaporation. When two atoms collide, they may merge. In this way, clusters are produced, which grow into separate islands. Eventually, the islands merge to form a continuous film of the material on the substrate. It is obvious that this growth process will depend on the temperature of the substrate. Controlling the temperature of the substrate at an optimal value helps in the growth of thin films.

6. Deposition of Thin Films by Sputtering

A second process for the deposition of thin films is sputtering. Figure 11 (from Ref. 9) illustrates this process.

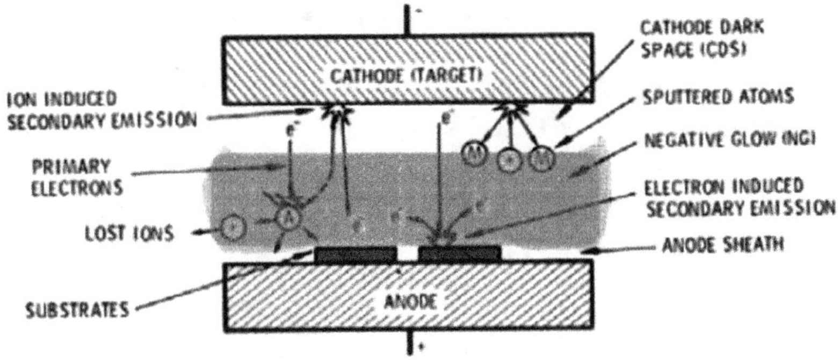

Figure 11. Schematic of sputter deposition (reprinted with permission from Ref.9).

The deposition is carried out inside a vacuum chamber filled with argon at a pressure of a few millibar. The material to be deposited, called the target, is the cathode itself. The substrate is mounted on the anode. A high voltage is applied between the anode and the cathode. A discharge is initiated in the argon gas at a pressure of a few millibar. The positively charged argon ions are attracted to the target, which is the cathode, and impinge on the target. The argon ions will transfer energy to the atoms in the target through elastic and inelastic processes. In elastic collisions, part of the kinetic energy of the argon ions is transferred to the target atoms. Those target atoms which acquire a larger kinetic energy than the binding energy of the atoms on the surface of the target will escape from the target. These are called sputtered atoms. All the sputtered atoms do not need to be neutral. Some of the sputtered atoms will be negatively charged, and some will be positively charged. The positively charged target atoms will fall back onto the target. The negatively charged target atoms will be attracted by the substrate which is maintained at ground potential. The neutral and negatively charged sputtered atoms will suffer collisions with the argon ions and will be scattered as they move toward the anode. After many such collisions, some of the sputtered atoms will get deposited on the substrate.

Secondary electrons produced by the inelastic collisions of the argon ions with the target help to maintain the plasma formed by ionizing the neutral argon atoms.

6.1. *Quantities which determine the rate of sputtering*

Only a fraction of the ions incident on the target will create target atoms for sputtering. The ratio of the number of target atoms for sputtering to

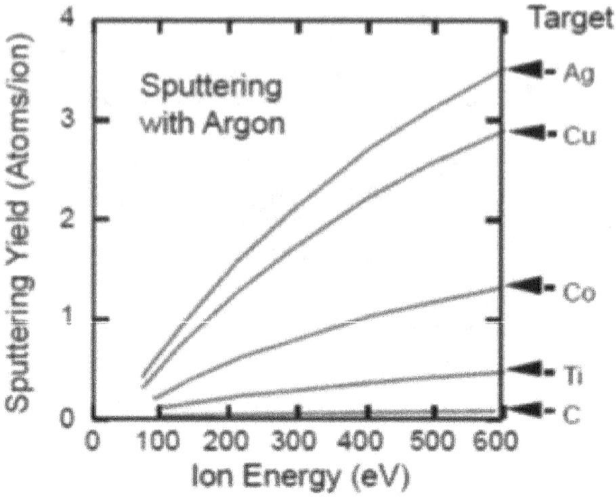

Figure 12. Sputtering yield as a function of ion energy for different targets with argon gas (reprinted with permission from Ref. 9).

the number of ions incident on the target is called the sputtering yield Y. The variation of sputtering yield as a function of the energy of the incident ions is shown in Figure 12 for different target materials, with argon as the rare gas.

Sputtering yield is observed to depend on (1) the energy E of the incident argon ions, (2) the surface binding energy U of the target atom, (3) the ratio of the mass of the target atom to the mass of the rare gas ion, M_t/M_i, and (4) the energy loss function $S_n(E)$ due to collisions of the argon ions with the target nuclei. There is a minimum energy of the ion, called the threshold energy E_{TH}, for sputtering target atoms. This threshold energy depends on the ratio of the mass of the target atom, M_t, to the mass of the rare gas ion, M_i, which is causing the sputtering. The variation of E_{th} with the mass ratio M_t/M_i is shown in Figure 13.

If J is the number of ions falling on the target in 1 s, the power in the ion beam is $P = JE$, where E is the energy of the incident ion. The J ions falling on the target produce $JY(E)$ sputtered target atoms in 1 s. The sputtering uses up an energy $JY(E)U$ to eject these atoms from the surface of the target. So, the energy efficiency of the sputtering process is

$$\eta = JY(E)U/JE = Y(E)U/E \tag{2}$$

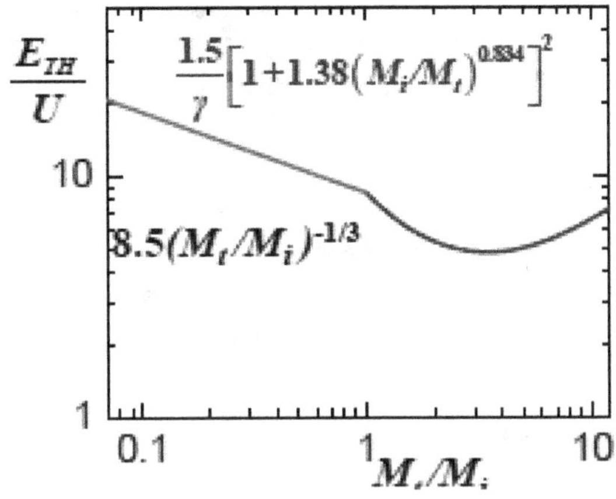

Figure 13. **Ratio of threshold energy E_{TH} to the surface binding energy U of the target atom plotted as a function of M_t/M_i (reprinted with permission from Ref. 9).**

The energy efficiency of sputtering is seen to reach a maximum, η_{max}, at an energy $E = 7E_{TH}$. In the range $3 < E/E_{TH} < 10$, the energy efficiency $\eta > 0.8\eta_{max}$. For copper, $U = 4$ eV and $E_{TH} = 30$ eV. So, efficiency is maximum at an energy E of 210 eV. Between 90 and 300 eV for the incident energy of the ions, the efficiency is greater than $0.8\eta_{max}$.

6.2. *Description of sputter deposition system*

A schematic diagram of a typical diode sputter deposition system is shown in Figure 14. The rare gas is channeled into the vacuum chamber in a continuous stream to maintain a pressure of about 50–80 mbar. The target is electrically insulated from the vacuum chamber. The vacuum chamber is earthed. The target is water cooled. The target is maintained at a negative voltage of a few kilovolts relative to the chamber. Surrounding the target but connected to the chamber is a dark shield. The argon ions are accelerated in the space between the dark shield and the target. The substrate on which the sputtered film is obtained is connected to a negative DC bias with respect to the chamber. The electrons created in the plasma will move towards the anode. They will also suffer collisions and be slowed down. To prevent the slow electrons from reaching the substrate, the substrate is maintained at a small negative potential. The substrate can be heated to maintain it at a high temperature.

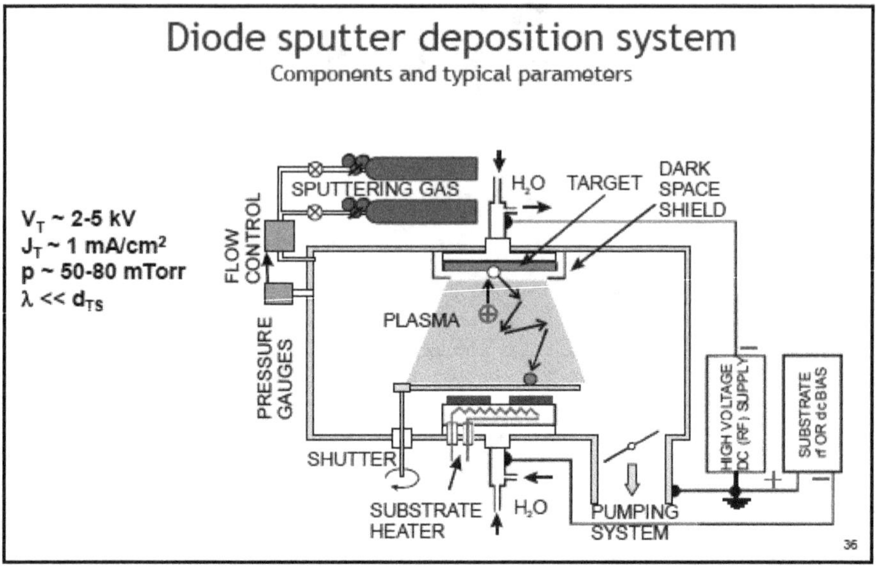

Figure 14. DC sputtering system (reprinted with permission from Ref. 9).

In a diode sputtering system, one needs to maintain a high argon pressure to have an appreciable flux of argon ions. This makes the mean free path of the target atoms low, so the film deposition rate is very low.

One can reduce the argon ion pressure and still maintain a satisfactory flux of ions by using a triode sputtering system. Here, we have a filament which, upon heating, emits electrons. These electrons get accelerated and produce more argon ions. The number of argon atoms which are ionized will depend on the filament current, which controls the output of the electrons. This decouples ion production from the energy of the ions which hit the target. Thus, the flux of argon ions and the energy with which the ions hit the target can be independently controlled. Since the pressure of argon gas in a triode sputtering system is much lower than the pressure in the diode sputtering system, the mean free path of the target atoms is longer in the triode sputtering system. This leads to a higher deposition rate of the target atoms on the substrate.

The ionization efficiency of the electrons can be increased by making them move along a spiral path. This is achieved by applying a magnetic field perpendicular to the electric field between the anode and the filament. The spiral path increases the ionization efficiency of the electrons, which increases the ion flux and reduces the potential to be applied to the target. This technique is called magnetron sputtering. A schematic illustration of this is shown in Figure 15.

The crossed electric and magnetic field (EXB) configuration can be achieved in several ways. Interested readers may consult the literature.

The target atoms may travel from the target to the substrate without collisions or may undergo many collisions with other atoms and diffuse to the substrate. Some of the sputtered atoms incident on the substrate may get reflected. The total deposition rate is mainly determined by the ballistic and diffusive processes.

DC sputtering has the disadvantage that it cannot be used with insulating materials as targets. As the ions impinge on the insulating target, the target becomes positively charged, repelling a further influx of ions. For insulating targets, RF sputtering is used. In RF sputtering, the

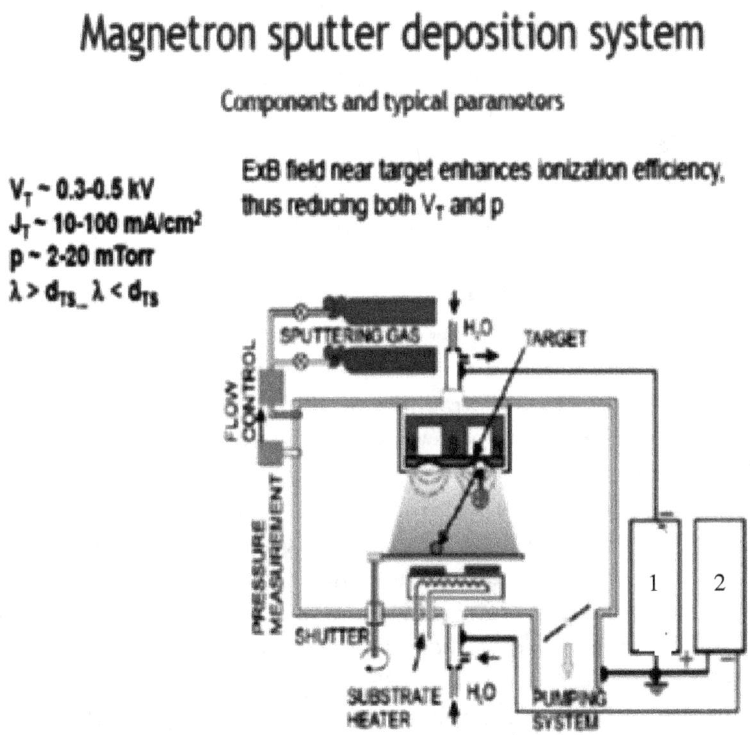

Figure 15. Magnetron sputtering system (reprinted with permission from Ref. 7): "1" denotes high-voltage DC power supply, and "2" is the substrate RF or DC bias voltage source.

potential difference between the cathode and the anode varies at radio frequencies.

RF sputtering has the following advantages. It can be used with metals, insulators, and composite targets. Charging effects on insulating targets are avoided. In this method, the operating gas pressure in the chamber is about 1–15 mbar. However, the deposition rates are low for some materials. Stray magnetic fields will disturb the sputtering process. Efficient heat removal from the target is necessary.

6.3. *Monitoring the thickness of the deposited film*

The thickness of the film during deposition can be monitored using a quartz crystal monitor. A thin quartz crystal with gold-plated electrodes is used for this purpose. The crystal vibrates mechanically with a natural frequency, f, when excited by an RF voltage applied across two opposite faces of the crystal. This frequency changes as the ions or atoms get deposited on the quartz crystal, thereby increasing its mass. The thickness of the film is monitored by measuring this frequency change. The mass deposited for the same thickness of the film depends on the density of the target material.

7. Molecular Beam Epitaxy

Molecular beam epitaxy (MBE) is a technique for growing thin films in ultrahigh vacuum with precise control of thickness, structure, and morphology. It permits the study of crystal growth in real time using reflection high-energy electron diffraction (RHEED), *in situ* X-ray diffraction, and scanning probe techniques. We can grow artificially layered crystalline films of great complexity with a high degree of control and reproducibility. This enables us to conduct studies on materials prepared as quantum dots (zero-dimensional), quantum wires (one-dimensional), and thin films (two-dimensional). We can prepare heterojunctions with improved performance. We can also prepare artificial materials even if there is a lattice mismatch or chemical incompatibility.

A schematic diagram of an MBE setup is shown in Figure 16 (from Ref. 11). An ultrahigh vacuum chamber is pumped by a turbopump (or cryopump) backed by a rotary pump. The pressure in the chamber must be about 10^{-9}–10^{-10} mbar. There are effusion cells, known as Knudsen cells (K-cells), attached to the vacuum chamber. These contain elements to be

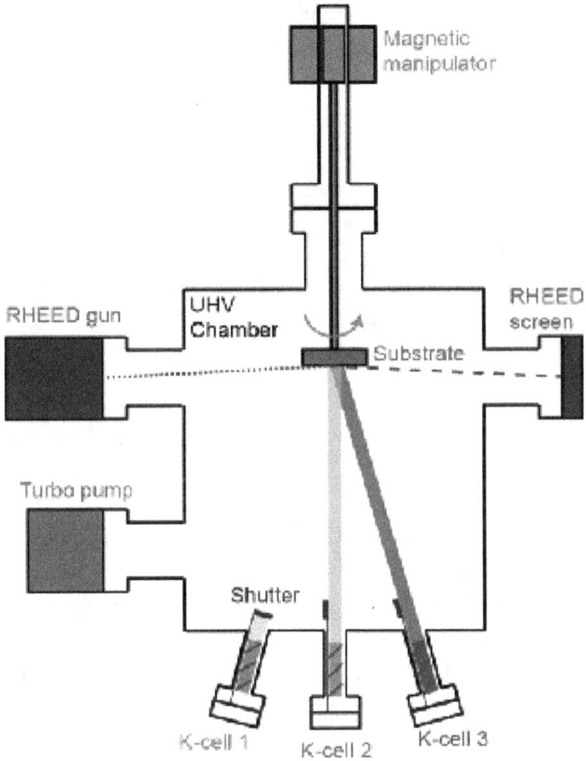

Figure 16. Schematic of an MBE setup (reprinted with permission from Ref. 11).

deposited. The materials used are in solid form and are of high purity (99.999% pure). The Knudsen cells can be heated so that the solids will sublime at low temperatures at the low pressure in the chamber. The mean free path for the atoms is very large. So, the atoms will travel in straight lines prior to being deposited on the substrate. Shutters in front of the Knudsen cells can be opened or closed to deposit the atoms of different elements in sequence. The rate of deposition is slow: about one atomic layer per second. The substrate can be heated and can also be rotated at a slow speed for a uniform deposition. Because of the low rate of deposition, the atoms have enough time to rearrange themselves into a perfect lattice structure.

The RHEED unit has an electron gun that produces high-energy electrons, which are incident on the film at a large angle of incidence. The

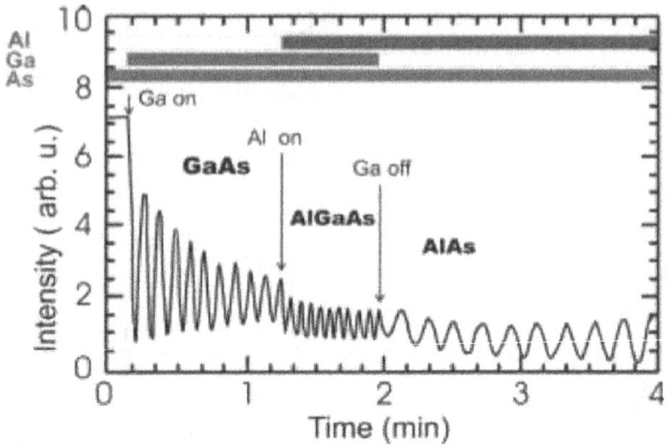

Figure 17. Growth of layers of GaAs, AlGaAs, and AlAs on top of one another (reprinted with permission from Ref. 10).

diffraction pattern is seen on a CCD camera, and the data are quickly analyzed on a computer. This enables monitoring the growth of the film *in situ* and in real time. One can also mount the scanning probes inside the vacuum chamber.

Figure 17 (from Ref. 10) shows the intensity of the RHEED pattern, where layers of AlGaAs are grown on layers of GaAs, and then, layers of AlAs are grown on layers of AlGaAs to form heterojunctions of III–V compounds.

MBE is a very powerful but expensive technique.

8. Conclusion

In this chapter, we have given a description of several deposition techniques for preparing thin films. The choice of the technique to be used will depend on the material of the film and the facilities available in the laboratory.

References

1. Dip Coating Theory | Dip Coating Thin Films, Complete Guide, www.ossila.com›pages›dip-coating-theory-film-thic.
2. Spin Coating: Complete Guide to Theory and Techniques, www.ossila.com›pages›spincoating.

3. Chapter 3: Preparative methods, www.unf.edu›chem4627›ch3_solid_state.
4. Wang, Q., Wu, G., Liu, S., Gan, Z., Yang, B., and Pan, J. (2019).*Crystals*, **9**(6), 320, https://doi.org/10.3390/cryst9060320.
5. Smith, D. (1995). *Thin Film Deposition — Principles and Practice*. McGraw Hill.
6. George, J. (1992). *Preparation of Thin Films*. CRC Press.
7. Srinivasan, R., Priolkar, K. R., and Ramesh, T. G. (2019). *A Manual on Experiments in Physics*. Indian Academy of Sciences, Bangalore.
8. Hurley, R. E. Vacuum Evaporation of Thin Films. School of Electrical and Electronics Engineering, Queen's University, Belfast, ftp://wsdetcp.upct.es/FelixM/...pdf/sc_Cartagena6a_vac-evap.pdf.
9. Petrov, I. (2011). Fundamentals of Deposition. Sputter Deposition, Nucleation and Growth 2011, www.mechanical.illinois.edu/.../sputterdepositionnucleationgrowth.20110316.
10. Ploog, K. Molecular Beam Epitaxy, www.aps.org>meetings.presentations>uploaf>ploog.
11. Molecular Beam Epitaxy (MBE) — Zeljkovic Lab, www.capricorn.bc.edu>zeljkoviclab>research.

Part II

Techniques for Materials Characterization

Chapter 3

X-RAY AND NEUTRON POWDER DIFFRACTION

1. Introduction

X-rays are produced when high-energy electrons are incident on a solid target made of metal. Figure 1 illustrates how X-rays are produced. Electrons are produced from a hot filament by thermionic emission. They are accelerated by the application of a potential difference of several kilovolts.

The incident high-energy electron ejects an electron (the small white circle in Figure 1) from the innermost shell ($n = 1$, K shell) of electrons of the target atom. An electron from the L shell can fill the hole created in the K shell. The difference in energy resulting from such a transition by the electron in the L shell emerges as a Kα X-ray photon. The Kα X-ray spectral line is a closely spaced doublet. On the other hand, if an electron in the M shell (the next higher shell) fills the hole in the K shell, the photon generated is called the K_β X-ray photon, and the spectral line is called the K_β line. The K_β photon has a higher energy than the K_α photon. The K_α and K_β X-ray lines have characteristic wavelengths that are slightly different from each other. If the target material is copper, the wavelength of the K_α photon is 1.541 Å and that of the K_β photon is 1.392 Å.

The wavelengths of the X-ray photons from some of the commonly used targets are given in Table 1. Here, $K_{\alpha 1}$ and $K_{\alpha 2}$ are the two closely spaced lines of the doublet.

The X-ray tube is a high vacuum sealed unit, normally operating at a power between 1.8 and 3 kW. The anode in the X-ray tube has to be water cooled to remove the heat generated in the anode when the

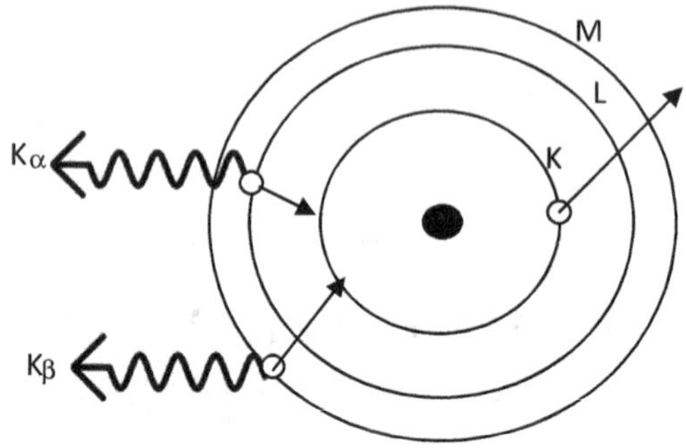

Figure 1. Production of K_α and K_β X-rays.

Table 1. Wavelengths of the X-ray lines from commonly used targets (Ref. 1).

Target	Line	Wavelength (Å)	Target	Line	Wavelength (Å)
Copper	$K_{\alpha1}$	1.540598	Cobalt	$K_{\alpha1}$	1.78901
	$K_{\alpha2}$	1.544426		$K_{\alpha2}$	1.7929
	K_β	1.39225		K_β	1.62083
Molybdenum	$K_{\alpha1}$	0.709319	Chromium	$K_{\alpha1}$	2.28976
	$K_{\alpha2}$	0.713609		$K_{\alpha2}$	2.293663
	K_β	0.632305		K_β	2.08492

high-energy electron beam is incident on it. The rotating anode X-ray tube is operated at a higher power (9–18 kW). By spinning the anode at a high speed of a few thousand RPM, the electron beam is made to fall on different spots in the target. So, any one spot in the target is not heated continuously. That is why one can achieve a higher power in the rotating anode X-ray tube.

X-rays come out of the tube through a beryllium window. **When the X-ray tube is in use, one should not expose oneself to this radiation. Suitable safety measures must be taken.**

2. Monochromatization and Collimation of the Radiation

For powder X-ray diffraction, one needs a beam of radiation of a single wavelength.

Monochromatization, i.e., isolating a radiation of a single wavelength from a mixture, can be achieved by eliminating the K_β radiation, which is usually weaker in intensity than the K_α radiation. This is done either by using an absorption filter or a monochromator. The filter is the cheaper solution.

Filtering by absorption is done by placing a foil of an element which is the one preceding the target material in the periodic table. For example, if Cu is the target material, we place a foil of Ni in the path of the X-ray beam from the tube. Ni is adjacent to Cu in the periodic table and has a lower atomic number. The K absorption edge of Ni is at 1.488 Å, which is in between the wavelengths of the K_α and K_β lines of copper. A Ni foil absorbs 50% of K_α and 99% of K_β radiation of copper (Figure 2). Thus, by using a foil of nickel as the filter, we can obtain almost pure K_α radiation of copper. However, the radiation absorbed by the filter is reemitted at longer wavelengths, which adds to the background radiation.

The radiation emerging from the X-ray tube is a diverging beam. This divergence should be reduced. The simplest way to limit divergence is to use a sheet of suitable absorber material with a rectangular slit (opening) in it. X-rays incident on the slit will pass through, whereas the rays falling on the solid portion of the sheet will be absorbed. This principle is illustrated in Figure 3.

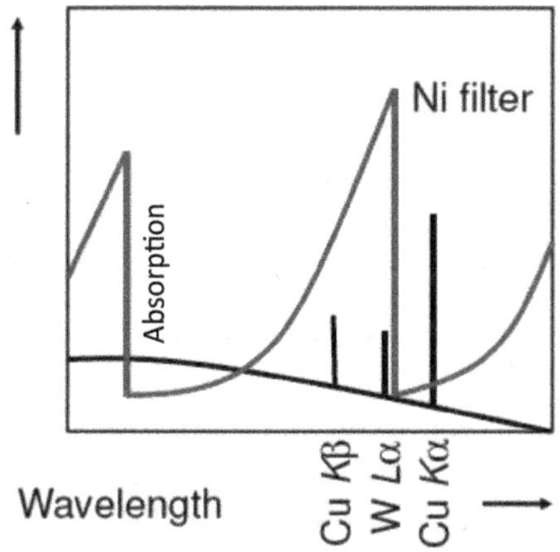

Figure 2. X-ray absorption versus wavelength for nickel (red graph) and X-ray emission versus wavelength for Cu K_α radiation (black graph) for the operation of a Ni filter for Cu K_α radiation (reprinted with permission from Ref. 1).

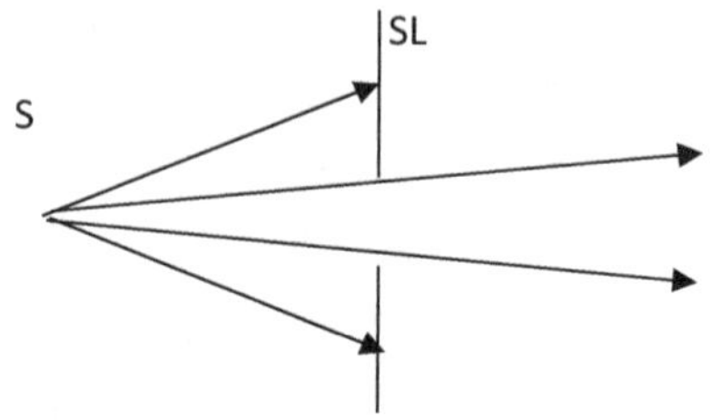

Figure 3. **Use of a slit (SL) to reduce the divergence of X-rays from a source (S).**

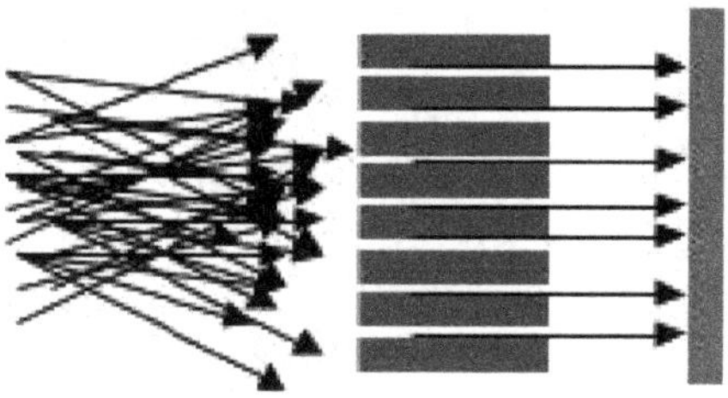

Figure 4. **Parallel-plate collimator.**

The narrower the slit, the more collimated the beam, but at the cost of reduced intensity of the beam. The width of the slit is determined (1) by the desired intensity of the beam at the test sample and (2) by the width of the sample.

Depending on the application, one may also use a parallel-plate collimator. Such a collimator is an array of open channels formed by a stack of absorber plates, as shown in Figure 4.

Unlike a single slit, which limits the width of the beam, a parallel-plate collimator does not limit the width. Such a collimator is called a Soller slit.

3. X-ray Detectors

In the early days, photographic film was used to detect the X-rays. Nowadays, a charge-coupled device (CCD) plate is used as the detector to count the number of X-ray photons incident on the surface of the detector.

4. Sample Preparation for a Powder Diffractometer

The powder sample should be densely packed in a sample holder so as to achieve a smooth flat surface. The grains in the packed sample should ideally have all possible orientations. The grains should be less than 10μm in size in any direction.

If the grain size is large or if the grains are not oriented randomly, the peak intensities may vary, and the diffraction pattern obtained will not be that of an ideal powder. In particular, the peak positions may not agree with the patterns given in the database.

Typically, we may fill a cylindrical well in the sample holder with finely ground powder and press it down with a glass rod or pestle. One may also mix the sample powder with a filler to randomize the orientation of the grains. Alternatively, one may sprinkle the sample powder on a glass plate and put this plate in the sample holder. One may also use a double-sided sticker tape, which is stuck to the sample holder, and on the exposed side of the tape, the powder is sprinkled.

5. Principle of Powder Diffraction of X-rays

Figure 5 illustrates the diffraction of X-rays by a set of parallel planes in a crystal. In each plane, atoms are identically arranged in a regular geometry.

The set of parallel planes in a crystal is specified by three indices (h, k, l). The unit normal to the crystal plane is indicated by the vector **s**. θ is the glancing angle made by the incident X-rays with the parallel

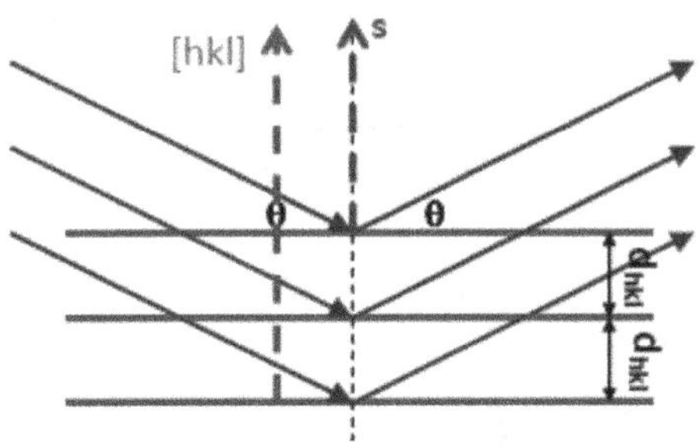

Figure 5. Diffraction of X-rays by a set of parallel planes in a crystal.

planes. The glancing angle of reflection is also θ. The rays reflected by the successive planes will be in phase if the Bragg's condition

$$2d_{h,k,l}\sin(\theta_{h,k,l}) = n\lambda \quad (1)$$

is satisfied, where λ is the wavelength of the incident X-rays and *n* is a positive integer. When *n* is greater than 1, the reflected X-ray beam is indexed as (*nh, nk, nl*). Thus, a (1, 1, 0) reflection will correspond to a reflection from the (110) planes with $n = 1$ in Equation (1), whereas a (2, 2, 0) reflection is the reflection from the same set of (110) planes with $n = 2$.

When the sample is a powder with random orientation of the grains, the diffracted X-ray beams from different grains will form a cone of diffraction angle 2θ, as shown in Figure 6(a).

The Ewald's sphere is a sphere of radius $2\pi/\lambda$, with its center at the spot where the incident X-ray falls on the powder specimen.

When the cone of rays is intercepted by a flat detector perpendicular to the incident X-ray direction, the diffraction pattern is recorded as a set

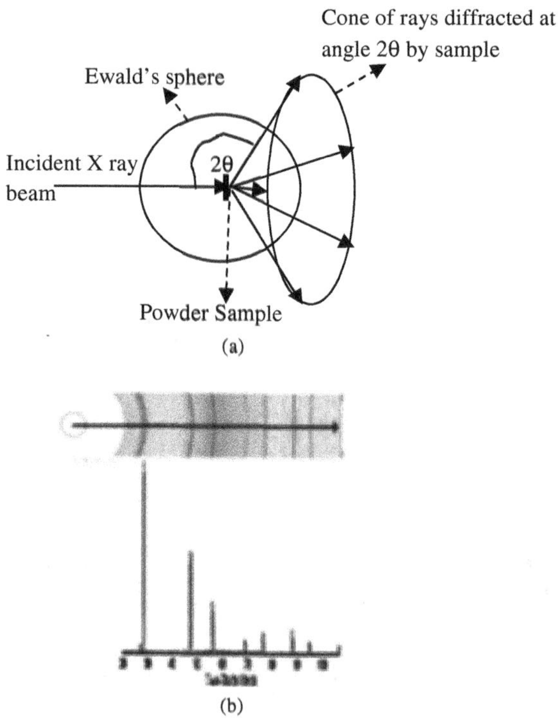

Figure 6. Diffraction pattern from a powder sample.

of rings, as shown in Figure 6(b). If the intensity of the diffraction pattern is plotted as a function of angle, we get a series of peaks. This is how a powder diffraction pattern appears.

6. Factors Affecting Powder Diffraction Patterns

Powder X-ray diffraction patterns are usually obtained experimentally in the Bragg–Brentano geometry. In this geometry, the glancing angle of the incident X-ray beam is changed by rotating the X-ray source relative to the sample by an angle Δ along a circle. Simultaneously, the detector is rotated by Δ in the opposite direction on the same circle relative to the sample to receive the diffracted beam. This is shown in Figure 7.

Whatever the glancing angle, the angle between the incident and diffracted beams is always twice the glancing angle of the incident beam, as shown in Figure 7. In actual experiments, the X-ray source and the detector are rotated simultaneously in the opposite sense at a constant rate. Furthermore, in Figure 7, $\boldsymbol{k}$ is a vector of length $2\pi/\lambda$ in the direction of the incident beam, and $\boldsymbol{k}'$ is a vector of the same length in the direction of the diffracted beam. In this geometry, the diffraction vector, $\boldsymbol{s} = (\boldsymbol{k}' - \boldsymbol{k})$, is always normal to the sample surface. We may note the following:

(a) As the incident beam is not exactly parallel but is slightly divergent, the width of the sample irradiated will decrease as the incident glancing angle is increased, as shown in Figure 8.

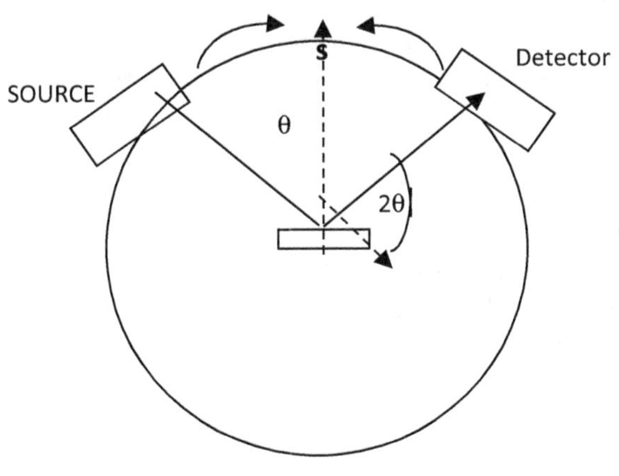

Figure 7. Bragg–Brentano geometry for powder X-ray diffraction.

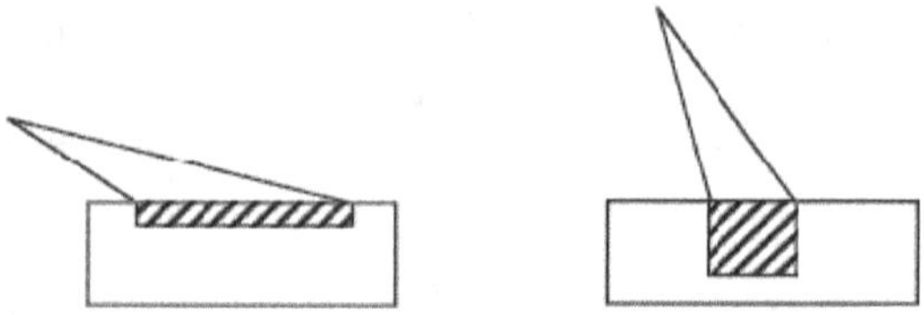

Figure 8. Change in the width of the irradiated spot on the sample as the glancing angle changes in a divergent beam.

Simultaneously with the change in width, the depth of penetration into the sample also changes, increasing as the glancing angle increases. The two contrasting effects will cause the volume of the sample diffracting the X-rays to be constant. If the sample is not homogeneous, this will cause the diffraction pattern to be inaccurate. For thin samples, the assumption of equal volumes for different glancing angles will not hold good.

(b) A divergent incident beam will give rise to a divergent diffracted beam, causing an increase in the width of the diffraction line and an uncertainty in the glancing angle. The smaller the divergence of the incident beam, the smaller this error. Reducing the divergence of the beam results in lowering the intensity of the incident beam, and it will take longer to record the diffraction pattern.

(c) If the sample is not a fine powder, the crystallites will not have all possible orientations. This causes a variation in intensity at different points on the diffraction ring.

(d) If we are not using a Soller slit to producc a parallel beam, then the source and detector are arranged such that the incident divergent X-rays are focused at the detector after Bragg reflection, as shown in Figure 9. If the detector is not located on the focusing circle, the diffraction rings will appear broad.

(e) The sample may not be located on the focusing circle but may be displaced relative to this circle. This is shown in Figure 10. The glancing angle will then be in error, which will increase as the displacement of the sample from the circle becomes more.

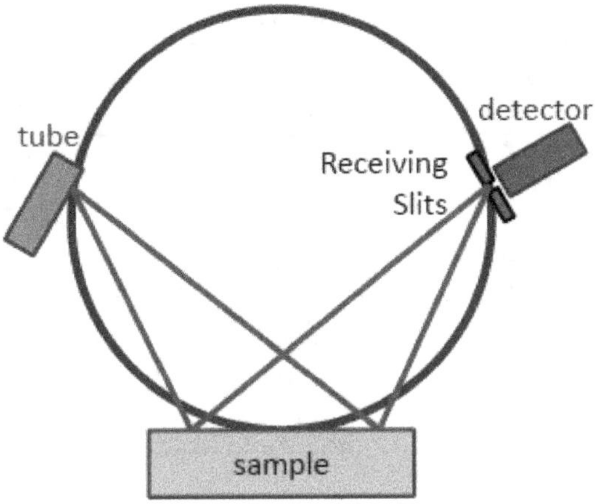

Figure 9. **Diffraction circle for focusing an incident divergent beam.**

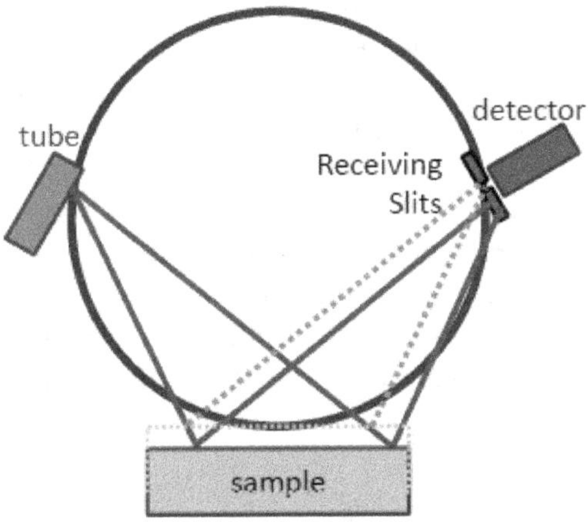

Figure 10. **Sample displacement error.**

7. Indexing a Powder Diffraction Pattern

Table 2 gives the seven crystal systems and the 14 Bravais lattices belonging to these systems in columns 1 and 2. The relations between the lattice parameters and the angles between the three primitive lattice vectors for the different crystal systems are listed in the last column.

Table 2. Crystal systems and Bravais lattices.

Crystal system	Bravais lattices	Symmetry	Axis system
Cubic	P,I,F	m3m	a = b = c; $\alpha = \beta = \gamma = 90°$
Tetragonal	P,I	4/mmm	a = b ≠ c; $\alpha = \beta = \gamma = 90°$
Hexagonal	P	6/mmm	a = b ≠ c; $\alpha = \beta = 90°$, $\gamma = 120°$
Rhombohedral	R	3m	a = b = c; $\alpha = \beta = \gamma \neq 90°$
Orthorhombic	P,C,I,F	mmm	a ≠ b ≠ c; $\alpha = \beta = \gamma = 90°$
Monoclinic	P,C	2/m	a ≠ b ≠ c; $\alpha = \gamma = 90°$, $\beta \neq 90°$
Triclinic	P	1	a ≠ b ≠ c; $\alpha \neq \beta \neq \gamma \neq 90°$

In column 2, P stands for the primitive lattice, I for the body-centered lattice, and F for the face-centered lattice. R stands for the rhombohedral lattice. C denotes a lattice with only the two faces, perpendicular to the unique axis C, having a lattice point at their centers.

Once the diffraction pattern is obtained, one will have to index the various lines, i.e., assign (h, k, l) values. First, we have to make a guess as to which of the seven crystal systems the sample belongs to and then look for systematic variations in the sin(θ) values. We illustrate the procedure for the case of the simple cubic system as follows.

In this system, the three primitive translations are equal in length, say a_0, and the three translations along the three crystal axes are mutually orthogonal. So, a set of planes (h, k, l) in this system will have an interplanar spacing, $d_{h,k,l}$, given by

$$d_{h,k,l} = a_0/(h^2 + k^2 + l^2)^{1/2} \tag{2}$$

So, from Equation (1), we get

$$\sin^2(\theta_{h,k,l}) = (\lambda/2a_0)^2 m \tag{3}$$

where

$$m = (h^2 + k^2 + l^2) \tag{4}$$

is an integer since h, k, and l are integers.

In the cubic system, we may have the following lattice types:

(a) simple cubic (SC) lattice, in which each of the eight corners has a lattice point;
(b) body-centered cubic (BCC) lattice, in which there is a lattice point at each of the eight corners of the cube and one lattice point at the center of the cube;
(c) face-centered cubic (FCC) lattice, in which there is a lattice point at each of the eight corners of the cube and at the center of each of the six faces of the cube; and
(d) the diamond lattice (diamond), which is FCC but in which every atom is surrounded by four identical atoms at the corners of a tetrahedron.

The possible values of m for various reflections in these four lattices are different, as shown in Table 3.

For the simple cubic lattice, the value of m mut be equal to the sum of the squares of three integers or zero. While the numbers from 1 to 6 can be represented this way (for example: $3 = (1^2 + 1^2 + 1^2)$ and $6 = (2^2 + 1^2 + 1^2)$), the numbers 7, 15, etc. cannot be so represented. So, these numbers are missing in the sequence for m for the simple cubic lattice. To understand the missing numbers in the sequences for the other lattices, one will have to delve into the theory of reciprocal lattice. This lengthy discussion is omitted here. For example, in addition to m being the sum of the squares of three integers h, k, and l, the sum $h + k + l$ must be an even number for BCC lattice. So, the reflections $m = 1$ or 3, for example, cannot occur in a

Table 3. Possible values of m in Equation (4) for the four different cubic lattices.

Lattice	Sequence of m values
Simple cubic	1,2,3,4,5,6,8,9,10,11,12,13,14,16,...
Body-centered cubic	2,4,6,8,10,12,14,16,...
Face-centered cubic	3,4,8,11,12,16,19,20,...
Diamond lattice	3,8,11,16,19,...

BCC lattice since $(h + k + l) = (1 + 0 + 0) = 1$ in the first case and $(h + k + l) = (1 + 1 + 1) = 3$ in the second, which violates the condition that $h + k + l$ must be even. For FCC, the indices h, k, and l must be all odd or all even. So, $m = 1$ ($h = 1, k = 0, l = 0$), $m = 2$ ($h = 1, k = 1, l = 0$) violate this condition. Hence, there are no reflections with $m = 1$ or $m = 2$.

Starting from the center of the diffraction pattern, the ratio of $\sin^2(\theta)$ values for the successive lines is calculated from the observed diffraction pattern. By comparing the ratios with the sequence of values of m in Table 3, one can identify the lattice system, the (h, k, l) values of the reflection planes, and calculate the lattice constant, a_0, of the crystal.

For other crystal systems, the procedure is more complicated. But in any case, from the diffraction pattern, the crystal system and the lattice parameters can be determined.

8. Powder X-ray Diffraction Data Bank

The powder X-ray diffraction data of all materials that have been studied so far are preserved in the International Centre for Diffraction Data (ICDD). A PDF 4+ file-2016 contains 384,613 entries. The file can be accessed on a computer using software provided by third-party vendors.

9. Applications of X-ray Powder Diffraction

9.1. *Phase identification*

A given material can be present in many different phases with different crystal structures. For example, silica can be in an amorphous phase or in two different crystalline phases, cristobalite and quartz. The amorphous phase will give a broad diffraction pattern without sharp peaks. The two crystalline phases are characterized by different diffraction patterns, comprising sharp peaks, with each peak corresponding to different glancing angles. If the sample is a mixture of all the three phases, then the diffraction pattern will be a sum of all these features, as shown on the right-hand side of Figure 11. From the positions of the peaks and the relative intensities, a quantitative estimation of the proportion of different phases in the mixture can be made.

Often, one tries to produce a new material, C, from a solid-state reaction of two other materials, A and B, mixed in the appropriate proportion. If the reaction is not complete, the diffraction pattern will show the peaks

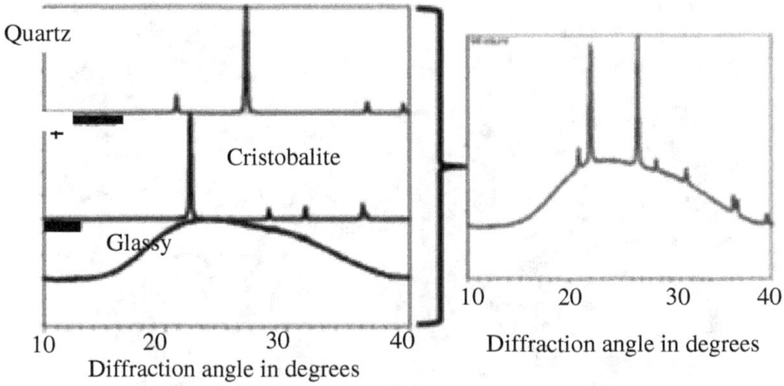

Figure 11. Left: Individual diffraction patterns of SiO_2 in the quartz, cristobalite and amorphous (glassy) phases. Right: Diffraction pattern when all three phases are present in the sample (extracted from Ref. 1).

due to materials A, B, and C. If the reaction is complete, only the peaks due to the material C will appear.

9.2. *Texture*

When the grains of the sample do not have a random orientation, the intensity will vary along the diffraction ring. This is shown in Figure 12.

The occurrence of such local regions with the preferred orientation is called "Texture."

By mapping the intensity of a single diffraction ring as a function of the position on the ring, one can determine the preferred crystal orientation in any part of the sample.

9.3. *Crystallite size*

When the crystallite size is less than 100 μm, the width of the diffraction peak in a powder sample increases as the grain size decreases.

After subtracting out instrumental broadening, the width β (in radian) of the diffraction peak at half the maximum intensity (FWHM) is determined. From this β value, the crystallite size τ is calculated from the Scherrer equation:

$$\tau = K\lambda/(\beta\cos(\theta)) \qquad (5)$$

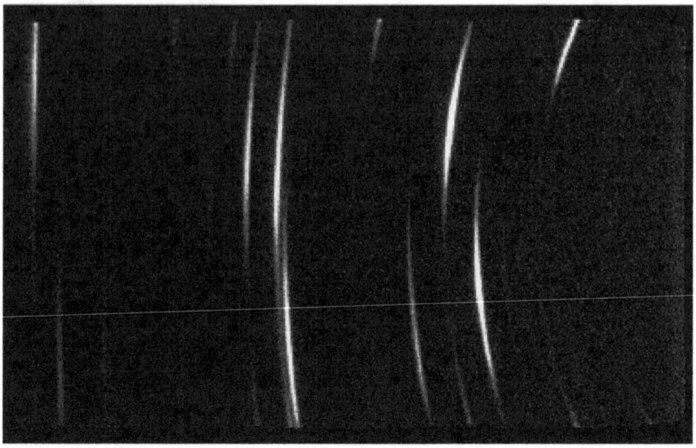

Figure 12. Diffraction pattern in the case of powders in which there is a preferred orientation (extracted from Ref. 1).

Here, K is a constant close to 0.9 and θ is the glancing angle of diffraction. One can determine τ from a number of prominent diffraction rings and calculate the average value of the crystallite size τ.

9.4. *Rietveld refinement for crystal structure determination*

In the previous sections, we have discussed the information to be obtained from a few of the diffraction peaks. Alternatively, one can fit the entire pattern over all angles θ using a procedure called Rietveld refinement with commercially available software. Using software, one can find the best fit for the entire diffraction pattern. If there is an unknown material in the sample, this refinement procedure gives information on the crystal structure of this unknown material.

Figure 13 shows a low-temperature, high-magnetic-field powder diffractometer available at the UGC-DAE Consortium for Scientific Research in Indore. This setup works on symmetric Bragg–Brentano geometry using a parallel X-ray beam from a rotating anode source working at 17 kW. Using this, one can carry out structural studies under nonambient conditions, i.e., at different temperatures (2–300 K) and different high magnetic fields (from +8 to −8 T). The available scattering angle ranges from $2\theta = 5°$ to $115°$ with a resolution better than

Figure 13. Powder X-ray diffractometer for low-temperature high-magnetic-field studies at UGC-DAE CSR, Indore (reprinted with permission from csr.res.in).

0.1° (Aga Shahee, Shivani Sharma, K. Singh, N. P. Lalla, and P. Chaddah, csr.res.in).

10. Other X-ray Diffraction Techniques

There are other X-ray diffraction techniques, such as:

(a) **Single-crystal X-ray diffraction:** For this, a single crystal of the material is required. Single-crystal diffractometers are more complicated in construction and operation. One can obtain complete structural information of the crystal through this technique.

(b) **Grazing angle X-ray diffraction:** In this technique, the incident glancing angle is made small. In such a situation, the X-rays penetrate a short thickness near the surface of the sample. This is a surface-sensitive technique and is used to study thin films.

(c) **Small-angle X-ray scattering:** In this technique, the intensity of X-rays scattered through a small angle (<10°) when the X-rays are transmitted through a material is measured. This technique is used to estimate the distribution in the nanoparticle and pore sizes in the material.

(d) **X-ray reflectivity:** In this technique, the intensity of X-rays reflected at various angles of incidence from a flat surface is measured. If the reflecting surface is not planar and smooth, the reflection coefficient calculated from the measured data will differ from the theoretical values obtained from the Fresnel coefficients of reflection. From these deviations, the inhomogeneities in the surface can be characterized.

A detailed discussion of the above techniques is beyond the scope of this book.

11. Limitations of X-ray Diffraction

X-rays are scattered by electrons in atoms. Only the charge on the electron influences the scattering and not the electron spin.

The limitations of X-ray diffraction techniques are:

(1) Atoms with a high atomic number Z have more electrons and scatter X-rays strongly since the scattering strength varies as Z^2. Atoms with a high atomic number are also heavier in mass. Consequently, it will be difficult to detect the presence of light atoms, such as hydrogen, in the presence of heavy atoms.

(2) When a magnetic material undergoes a transition from a disordered paramagnetic state to an ordered magnetic state, the orientation of electron spins in the material gets ordered. Since the scattering of X-rays is independent of the spin orientation, magnetic transitions cannot be detected by X-rays.

12. Neutron Diffraction

12.1. *Coherent and incoherent scattering cross-sections*

Neutrons, produced by nuclear reactions in the core of a nuclear reactor, have very high energies. These neutrons are slowed down by collisions with deuterons in heavy water or with carbon atoms in a graphite moderator. Thereby, they acquire a Maxwellian distribution of energies at the temperature of the moderator. The average kinetic energy of a neutron at a temperature of 300 K is $(3/2)kT$, where k is the Boltzmann's constant. Such a neutron will have a momentum $p = (3mkT)^{1/2}$, where m is the mass of the neutron. The de Broglie wavelength of the neutron will be h/p. So, a neutron having an average thermal energy at 300 K will have a wavelength of 0.146 nm. This is similar in magnitude to X-ray wavelengths. Hence, a beam of such neutrons will be diffracted by crystal planes, similar to X-rays.

There is a major difference between X-ray and neutron diffractions. While X-rays are scattered by electrons due to the electromagnetic interaction, the neutron, which is electrically neutral, is scattered by the strong forces within the nucleus. The strength of the scattering force is indicated by the scattering length b. This is the radius of a hard sphere, which will scatter the neutrons in equal measure as the nucleus. The incident neutron beam has a spin $\pm 1/2$. All nuclei of atoms have spin angular momentum, characterized by a quantum number I. In the incident beam, the neutrons are distributed randomly, with 50% of neutrons having spin $s = +1/2$ and 50% of neutrons having spin $s = -1/2$. When a neutron is scattered by the nucleus of an atom, the total spin of the nucleus and the neutron can be $M = I + ½$ or $M = I - ½$. The scattering length, in the first case, will be denoted by b^+ and in the second case by b^-. The state M has a degeneracy equal to $(2M + 1)$. So, the number of states for which $b = b^+$ is $2(I + 1)$, and the number of states for which $b = b^-$ is $2I$. The total scattering cross section σ is related to the scattering length b by

$$\sigma = 4\pi b^2 \tag{6}$$

The scattering lengths b^+ and b^- are different and have weightages $2(I + 1)$ and $2I$. So, the average scattering length for the given nucleus is

$$\langle b \rangle = [2(I + 1)b^+ + 2Ib^-]/2(2I + 1) \tag{7}$$

The average of the square of the scattering length is

$$<b^2> = [2(I+1)b^{+2} + 2Ib^{-2}]/2(2I+1) \quad (8)$$

The average value of $<b>$ is a measure of **coherent scattering**, which gives rise to diffraction peaks. This is called the **coherent scattering length** b_{COH}. The root mean square fluctuations in the value of b are a measure of **incoherent scattering of the neutrons, leading to a broad background**. The scattering length b_{INCOH} is given by

$$b_{INCOH} = [<b^2> - <b>^2]^{1/2} \quad (9)$$

For the hydrogen nucleus, the value of $I = ½$ and b^+ is 10.8 fm and b^- is −47.4 fm. One femtometer (fm) is 10^{-15} m.

So,

$$b_{COH} = -3.75 \text{ fm} \quad (10a)$$

and

$$b_{INCOH} = 25.2 \text{ fm} \quad (10b)$$

The corresponding cross-sections are

$$\sigma_{COH} = 4\pi<b_{COH}>^2 = 176 \text{ fm}^2 = 1.76 \text{ barn} \quad (11a)$$

and

$$\sigma_{INCOH} = 4\pi<b_{INCOH}>^2 = 7977 \text{ fm}^2 = 79.77 \text{ barn} \quad (11b)$$

One barn is 100 fm^2. Thus, hydrogen atoms in the sample will produce much more incoherent scattering than coherent scattering. Hence, the neutron diffraction peaks due to hydrogen will be submerged in the incoherent background.

On the other hand, for deuterium, the coherent scattering cross-section is 5.59 barns, while the incoherent scattering cross-section is 2.05 barns. In this case, the diffraction peaks will stand out against a much reduced background. The advantage of neutron scattering is that we can substitute one isotope for another to increase coherent scattering without affecting the crystal structure. There are other isotopes of elements for which the incoherent scattering cross-section will be much smaller than the coherent scattering cross-section, and the samples containing these isotopes will yield prominent diffraction peaks.

12.2. *Advantages of neutron diffraction*

(a) Ability to distinguish light elements in the presence of heavy elements:

In Figure 14, we plot the neutron coherent scattering length as a function of atomic number for various elements in the periodic table.

We note that the coherent scattering lengths for light elements are comparable in magnitude to the coherent scattering lengths of heavy elements. On the other hand, for X-ray scattering, the scattering cross-section varies as Z^2, where Z is the atomic number of the element. Thus, heavy elements will scatter X-rays much more strongly than light elements. One cannot detect the signature of light elements in X-ray scattering data in the presence of heavy elements. On the contrary, one can determine from neutron diffraction data the position of both light and heavy elements.

(b) Ability to distinguish between different types of magnetic ordering:

Magnetic materials contain atoms which have permanent magnetic moments because of the orbital and spin angular momenta of the

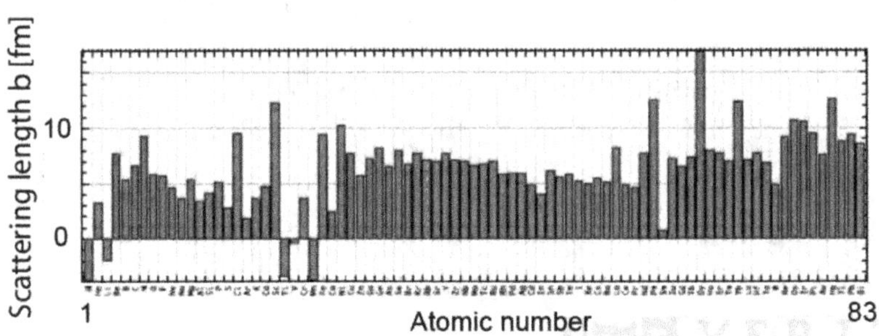

Figure 14. Neutron coherent scattering length as a function of atomic number for various elements.

electrons in the atoms. At high temperatures, the orientation of the magnetic moments of the atoms is disordered. This phase is called the paramagnetic phase. For a one-dimensional chain of atoms, this is schematically indicated in Figure 15, assuming only up and down orientations for the magnetic moment.

In Figure 15, the orientation of the magnetic moments is random. The net magnetic moment in the infinite chain is zero, and hence, this is the paramagnetic phase.

When the material is cooled, the magnetic moments can become ordered. Two types of ordering are shown in Figures 16(a) and (b).

In Figure 16(a), the magnetic moments of all atoms are parallelly aligned. So, there is a large net magnetic moment for this chain. This is called the ferromagnetic phase. In Figure 16(b), the adjacent atoms have antiparallel alignment of the magnetic moments. In this case, the system is highly ordered, but the net magnetic moment is zero. This phase is called the antiferromagnetic phase.

X-rays do not interact with the magnetic moments of the atoms. So, for X-rays, all three arrangements will look alike. The lattice constant determined by X-rays will be the distance between adjacent atoms in all three cases. The X-ray diffraction pattern will not show any difference.

Neutrons have a magnetic moment. So, a neutron can interact with the magnetic moment of an electron. In the paramagnetic phase, due to the disordered arrangement, the magnetic moments of the atoms will contribute only to an incoherent background scattering. The neutron

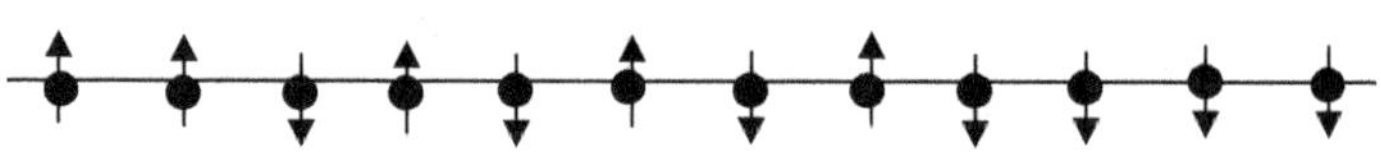

Figure 15. One-dimensional infinite chain of atoms with a random arrangement of magnetic moments in the paramagnetic phase.

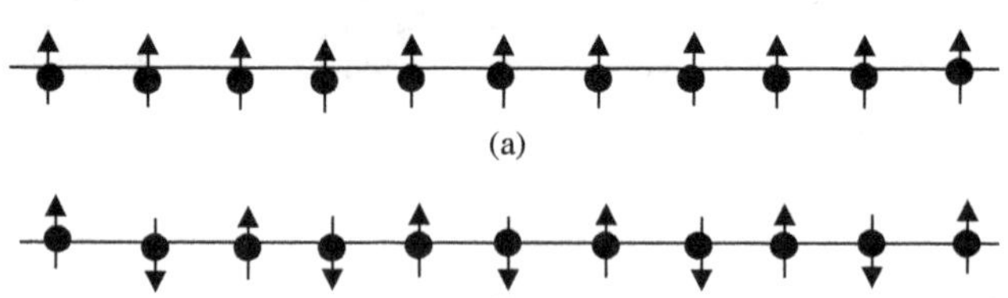

Figure 16. (a) Ferromagnetic and (b) antiferromagnetic orderings of the magnetic moments.

diffraction pattern will look similar to the X-ray diffraction pattern, and the measured lattice constant will just be the interatomic distance.

When the paramagnetic phase changes to the ferromagnetic phase, the incoherent scattering will reduce to zero, but the diffraction pattern will not change. The lattice constant measured will be the interatomic distance.

When the paramagnetic phase changes to the antiferromagnetic phase, the neutron diffraction pattern will change. The neutron will now indicate a lattice with twice the interatomic spacing as the lattice parameter.

As an example, we show in Figure 17 the neutron powder diffraction pattern of manganese telluride in the paramagnetic and antiferromagnetic phases studied by Kunitomi and Hamaguchi (6).

This material is paramagnetic above 50°C. Magnetic ordering takes place below 50°C. In planes perpendicular to the *c* axis, the atoms are coupled ferromagnetically, but the alternate sheets along the *c* axis are coupled

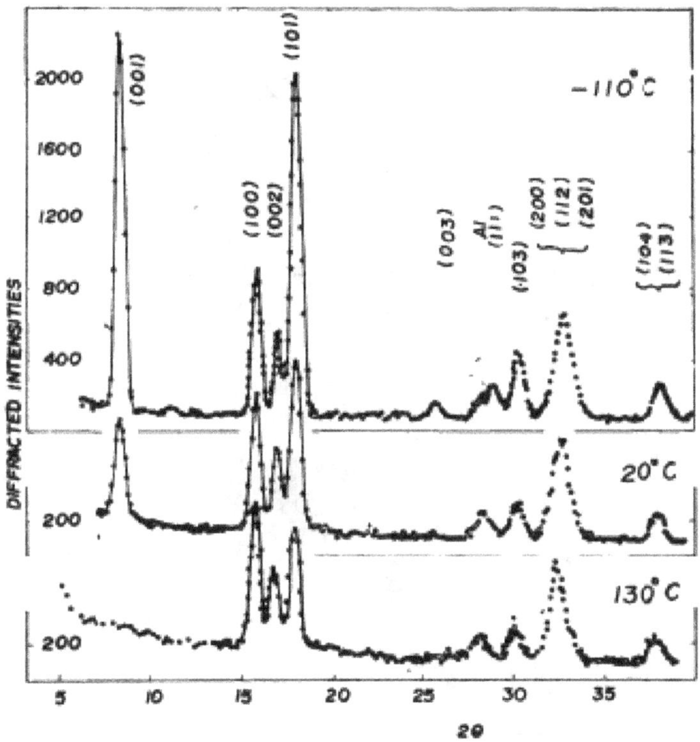

Figure 17. Neutron diffraction of MnTe at 130°C, 20°C, and −110°C (reprinted with permission from Ref. 6).

antiferromagnetically. This antiferromagnetic coupling gives rise to the development of the (001) reflection below 50°C. Thus, neutron diffraction will be able to distinguish between different types of magnetic ordering.

There are three neutron powder diffractometers around the Dhruva reactor in BARC, Trombay, Mumbai. They are described in Ref. 4. The layout of powder diffractometer 2 is shown in Figure 18.

The polychromatic thermal neutrons from the moderator of the Dhruva reactor in BARC come out through the biological radiation shield. This beam of polychromatic neutrons is monochromatized by Bragg reflection at the (3,3,1) face of a germanium single crystal. The monochromatized neutron beam has a wavelength of 1.244 Å. The monochromatized beam of size 4 × 1.5 cm falls on the powder sample. From the beam size, one realizes that the sample used in neutron diffraction is larger, in general, compared to the sample used in X-ray diffraction. The neutron

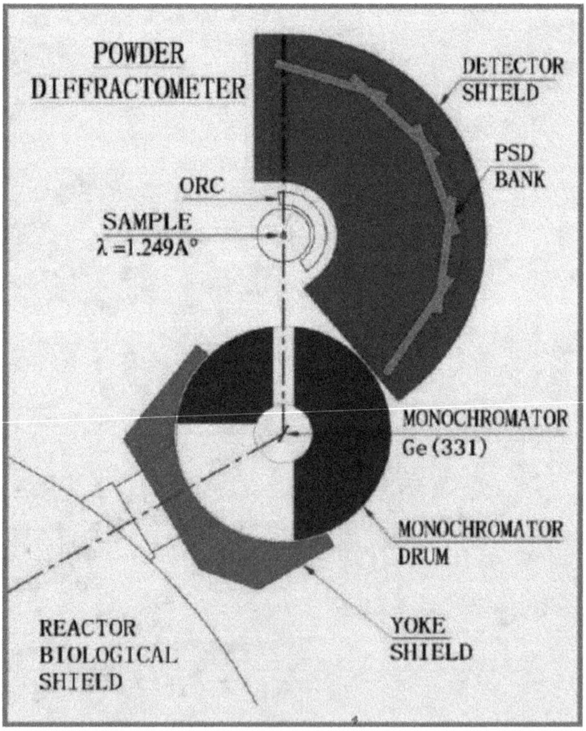

Figure 18. Layout of powder diffractometer 2 in BARC, Trombay, Mumbai (reprinted with permission from Ref. 4).

flux is 8.5 × 10^5 neutrons/cm^2/s. One can record the powder spectrum over a range of 2θ values from 4° to 140° using a bank of five position-sensitive detectors. The sample temperature can be maintained in the range from 5 to 2000 K. This spectrometer is used to study magnetic structures in a large variety of systems, such as titanates and zirconates, alloys of transition metals, and rare earth metals.

The powder diffraction of magnetite (Fe_3O_4) recorded at 300 K with this diffractometer is shown in Figure 19.

Figure 20 shows a photograph of a neutron powder diffractometer for low-temperature studies under high magnetic fields developed by the UGC-DAE Consortium for Scientific Research (Ref. 5).

This is a multi-PSD high-throughput diffractometer. This uses a bent perfect Si crystal monochromator to get a highly focused neutron beam of size 15 × 25 mm^2. The flux is 7 × 10^7 neutrons/cm^2/s. The neutron wavelength is 1.48 Å. The monochromator can be flipped to give neutrons of wavelength ranging from 1.1 to 2.3 Å. The scattering angle ranges from 6° to 123°. Samples of about 1 cc in volume may be used. The precision in measuring the separation of lattice planes, i.e., $\Delta d/d$ is approximately 0.3%. Measurements can be done down to 4 K in a magnetic field up to 7 T.

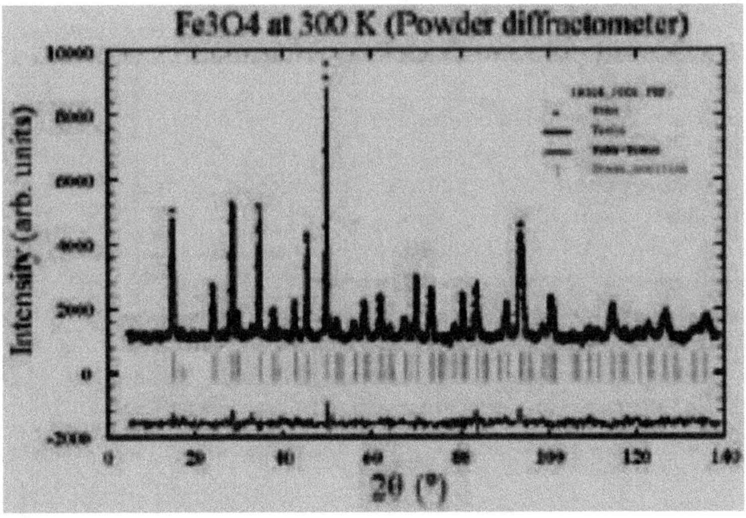

Figure 19. Neutron powder diffraction spectrum of Fe_3O_4 (reprinted with permission from Ref. 4).

Figure 20. Photograph of the powder diffractometer for low-temperature studies under high magnetic fields, developed and installed by UGC-DAE CSR on the Dhruva reactor in Trombay (reprinted with permission from Refs. 4 and 5).

A more detailed discussion of neutron diffraction is beyond the scope of this book.

References

1. Speakman, S. A. Basics of X ray Powder Diffraction, http//prism.mit.edu/xray.
2. Diffraction: Powder Method-Properties of X rays, https://web.stanford.edu › glam › xlab › LectureNotes ›.
3. An Introduction to Neutron Scattering, https://neutrons.ornl.gov>sites>default>files.
4. Aswal, V. K., Sen, D., Mukhopadhyay, R., and Chaplot, S. L. (2013). National Facility for Neutron Beam Research, BARC.
5. Pimpale, A. V., Dasannacharya, B. A., Siriguri, V., Babu, P. D., and Goyal, P. S. (2002). *Nucl. Instrum. Methods* **A481**, 615.
6. Kunitomi and Hamaguchi (1964). *Journal de Physique* **25**, 568.

Chapter 4

ELECTRON SPECTROSCOPY FOR CHEMICAL ANALYSIS

1. Introduction

Photoelectron spectroscopy using photons in the far UV or X-ray region is better known by the name Electron Spectroscopy for Chemical Analysis (ESCA). Before we discuss this important technique, let us take a brief digression into the energy states of electrons bound to a free atom and how these are affected by the state of ionization of the atom and by the type and number of its neighbors.

2. Binding Energy of an Electron in an Atom

Let us consider, for example, a sodium atom. A free sodium atom has 11 electrons. The electrons are arranged in shells corresponding to their total quantum number n. The shell with the lowest energy is the K shell, which corresponds to $n = 1$. For $n = 1$, the angular momentum quantum number $\ell = 0$. When $\ell = 0$, it is called the s state. The K shell has two such s states, which can be occupied by two electrons, and such a filled shell is designated as $1s^2$. The next shell is the L shell, for which $n = 2$. For $n = 2$, ℓ can take two values: 0 and 1. The electrons in an atom see the attractive Coulomb potential due to the positively charged nucleus, which can be considered a point charge. The electrons also experience a repulsive potential due to the negatively charged cloud of other bound electrons which is smeared all over the atom. While the potential due to a point charge varies as $1/r$, the potential due to the distributed charge cloud contains terms that vary as $1/r^p$, where p takes values different from 1.

Due to this reason, the energy of the electron not only depends on n but also on ℓ. In the 2s state ($n = 2$ and $\ell = 0$), there will be two electrons. In the 2p state ($n = 2$ and $\ell = 1$), one can have a maximum of six (i.e., $2(2\ell+1)$) electrons. The energy of electrons in the 2p state is greater than that in the 2s state.

The 2p state itself is split into two closely spaced states due to spin–orbit interactions. The electron has an intrinsic spin angular momentum characterized by a spin quantum number s, which is equal to ½. Due to this angular momentum, the electron, which carries a negative charge, has a magnetic moment $\boldsymbol{\mu}$ proportional to its angular momentum $sh/2\pi$. A charged particle moving with an orbital angular momentum $\ell\hbar$ is equivalent to a current and produces a magnetic field, **B**, proportional to its orbital angular momentum. The magnetic moment $\boldsymbol{\mu}$ in the presence of the magnetic field **B** has an energy $-\boldsymbol{\mu}\cdot\mathbf{B}$. So, we have an additional term in the Hamiltonian, called the spin–orbit Hamiltonian, H_{so}, which is proportional to $-\boldsymbol{\ell}\cdot\mathbf{s}$. This is the origin of spin–orbit splitting of the energy levels, with $\ell \neq 0$ for an atom. The energy level is split into two close energy levels, characterized by the quantum numbers $j = \ell + s$ and $j = \ell - s$. Thus, for p states, j takes the two values 3/2 and 1/2. For $n = 2$, these levels are designated as $2p_{3/2}$ and $2p_{1/2}$.

The 1s, 2s, and 2p levels of a sodium atom accommodate a total of 10 out of the 11 electrons. The 11th electron goes into the $n = 3$ shell. In this shell, ℓ can take the values 0, 1, and 2. Correspondingly, we have 3s, 3p, and 3d states. The 3s state has the lowest energy and can take a maximum of two electrons. The 11th electron of sodium goes into this 3s state. The energy of the 3s state is less than the energy of the 3p state, whose energy is less than the energy of the 3d state. Spin–orbit splitting causes the 3p state to be split into two closely spaced states: $3p_{1/2}$ and $3p_{3/2}$. The 3d state is split by spin–orbit splitting into two states: $3d_{3/2}$ and $3d_{5/2}$. There are no more electrons in the free sodium atom to fill these states.

There are higher energy shells, $n = 4$, $n = 5$,..., till n goes to infinity. When n is infinity, the energy of the state is zero, and the states beyond this form a continuum of positive energy. If a bound electron acquires such positive energy, it escapes from the atom, which is then said to be ionized. The energy required to remove an electron from the K shell and put it in the state $n = \infty$ is called the binding energy of the K-shell electron. The binding energy of the electron decreases as the value of n increases. Thus, the binding energy of the electron in the $n = 2$ shell is less than the binding energy of the electron in the $n = 1$ shell.

3. Influence of the Valence State of the Atom and its Environment on the Binding Energy

Let us take the singly ionized sodium ion Na^+. This ion has only 10 bound electrons since one of the 11 electrons in the sodium atom is removed. Consequently, the repulsive energy due to the electron cloud is less in the ion Na^+ than in the neutral atom. The attractive energy due to the nucleus remains the same since the number of protons in the nucleus remains unchanged when the sodium atom is ionized. So, the binding energy of the electrons in all the shells will be increased when sodium atom is ionized.

Suppose we add one electron to a neutral sodium atom to make the negative ion Na^-. In the Na^- ion, there will be 12 bound electrons instead of 11 electrons in the neutral sodium atom. So, the repulsive energy due to the charge cloud will increase, reducing the binding energy of an electron in all the shells.

In a molecule of sodium chloride, an electron is transferred from the sodium atom to the chlorine atom. In the molecule, we have a positively charged sodium ion and a negatively charged chlorine ion, and the Coulomb interaction between the oppositely charged ions binds them to form the molecule of sodium chloride. Such a transfer of electrons is called oxidation or reduction. When electrons are lost, the atom is said to be oxidized. When electrons are gained, the atom is said to be reduced. Thus, when a neutral manganese atom progressively loses electrons in going from Mn^0 to Mn^{2+} to Mn^{3+} to Mn^{4+}, we say that its oxidation state is increasing. In $MnCl_2$, manganese is in the 2+ state; in $MnCl_3$, Mn is in the 3+ state; and in $MnCl_4$, Mn is in the 4+ state. We say that the oxidation or valence state of Mn increases from 2 to 4 when going from $MnCl_2$ to $MnCl_4$. This will cause the binding energy of the electron in the K, L,... shells of Mn to increase as we go from $MnCl_2$ to $MnCl_4$.

Carbon and its compounds form covalent bonds. Each carbon atom has four electrons in its outer shell. The s and p atomic orbitals hybridize, forming four sp^3 orbitals directed along the four corners of a tetrahedron. Each of these orbitals can be filled with two electrons. In methane, CH_4, hydrogen contributes one electron to each sp^3 orbital, while carbon contributes another to fill the orbital. Such a bond is called a covalent bond. In methane, four hydrogen atoms are attached to carbon with four covalent single bonds.

In many organic compounds, there are different carbon atoms, with some forming a single bond, some a double bond, and some a triple bond. Figure 1 shows some typical organic compounds. It is seen that different carbon atoms can be attached to different atoms in a large organic compound.

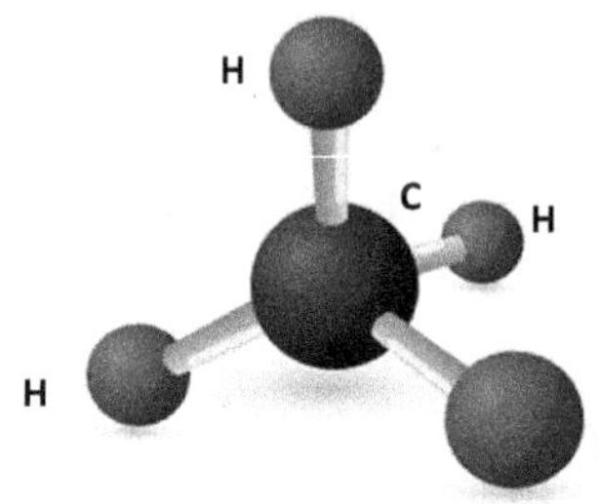

(a)

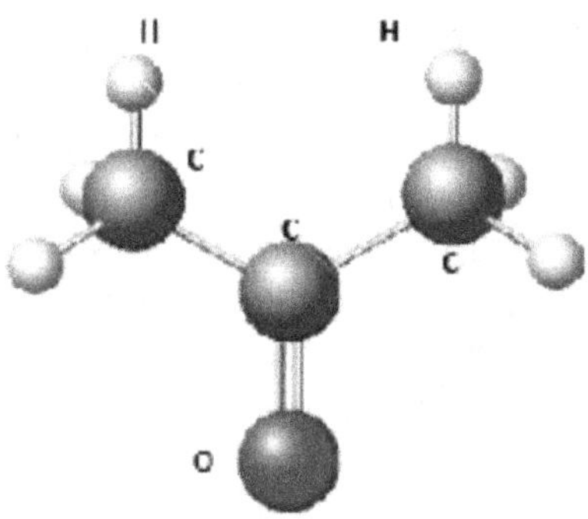

(b)

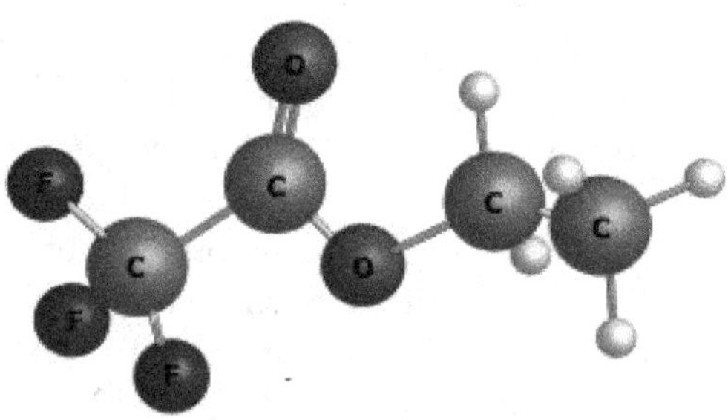

(c)

Figure 1. (a) Methane molecule, CH_4; (b) acetone molecule, $H_3C–C–O–CH_3$; (c) ethyl trifluoro acetate, $F_3C–C–O–CH_2–CH_3$.

In the methane molecule, the carbon atom is surrounded by four hydrogen atoms, each connected by a single bond. In acetone, there is a central carbon atom connected by a double bond to one oxygen atom and also connected to two CH_3 groups with a single bond. In ethyl trifluoro acetate, the different carbon atoms have different environments. As seen in Figure 1(c), the carbon atom on the extreme right has three hydrogen atoms and another carbon atom as its neighbors. The next carbon atom has two hydrogen atoms, another carbon, and an oxygen atom as its neighbors. The third carbon atom has two oxygen atoms and one carbon atom as neighbors. It is bound by a double bond to one of the oxygen atoms. The carbon atom on the extreme left has one carbon and three fluorine atoms as neighbors. The different environments and the difference in the nature of the chemical bond cause slight differences in the binding energy of the electron in the 1s shell of different carbon atoms.

4. Energy Bands in a Solid

A free atom has a size of the order of a few angstroms. This means that, beyond a distance of this order from the nucleus of this atom, its electron charge density becomes negligibly small. Thus, for sodium in the vapor state, any two sodium atoms are separated by about a micrometer or more; therefore, the electron energy level in each sodium atom is the same as that in a free atom. So, for a pair of such sodium atoms, the total number of energy levels is twice that of a single sodium atom. It should be noted that all the energy levels are discrete. In each atom, the 1s level is filled with two electrons, the 2s and 2p levels with eight electrons, and the 3s level with one electron. We may label the energy levels of the pair by attaching an extra symbol to identify each atom.

If the two atoms are closer together, as in a liquid or solid phase, the electronic charge clouds of the outer electrons overlap. Then, the electron attached to nucleus 1 sees a small potential, v, due to the nucleus and charge cloud of the neighboring atom, in addition to the potential V due to its own nucleus and electronic cloud. Due to this interaction, the degenerate 3s energy levels of the two atoms will form a pair of energy levels, separated by a small difference in energy. This separation will be very small for the 1s energy level and will progressively increase for the higher energy levels.

In Na metal, each sodium atom is surrounded by other sodium atoms at different distances. Sodium crystal has a body-centered cubic structure. Each sodium atom is surrounded by eight nearest neighbors at a distance of 0.371 nm, six next-nearest neighbors at a distance of 0.428 nm, and so on. The first eight nearest neighbors will interact to create a set of eight closely spaced energy levels for each of the atomic levels 1s, 2s, 2p, and 3s. The second neighbor interaction will split these energy levels further. This kind of splitting of energy levels will go on as we include more distant neighbors in the calculations. So, if we take one cubic centimeter of sodium metal, each of the energy levels of an isolated sodium atom will be split into N levels, where N is the number of sodium atoms in the entire sample. The N split levels have energy values over a finite energy range. This range will be small for the inner electronic shells. It will progressively increase as we go to the outer shells. This is illustrated in Figure 2.

Each energy level in the free atom spreads out into a band in the solid phase. The bands of some energy levels may overlap, as shown in Figure 3. For example, in sodium metal, the bands arising from 2s and 2p levels overlap. Each band arising from an atomic energy level can contain a maximum of εN electrons, where ε is the maximum number of electrons that the atomic energy level can have. Thus, the 1s band can have a maximum of $2N$ electrons. The overlapping 2s and 2p bands can have a maximum of $8N$ electrons. The 3s band can have a maximum of $2N$ electrons. Since the 1s, 2s, and 2p shells in the isolated sodium atom are filled, the

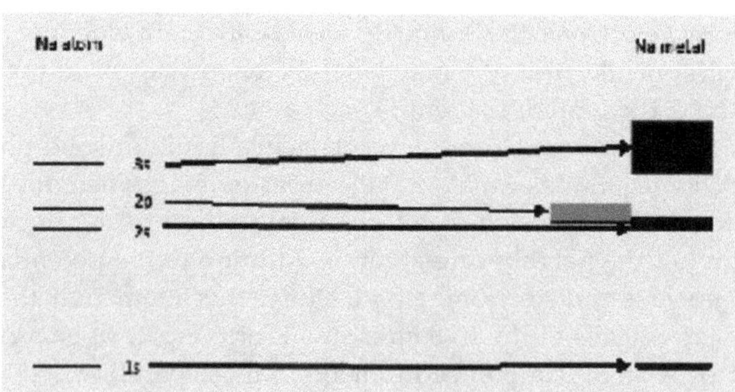

Figure 2. Discrete energy levels of a single sodium atom (left) spreads out into energy bands in sodium metal (right).

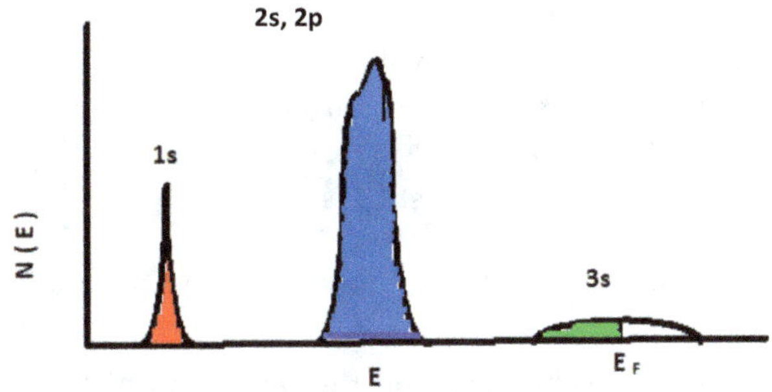

Figure 3. Schematic plot of *N*(*E*) vs. *E* in sodium metal.

bands arising from these levels in the metal are filled completely too. In the atom, the 3s shell has one electron. So, the 3s band in the metal is half filled with N electrons, and there are N unoccupied energy states.

In one cubic centimeter of sodium, there are 2.54×10^{22} atoms. The width of the 3s band is of the order of a few electron volts. For the inner bands, the energy width is even smaller. Thus, one cubic centimeter of sodium metal has 2.54×10^{22} closely spaced energy levels in the 3s band. So, we consider the energy levels to be continuously distributed in the 3s band and describe them in terms of the density of states $N(E)$. The number of energy levels between E and $E + \mathrm{d}E$ in a unit volume of the material is $N(E)\mathrm{d}E$. This number density will go to zero at the two ends of an energy band and will reach a maximum at an energy somewhere near the middle of the band. This is shown in Figure 3.

The 1s energy band has a narrow width. The outer bands have larger band widths. The 1s, 2s, and 2p bands are filled. The 3s band is half full. The area of each of the colored regions must be proportional to the numbers 2, 8, and 1 electron per atom in these bands; this proportionality is not evident in the figure as it is not drawn to scale.

Thus, in a metal, the outermost band, called the conduction band, is only partially filled. This is the feature which is responsible for the very high electrical and thermal conductivities of metals.

At absolute zero, each energy state in the conduction band is filled with two electrons, up to a maximum energy E_F (0), called the Fermi energy. This energy is calculated from the relation

$$2\int_{0}^{E_{F0}} N_c(E)\,dE = n_c \tag{1}$$

where n_c is the number of electrons per unit volume in the conduction band and $N_C(E)$ is the number density of states in the conduction band. As the temperature increases, the Fermi energy value decreases slowly because the occupation probability $f(E, T)$ of a state with energy E at a temperature T in Fermi–Dirac statistics is

$$f(E, T) = 1/[\exp((E - E_F(T))/kT) + 1] \tag{2}$$

We must introduce this factor in the integrand of Equation (1) and change the upper limit of the integral to infinity to calculate $E_F(T)$.

In a solid, one measures the energy of a state, taking the zero of energy at the Fermi level. So, the binding energies of all the inner levels have a negative sign. Energies above the Fermi level have a positive sign.

In a semiconductor, such as silicon, the valence band is the topmost filled band, and the conduction band is the next higher unfilled band. There is a gap in energy between the maximum of the valence band and the minimum of the conduction band. This energy difference is called the band gap. In an intrinsic semiconductor, the Fermi level E_F lies exactly halfway between the top of the valence band and the bottom of the conduction band.

5. Photoelectric Effect

When a photon of energy $h\nu$ is incident on a metallic material, it can be absorbed by an electron in the material. If the energy of the photon is greater than a threshold value, an electron can come out of the material. This threshold energy, Φ, is called the work function of the material, and it is the energy required to pull out an electron at the Fermi level in the material to a state of zero kinetic energy outside the material. This state of zero kinetic energy outside the material is called the vacuum state. If the electron is in an inner shell with energy $-E_i$, the minimum energy of the photon to pull out an electron from this inner shell is $(E_i + \Phi)$. A photon with this energy can pull out electrons from all the inner shells with a binding energy of $-E_k$ that is greater than $-E_i$. The electron pulled out from a shell with binding energy $-E_k$ by a photon of energy $h\nu$ comes out with a kinetic energy

$$E_{KE} = h\nu - (E_k + \Phi) \tag{3}$$

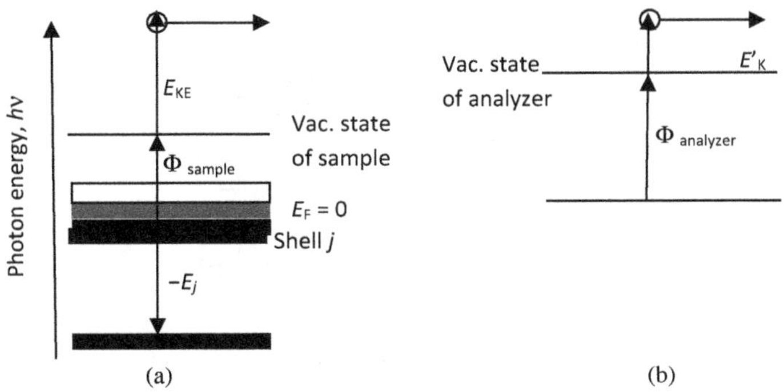

Figure 4. Schematic of the energy level diagram in a metal pertaining to the photoelectric effect.

This is shown in Figure 4(a).

Thus, an analysis of the spectrum of the kinetic energies of emitted electrons will give information about the number density of states of the various inner energy bands in a material.

To analyze the kinetic energy of the photoelectron, we use an analyzer. The work function of the material of the analyzer, $\Phi_{Analyzer}$, will be different from Φ_{Sample}, the work function of the test sample. If the sample and analyzer are connected by a conducting wire, the Fermi levels of the sample and the analyzer will become aligned. There will be a contact potential difference, $(\Phi_{Analyzer} - \Phi_{Sample})$, which will decelerate (or accelerate, depending on the sign of the contact potential difference) the electron. The electron kinetic energy measured by the analyzer will be E'_K, as shown in Figure 4(b).

When an incident photon of large energy falls on a material, a photoelectron may be ejected from an inner shell, say the K shell. Then, two processes may take place, as shown in Figure 5. In the first process, the hole in the K shell is filled by an electron in the L shell falling into the hole. The energy difference between the L and K shells is emitted as a photon of energy $h\nu_F$. This is called fluorescence and is shown in Figure 5(a). A second possibility is shown in Figure 5(b). Following the transition of an electron from the L shell to fill the hole in the K shell, a second electron is ejected from a shell of higher energy than L. This second electron is called the Auger electron. The energy of the fluorescence photon in Figure 5(a) and the Auger electron in Figure 5(b) are characteristic of the target atom and do not depend on the energy of the incident photon. This is how

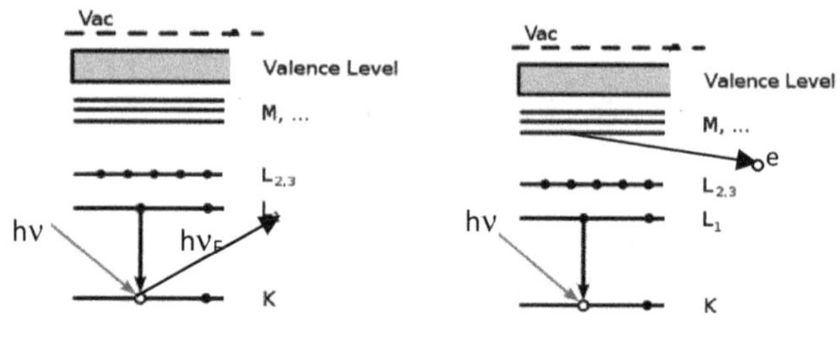

(a) (b)

Figure 5. (a) Fluorescence; (b) Auger process.

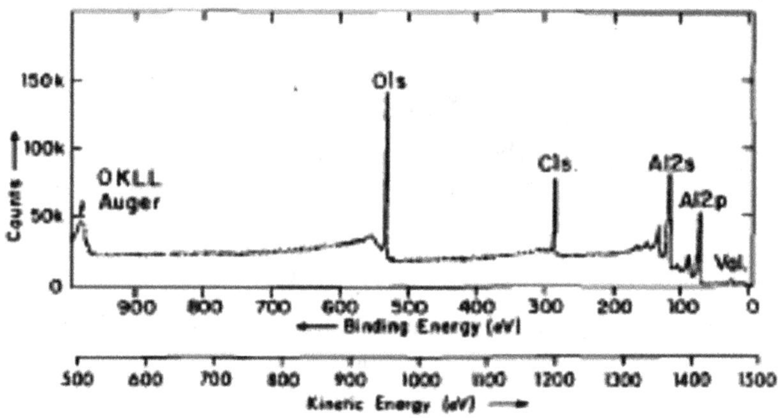

Figure 6. Photoelectron spectrum of an oxidized surface of an aluminum sample (reprinted with permission from Ref. 1).

Auger electrons are distinguished from photoelectrons. However, the energy of the incident photon must be sufficient to cause the ejection of an electron from the inner shell (K shell in the current case).

Both X-ray fluorescence and Auger electrons can be used for chemical analysis. X-ray fluorescence was discussed in the previous chapter.

The incident radiation has a penetration depth in the solid of the order of a few micrometers. The electrons ejected from deep within this penetration depth suffer inelastic collisions on their way out and come out with a range of kinetic energies below the value given by Equation (2). These are called secondary electrons. Only electrons coming from about the top 10 atomic layers from the surface do not undergo such inelastic collisions. They produce prominent no-loss peaks in the measured spectrum, from which one can calculate the binding energy of the electron using Equation (2). So, photoelectron spectroscopy gives information about the surface layers of the material and not about its bulk.

This is shown in the kinetic energy distribution of electrons from an aluminum sheet with a thin oxide layer on top, when X-ray photons of energy 1487 eV are incident on it (Fig.6).

Marked on the spectrum are the no-loss peaks generated by the oxygen 1s, carbon 1s, and aluminum 2s and 2p states. The carbon peak comes from the contamination of the surface with some carbonaceous material. We see that the oxygen and carbon peaks are comparable in height with the aluminum peaks. This is a clear indication that the photoelectrons are emitted by the surface layers and do not come from the bulk. We also see, on the low kinetic energy side of the oxygen 1s peak, a broad background of electrons arising from the scattering of the photoelectrons.

At a pressure of 10^{-9} mbar, it will take about an hour to form a monolayer of oxygen on the specimen surface. Since an XPS measurement may take about an hour or more, it is necessary to work at a pressure lower than this value. Furthermore, it will be necessary to prepare a clean surface of the material *in situ* by cleaving or thin-film deposition in vacuum. In addition, cleaning the surface with ion bombardment is often resorted to.

6. ESCA Spectrometer

Figure 7 shows the schematic arrangement of X-ray photoelectron spectroscopy.

For probing the inner shells of atoms in a material, one needs to use a monochromatic X-ray source that will produce X-ray photons of energy considerably more than the binding energy of the electron in an inner shell of the atom present in the material to be studied.

Monochromatic X-rays characteristic of the target (Mg or Al) in the source are generated when electrons accelerated to energies of several keV are focused on the target. The common targets used for X-ray emission are Mg or Al. The photons of the K_α line of Mg have an energy of 1253.6 eV, and the photons of the K_α line of Al have an energy of 1486.6 eV.

The radiation from these sources has a natural line width, which will cause a width in the measurement of the kinetic energy of the photoelectron. The resolution in the measurement of the energy of the

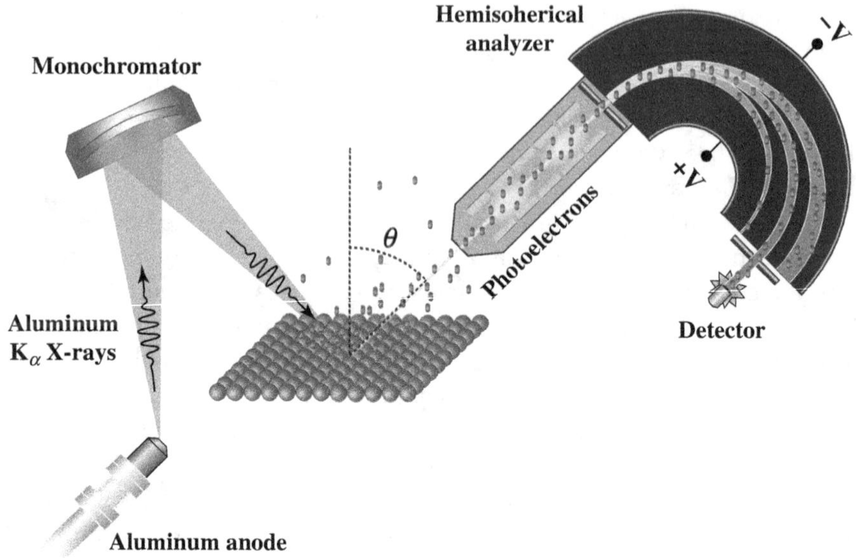

Figure 7. Schematic arrangement of an ESCA spectrometer (Redrawn from https://grimmgroup.net>research>xps>background).

photoelectron can be improved to about 0.5 eV if a monochromator is used. The use of a monochromator results in a reduction in the intensity of the X-radiation.

A synchrotron provides a powerful source of X-rays, tunable with a grating over a wide range of wavelengths. The advent of powerful synchrotron radiation sources has proved, in general, a boon for research in materials science.

One can study the XPS of materials in the gas, liquid, or solid phase. For studying the material in the gaseous state, one must have a controlled gas inlet system. The pressure of the gas must be around 10^{-4} mbar so that there are sufficient molecules in the sampling volume to produce photoelectrons. One must use differential pumping to maintain the sample gas volume at this pressure, while the X-ray source and the electron energy analyzer are maintained at a much lower pressure in the same chamber.

The electron energy analyzer and multichannel detector must satisfy the following criteria: (1) They should permit easy access to the specimen and detector regions; (2) they should be able to collect a large fraction of the photoelectrons emitted from the specimen; and (3) the energy resolution must be about 0.1– 0.5 eV at a kinetic energy of 1000 eV.

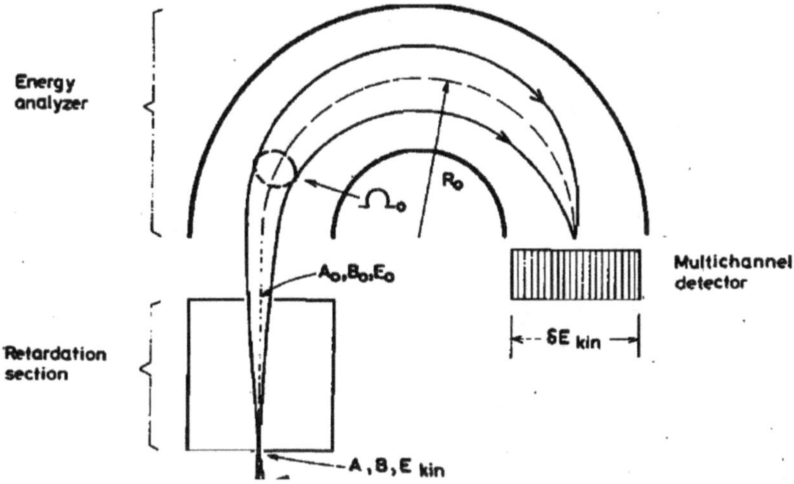

Figure 8. A schematic diagram of the electron energy analyzer (reprinted with permission from Ref. 1).

A schematic diagram of the electrostatic hemispherical analyzer with a retarding section and a multichannel detector is shown in Figure 8.

The beam of electrons coming from the sample is focused on the slit at the entrance of the retarding section. The retarding grid analyzer reduces the kinetic energy, E_{kin}, of the electron to a fixed value E_0, which is analyzed by the hemispherical analyzer. The hemispherical analyzer has two concentric metal hemispheres. A positive potential difference ($V_{inner} - V_{outer}$) is maintained between the two hemispheres, which produces a radial electric field. This field bends the electron beam into a circular orbit and focuses the electron beam onto the detector. In the detector, each photoelectron produces many secondary electrons, resulting in an amplified output signal. The analyzer is surrounded by a mu-metal shield to protect it from stray magnetic fields.

The photoelectrons emitted by the sample have a very wide range (several hundred eV) of kinetic energy. This large energy range is reduced to a narrow range, centered around a fixed value of E_0, by using the retardation grid section. This can be achieved by suitably adjusting the retarding potential on the grids. The electrons emerging from the retardation section enter the hemispherical analyzer with kinetic energies close to E_0. The big advantage of this reduction of energy range is that the kinetic energy of electrons can be measured, while keeping the resolution in measurement a constant, over the entire range of energies of the emitted photoelectrons.

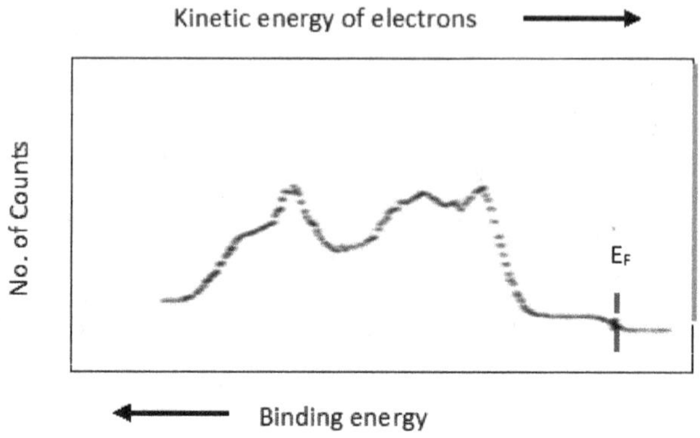

Figure 9. Photoelectron spectrum of a gold film irradiated with Mg K_α radiation.

In a gas of molecules, the binding energy of an electron is measured relative to the vacuum state. In a solid metallic sample, the binding energy is measured with reference to the Fermi energy in the metal. We need to identify the Fermi level in the photoelectron spectrum of a metal sample. The photoelectron spectrum from a gold film, irradiated with K_α X-rays from a Mg source, is shown in Figure 9. This is a typical spectrum for metals.

We see peaks in the spectrum arising from photoelectrons coming from the core levels of gold. The kinetic energy of the electrons increases as we move to the outer shells until we reach the conduction band. This band is wide. So, the number of photoelectrons emitted with a kinetic energy between E and $E + \Delta E$ becomes small. The levels above the Fermi level are sparsely filled with electrons. So, the intensity of the electrons coming from the sample shows a sudden drop in the conduction band. The energy at which this sudden drop occurs is the position of the Fermi level. Note that the binding energy of the core-level electrons increases to the left relative to E_F

7. Examples of ESCA Spectra

7.1. *Core-level shifts in binding energy*

The core-level binding energies of Mn in different oxides are shown in Figure 10. The 3s spectrum shows two peaks. The splitting of the 3s core level arises from the interaction of the 3s electron with the 3d valence

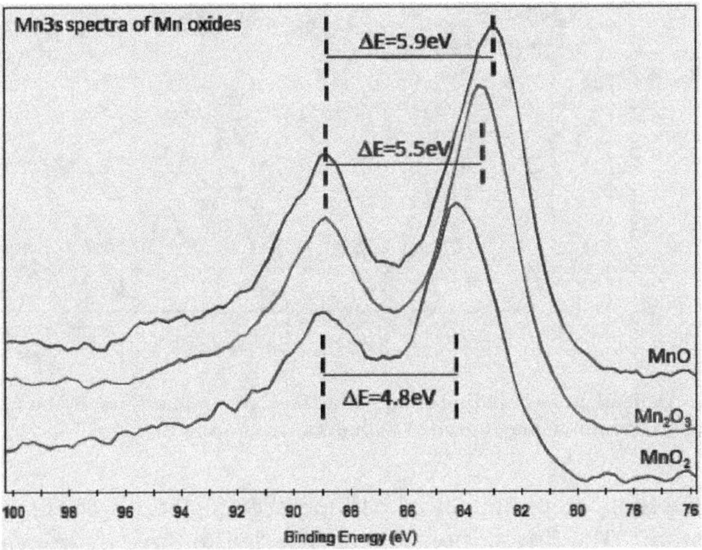

Figure 10. The binding energy of 3s core level of manganese in different oxides (reprinted with permission from Ref. 3).

band electron. From the figure, we see that one can get a resolution of 0.1 eV in measuring the binding energy.

We note that Mn exists in the 2+ state in MnO, the 3+ state in Mn_2O_3, and the 4+ state in MnO_2. Note that the binding energy of 3s core electron of Mn shifts to higher values as the valence state of Mn increases. This is because the number of electrons decreases as the valence state increases, resulting in a reduction in the repulsive energy of the electrons and an increase in the binding energy of the 3s core electron.

The separation ΔE between the two 3s electron peaks decreases as the valence state of Mn increases. This splitting is characteristic of the valence state of Mn and can be used to diagnose its valence state.

Figure 11 shows the binding energy of the core 1s level of carbon in two different molecules.

As mentioned above in Section 3, carbon can be in the single-, double-, or triple-bonded state, and it may be connected to different atoms in organic compounds. The electronegativity values for the different atoms are as follows: hydrogen: 2.20, carbon: 2.55, oxygen: 3.44, and fluorine: 3.98. In a covalent bond, the atom with the larger electronegativity pulls the electrons in the bond closer to itself.

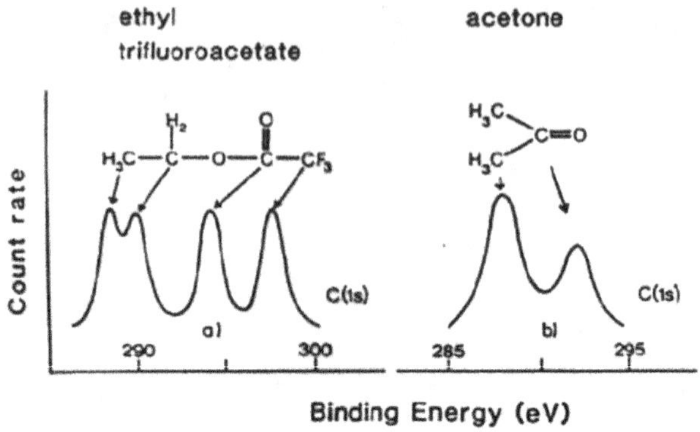

Figure 11. Dependence of binding energy of C(1s) electron in differently bonded states with different environments (reprinted with permission from Ref. 2).

In the first compound, ethyl trifluoroacetate, there are four different carbon atoms. The first carbon atom is bonded to three hydrogen atoms and another carbon atom. Since the difference in the electronegativities of hydrogen and carbon is small, the electrons are almost equally shared by carbon and hydrogen. In the case of the second carbon atom, one of its neighbors is oxygen, which is more electronegative than carbon. Hence, in the C–O bond, the charge cloud moves away from carbon toward oxygen. So, the repulsive energy of the electronic cloud is less in the case of the second carbon atom than in the first carbon atom. Hence, the binding energy of the 1s electron shell in the second carbon atom is greater than that of the same in the first carbon atom. The fourth carbon atom is bonded to three fluorine atoms. Fluorine has an even larger electronegativity than oxygen, and there are three atoms of fluorine attracting the electrons in the three bonds with the fourth carbon atom. So, the repulsive energy of the electronic cloud is reduced much more, and the binding energy of the 1s core level of this carbon atom is the maximum in this compound. The third carbon atom is attached to oxygen with a double bond, which carries more charge than a single bond. So, the binding energy of the 1s core electron of the third carbon atom is intermediate between that of the second and fourth carbon atoms. The two C(1s) peaks in acetone can be understood using similar arguments.

Figure 12 shows a high-resolution spectrum of the binding energy of 1s core shell of carbon in CH_4.

In a molecule, there are different vibration states due to the interatomic vibrations. In the photoionization of an atom in a molecule, the

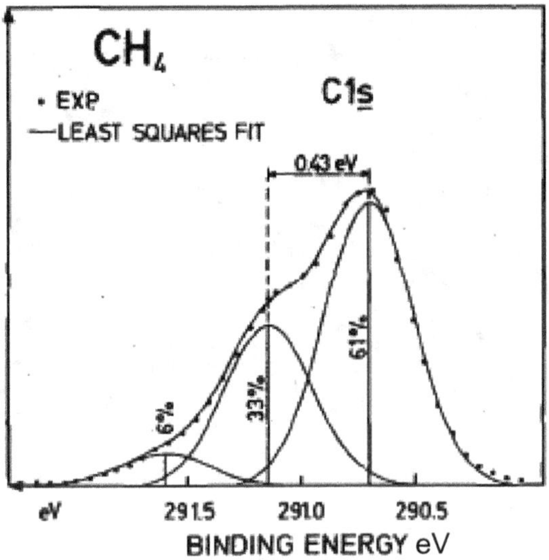

Figure 12. Vibration broadening of C(1s) peak in methane (reprinted with permission from Ref. 1).

vibration state of the molecule may change. This results in a multiplet structure in the spectrum of binding energy of the core shell of an atom, as determined by XPS. Vibration frequencies are high if the molecule has light atoms. The energy associated with vibration even in such molecules is a fraction of an eV. To see the multiplet structure due to vibrations, one must have an analyzer with the highest possible resolution.

The C(1s) peak broadening shown in Figure 12 is analyzed in terms of three Lorentzian profiles. The peaks of the three Lorentzian profiles correspond to three different vibration states of the symmetric C–H stretching vibration.

7.2. *Fixing the Fermi level and measuring the density of states in a metal*

Figure 13 shows an extended XPS spectrum of polycrystalline silver taken with K_α X-rays from aluminum.

The electronic configuration of silver is given at the top of Figure 13. The peaks in the photoelectron spectrum due to the electrons ejected from core levels 3s, 3p, 3d, 4s, 4p, and 4d are seen clearly.

The areas under the spin–orbit split $j = \ell + 1/2$ and $\ell - 1/2$ levels must be in the ratio of $(2\ell + 2)/(2\ell)$. Thus, for the $p_{1/2}$ and $p_{3/2}$ levels, the ratio must be 1:2. For $d_{3/2}$ and $d_{5/2}$ levels, the ratio must be 2:3. This is seen

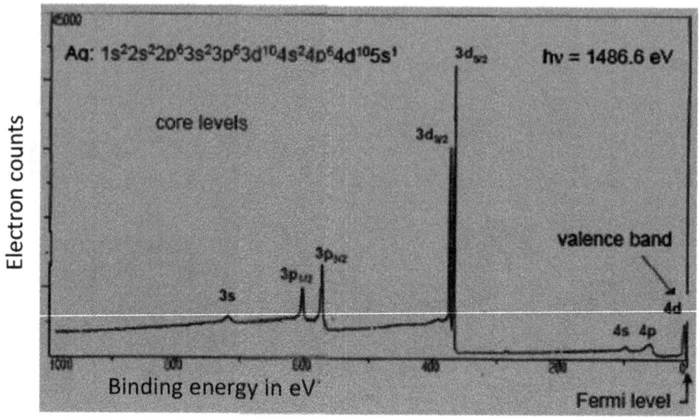

Figure 13. Extended XPS spectrum in polycrystalline silver (reprinted with permission from Ref. 2).

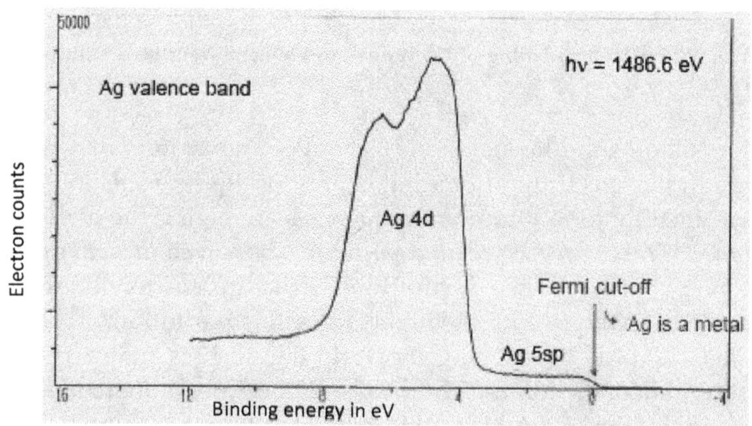

Figure 14. XPS spectrum of silver covering the Ag 4d valence band and Ag 5sp conduction band (reprinted with permission from Ref. 2).

to be true in Figure 13 for the $3p_{1/2}$ and $3p_{3/2}$ levels and the $3d_{3/2}$ and $3d_{5/2}$ levels. This ratio helps in identifying the spin–orbit split levels.

In Figure 14, the photoelectron spectrum of the same specimen of silver over a much narrower range of 20 eV in energy covering the 4d and 5s bands is shown.

The overlapping 5s 5p bands (shown as 5sp in the figure) are the partially filled conduction bands of metallic silver. The sharp drop in the number density of states at the Fermi level is clearly seen in the extended spectrum. This fixes the position of the Fermi level. The binding energies of the core shells relative to the Fermi level have a negative sign, which is not shown in Figure 14. The conduction band energy levels above the Fermi level have a positive sign. The occupation probability for levels above the Fermi level is small.

The 4d band is the highest fully occupied band in silver. This is called the valence band. It extends over nearly 4 eV of energy. The graph of electron counts versus energy reflects the density of states N(E) at different energies within this band.

7.3. *Plasmon peaks in metals*

Plasmon is a collective excitation of electrons in the conduction band. These electrons are treated as a gas. The photoelectron arising from a core shell may excite a quantum of plasma oscillations in the electron gas and come out with a lower kinetic energy. This gives rise to peaks in the XPS spectrum at binding energies higher than the binding energy of the shell from which the photoelectron was ejected. This is shown for aluminum in Figure 15.

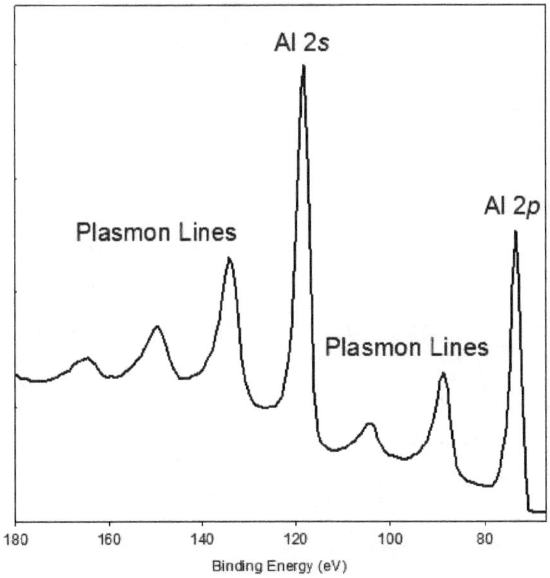

Figure 15. Plasmon peaks in XPS of aluminum (reprinted with permission from Ref. 4).

8. Conclusion

Table 1 taken from the work of Fadley (Ref. 1) gives a summary of the information that can be obtained about a material from the various features of its ESCA spectrum.

There is vast literature on ESCA. It is a very powerful technique and is used extensively in materials research. In this chapter, we have attempted to give a brief introduction to this important technique. Interested readers may look into the references given at the end of this chapter for further details.

Table 1. Schematic illustration of the interrelationships between various observable XPS spectral features or their associated effects and the basic system properties potentially derivable from an analysis of such observations (Ref. 1).

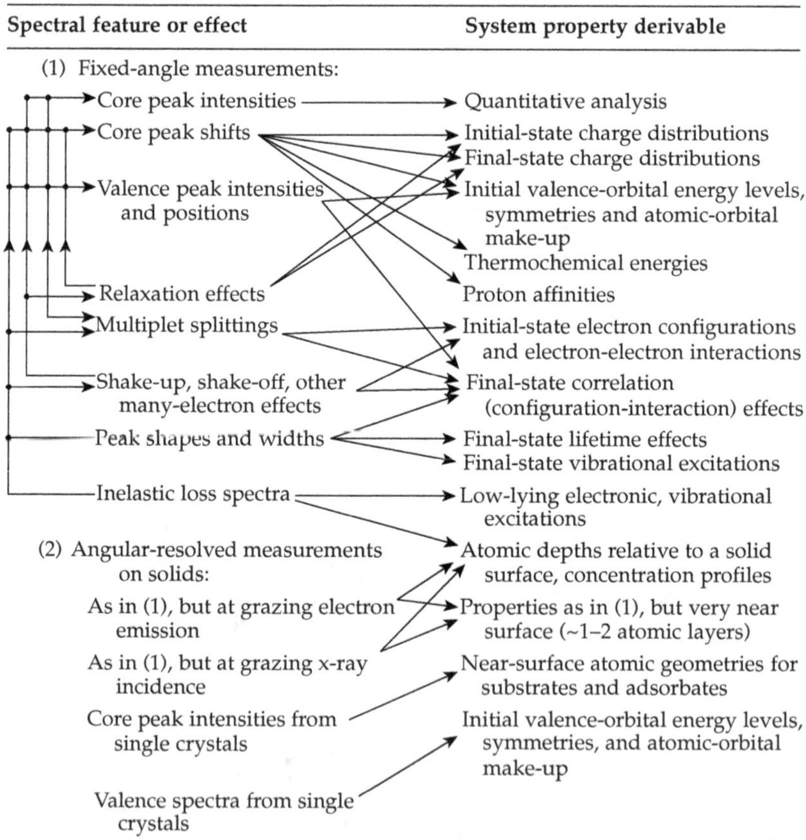

References

1. Fadley, C. S. (1978). Chapter 1: Basic concepts of X ray photoelectron spectroscopy. In C. R. Brundle, and E. D. Baker (eds.) *Electron Spectroscopy: Theory, Techniques and Applications*, vol. 2. Academic Press.
2. Electron Spectroscopy for Chemical Analysis, https://www.cpfs.mpg.de›5_15_05_2018-pqm_.
3. www.xpssimplified.com/elements/manganese.php.
4. Smart, R. *et al.* X ray Photoelectron Spectroscopy- Caltech MMRC, Department of Physics and Materials Science, City University of Hongkong, www.mmrc.caltech.edu>SS_XPS>XPS_PPT>XPS_slide.

Chapter 5

ELLIPSOMETRY FOR THIN-FILM ANALYSIS

1. Introduction

The preparation of materials in the form of thin films has become a standard process in almost all research laboratories and many industries. This has led to the development of a variety of measurement techniques to characterize such thin films. Among them, ellipsometry has become a key technique for the accurate measurement of the optical parameters of a thin film in a nondestructive way. Ellipsometry involves the determination of a change in the state of polarization of reflected light in comparison to that of an incident light beam. In this sense, ellipsometry is very different from other optical techniques, wherein only the intensity of the light beams is measured. By analyzing the measured data, we can determine optical parameters such as refractive index (n), the extinction coefficient (k) as a function of λ, the wavelength of light, and the thickness (d) of a thin-film sample. By a more elaborate analysis, one can also determine the chemical composition of the thin film, crystallinity of the film, doping concentration, and band gap energy of semiconductor thin films. In fact, ellipsometry is ideal for measuring the thickness of films in the range of one nanometer to one micrometer. Ellipsometry carried out over a range of wavelengths, spanning from the UV to the NIR regions (200 nm – 2000 nm), is called spectroscopic ellipsometry (SE). This has become a widely used technique in both research laboratories and industry for material characterization and thin-film device fabrication. All modern-day ellipsometers are interfaced to desktop computers to control their operation as well as for data analysis. Thus, ellipsometry has emerged as a technique for fast measurements and, hence, can be employed for online monitoring of the physical properties of thin films while they are being deposited. In this chapter, we confine ourselves to the determination of the optical constants and film thickness.

2. Fresnel's Equations for Reflection and Transmission

In ellipsometry, the basic physical phenomenon of interest is the reflection and transmission of a beam of polarized light at the surface of a transparent solid. Electromagnetic waves, such as light, are transverse waves and are hence characterized by a "state of polarization" (SoP), which represents the orientation of the electric field vector $\boldsymbol{E}$ of the wave and its evolution with time as the wave travels along the propagation vector $\boldsymbol{k}$ of the wave. We are interested in the three specific states of polarization shown in Figures 1(a)–(c).

Figure 1 shows, qualitatively, the pattern of the electric field vector $\boldsymbol{E}$ due to the mixing of two coherent orthogonally plane-polarized electromagnetic waves of the same frequency and wavelength, wave 1 and wave 2, propagating together along the z-axis. Figure 1(a) represents the generation of a new plane-polarized wave due to the mixing of the two coherent waves traveling in phase. In this case, the projection of the electric field on the x–y-plane, which is perpendicular to the z-axis, the propagation direction, is a straight line as shown. The angle made by this straight line with the x-axis is determined by the ratio of the amplitudes of wave 1 and wave 2. If the phase difference between the two propagating waves is 90° (or $\pi/2$) and they have equal amplitudes, then the resulting wave is said to be circularly polarized, as shown in Figure 1(b). The projection of the tip of the electric field vector on the x–y-plane is a circle. If the phase difference between the two propagating waves is other than zero or 90°, the wave is said to be elliptically polarized, as shown in Figure 1(c), and the projection of the tip of the electric field vector on the x–y-plane is an ellipse. These three states of polarization are called "pure states" because of a constant phase difference between the component waves wave 1 and wave 2 in each case. If the phase difference is a random variable, the resulting wave is called "mixed state" or unpolarized light. Such light is emitted by a hot body (a thermal source of light) and is a superposition of a large number of waves with completely random phase differences or SoPs.

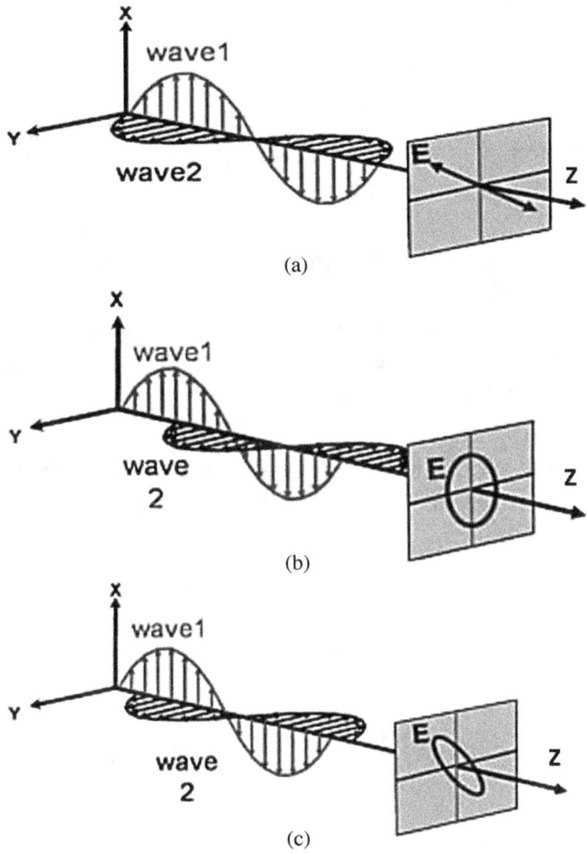

Figure 1. Schematic diagrams of the electric field pattern of two coherent electromagnetic waves (wave 1 and wave 2) traveling together along the z-axis. The field vectors are shown for one complete cycle of each wave for the case of (a) plane-polarized, (b) circularly polarized, and (c) elliptically polarized electromagnetic waves (from the website of J. A. Woollam Co.).

In ellipsometry, we basically measure the reflection of a highly monochromatic and well-collimated beam of light from the plane surfaces or interfaces of a transparent plate or a thin film deposited on a substrate. For this purpose, it is necessary to consider the following two specific cases:

- The incident light beam is plane polarized such that the electric vector E is entirely confined to the plane of incidence. **This is referred to as**

p-polarized state since *E* is parallel to the plane of incidence all along the light beam.

- The incident light beam is plane polarized such that the electric vector *E* is confined to a plane perpendicular to the plane of incidence. **This case is referred to as s-polarized state** since *E* is perpendicular to the plane of incidence all along the light beam.

If the plate, or the thin film on a substrate, is made of optically isotropic homogeneous material, it is seen that a p-polarized incident beam is reflected as a p-polarized beam, and if the incident beam is s-polarized, the reflected beam is also an s-polarized beam. The reflection and transmission coefficients for these two cases are different. These coefficients are the ratios of the reflected or transmitted wave amplitude to the incident wave amplitude. They are represented by equations called the Fresnel equations (Ref. 1). For the p-polarized state, the Fresnel equations are

$$r_p = \frac{E_{pr}}{E_{pi}} = \frac{n_t \cos\theta_i - n_i \cos\theta_t}{n_t \cos\theta_i + n_i \cos\theta_t} \tag{1a}$$

$$t_p = \frac{E_{pt}}{E_{pi}} = \frac{2n_i \cos\theta_i}{n_t \cos\theta_i + n_i \cos\theta_t} \tag{1b}$$

where r_p is the amplitude reflection coefficient, n_i is the refractive index of the medium in which the incident beam exists, n_t is the refractive index of the medium into which the beam is transmitted, θ_i is the angle of incidence, θ_t is the angle of refraction, and t_p is the amplitude transmission coefficient. E_{pi}, E_{pr}, and E_{pt} are the electric field amplitudes of the incident, reflected, and transmitted light beams, respectively.

Similarly, the Fresnel equation for the s-polarized state are as follows:

$$r_s = \frac{E_{sr}}{E_{si}} = \frac{n_i \cos\theta_i - n_t \cos\theta_t}{n_i \cos\theta_i + n_t \cos\theta_t} \tag{2a}$$

$$t_s = \frac{E_{st}}{E_{si}} = \frac{2n_i \cos\theta_i}{n_i \cos\theta_i + n_t \cos\theta_t} \tag{2b}$$

where E_{si}, E_{sr}, and E_{st} are the electric field amplitudes of the incident, reflected, and transmitted light beams, respectively.

The coefficients r_p, t_p, r_s, and t_s are complex numbers. We define the reflection and transmission coefficients in terms of intensities as follows:

$$R_p = r_p{}^* r_p,\ R_s = r_s{}^* r_s \tag{3a}$$

$$T_p = t_p{}^* t_p,\ T_s = t_s{}^* t_s. \tag{3b}$$

Equation (3a) defines the intensity reflection coefficients, and Equation (3b) defines the intensity transmission coefficients for the p-polarized and s-polarized light beams, respectively, and these are the quantities measured through experiments.

Figure 2(a) shows the variation in amplitude reflection coefficients r_p and r_s and transmission coefficients t_p and t_s, also called **Fresnel coefficients**, as a function of the angle of incidence for the light incident from air on glass. The angle θ_B, at which r_p becomes zero, is called Brewster's angle of incidence. It is given by

$$tan(\theta_B) = \left(\frac{n_t}{n_i}\right) \tag{4}$$

Figure 2(b) shows the variation in the intensity reflection coefficients R_p, R_s and T_p, T_s as a function of the angle of incidence, corresponding to Figure 2(a).

The graphs shown in Figures 2(a) and 2(b) correspond to the case of a light beam traveling from a medium of lower refractive index (optically less dense medium) to a medium of higher refractive index (optically denser medium). In the graphs, the Brewster angle is the angle of incidence at which r_p and hence R_p are equal to zero. In other words, a light beam reflected at the Brewster angle is completely s-polarized, i.e., the reflected light beam is perfectly plane polarized. The transmitted light beam is a mixture of both p-polarized and s-polarized components. The graphs also indicate that, in general, the fraction of the s-polarized light is greater than the fraction of the p-polarized light in the reflected light beam, except at an angle of incidence equal to 0° or 90°.

In contrast, the transmitted light beam contains a greater proportion of the p-polarized light compared to the s-polarized light. The difference

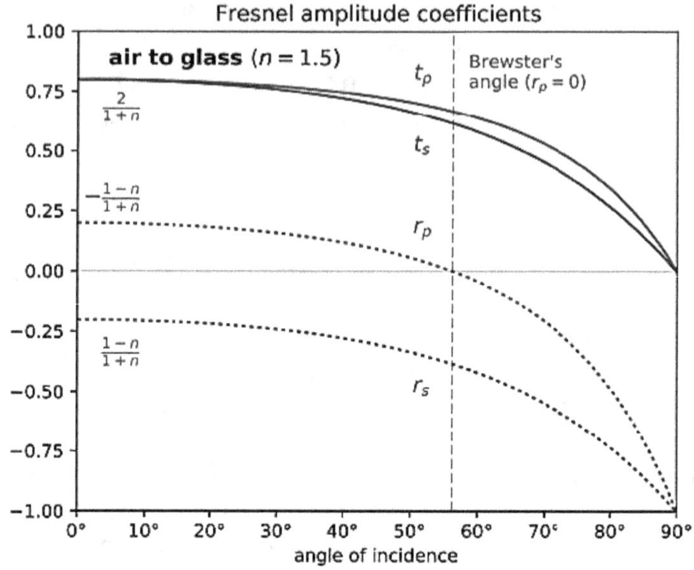

(a)

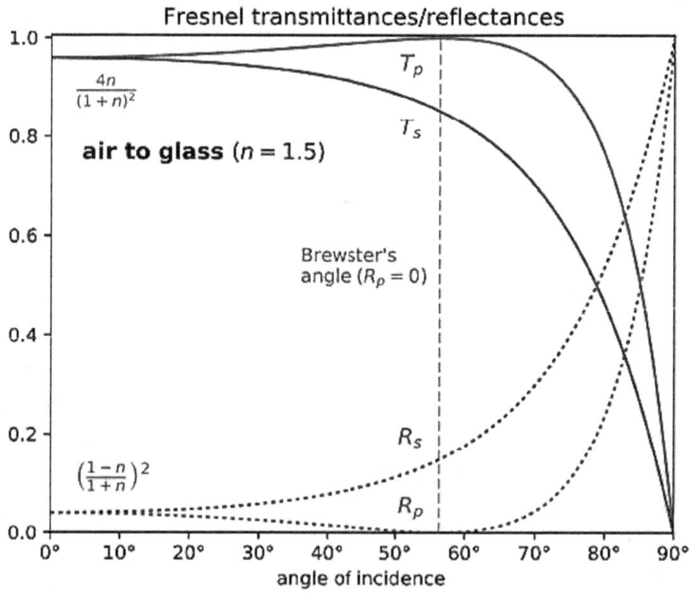

(b)

Figure 2. Variation in (a) the amplitude reflection coefficients r_p and r_s and transmission coefficients t_p and t_s and (b) the intensity reflection coefficients R_p, R_s and intensity transmission coefficients T_p, T_s with the angle of incidence (from Wikipedia).

in the magnitudes of E_{rp} and E_{rs} leads to a rotation of the polarization ellipse of the reflected light beam as compared to that of the incident beam. Furthermore, if the plate or thin film, which reflects the light, is an absorbing medium, its refractive index is a complex number. This leads to the addition of an extra phase difference between the s-polarized and p-polarized components of the reflected light beam. Thus, if the incident light beam is plane polarized, then after reflection from an optically dense medium, the reflected light beam becomes elliptically polarized. **In ellipsometry, we basically measure the phase change, Δ, between the incident light beam and the reflected beam and the angle ψ, where $\tan(\psi)$ is equal to the square root of the ratio of the intensity reflection coefficients for the p-polarized to that of the s-polarized component of the beam.**

If a beam of light is incident on a transparent thin film deposited on a large substrate, multiple partial reflections and transmissions of the beam will occur at the top and bottom surfaces of the film, as shown in Figure 3. Thus, a single incident beam of light is split into a very large number of beams, which emerge from the top and bottom surfaces of the thin film. The multiple rays emerging from the top surface are all parallel to each other. The same is true for the rays emerging from the bottom surface of the thin film.

The reflected light beam will be a superposition of all the ray components emerging from the top surface due to multiple reflections and transmissions. The thin film and the substrate are assumed to be made of

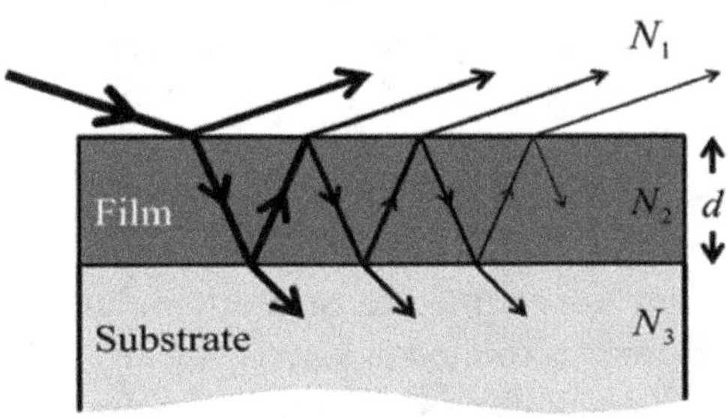

Figure 3. Schematic diagram of light beam reflection from a thin film of thickness d deposited on a substrate of refractive index N_3. The refractive index of the thin film is N_2 and that of the medium above the film is N_1. Usually, the medium above the thin film is air, for which $N_1 = 1.0$.

isotropic materials. In such cases, the amplitude reflection and transmission coefficients are given, respectively, by

$$r_p = \frac{r_{p,12} + r_{p,23}e^{-i2\beta}}{1 + r_{p,12}r_{p,23}e^{-i2\beta}} \tag{5}$$

$$r_s = \frac{r_{s,12} + r_{s,23}e^{-i2\beta}}{1 + r_{s,12}r_{s,23}e^{-i2\beta}} \tag{6}$$

where

$$\beta = 2\pi\left(\frac{d}{\lambda}\right)N_2 \cos\theta_2 \tag{7}$$

β is the phase difference between any two neighboring rays emerging out of the top surface of the thin film, and θ_2 is the angle of refraction in the film. In the above equations, $r_{p,12}$ is the amplitude reflection coefficient for the top surface of the thin film and $r_{p,23}$ is that at the lower surface of the thin film, i.e., the interface between the film and the substrate. A similar interpretation applies to $r_{s,12}$ and $r_{s,23}$. Equations may also be written for the transmission coefficients, but for our discussions, the equations for reflection coefficients will suffice.

3. Ellipsometer and its Main Components

As mentioned above, the primary task in ellipsometry is to measure the parameters characterizing the state of polarization of monochromatic light reflected from planar surfaces. The instrument used for this purpose is called an **ellipsometer**.

Figure 4 shows a commercial ellipsometer. The instrument consists of two side arms oriented at exactly the same angle with respect to the horizontal plane. They are coupled to each other and attached to a rotary stage such that if one arm is rotated clockwise by an angle θ, the other arm gets rotated anticlockwise by the same angle, θ, simultaneously. Thus, the two arms make equal angles with the horizontal plane even after a small rotation. In one arm, a light source, usually a laser, is mounted, and the other arm carries a photodetector. In between the two arms, a plate, in the shape of a disk (like a prism table in spectrometers), is fixed so that its

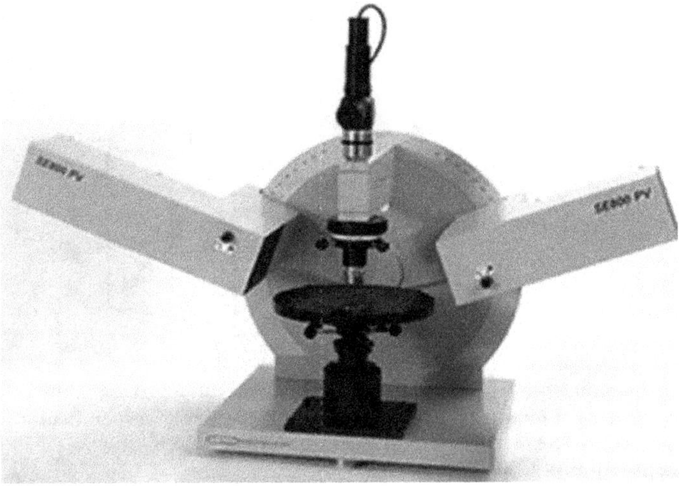

Figure 4. A photograph of a commercial ellipsometer (Model SE 800PV; from the website of Sentech, Germany).

surface is horizontal. This is the sample holder. Thus, the laser beam incident on the horizontal surface of the sample should enter the photodetector in accordance with the law that the angle of reflection is equal to the angle of incidence. The instrument must be set up such that these conditions are satisfied; this is ensured in all commercial ellipsometers. In the ellipsometer shown in Figure 4, the plane of incidence will be the vertical plane containing the incident and reflected laser beams and the normal to the reflecting surface at the point of incidence of the laser beam on the sample. Figure 4 also shows a microscope, mounted vertically in the instrument, for viewing the sample surface.

Apart from the light source and the photodetector, the other main components of the ellipsometer are a polarizer, a compensator, and an analyzer (essentially the same as a polarizer).

A **polarizer** is a highly transparent optical component and is available in the form of a film (sheet polarizers), plate, or cube, which is made using crystalline solids. Some examples of the cube types are the Glan–Thomson polarizer, Wollaston polarizer, Rochon polarizer, and Nicol prism. It is used to produce linearly polarized light and also to determine the state of polarization of a beam of light. It is characterized by a pass axis, which is a specific direction in the device. The direction perpendicular to the pass axis is

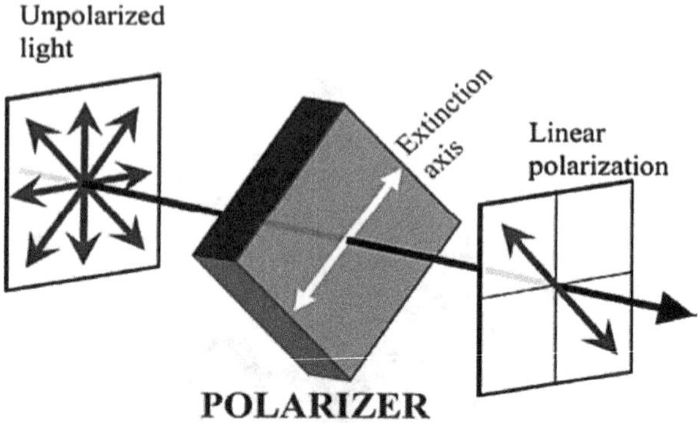

Figure 5. **Schematic illustration of the conversion of unpolarized light into linearly polarized light using a polarizer. The extinction axis shown is perpendicular to the pass axis of the polarizer. The electric field of the light transmitted by the polarizer is parallel to the pass axis (from Wikipedia).**

called the extinction axis. A plate polarizer is usually a rectangular or circular plate of about 2–3 mm thickness. The pass axis is generally marked on the plate holder. This device completely transmits the component of the wave electric field vector that is parallel to the pass axis of the polarizer and totally blocks the component parallel to the extinction axis. Figure 5 illustrates the generation of plane-polarized light from unpolarized light.

Furthermore, when a polarized light beam is passed through a polarizer at normal incidence and if the polarizer is rotated gradually from 0° to 360° about the beam as the axis, the transmitted light intensity will be:

- maximum twice and zero twice if the light is plane polarized;
- constant if the light is circularly polarized;
- maximum twice and minimum twice and will vary smoothly between these two values if the light is elliptically polarized.

This procedure is used to detect the SoP of a light beam, or any electromagnetic wave, in general. This is one of the basic steps in any ellipsometry measurement.

An **analyzer** is just another polarizer fixed in the arm containing the photodetector. It is named an analyzer simply because it is used to determine the SoP of the reflected light beam.

Apart from the polarizer, the other important optical component used in ellipsometry is the **compensator**. This is a transparent component made using appropriate crystals and is characterized by two special directions called "fast axis" and "slow axis"; these are mutually perpendicular directions. Using this device, one may introduce an extra phase shift between any two plane-polarized waves. If plane-polarized light is passed through a compensator, the emerging light beam will be elliptically polarized, in general. More specifically, if the compensator introduces a phase difference of 90° ($\pi/2$ radians) between wave 1 and wave 2 (see Figure 1(b)) and if the amplitudes of the two waves are equal, the light emerging from the compensator is seen to be circularly polarized. Such a compensator is also called a quarter wave plate (QWP) and is just a plate of a crystal, such as quartz or calcite, of appropriate thickness, which is dependent on the wavelength of the light beam.

Figure 6 illustrates the use of QWP for the generation of circularly polarized light. For this, it is essential that the electric field of the incident light be oriented at 45° to the fast axis of the QWP. For other orientation angles, the output is elliptically polarized.

Another such device is the half wave plate (HWP), which introduces a phase difference of 180° (π radians) between the two waves. More versatile compensators, such as the Babinet's compensator, are used to generate variable phase shifts between two waves. These devices are made using a pair of crystals in the shape of prisms (for details, see Ref. 1). However, it must be noted that the phase shift introduced by a QWP, HWP, or a compensator varies with the wavelength of light. Hence,

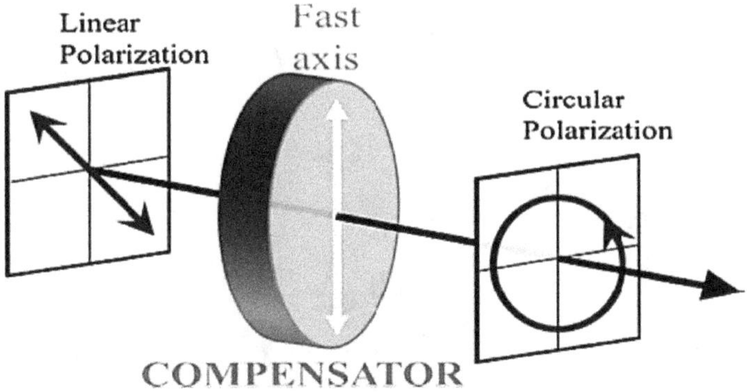

Figure 6. **Schematic illustration of the conversion of plane-polarized light into circularly polarized light using a QWP shown as a compensator. Note that the electric field of the incident light beam is oriented at 45° to the fast axis of the QWP. The slow axis of the QWP is perpendicular to its fast axis (Wikipedia).**

compensators must be selected appropriately for spectroscopic ellipsometry for measurements at different wavelengths.

4. Basics of Elliptically Polarized Light

The most general pure state of polarization of perfectly monochromatic light is elliptical polarization. The ellipse, shown in Figure 7, is generated in the x–y-plane by the tip of the electric vector of the wave traveling along the positive z-axis.

The electric field of the light wave may be expressed as

$$\boldsymbol{E}(z, t) = \{E_x \boldsymbol{x} + E_y \boldsymbol{y}\} e^{i(kz-\omega t)} \tag{8}$$

where $\boldsymbol{x}$ and $\boldsymbol{y}$ are unit vectors along the x- and y-axes, respectively, and E_x and E_y are complex amplitudes expressed as

$$E_x = Xe^{i\alpha} \text{ and } E_y = Ye^{i\beta} \tag{9}$$

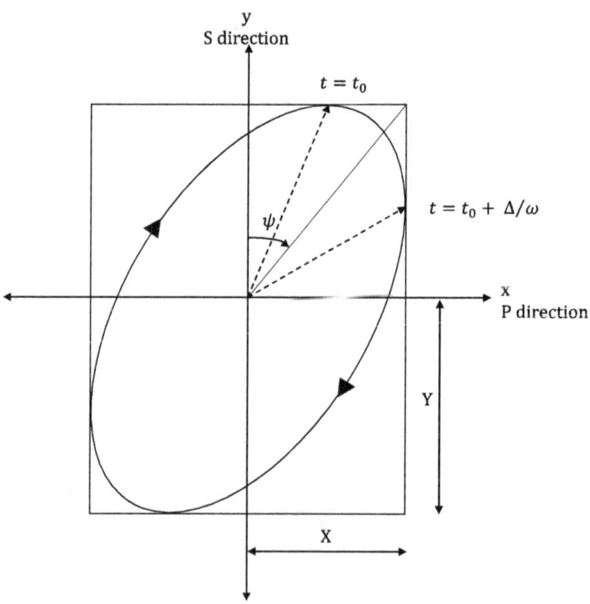

Figure 7. The ellipse traced out by the tip of the electric field of a monochromatic light wave traveling along the positive *z*-axis. The electric field attains its maximum value denoted by *X* along the *x*-axis and *Y* along the *y*-axis. The angle Ψ is determined by the ratio *X*/*Y*. It is one of the two parameters measured in ellipsometry.

Combining Equations (8) and (9), we get

$$\boldsymbol{E}(z, t) = \{Xe^{i\Delta}\, \boldsymbol{x} + Y\, \boldsymbol{y}\}e^{i\beta}e^{\, i(kz-\omega t)} \tag{10}$$

where $\Delta = (\alpha - \beta)$ is the relative phase difference between the amplitudes E_x and E_y. Alternatively, the wave electric field $\boldsymbol{E}$ can be expressed as a two-component column matrix:

$$\boldsymbol{E}(z,t) = \left\{ \begin{bmatrix} X & e^{i\Delta} \\ & Y \end{bmatrix} e^{i(kz-\omega t+\beta)} \right\} \tag{11}$$

In ellipsometry, we may ignore the constant phase factor $e^{i\beta}$. The ellipse shown in Figure 7 represents the time evolution of the electric field in the plane $z = 0$. The ellipse is fully described by the column vector

$$\boldsymbol{E}(t) = \mathrm{Re}\left\{ \begin{bmatrix} X & e^{i\Delta} \\ & Y \end{bmatrix} e^{-i(\omega(t-t_0))} \right\} \tag{12}$$

where t_0 is the initial time of observation. From Equation (12), it follows that at $t = t_0$, the y-component of $\boldsymbol{E}(t)$ attains the maximum magnitude Y, whereas the x-component is less than X, its maximum value. This electric vector is shown as a dashed line in Figure 7. At $t = (t_0 + \Delta/\omega)$, the x-component attains the maximum magnitude X, whereas the y-component is smaller than Y, its maximum value. The corresponding electric field vector is shown as a dotted line in Figure 7. The angle Ψ, which is a characteristic parameter of the ellipse shown in the figure, is given by

$$\tan(\Psi) = \frac{X}{Y} \tag{13}$$

where X and Y are the maximum electric field amplitudes of the wave along the x- and y-axes, respectively. In the experimental setup, the electric field of the p-polarized wave is oriented along the x-axis, and the electric field of the s-polarized wave is along the y-axis.

The ratio X/Y and the phase difference Δ are the two main quantities measured in ellipsometry. The angle Δ is determined by the ellipticity of the ellipse in Figure 7. If the angle Δ is a positive quantity, then the ellipse is traced in the clockwise sense, and it corresponds to right-handed polarization. If the angle Δ is a negative quantity, the ellipse is traced in the anticlockwise sense, and the corresponding polarization state is left-handed polarization. However, this convention is not unique. The range of values of Δ is 0–2π or, equivalently, $-\pi$ to $+\pi$. Furthermore, it is convenient to determine the ratio (X/Y) in terms of the angle Ψ. From Equation (12), it is easy to see that the range of values of Ψ is 0 to $\pi/2$. We note that Equation (12) fully represents the SoP of the light beam. In ellipsometry, the incident light beam has a certain SoP, which gets altered due to reflection, and hence, the reflected beam has a different SoP. The task in ellipsometry is to measure the change in the SoP due to reflection, i.e., the angles Ψ and Δ.

From the above discussions, we may now state that for ellipsometry, the electric vector of an elliptically polarized light may be adequately represented by the column vector

$$J_1 = \begin{bmatrix} \sin(\Psi)e^{i\Delta} \\ \cos(\Psi) \end{bmatrix} \tag{14}$$

This column matrix is completely determined by the real angles Ψ and Δ and is called the **Jones vector** for elliptically polarized light. One can construct a vector orthogonal to this Jones vector, which is given by

$$J_2 = \begin{bmatrix} -\cos(\Psi)e^{i\Delta} \\ \sin(\Psi) \end{bmatrix} \tag{15}$$

These two column matrices satisfy the orthogonality conditions

$$J_1^* J_2 = J_2^* J_1 = 0 \tag{16}$$

where the * symbol indicates taking the Hermitian adjoint of the column matrix, i.e., taking the complex conjugate of each element of the column matrix and taking its transpose, and writing it as a row matrix. Such orthogonal Jones vectors may be used as basis vectors for the analysis of the general polarization states of light beams.

Two special cases of the Jones vector J_1 given by Equation (14) are as follows:

- **Linearly polarized light:** This is obtained for $\Delta = 0$ or π. Furthermore, for $\Psi = 0$, the light is polarized in the y-direction, and for $\Psi = (\pi/2)$, it is polarized in the x-direction. If the electric field amplitude is unity, then the corresponding Jones vectors are given by

$$x\text{-polarized light: } \begin{bmatrix} 1 \\ 0 \end{bmatrix} \text{ and } y\text{-polarized light: } \begin{bmatrix} 0 \\ 1 \end{bmatrix} \tag{17}$$

These two Jones vectors can be considered unit basis vectors for the analysis of polarized light.

- **Circularly polarized light:** This is obtained for $\Psi = \pi/4$ and $\Delta = +\pi/2$ or $-\pi/2$. It is seen that $\Delta = +\pi/2$ corresponds to right circularly polarized (RCP) light, and for $\Delta = -\pi/2$, we obtain left circularly polarized (LCP) light. The corresponding Jones vectors of unit magnitude are:

$$\text{RCP light: } \frac{1}{\sqrt{2}}\begin{bmatrix} 1 \\ +j \end{bmatrix} \text{ and LCP light: } \frac{1}{\sqrt{2}}\begin{bmatrix} 1 \\ -j \end{bmatrix} \tag{18}$$

Here, $j = \sqrt{-1}$.

These unit vectors may also be used as basis vectors. We may generate a circularly polarized light by passing an unpolarized light through a polarizer and next through a QWP, aligning the fast axis of the QWP at an angle of $\pm\pi/4$ with respect to the pass axis of the polarizer, as shown in Figure 6.

5. Basic Principles of Ellipsometry

The electric field of a plane electromagnetic wave propagating along the positive z-axis in an infinite medium is given by

$$\boldsymbol{E} = \boldsymbol{E}_0\, exp[i(\omega t - Kz + \delta)] \tag{19}$$

where $\boldsymbol{E}_0$ is the wave amplitude, K is the propagation constant, ω is the wave frequency, and δ is the initial phase of the wave. The propagation constant K may be written as

$$K = \frac{2\pi}{\lambda} N \tag{20}$$

where N is the complex refractive index of the medium given by

$$N = n - ik \tag{21}$$

Using Equations (20) and (21) in Equation (19), we may write the electric field of the wave as

$$\boldsymbol{E} = \boldsymbol{E}_0 \exp\left[-\frac{2\pi k}{\lambda}z\right]\exp\left[i\left(\omega t - \frac{2\pi n}{\lambda}z + \delta\right)\right] \tag{22}$$

where λ is the free space wavelength of the wave. It is clear from this expression that the attenuation of wave amplitude depends on k and the phase change depends on n. The intensity of the wave is given by

$$I = I_0 \exp[-\alpha z] \tag{23}$$

which is called Beer's law in optics. Here, $I_0 = E_0^2$, and the absorption coefficient α is given by

$$\alpha = \frac{4\pi k}{\lambda} \tag{24}$$

From the above discussion, it is clear that the electric field of the wave depends on both n and k, the real and imaginary parts of the complex refractive index N of the medium, respectively.

We may note that the refractive index N is related to the complex dielectric permittivity ε by

$$\varepsilon = N^2 = (n - ik)^2 \tag{25a}$$

where

$$\varepsilon = \varepsilon_1 - i\varepsilon_2 \tag{25b}$$

From Equations (25a) and (25b), we obtain

$$\varepsilon_1 = n^2 - k^2 \text{ and } \varepsilon_2 = 2nk \tag{25c}$$

Thus, for the sample material, we can also calculate ε_1 and ε_2 from n and k. If k is very small for a material, we see from Equation (25c) that

$n = \sqrt{\varepsilon_1}$, which is commonly used for calculating the refractive index when absorption is negligible.

When a plane electromagnetic wave is reflected by the plane surface of a solid material, the electric field of the reflected wave is obtained by multiplying the electric field of the incident wave by the amplitude reflection coefficient r_p or r_s, depending on the SoP of the incident wave. Since the reflection coefficient is a complex number of the type $re^{i\theta}$, the reflected wave field acquires an added phase change of θ due to the reflection process. This phase change will depend on both n and k, which, in turn, depend on the wavelength λ.

In ellipsometry, the light beam incident on the sample is plane-polarized light, which is prepared as a superposition of a p-polarized and an s-polarized plane wave. These two plane waves are in phase in the incident light beam. After reflection from the sample surface, the two plane waves acquire different phase shifts because $r_p \neq r_s$. Therefore, the reflected beam becomes an elliptically polarized wave. Thus, in ellipsometry, we determine ψ, given by Equation (13), as well as the phase shift Δ between the two reflected s- and p-polarized wave components. **The values of both ψ and Δ depend on the wavelength λ of the light beam and the angle of incidence of the light beam at the sample surface.**

6. Experimental Techniques in Ellipsometry

The main types of ellipsometry techniques are:

- **Null ellipsometry:** In this method, either the polarizer or the analyzer is rotated to reduce the light intensity detected by the photodetector to zero (or negligibly small). Alternatively, the compensator may be rotated to accomplish the same result. The angle by which the polarizer, analyzer, or compensator is rotated relative to the initial setting is one of the measured data.
- **Photometric ellipsometry:** In this case, the compensator is not required. The polarizer may be held fixed, and the analyzer is rotated continuously. The detected light intensity varies periodically (approximately sinusoidal variation) with the angle of rotation of the analyzer. From the Fourier analysis of the data, the two parameters Ψ and Δ can be recovered.

- **Spectroscopic ellipsometry:** Measurements of the above types are carried out at several wavelengths of the light beam. For this, one may use a broad-spectrum light source together with a monochromator or lasers at several wavelengths. In modern ellipsometers, measurements are carried out at 50–300 different wavelengths in the IR–UV range simultaneously by using 1D or 2D array photodetectors. Such measurements enable us to determine the wavelength dependence of the refractive index, film thickness, crystallinity of the film, composition of the thin film, energy band gap, etc.

Some **advantages of ellipsometry** over other techniques are as follows:

- It is a nondestructive and noninvasive method. The sample is not damaged in any way.
- Measurement over a range of wavelengths takes less than a minute.
- It yields values for several physical parameters of the thin-film sample and has become a standard technique for material characterization.
- Measurements can be made at several wavelengths of light (from 200 to 2000 nm). Thus, we can obtain refractive index as a function of wavelength and, hence, information about material properties, such as composition and energy band gap.
- The sample may be cleaned by washing it with acetone and alcohol and then drying in air, which is a simple laboratory process. This removes any organic material and dust particles from the surface of the test sample. If the surface is rough, the reflected light would be diffuse rather than specular. In such cases, the sample surface may first have to be polished by standard laboratory procedures. Additionally, one may include the surface roughness in the computation by adopting different models for the surface roughness.

Some of the **difficulties with this technique** are as follows:

- The accuracy of the physical parameters determined depends on the theoretical model of the multilayered structure and the model selected for the wavelength dependence of the refractive index $n(\lambda)$.

- The value of phase shift Δ depends strongly on the roughness of the sample surface. Hence, the sample surface must be smooth, and all surfaces/interfaces must be highly parallel.
- The spatial resolution of the measurements is limited by the spot size of the light beam on the sample surface, i.e., by the diffraction of light beam.
- The thin film must possess a small but finite absorption coefficient.
- The analysis of crystalline thin films is quite intricate due to their anisotropy.
- The depolarization of the light beam due to the poor quality of the optical components in the ellipsometer reduces the accuracy of the results.

The schematic diagrams of the setup for ellipsometry measurements is shown in Figure 8. The arrangement of the main optical components in Figure 8 is called the **PCSA** layout since the light beam from the source passes first through the polarizer, then the compensator, and is reflected by the sample surface before it passes through the analyzer. The solid line starting from the source represents the incident ray of light. The ray is incident on the flat surface of the sample at an angle of incidence φ with the normal to the sample surface (shown as a dot–dash line). The incident ray passes through a polarizer and a compensator (phase retarder). The solid line from the sample to the detector is the reflected ray, which makes the same angle φ with the normal. The reflected ray passes through the analyzer, which is also basically a polarizer. The incident ray, the normal

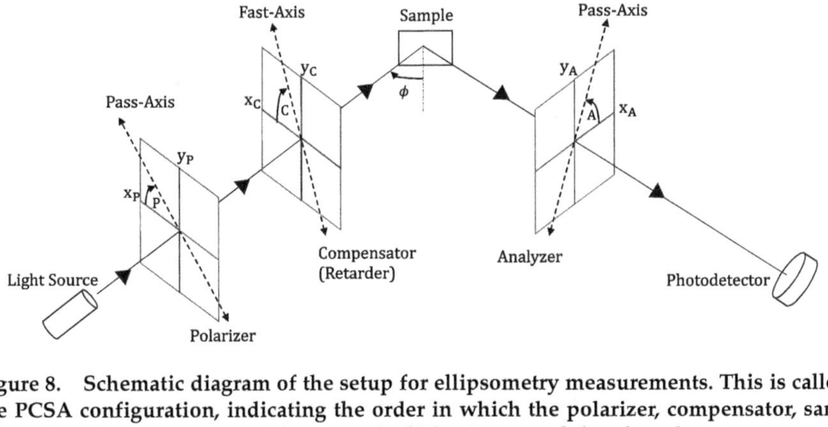

Figure 8. Schematic diagram of the setup for ellipsometry measurements. This is called the PCSA configuration, indicating the order in which the polarizer, compensator, sample, and analyzer are arranged between the light source and the photodetector.

at the point of incidence, and the reflected ray define the plane of incidence, which is treated as a horizontal plane in Figure 8. The sample surface is in the vertical plane. The dotted line, with arrow heads at both ends, represents (i) the pass axis for the polarizer and analyzer and (ii) the fast axis for the compensator.

In Figure 8, the light beam from the source may be unpolarized. The beam transmitted by the polarizer is plane polarized, with its electric field vector oriented at an angle P with the X_p axis. As this plane-polarized light passes through the compensator, it is split into two components, with electric field E_f parallel to the fast axis and E_S parallel to the slow axis of the compensator. These components are incident on the sample and are given by

$$E_f = t_f E_p \cos(P - C) \text{ and } E_s = t_s E_p \sin(P - C) \tag{26}$$

where t_f and t_s are the transmission coefficients of the compensator corresponding to the fast and slow axes, respectively. These two wave components are plane-polarized waves. Hence, each one can be decomposed into a p-polarized state and an s-polarized state with reference to the plane of incidence. Thus, the wave electric field incident on the sample may be expressed as

$$p\text{-polarized state: } E_f \cos(C) - E_s \sin(C) \tag{27a}$$

$$s\text{-polarized state: } E_f \sin(C) + E_s \cos(C) \tag{27b}$$

The reflection process is characterized by the amplitude reflection coefficients r_p and r_s, defined by Equations (1) and (2), respectively. Hence, the p- and s-polarized components after reflection are given by

$$E_{rp} = r_p[E_f \cos(C) - E_s \sin(C)] \tag{28a}$$

and

$$E_{rs} = r_s[E_f \sin(C) + E_s \cos(C)] \tag{28b}$$

Finally, the analyzer transmits only the components of the fields E_{rp} and E_{rs} along its pass axis, and the total transmitted field is

$$
\begin{aligned}
E_A &= E_{rp}\cos(A) + E_{rs}\sin(A) \\
&= E_p r_p \cos(A)\{t_f \cos(P - C)\cos(C) - t_s \sin(P - C)\sin(C)\} \\
&\quad + E_p r_s \sin(A)\,\{t_s \sin(P - C)\cos(C) + t_f \cos(P - C)\sin(C)\}
\end{aligned} \tag{29}
$$

The light intensity measured by the photodetector is proportional to the square of the amplitude $E_A E_A^*$, where E_A is a complex quantity. Equation (29) gives the light amplitude incident on the photodetector **in the PCSA configuration.**

A simpler configuration for ellipsometry is the **PSA configuration,** in which the compensator is removed. This is achieved, effectively, by putting $t_f = t_s = 1$ and $C = P$ in Equation (29). Thus, the final output amplitude in this case becomes

$$
\begin{aligned}
E_A &= E_p\{r_p \cos(P)\cos(A) + r_s \sin(P)\sin(A)\} \\
&= E_p r_s\{\rho\cos(P)\cos(A) + \sin(P)\sin(A)\}
\end{aligned} \tag{30}
$$

where ρ is a complex quantity defined as

$$
\rho = \frac{r_p}{r_s} = \tan(\Psi)e^{i\Delta} \tag{31}
$$

The quantities Ψ and Δ are determined by ellipsometry measurements.

In the matrix formalism, the result represented by Equation (30) may be expressed as

$$
E_A \begin{bmatrix} 1 \\ 0 \end{bmatrix} = \begin{bmatrix} 1 & 0 \\ 0 & 0 \end{bmatrix} \begin{bmatrix} cosA & sinA \\ -sinA & cosA \end{bmatrix} \begin{bmatrix} r_p & 0 \\ 0 & r_s \end{bmatrix} \begin{bmatrix} cosP \\ sinP \end{bmatrix} E_P \tag{32}
$$

Here, the column vector on the right-hand side gives the x- and y-components of the plane-polarized light transmitted by the polarizer. This is multiplied by the diagonal matrix, with elements r_p and r_s to account for the reflection by the sample. The reflected light consists of x- and y-polarized components. These components are multiplied by the rotation matrix, which results in aligning the two field components parallel and perpendicular to the pass axis of the analyzer. After this transformation, we are left with calculating the effect of the analyzer. This is achieved by multiplying the 2×2 matrix with elements 1 and 0, which represents the action of the analyzer on field components parallel and perpendicular to the pass axis; this matrix is said to be the Jones matrix of the analyzer in its eigen-polarization basis system. We note that Equation (32) yields the same results as Equation (30).

7. Null Ellipsometry Method

The apparatus is set up in the PCSA configuration. The polarizer azimuthal angle is set at P_0, and the compensator azimuthal angle is adjusted to a value of C_0 such that the reflected beam is plane polarized; this can be verified by rotating the analyzer. Finally, the azimuthal angle of the analyzer is adjusted to a value of A_0 such that the transmitted light intensity is reduced to zero (i.e., negligibly small). Thus, the photodetector records zero intensity. Hence, putting $E_A = 0$ in Equation (29) and after some manipulations, we obtain

$$\rho = \frac{\tau_c tan(P_0 - C_0) + tan(C_0)}{\tau_c tan(P_0 - C_0) tan(C_0) - 1} \tan(A_0) \tag{33}$$

where τ_c is the complex transmission ratio of the compensator defined by

$$\tau_c = \frac{t_s}{t_f} = tan\Psi_c e^{i\Delta_c} \tag{34}$$

This quantity should be known for a given compensator. Equation 33 gets simplified considerably if the compensator is a QWP, for which $\tau_c = i$, and it is oriented at angle $C = \pi/4$ radians. With this choice, we may achieve zero intensity at the detector for angles P_1 and A_1 of the polarizer and analyzer, respectively. It can be shown that a zero output intensity can also be achieved at angles $-P_1$ and $-A_1$. We can show further that

$$\Psi = A_1 \quad \text{and} \quad \Delta = 2P_1 + (\pi/2) \tag{35}$$

Thus, the experimentally measured angles P_1 and A_1 directly yield the values of the ellipsometry parameters Ψ and Δ at the specific wavelength at which the compensator works as a QWP. We must ensure that

$A_1 > 0$. The measurements are repeated for $C = -\pi/4$ radians for the QWP. From these two experimental results, the average values of Ψ and Δ are calculated. Measurements at multiple angles of incidence enable us to determine *n*, *k*, and *d*.

8. Photometric Ellipsometry Method

For this method, the compensator is not necessary, and hence, the ellipsometer is set in the PSA configuration. The polarizer is set at some orientation *P*, and the analyzer is rotated continuously about the reflected beam as the axis. The output intensity *I* measured by the photodetector varies periodically with the analyzer angle *A*. The graph of *I*(*A*) as a function of *A* is Fourier analyzed to obtain the parameters Ψ and Δ. Using Equation (30), we may obtain the equation for the output intensity as follows:

$$I(A) = I(P)\ |r_s|^2 \cos^2 P\{\tan^2\Psi\cos^2 A + \tan^2 P\sin^2 A + 2\tan\Psi\cos\Delta\tan P\cos A\sin A\} \tag{36}$$

where *I*(*P*) is the intensity of the light beam passed by the polarizer and, hence, incident on the sample. The measurements are carried out for a fixed value of angle *P* and for a large number of values of angle *A* by rotating only the analyzer. Since only the analyzer is rotated, this method is called rotating analyzer ellipsometry (RAE). Alternatively, one may fix angle *A* and make the measurements for several values of angle *P*. This method is called rotating polarizer ellipsometry (RPE). We can obtain the parameters Ψ and Δ by either the RAE or RPE method. For further analysis, we may rewrite Equation (36) as follows:

$$I(A) = \frac{I(P)|r_s|^2 \cos^2 P}{2(\tan^2\Psi + \tan^2 P)}\quad [1 + \alpha\cos 2A + \beta\sin 2A] \tag{37}$$

$$\alpha = \frac{(\tan^2\Psi - \tan^2 P)}{(\tan^2\Psi + \tan^2 P)} \quad \text{and} \quad \beta = \frac{2\tan\Psi\tan P\cos\Delta}{\tan^2\Psi + \tan^2 P} \tag{38}$$

α and β are the cosine and sine Fourier coefficients of the periodically varying output intensity, respectively, normalized to the constant background intensity, which are measured using the photodetector. Thus, the measured output light signal has to be Fourier analyzed to obtain α and β. The parameters Ψ and Δ are related to α and β by the equations

$$tan\Psi = |tanP|\sqrt{\frac{1+\alpha}{1-\alpha}}, \qquad cos\Delta = sgn(P)\frac{\beta}{\sqrt{1-\alpha^2}} \tag{39}$$

From the above equations, we can obtain Ψ and Δ. The RAE method is schematically illustrated in Figure 9.

This measurement process is repeated for different values of angle P to evaluate the average values of Ψ and Δ at a specific wavelength. One precaution to be taken in the RAE method is that angle P of the polarizer should not be close to zero or $\pm\frac{\pi}{2}$ to avoid large errors. In this method, errors can also occur due to the sensitivity of the photodetector to the different orientations of the electric field of the plane-polarized light beam passed by the rotating analyzer. This problem is avoided in the RPE method since the analyzer is kept fixed, but we have to ensure that the primary light source (laser or lamp) produces either completely unpolarized light or circularly polarized light. In this method too, measurements at multiple angles of incidence enable us to determine all three parameters n, k, and d. Spectroscopic ellipsometry can also be carried out by this method.

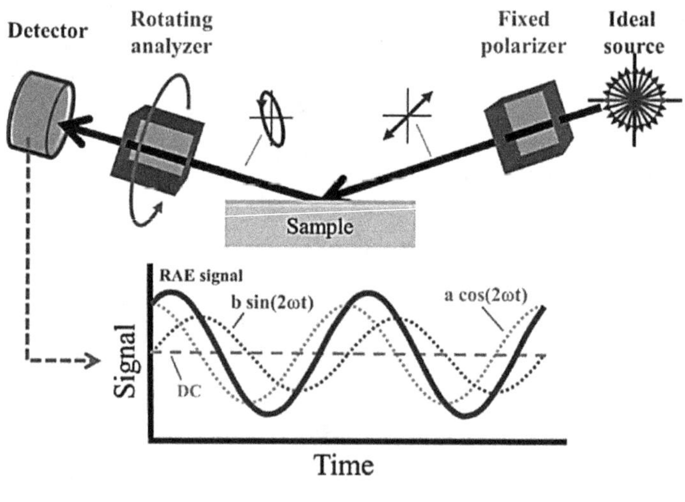

Figure 9. Schematic diagram of the RAE photometric ellipsometry and Fourier analysis of the output signal. The RAE signal has a large DC component and both sine and cosine components at 2ω frequency, where ω is the frequency of rotation of the analyzer (from Wikipedia).

We may carry out photometric ellipsometry with the apparatus in the PCSA or PSCA configuration. The compensator may be a QWP. It has become standard practice to fix both angles P and A and rotate the compensator. The advantage of this procedure is that the measured data are not affected by the SoP of the source light or the polarization sensitivity of the photodetector. In this case too, the output light intensity is described by Equation (36) or (37).

The measurements may be carried out using a thin collimated light beam or a beam focused on the sample surface. In such cases, a single photodetector is used. For measurements over large sample surface areas, one may scan the beam over the surface or use light beams with a larger cross-section. In such cases, one must use a linear or 2D array photodetector. For measurements over a range of wavelengths, i.e., spectroscopic ellipsometry, too, one may use array detectors so that the intensities at all wavelengths are recorded simultaneously in a single measurement. The use of such detectors greatly reduces the time required for the acquisition of data.

9. Calibration of the Ellipsometer and Sample Preparation

Before any measurement, we must calibrate the ellipsometer. Calibration means the measurement of angles P, C, and A with reference to the x-axis, as shown in Figure 8. Note that the x-axis lies in the plane of incidence of the light beam. This implies that we first determine the reading on the circular scale of the polarizer when its pass axis is oriented parallel to the plane of incidence. Using this setting, one can determine the reading corresponding to the orientation of the fast axis of the compensator and the pass axis of the analyzer. These three readings constitute the initial readings with respect to which the angles P, C, and A are to be measured. Calibration may be achieved by using a standard sample for which we know the values of n, k, and d. The procedure normally used for calibration is to carry out the measurements using the standard sample and adjust the angles P, C, and A until the experimental values match the known values of n, k, and d. Once this is achieved, the standard sample is replaced by the sample to be studied, and measurements are performed.

Alternatively, the initial readings can be determined using the Brewster angle phenomenon. Accordingly, if a p-polarized light is reflected at the Brewster's angle using the standard sample, there will be no reflected beam. First, we remove the compensator and analyzer and rotate the polarizer until the photodetector shows zero reading. The orientation of the polarizer would then be parallel to the plane of incidence, and the corresponding reading, P_0, on its circular scale would be the initial value of P. Next, we place the compensator on its mount and approximately align its fast axis parallel to the pass axis of the polarizer. We now slowly rotate the compensator until the photodetector shows a zero reading, which implies that the fast axis of the compensator is perfectly aligned parallel to the plane of incidence. The reading, C_0, on the circular scale of the compensator would be the initial value of the angle C. Finally, the analyzer is placed on its mount, and both the polarizer and compensator are rotated by 90° relative to their initial orientations P_0 and C_0.The light beam now incident on the sample is s-polarized, and hence, the reflected beam is also s-polarized. We now rotate the analyzer until the photodetector reads zero. In this situation, the pass axis of the analyzer would be parallel to the plane of incidence, and the reading on the circular scale of the analyzer will be the initial value A_0 of the analyzer. The ellipsometer is now calibrated and ready for measurements.

The sample to be studied should have as smooth a surface as possible so that the light beam undergoes specular reflection, i.e., the reflected beam should also be a well-collimated beam. If the surface of the sample is very rough, diffuse reflection occurs, and the light is scattered. Such samples will first have to be polished through standard laboratory procedures and then cleaned. All samples should be cleaned through standard procedure, i.e., washed using acetone and then alcohol, to remove any organic deposits. The cleaned sample is dried by blowing warm air on the surface to remove residual molecules of the cleaning liquids and any dust particles. The sample is now ready for ellipsometry measurements.

10. Analysis of Ellipsometry Data

We discuss the procedure to calculate the optical parameters n, k, and d from the values of Ψ and Δ obtained as functions of the wavelength of light. The spectra of Ψ and Δ are the outcomes of the spectroscopic ellipsometry of the given sample at a specific angle of incidence. The direct calculation of n, k, and d for the film by using the Fresnel equations is too intricate due to multiple interfaces. The procedure adopted is called "regression analysis," and it consists of the following steps:

- **Sample model:** The geometry of a typical device is shown in Figure 3. We must know the optical properties of the surrounding medium (air in most cases) and the substrate. We must also have an approximate estimate of the values of *n*, *k*, and *d* for the thin film. For transparent materials with negligible absorption of visible light, the refractive index *n*, which is a function of the wavelength λ, is accurately expressed by Cauchy's dispersion formula:

$$n(\lambda) = A + \frac{B}{\lambda^2} + \frac{C}{\lambda^4} \tag{40}$$

where A, B, and C are constants specific to the material. We arbitrarily fix the values of these constants and evaluate *n* for different values of λ. The constants A, B, C, and *d* (film thickness) are the parameters to be adjusted iteratively until the computed graphs of Ψ and Δ as functions of λ match to the desired accuracy with the graphs of Ψ and Δ determined by ellipsometry measurements. For measurements extending into the IR wavelengths, the Sellmeier's formula gives more accurate values for $n(\lambda)$. Materials may absorb light at UV and IR wavelengths. In such cases, the Tauc–Lorentz formula may be used instead of Equation (40).

- Using the estimated (or guessed) values of the optical parameters of the thin film and the known precise values of the optical constants of the substrate and the medium above the thin film, we evaluate the reflection coefficients R_p and R_s by substituting Equations (5) and (6) in Equation (3a). The values of Ψ and Δ can be calculated theoretically as functions of λ using Equation (31) along with Equations (5) and (6). Thus, the graphs of Ψ and Δ versus λ are plotted. If these graphs show deviations from the experimentally determined graphs, we must change the values of the parameters A, B, C, and *d*, and repeat the theoretical computations based on Equation (31). The difference between the two graphs is subjected to a mean square error (MSE) analysis. The difference between the two graphs at different values of λ gives the error, and the mean square of this error is the value of the MSE. The iteration is repeated until the MSE is smaller than a desired value. In other words, a good match is obtained between the experimental graph and the theoretical graph. The corresponding values of A, B, C, and d are used to compute $n(\lambda)$, $k(\lambda)$, and thickness *d*, and these are taken to be the precise values for the thin film being investigated.

- The regression analysis described above is the process of fitting a theoretical function to the experimental data by successive corrections of the parameters describing the theoretical function. The goal is to minimize the MSE by adjusting these parameters for the sample material. This is illustrated in Figure 10 for a sample for which only the thickness of the thin film was to be determined. For this, the function $\Psi(\lambda)$ is evaluated for a range of thickness values. In the first step, the film thickness was assumed to be 700 nm. For this film thickness, the spectrum of Ψ was evaluated and the MSE was calculated. Next, the thickness was increased by a small amount, and the MSE was evaluated, which was found to be lower than the previous value. Hence, in the succeeding steps, the film thickness was progressively increased in small steps until the MSE reaches a minimum value. Instead, if the MSE were to increase in the second step, we would reduce the thickness in steps and repeat the calculations. Figure 10 shows the case where successive iterations, with increasing thicknesses, lead to the minimization of the MSE after nine iterations. The thickness assumed for the last iteration was 749.18 nm, and this was taken to be the real thickness of the film.

The iterative regression analysis algorithm is illustrated in Figure 11. This analysis yields a unique set of sample parameters (film thickness and constants A, B, and C) for which the measured spectra of Ψ and Δ closely match the computed spectra. The calculated values of n, k, and d, which

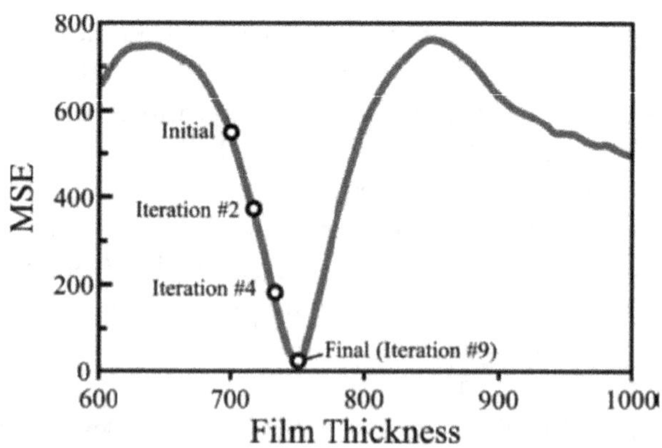

Figure 10. MSE profile as a function of film thickness, the unknown parameter of the sample device (from Wikipedia).

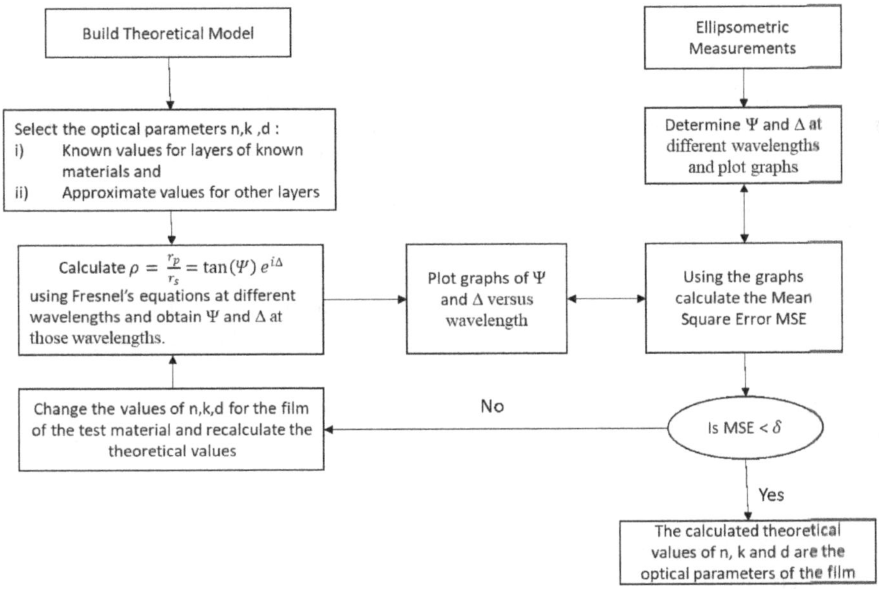

Figure 11. Flowchart for the evaluation of optical constants, *n* and *k*, and film thickness, *d*, by the iterative regression analysis. Here, δ is the prescribed upper limit of the MSE. The iteration is stopped once the MSE is less than δ.

correspond to these unique values of the parameters, are taken to be the most accurate values of the optical properties of the thin film.

The regression analysis described above refers to the device shown in Figure 3 in which a single thin film of an isotropic material is deposited on a substrate of an isotropic material. The optical parameters of the substrate and the surrounding medium are assumed to be known *a priori*. For samples with multiple thin-film layers or crystalline thin films, the computation process is similar but more elaborate, requiring more extensive measurements and data analysis. One may determine the optical constants and thicknesses of different layers. From the graphs of n and k and, therefore, the absorption coefficient α as a function of wavelength, one may obtain the band gap of the thin film, especially if it is made of a semiconductor. If the thin film is made of an alloy, the composition of the alloy can also be evaluated. By employing focused light beams, we can determine the spatial variation of the composition of the material on the film surface. These aspects are beyond the scope of this book. The reader may consult the references (Refs. 1–6) listed at the end of this chapter for more details.

11. Conclusion

We have presented an elementary account of ellipsometry as a technique for the determination of the optical parameters of highly transparent dielectrics, oxides, and semiconductor materials with very high accuracy. Spectroscopic ellipsometry is a very versatile technique which can be employed to determine:

(i) the optical constants (n, k) and thickness d of a thin film;
(ii) the spatially resolved composition of the material on the surface of a thin-film sample;
(iii) the band gap energy of the film material;
(iv) the optical constants and film thickness of multilayered optical components and devices.

Ellipsometry is discussed in greater depth in the following references.

References

1. Hecht, E. (2019). *Optics* (5th edition). Pearson, New York.
2. Tompkins, H. G., and Hilfiker, J. N. (2016). *Spectroscopic Ellipsometry: Practical Applications to Thin Film Characterization*. Momentum, New York.
3. Fujiwara, H. (2007). *Spectroscopy Ellipsometry: Principles and Applications*. John Wiley & Sons, UK.
4. "Ellipsometry Tutorials", from website of J.A. Woollam Co., www.jawoollam.com/resources/ellipsometry-tutorial.
5. "Spectroscopic Ellipsometry- Basic Concepts", from website of Horiba Scientific, www.horiba.com/int/scientific/technologies/spectroscopic-ellipsometry.
6. Tompkins, H. G., and Irene, E. A. (eds.). (2005). *Handbook of Ellipsometry*. William Andrews, New York.

Chapter 6

ELECTRON MICROSCOPY

1. Introduction

The crystal structure of a material refers to the orderly arrangement of atoms in a perfectly periodic structure. The crystal structure is determined using X-ray and neutron diffraction techniques. Materials, whether naturally occurring or prepared in the laboratory or industry, have many other features which need to be studied, such as surface morphology, microstructure, and composition. The determination of such aspects is also equally important. Electron microscopy has become one of the preferred tools for these studies.

In an actual crystal, there are defects. These defects may be impurities, vacancies, interstitials, or dislocations. A vacancy is an atomic site at which an atom is not present. An interstitial defect refers to the presence of an atom of small size in between the lattice sites occupied by the larger atoms of the material. Impurities play an important role. It is well known that dopant atoms, which are basically impurities, drastically alter the electrical properties of semiconductors.

Dislocations are of two types: (1) edge dislocation and (2) screw dislocation. These are described as follows.

The edge dislocation is formed by inserting an extra half plane of atoms in the lattice, as shown in Figure 1a. This causes a distortion of the lattice around the edge of the half plane of atoms. Consider a closed loop MNOP in the perfect crystal, as shown in the upper-left panel of Figure 1, which consists of three steps down from M to N, four steps to the left from N to O, again three steps up from O to P, and four steps to the right from P to M. Each step is a translation from one lattice site to the next. Now, perform an identical number of steps around an edge dislocation, starting from the lattice site M. We find that we end up at the site R, and the loop is not closed. We need to take one more step from R to M to close the loop. The vector RM is called Burger's vector. The plane of the half plane of atoms is called the slip plane, and the Burger's vector ***b* is perpendicular to the slip plane**.

A screw dislocation is shown in the right panel of Figure 1b. One part of the lattice is sheared with respect to the other part by one lattice spacing. In the perfect crystal, shown in the left panel of Figure 1b, we go from M to N by four steps down, from N to O by five steps left, O to P four steps up, and from P to M by five steps to the right. The loop is closed. However, if we perform the same operations on the figure at bottom right, we end up at the lattice point R, and to close the loop, we have to add the lattice vector from R to M. This vector RM, which is the Burger's vector ***b***, is now **parallel to the slip plane,** which is the plane between the two parts of the sheared lattice.

The above two examples show that as long as we take a rectangular loop around a dislocation involving an integral number of steps along each side of a rectangle, the loop will not be closed. It will need another step to close the loop, and this step, called the Burger's vector, is perpendicular to the slip plane in an edge dislocation and parallel to the slip plane in a screw dislocation.

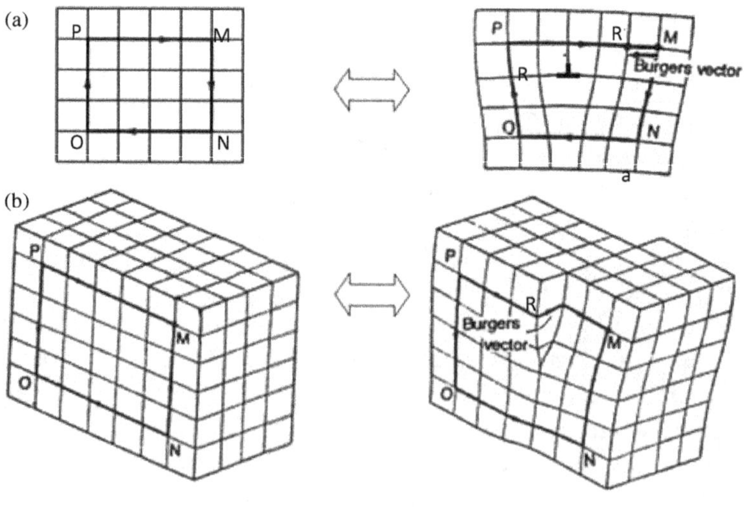

Figure 1. (a) Edge and (b) screw dislocations.

Due to the migration of atoms and vacancies, dislocations can move. One may have dislocations of mixed types. The interactions between the dislocations are responsible for the difference between the mechanical properties of normal and strain-hardened materials. Screw dislocations are involved in the growth of crystals.

A Stacking fault is a planar defect arising from an error in the sequencing of different planes of atoms. This is shown in Figure 2.

In the perfect crystal, shown on the left in Figure 2, the atomic planes are sequenced as ABC, ABC,.... The stacking fault occurs when this sequence is broken. At the stacking fault, shown on the right of Figure 2, the sequence is ABC, AB, ABC,.... This defect is common in close-packed crystal structures, such as face-centered cubic or hexagonal close packing. A stacking fault is caused by a moving dislocation.

When we grow crystals from the melt, there are several nucleation centers that give rise to the growth of grains. The size of any grain will depend on the conditions of growth. When two grains meet, the lattice planes in the two grains will be misoriented. The grain boundary is the region where the orientation of the lattice planes in one grain slowly changes in space to the orientation in the other grain. When the angle of misorientation is small, we call the boundary a small-angle grain boundary. When the angle of mis-orientation is large, it is called a large-angle grain boundary.

A material can exist in different phases, such as ice floating on water. When an alloy of metals is cooled from a high temperature, there can be local regions with different phases of the alloy with different compositions, all segregated but coexisting. So, the material will consist of these different phases with different compositions.

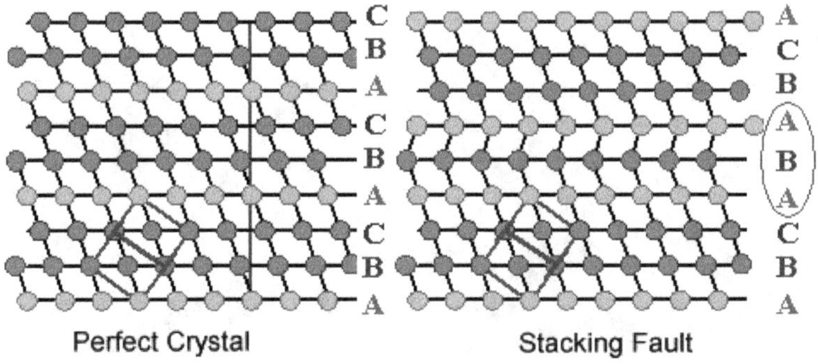

Figure 2. Illustrating a stacking fault.

The mechanical properties of a material will depend on the grain structure.

There could be micrometer-sized voids in the material. The failure of a material occurs under stress at these voids.

The presence of defects, dislocations, grains, grain boundaries, and different phases in the material is collectively called the microstructure of the material.

The length scale of these different structures is illustrated in Figure 3.

The size of the structures spans a wide range, from nanometer to millimeter. As they affect the properties of the material, the characterization of these structures is important. One can examine these structures using a variety of microscopes.

An optical microscope is often used to study microscopic objects. The schematic of an optical microscope in the transmission mode of operation is shown on the left in Figure 4.

Light from a source is focused on the sample stage on which the object is mounted. The light, scattered by the sample, is collected by the objective lens, which forms a real image of the object in the intermediate image plane. This is magnified by the eyepiece, or ocular, to form a final image, which can be seen with the eye. The image can also be photographed. Microscopes can also operate in the reflection made.

Two important parameters of such a microscope are (1) magnification and (2) resolution. Magnification is the ratio of the dimension of the image to the dimension of the object. Magnification up to 1000 can be realized in an optical microscope. Resolution is a measure of the smallest separation between any two features in an object which can be discerned

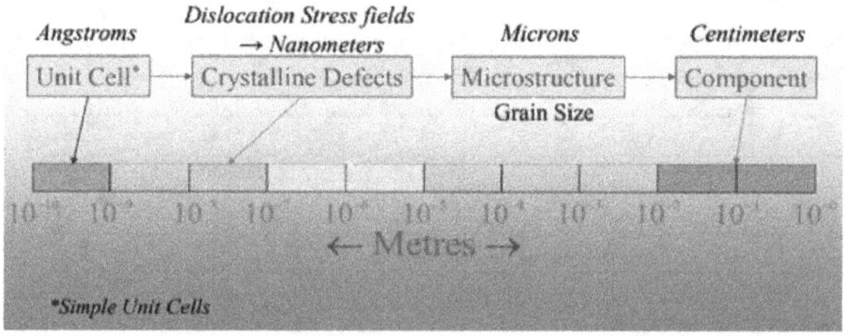

Figure 3. Length scale of different structures in a material.

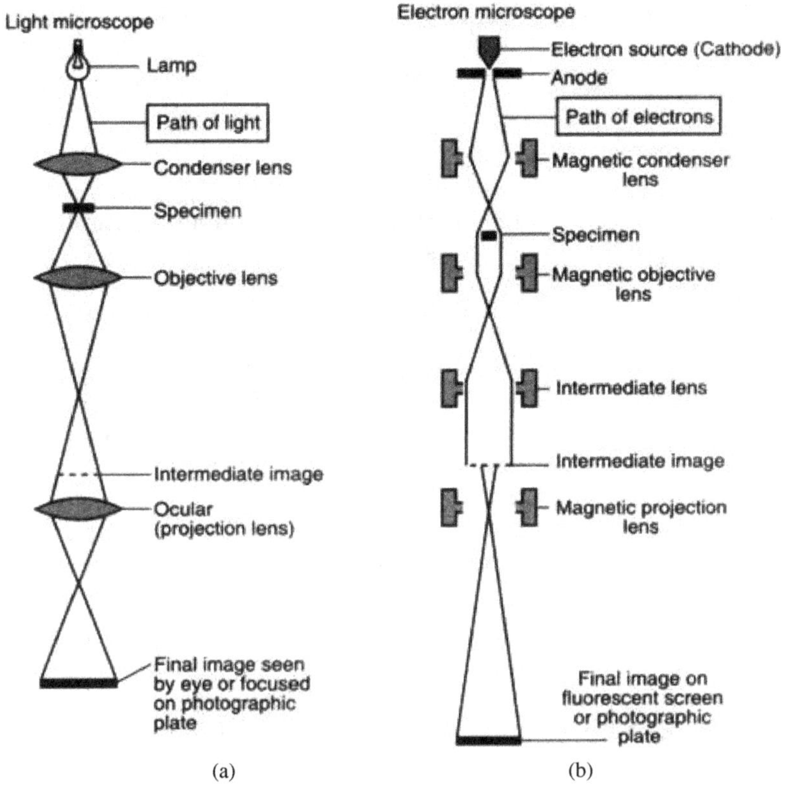

Figure 4. Schematic representations of (a) optical and (b) electron microscopes in the transmission mode.

as distinct in the final image. This depends on (1) the wavelength of light and (2) the numerical aperture (NA) of the objective.

The smaller the wavelength, the higher the resolution and the closer the features which can be seen distinctly. One may use an ultraviolet source of light to get higher resolutions. The NA is the product of the refractive index of the medium between the object and the objective lens, and sin(α), where α is the semi-angle of the cone of rays reaching the objective from a point on the object directly below its center. For a given α, one can increase the resolving power by filling the space between the object and the objective lens with oil, which has a refractive index greater than unity. At best, the smallest distance between two features which can be seen as distinct is of the order of one μm in an optical microscope.

2. Principle of Operation of an Electron Microscope

An electron, accelerated by a voltage V, has a kinetic energy given by

$$E = p^2/2m_0 = eV \tag{1}$$

In Equation (1), E is the kinetic energy, p the momentum of the electron, e the magnitude of the charge on the electron, and m_0 the rest mass of the electron.

The above relation is valid when the velocity of the electron is small compared to the velocity of light, which is the case when the electron is accelerated to a voltage less than 100 V. If the accelerating voltage is in kilovolts, then p and E are related by the relativistic equation given by

$$p^2c^2 + m_0^2c^4 = (m_0c^2 + eV)^2 \tag{2}$$

According to Louis de Broglie, the probability wave associated with the electron has a wavelength

$$\lambda = h/p \tag{3}$$

In the nonrelativistic case, the wavelength and energy are related as

$$\lambda = h/(2m_0eV)^{1/2} \tag{4}$$

Here, h is the Planck's constant. Substituting for h, e, and m_0, we have

$$\lambda \text{ (in nm)} = 1.23/\sqrt{V} \tag{5}$$

Here, V is in volts. If V is 100 V, the de Broglie wavelength λ will be 0.123 nm. Due to such a small wavelength, the electron microscope offers a much higher resolution than an optical microscope, wherein the wavelength is about 500 nm. In addition, the magnification achievable with an electron microscope is about 10,000, which is much higher than in an optical microscope. These are the main advantages of an electron microscope over an optical microscope.

However, the electron microscope can only operate in an ultrahigh vacuum. The presence of even a minute amount of air will result in the electrons getting scattered and, also, losing energy in ionizing the air molecules, and no image will be formed.

Electrons have charge. They are absorbed heavily by materials. In the transmission mode, the electrons have to pass through the sample. So, the sample will have to be thin so that enough electrons pass through it to create the image. In the transmission electron microscope (TEM), the sample should also be mounted in a vacuum.

On the right in Figure 4, the path of the electron beam in a transmission electron microscope is shown. The schematic layout of the transmission electron microscope is very similar to that of an optical microscope, except that in a TEM, magnetic lenses are used for focusing the electrons. Magnetic lenses are basically coils carrying current that generate a magnetic field for bending the electron paths.

3. Diffraction of Electron Beams

We see that electrons accelerated to 100 V and above have wavelengths comparable to, or less than, the interatomic distances in crystals. Therefore, they can also be used in the diffraction mode to study the lattice spacing in the crystallites.

If a collimated beam of electrons of de Broglie wavelength λ is incident at a glancing angle θ on a set of parallel atomic planes with spacing d, the beam will be reflected if the Bragg condition

$$2d\sin(\theta) = n\lambda \tag{6}$$

is satisfied.

Such diffraction by crystals is not possible with visible light since it has a wavelength which is about hundred times the atomic spacing.

If the material is multi-grained with grains oriented randomly in all directions, then the diffracted rays will come out as a cone. They will be recorded as rings on a flat photographic film placed perpendicular to the incident beam.

Figure 5 shows the diffraction images from a foil of aluminum, with both X-rays and an electron beam. The similarity of both patterns is striking.

Thus, electrons can be used for both microscopy and diffraction.

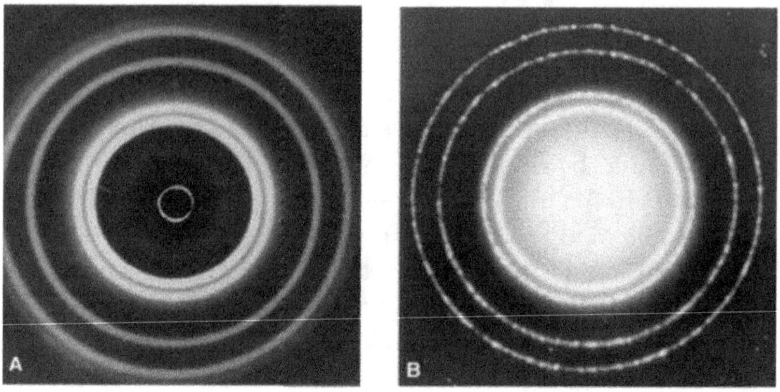

x-ray diffraction of aluminum foil — electron diffraction of aluminum foil

Figure 5. X-ray and electron diffraction from an aluminum foil.

Figure 6. Photograph of a TEM.

4. Transmission Electron Microscopy

Figure 6 shows a photograph of a TEM. At the top, an electron gun consisting of a tungsten filament produces electrons by thermionic emission. The beam is accelerated by applying a few kilovolts to an anode.

The beam is focused by a magnetic lens onto a sample stage. This lens should be carefully designed to avoid astigmatism and spherical aberration, which will spoil the quality of the image.

The sample should be very thin. If the sample is a solid material, such as a metal or a semiconductor, the sample has to be ion-milled so that it has a very thin central region and should be held by a rim which is sufficiently thick. If it is a biological sample, it is either fixed with a chemical or rapidly frozen to avoid damage to the cell wall through nucleation and growth of ice. The chemical produces cross-links between the proteins or lipids in the biological sample, which prevents any change in the morphology of the sample. Biological samples, such as tissues, are potted in epoxy and then sectioned into thin films (about a few microns thick) using a microtome.

The contrast in the image is produced by the difference in the number of electrons transmitted through the different parts of the sample. It is this contrast which enables one to study the texture and morphology of the samples. A sample may contain different chemical elements with different atomic numbers. Heavy elements absorb electrons more than light elements. In biological samples, which contain mainly light elements, the contrast will not be good. The contrast can be improved by staining the biological sample with heavy elements.

The microscope can be operated in the following two modes: (a) the imaging mode, which produces an enlarged image of the object, and (b) the diffracting mode, in which all rays coming in the same direction from different points in the object are brought to a focus at a single point in the image plane. The difference between the two modes is shown in Figure 7.

For imaging purposes, the electron beam may be treated as a bundle of rays, just as in the case of a light beam. In the imaging mode, the incident electron beam falls on the specimen as a convergent beam. The distance of the objective lens from the specimen is adjusted so that the rays, denoted by 1 and 1′ on the left-hand side of Figure 7, coming from a given point in the object are focused at one point in the image plane. Similarly, rays 2 and 2′ are focused at a different point in the image plane. If the distance between the lens and the image plane is more than the focal length of the lens, a magnified image of the object is produced. On the other hand, in the diffraction mode, a parallel beam of electrons falls on the specimen. Parallel rays, such as those denoted by 1 and 2 (or 1′ and 2′), from different points of the object are focused at a point in the image plane. In this case, the distance between the lens and the image plane is

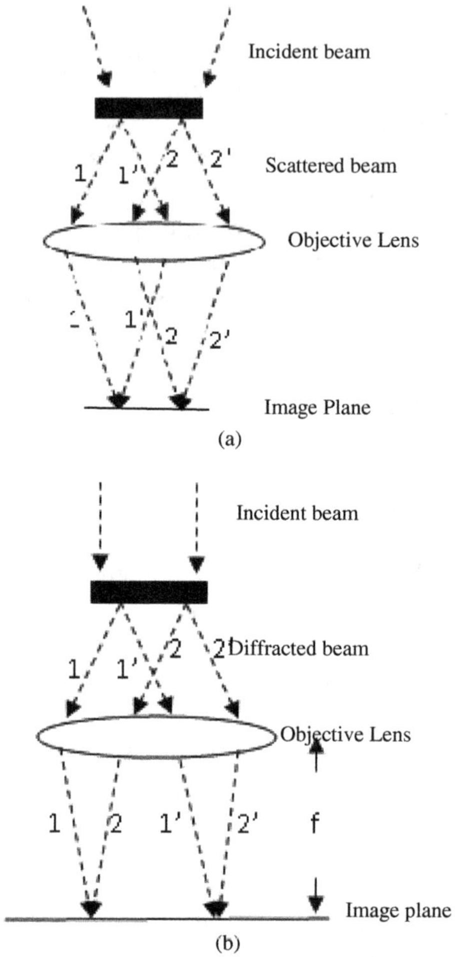

Figure 7. Path of the rays in the (a) imaging and (b) diffraction modes.

kept equal to the focal length, *f*, of the lens. The final image is projected onto a CCD plate.

One may improve the contrast of the image by using an aperture, which will prevent electrons that do not pass through the specimen from reaching the image plane. This is called dark field imaging.

Figure 8(a) shows the improvement in contrast in the imaging mode when using dark field imaging. Note the length scale. The dark features in image (a) and the bright features in image (b) are of nanometer size.

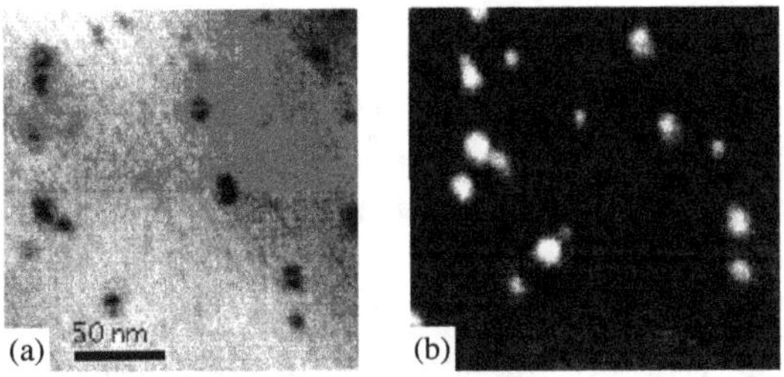

Bright Field | Dark Field

Figure 8. Improvement in contrast using dark field imaging.

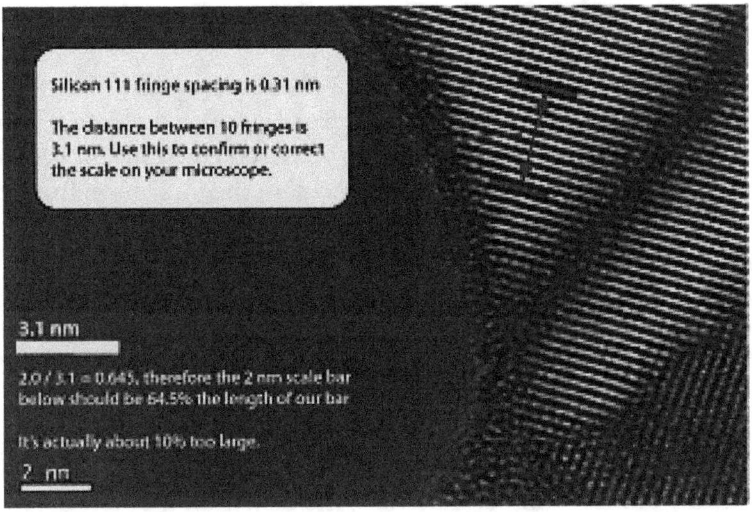

Figure 9. Lattice fringes formed by (111) planes of silicon (reprinted with permission from Steve Cham, EM Resolutions Limited, February 8, 2016).

In the diffraction mode, high-resolution imaging is possible using the phase contrast technique. In this technique, the wave transmitted through the crystalline sample and the diffracted waves produced by the lattice planes interfere, producing fringes. The changes in the phase of electrons passing through the material are converted into changes in brightness. The spacing of the fringes gives the spacing of the lattice planes. Figure 9 shows the lattice fringes formed by Si (111) planes. The spacing between the (111) planes in silicon is 0.31 nm, which is equal to the fringe width determined from Figure 9, which shows that the width of 10 fringes is 3.1 nm.

4.1. *Some examples of TEM images*

The following are some examples of TEM images.

In Figure 10(a), we show an image of a stacking fault produced by a moving dislocation. There are two locations where a break in the sequence ABC, ABC,... is seen to occur.

In Figure 10(b), we show a small-angle grain boundary. Note the change in the inclination of the lattice planes in two neighboring grains.

In Figure 10(c), we show the image of nanorods about 10 nm in diameter, and in Figure 10(d), we show the image of a biological (liver) cell.

5. Scanning Electron Microscope

Figure 11 is a schematic of a scanning electron microscope (SEM).

In a SEM, the incident beam of electrons is focused at a fine point on the specimen. **Here, we look at the backscattered electrons, or secondary electrons, scattered at an angle to the incident beam. Hence, the sample does not need to be thin.**

The accelerated electron beam is focused on a specimen that can be tilted. The beam can be deflected by deflection coils so that the sample can be scanned.

The beam penetrates the sample up to a small thickness. The depth of penetration depends on the energy of the electron beam and the atomic number of the elements in the sample. This is shown in Figure 12 (from Ref. 5).

For low-atomic-number materials, the penetration into the sample is deeper, and the penetration region takes the shape of a drop. For materials with a high atomic number, the penetration region is cylindrical in shape. The higher the energy of the beam, the deeper the penetration.

As the electron beam penetrates the material, different kinds of secondary emissions take place, as shown in Figure 13. The convergent incident electron beam (denoted by "a") falls on a narrow spot on the specimen. These high-energy electrons enter the specimen and create various emissions before they lose their energy and get absorbed. Secondary electrons ("b") are emitted at a small angle with respect to the surface. The secondary electrons are valence electrons knocked off from the atoms inside the specimen by the incident electrons. They have a low energy of up to 50 eV. These electrons come from a very thin layer of the penetrated region near the surface of the specimen.

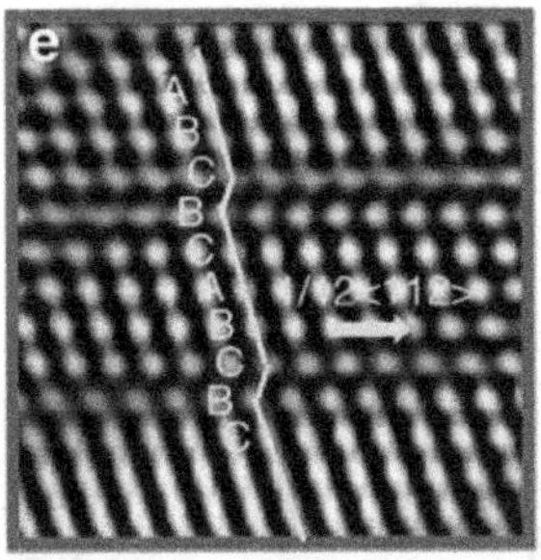

(a)

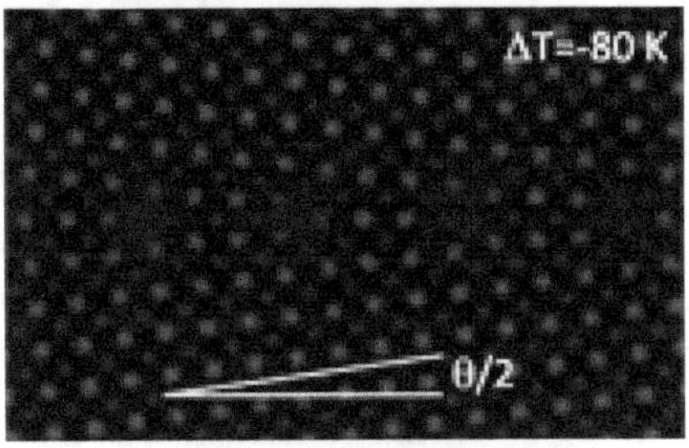

(b)

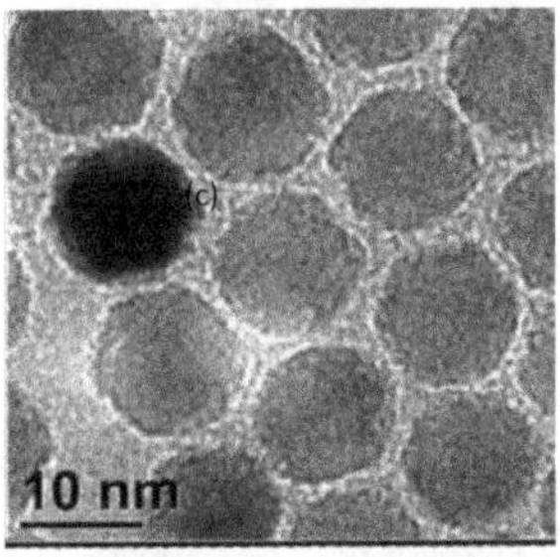

(c)

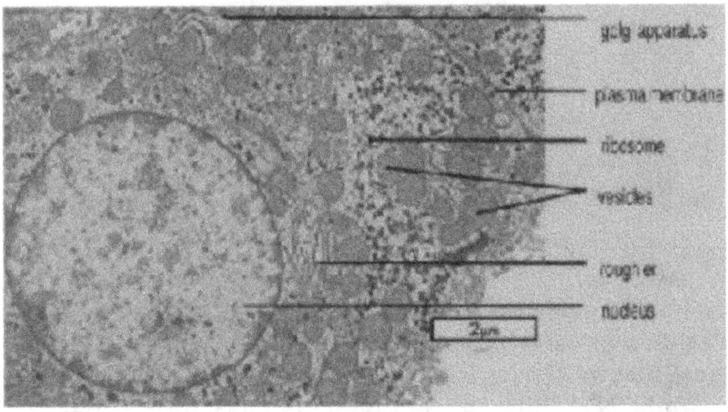

(d)

Figure 10. **(a) Stacking fault; (b) small-angle grain boundary; (c) nanorods about 10 nm in diameter; (d) image of a liver cell.**

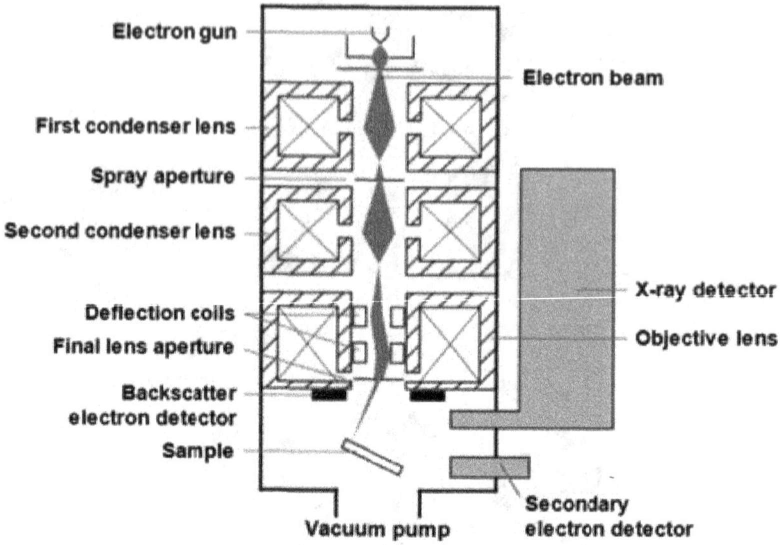

Figure 11. Schematic diagram of a SEM.

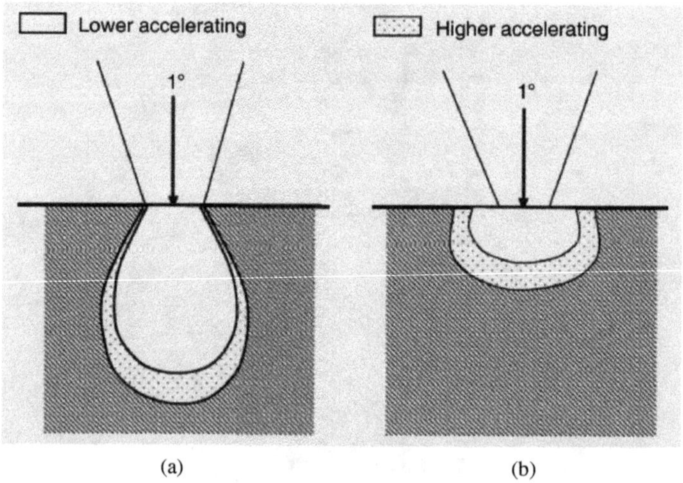

Figure 12. Penetration region for (a) low-atomic-number materials and (b) high-atomic-number materials (Reprinted with permission from Zhou, W., Apkarian, R., Wang, Z.L., Joy, D. (2006). In: Zhou, W., Wang, Z.L. (eds) Scanning Microscopy for Nanotechnology. Springer, New York, NY. https://doi.org/10.1007/978-0-387-39620-0_1). The region for higher acceleration cases is shown as a dotted area.

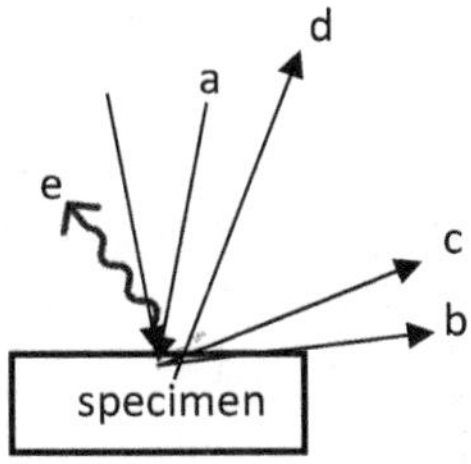

Figure 13. Incident beam of electrons "a", and various emissions from the specimen: secondary electrons "b," Auger electrons "c," backscattered electrons "d," and X-rays "e."

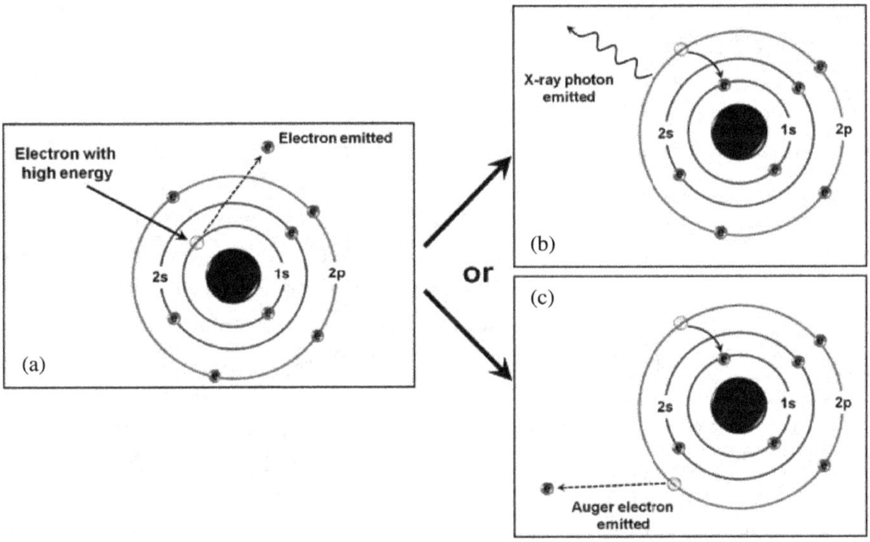

Figure 14. An illustration of the Auger process.

The Auger process, which leads to the generation of Auger electrons ("c"), is illustrated in Figure 14.

The incident electron ionizes the atom by removing an electron from the inner 1s shell. This is shown in Figure 14(a). An electron from an outer 2p shell occupies the vacancy in the 1s shell, and an X-ray photon is emitted. This is shown in Figure 14(b). In the Auger process, a second electron from the 2p shell is knocked out, instead of an X-ray photon being emitted. The energy of the Auger electron is characteristic of the atom.

The Auger electrons have low energy. They come out of a thin layer at the top of the penetrated region.

A part of the incident electron beam is scattered at a large angle from elastic collisions with an atom or inelastic collisions after producing secondary electrons. The backscattered electrons ("d") have a large spread in energy, ranging from a few hundred electron volts to a few thousand electron volts, i.e., the energy of the incident electrons.

Furthermore, the emission of X-ray photons ("e"), which are characteristic of the elements in the sample, also occurs, as illustrated in Figure 14(b). In addition, we have X-rays due to bremsstrahlung emitted as a result of the deceleration of the incident electrons. The bremsstrahlung is spread continuously over a range of wavelengths and occurs as background radiation.

One can use the secondary and Auger electrons for characterizing the surface layers and their composition. The backscattered electrons and X rays yield information about the surface morphology and chemical composition.

Detectors collect the secondary and Auger electrons coming at small angles with respect to the surface. The detector for secondary electrons comprises a Faraday cage and a scintillator. The Faraday cage is maintained at a small positive potential of a few hundred volts, and the scintillator is at a large positive potential in the kilovolt range. The secondary electrons coming from a point on the surface of the sample in the penetrated region will be accelerated and will fall on the scintillator to produce photons. These are converted into electronic signals, which, after further processing, form an image of the surface around the penetration region in the sample. The contrast will depend on the surface topography of the sample, as shown in Figure 15.

In Figure 15, we show secondary electrons being emitted from two points, A and B, on the surface of the sample when the primary electron beam is focused on the surface of the sample. Between A and B, there is a small hump on the material projecting out of the surface. When the detector is in the position shown, more secondary electrons reach the detector from point B than from point A because the hump will block some of the secondary electrons from point A from reaching the detector. So, point B will appear brighter than point A.

The dependence of brightness, in the secondary electron image, on surface topography for needle-shaped objects is due to a different reason. At the tips of needle-shaped objects, there is a larger volume giving rise

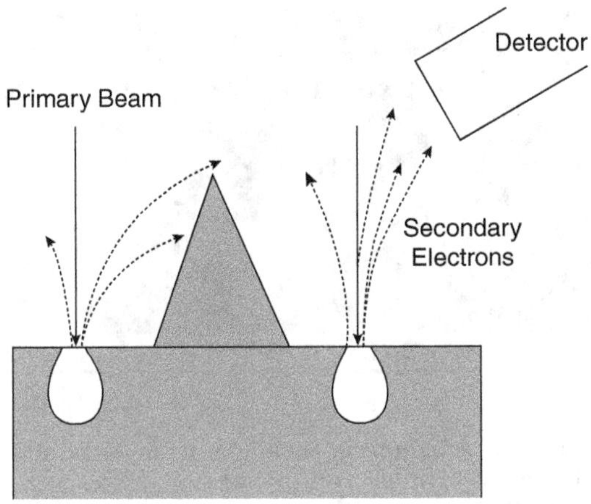

Figure 15. Influence of surface topography (redrawn from Ref. 3).

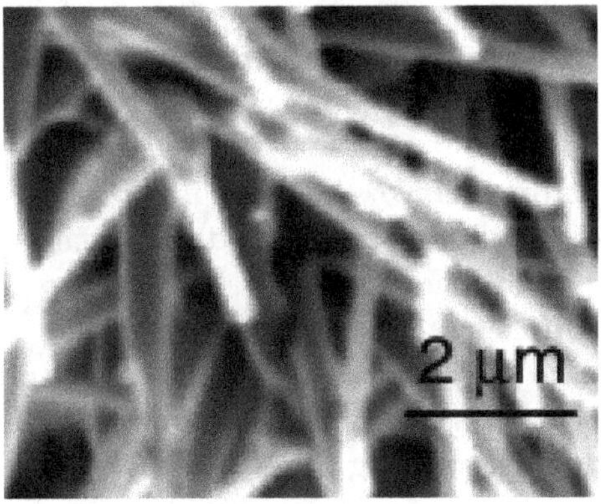

Figure 16. SEM image of ZnO nanoneedles (reprinted with permission from Ref. 5).

to secondary electron emission than from a flat surface. So, secondary electron emission is enhanced at sharp tips. Figure 16 (from Ref. 5) shows an SEM image of zinc oxide nanoneedles, where the emission is seen to be enhanced at the tips.

The sample is mounted on a stage which can be tilted. Tilting of the specimen increases the secondary emission area. So, tilting generally enhances secondary electron emission. It also changes the contrast. The sample can be scanned in the plane of the tilted surface.

The backscattered electrons have much higher energy than the secondary electrons. So, in the detector used for backscattered electrons, the

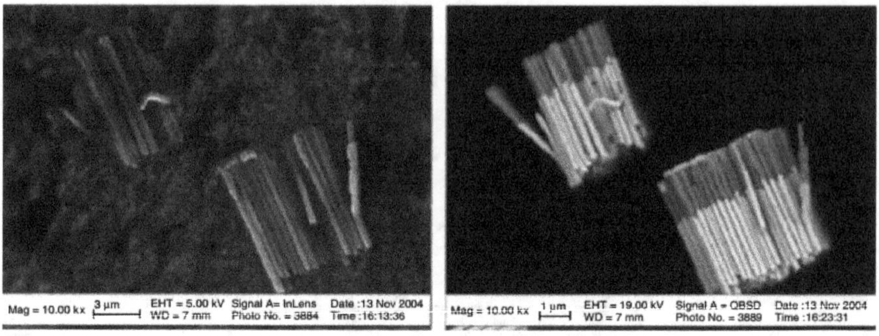

(a) (b)

Figure 17. SEM images of a bundle of nanorods with (a) secondary electrons and (b) backscattered electrons (Reprinted with permission from Zhou, W., Apkarian, R., Wang, Z.L., Joy, D. (2006). In: Zhou, W., Wang, Z.L. (eds) Scanning Microscopy for Nanotechnology. Springer, New York, NY. https://doi.org/10.1007/978-0-387-39620-0_1).

Faraday cage is maintained at a negative potential of a few hundred volts. This prevents the secondary electrons from reaching the detector. High-atomic-number elements yield more backscattered electrons than secondary electrons.

In Figure 17 (from Ref. 3), the SEM images of a bundle of nickel nanorods, recorded with secondary and backscattered electrons, are shown. The images with backscattered electrons are brighter and show the details of the sample to a greater extent.

As the accelerating voltage increases, the penetration depth of the incident electrons increases. So, the features at different depths within the specimen get overlapped, making the image more blurred.

If the specimen is nonmetallic, it may get charged. Then, the incident electron will get deflected, and this will cause distortion in the image. If the charge is not large, it does not affect the incident electron beam. However, the secondary electrons emitted get deflected by the charge because of their low energy, and this will result in reduced contrast in the image of the charged spot. To prevent charging, the specimen is coated with a thin (10 nm) film of gold or platinum to make the surface electrically conducting and then placed inside the SEM.

In tissue samples, the cells may die, and the sample may get distorted. To prevent this distortion, biological samples are chemically fixed with glutaraldehyde or formaldehyde. The tissue can also be fixed by dipping it in an ethanol solution for a certain duration before dehydrating it. It is important to ensure that the sample does not degas inside the SEM.

After the specimen is prepared, it is mounted on the sample mount in the SEM. Bulk specimens can be mounted using conductive paste. Powders are either dusted onto a layer of conductive paste or fixed using double-sided sticky tape.

5.1. *Some examples of SEM images*

Figure 18 shows some examples of SEM images. An image of nanofibers is shown in Figure 18(a). The hexagonal cells of graphene are shown in Figure 18(b). In Figure 18(c), we show the grain boundaries in a multi-grained specimen of $SrTiO_3$. Figure 18(d) shows the image of two phases, Mn_2Sb (dark grains) and MnSb (bright needles), when $Mn_{59.8}Sb_{40.2}$ is quenched from its melt.

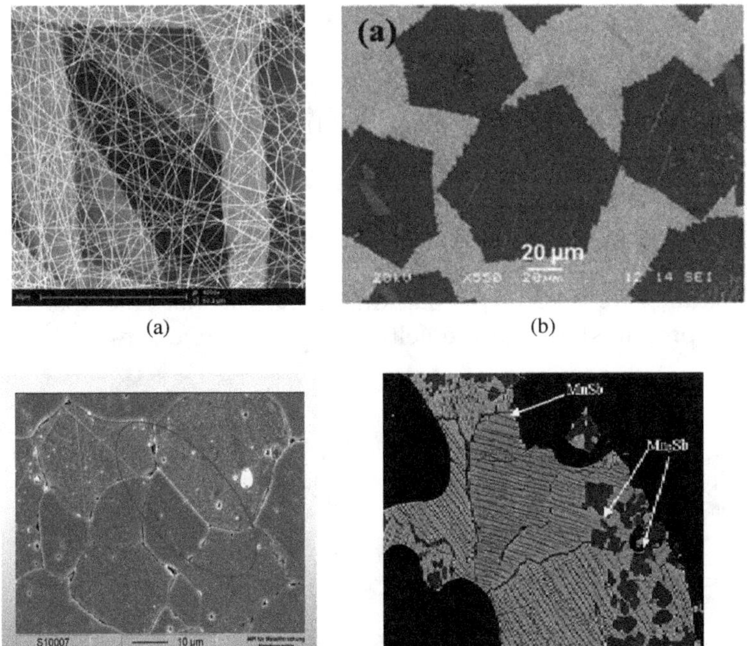

(a) (b) (c) (d)

Figure 18. **SEM Images of (a) nanofibers, (b) hexagonal cells of graphene, (c) grain boundaries in $SrTiO_3$, (d) Mn_2Sb and MnSb phases when $Mn_{59.8}Sb_{40.2}$ is quenched from its melt.**

5.2. *Energy-dispersive X-ray analysis*

The energy-dispersive X-ray analysis technique is more often called by the acronym EDAX. As shown in Figure 13, the incident electrons also cause X-ray photons to be emitted. These are the K and L X-ray lines, the frequencies of which are characteristic of the elements present in the sample. An analysis of the energies and intensities of these characteristic X-ray photons gives information about not only the elements present but also their relative proportions.

An example of an EDAX spectrum (Figure 20) is shown below the corresponding SEM photograph (Figure 19) of the surface of the brain tissue taken from a sample damaged by a bullet from a pistol (Ref. 6).

The EDX analysis in Figure 20 shows the presence of Pb, Si, Ca, and Ba, which is indicative of the materials deposited at the wound site by the bullet.

6. Applications of Electron Microscopes

Electron microscopy (both TEM and SEM) has a wide variety of applications in:

(a) materials science for elemental analysis, detecting mixed phases, and the study of nanomaterials;
(b) metallurgy for the study of stacking faults, the motion of dislocations, crack propagation, and premelting at grain boundaries;

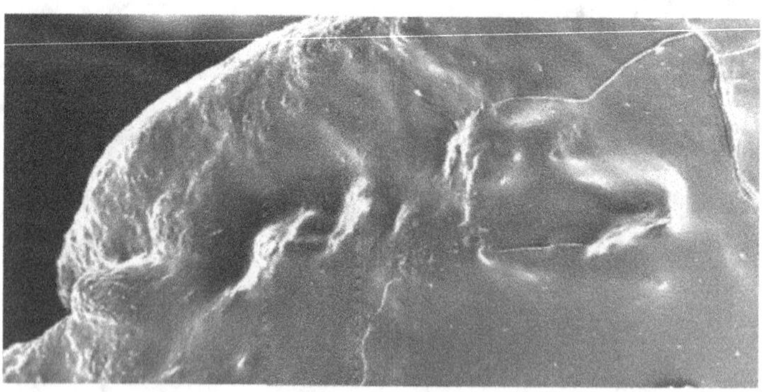

Figure 19. **SEM of the brain tissue taken from a sample with a bullet wound (reprinted with permission from Ref. 6).**

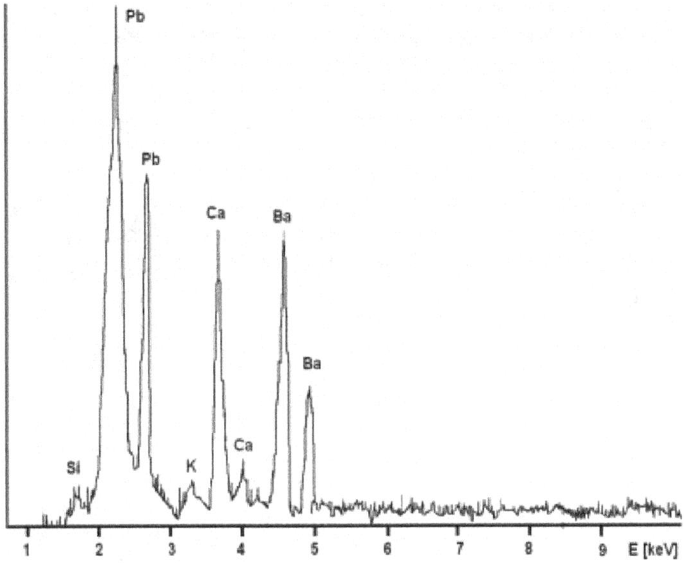

Figure 20. EDX analysis of the brain tissue damaged by the bullet (reprinted with permission from Ref. 6).

(c) the study of semiconductor surfaces in the manufacturing of integrated electronic chips;
(d) biology for the study of cells and tissues; and
(e) forensic science.

7. Conclusion

Electron microscopy is an extensive field of study. We have only given a brief overview of the subject. SEM (and, to a lesser extent, TEM) are available in all laboratories involved in research on materials science. It is necessary to have knowledge about the operating principles of these instruments and the analysis of the data obtained from them.

References

1. Egerton, R. F. (2005). *Physical Principles of Electron Microscopy- An Introduction to TEM, SEM and AEM*. Springer Verlag.
2. Ul-Hamid, A. (2018). *A Beginners Guide to Scanning Electron Microscopy*. Springer.

3. Zhou, W., Apkarian, R. P., Wang, Z. L., and Joy, D. (2006). Chapter 1: Fundamentals of scanning electron microscopy. In W. Zhou and Z. L. Wang (eds.) *Scanning Microscopy for Nanotechnology and Applications*. Springer Verlag.
4. Transmission Electron Microscopy, www.fisica.unige.it~rocca>Didattica>14TEM.
5. Scanning Electron Microscopy-Nanoscience Instruments, https://www.nanoscience.com/techniques/scanning-electron-microscopy/.
6. Biro, C., Kovac, P., Palkovic, M., El-Hassoun, O., Caplovicova, M., Novotny, J., and Jakubovsky, J. (2010). *Rom. J. Leg. Med.* **18**, 225.

Chapter 7

SURFACE PROBE TECHNIQUES

1. Introduction

Study of the surface topography of materials is of interest in many areas. Surfaces provide a hint towards the geometric and electronic properties of the bulk. Interesting physics and chemistry can be learned from the study of surfaces in the fields of heterogeneous catalysis, self-assembly, surface reconstruction, surface states, charge density waves, and adsorption.

Binnig and Rohrer developed the scanning tunnelling microscope (STM) in 1981 and produced a stunning image of the surface of a silicon film with **atomic resolution**. Since that time, a variety of surface probes have come into vogue.

The general layout of a scanning probe microscope is shown in Figure 1.

There is a probing tip on a reed which can be positioned roughly on the surface of a sample to be studied. The sample is mounted on a piezo actuator. The sample can be moved in a raster pattern in the *XY* plane i.e., plane of the surface, and translated in the *Z* direction. A computer system

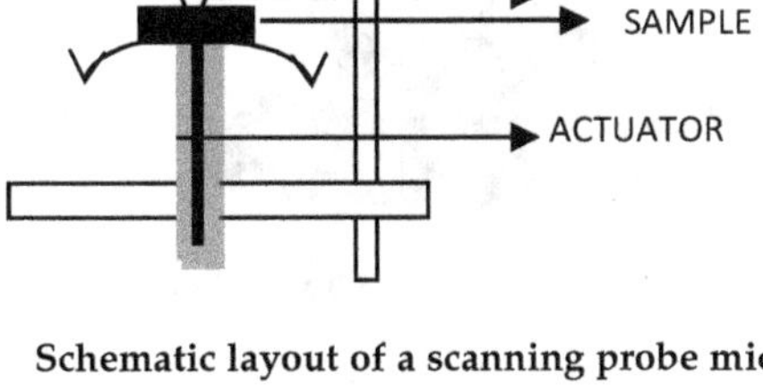

Figure 1. Schematic layout of a scanning probe microscope.

actuates the piezoelectric actuators and acquires data. The data is projected as a two-dimensional image on the computer monitor screen.

Since we are talking about nanometer scale resolution, all scanning probe systems must be mounted on vibration isolation mounts.

In this chapter, we give a brief description of the principle of operation of the scanning tunnelling, atomic force, and magnetic force microscopes.

2. Scanning Tunnelling Microscope

This microscope operates on the quantum mechanical tunnelling effect (Figure 2). A fine tip of a tungsten probe is brought within a few angstroms of the surface of **a metal or semiconductor** specimen.

The distance between the tip and the surface of the sample is *d*. The tip is biased to a voltage, *V*, relative to the surface of the sample. Then a tunnelling current, *I*, can flow from the tip to the sample or vice versa.

Figure 3 explains the tunnelling phenomenon. E_F is the Fermi level in the sample. In the absence of a bias voltage *V*, the Fermi level in the sample is aligned with the Fermi level in the tip. When a bias potential *V* is applied, the Fermi level of the tip is lowered (raised) relative to the Fermi level in the sample by *eV*, when *V* is positive (negative). An electron approaches the barrier from one side with energy, *E*, less than the barrier height. In classical mechanics, this electron cannot climb the barrier and is fully reflected. In quantum mechanics, this electron is accompanied by a de Broglie wave of probability amplitude. At the barrier, the wave is partially reflected and travels back. A part of the wave propagates through

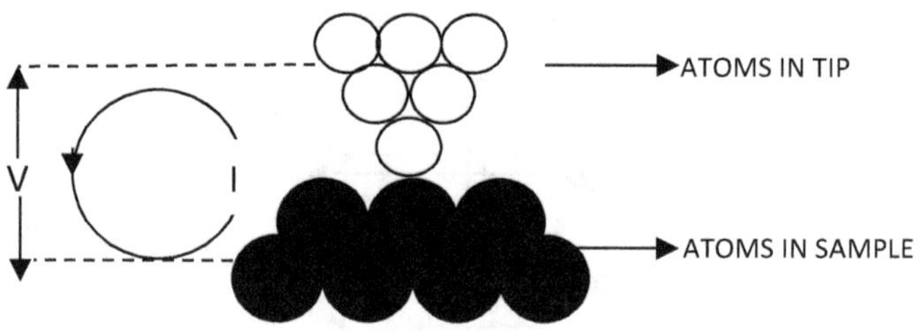

Figure 2. Schematic diagram of an STM. The circles represent atoms.

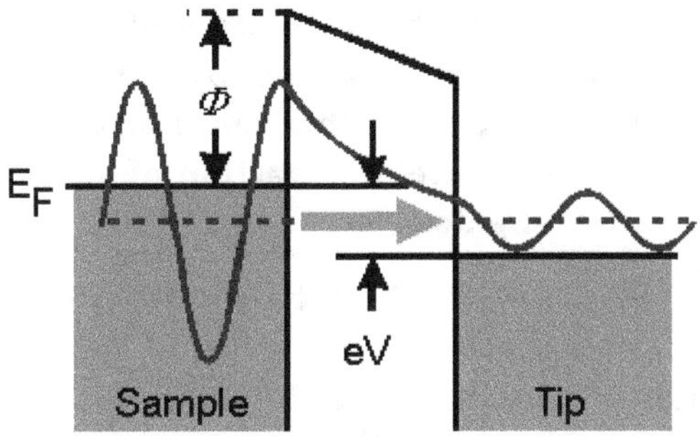

Figure 3. Tunnelling phenomenon.

the barrier with exponentially decaying amplitude and comes out on the other side as a travelling wave. This indicates that there is a finite probability of the electron to be found on the other side of the barrier with the same energy, provided there is a vacant energy level on this side. This phenomenon is called tunnelling. The tunnelling current, I, is proportional to

$$I \propto V \rho \exp(-\kappa d) \tag{1}$$

Here, ρ is the electron density in the sample at the point of the tip. κ, called the inverse decay length, is related to the average work function, Φ, between the sample and tip by

$$\kappa = (2m\Phi)^{1/2}/\hbar \tag{2}$$

when the bias V is small compared to Φ/e.

The tunnelling current is in the range of picoampere to nanoampere.

The tunnelling conductance, I/V, at a point gives the electron density at that point on the sample. Thus, a measurement of the conductance, keeping d constant at different points on the surface, gives the electron density distribution on the surface. **The STM does not locate the position of the nucleus of the atom. It only measures the electron density distribution on the surface.**

If we vary d at a given point and measure I at a given voltage, V, then $\partial \ell n I/\partial d$ measures the work function, Φ, at that point on the sample. For a clean surface, Φ will be uniform all over the surface.

Taking a value of $\Phi = 4$ *eV*, which is the order of the work function for metals, the tunnelling current decreases by a factor of 10 for every increase of *d* by 0.1 nm. The exponential variation of the tunnelling current allows a change in distance of a fraction of an angstrom between the tip and the sample to be detected.

There are two modes in which the STM can operate:

1. **Constant height mode:** In this mode, the tip travels in a horizontal plane at a given height above the surface.
2. **Constant current mode:** In this mode, the tunnelling current is kept constant by a feedback mechanism involving a piezo actuator. This mechanism adjusts the distance *d* between the sample and the tip by moving the position of the sample in the vertical direction by a change in voltage applied to the piezo actuator. The variation of the voltage on the piezo actuator, as the tip is moved across the sample, yields the desired information.

The constant height method takes less time than the constant current method to scan a given area of the surface.

The tip in a STM need not be as sharp as in an atomic force microscope.

The limitation of STM is that it can only be used with metallic or semiconducting surfaces and not with insulating surfaces. It also requires ultrahigh vacuum so that the surface is not contaminated.

In Figure 4 are shown 4-nm clusters of gold atoms on a GaAs surface. This gives an idea of the resolution achievable with an STM.

When a crystal is cleaved, the atoms at the surface have some dangling bonds. The equilibrium arrangement of atoms on the surface plane departs from the arrangements of atoms of a parallel plane in the interior of the sample. This is called reconstruction. The reconstructed Si(111) face has a two-dimensional unit cell (with 49 atoms and 19 dangling bonds), which is described on the dimer–adatom–stacking fault (DAS) model. All the atoms are not in the same plane. This structure is called Si(111) 7×7.

In Figure 5(c) is shown the schematic representation of the DAS 7×7 structure of reconstructed Si(111) face in plan and side views (Ref. 3). This model is constructed from dimer–adatom–stacking fault and hence the acronym DAS. The yellow-coloured atoms represent Si-adsorbed atoms,

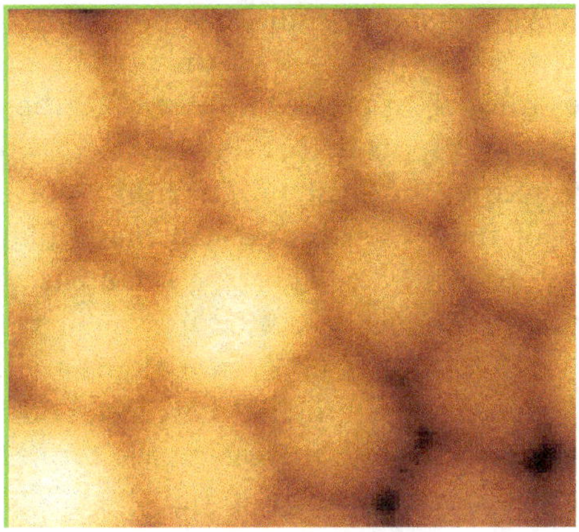

Figure 4. 4-nm clusters of gold on GaAs (reprinted with permission from Ref. 2).

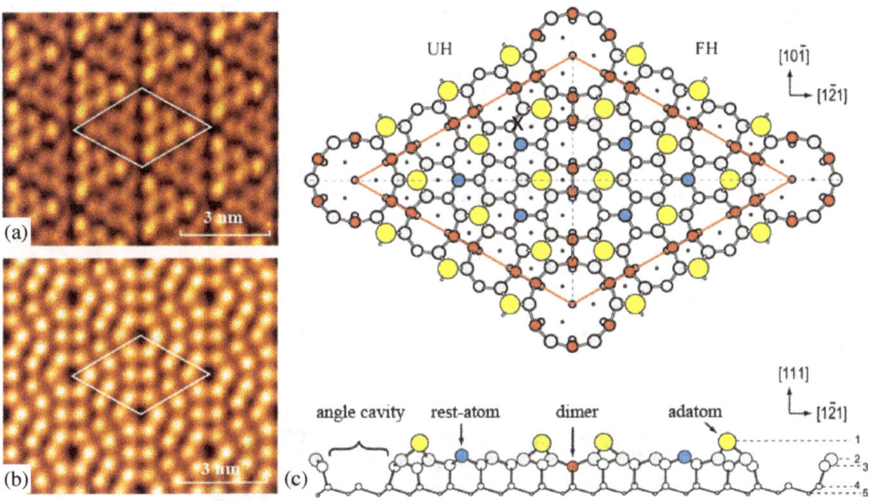

Figure 5. DAS model for the reconstruction of the Si(111) face 7 × 7 (reprinted with permission from Ref. 3).

the red circles dimerized Si atoms, and the blue circles the second layer Si rest atoms. The closed parallelogram (in red) in (c) represents the two-dimensional unit cell of the reconstructed surface. This unit cell has two triangular half cells, separated by a line of dimers. The faulted right half of the figure (c) is marked FH. The unfaulted left half of the unit cell is marked UH. The STM image of the filled states is shown in (a) and that of the unfilled states is shown in (b). The unit cell is marked in (a). The faulted half of the unit cell appears brighter than the unfaulted half in STM image (a). The bright spots in (b) represent the Si ad atoms. Thus, the STM image clearly verifies the structure derived theoretically.

With the help of the STM tip, one can manipulate the position of individual atoms on a substrate. The IBM scientists were able to form atomic patterns thus. Figure 6 shows 48 iron atoms arranged in a perfect circle on a copper (111) surface at 4.2 K. Iron atoms are first physiosorbed on the Cu surface. Then the tip of the STM is placed directly over a physiosorbed atom. The tip is lowered, increasing the tunnelling current. This increases the attractive force of the tip to the atom. The atom is dragged by the tip and is moved across the surface to a desired position. Then, the tip is withdrawn by lowering the tunnelling current. The image of the corral shows the electron density distribution.

Figure 7 shows a "carbon monoxide molecule man" made by aligning 28 CO molecules. The height of the figure is 50 Å. These figures are taken from the IBM gallery of STM images.

Figure 8 shows a photograph of an STM system operating at 4.2 K.

3. Atomic Force Microscope

The atomic force microscope is based on the van der Waals interaction between an atom on the tip and an atom on the surface. **It has the advantage that it can be used with all materials unlike the STM, the use of**

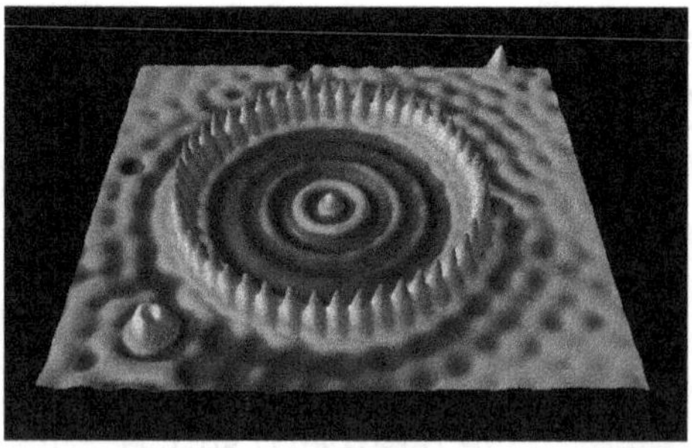

Figure 6. **48 iron atoms trapped in a perfect circle.**

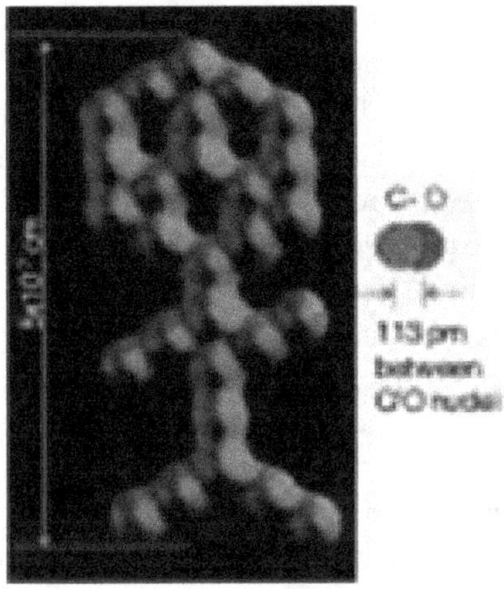

Figure 7. Carbon monoxide molecule man made of 28 CO molecules.

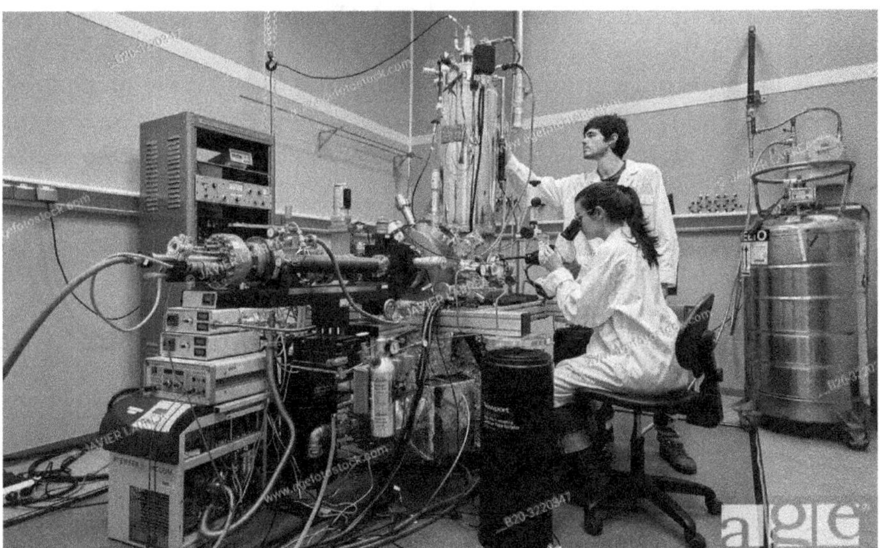

Figure 8. Photograph of an ultralow, UHV DTM system (from Alamy Stock Photo).

which is limited to metals and semiconductors. Also, the atomic force microscope does not need vacuum to operate.

To understand the working of the atomic force microscope, we plot in Figure 9 how the van der Waals interaction varies with the distance between the tip and the surface under study.

At a large distance between the tip and the surface, the force is attractive and small. As the tip comes closer to the surface, this attractive interaction increases till it reaches a maximum (negative) value. Now, the electronic clouds of the atoms in the tip start to overlap with the electronic clouds of atoms on the surface of the sample. Pauli exclusion principle leads to a repulsive interaction. As the tip comes closer to the surface, this repulsive force cancels part of the attractive force. The total force becomes zero at a certain distance between the tip and the surface. If the tip comes still closer to the surface, the repulsive force is dominant and increases rapidly as the distance becomes less.

There are two principal modes in which the atomic force microscope is used. The first is the contact mode, in which the tip comes very close to the surface so that the force is repulsive. The force changes by a large amount when the distance of the tip from the surface is changed. The other is the non-contact mode. Here, the distance between the tip and the surface is large, so that the force is attractive and changes slowly with the distance. A third mode, called the intermittent mode, is used sometimes. This will be discussed later.

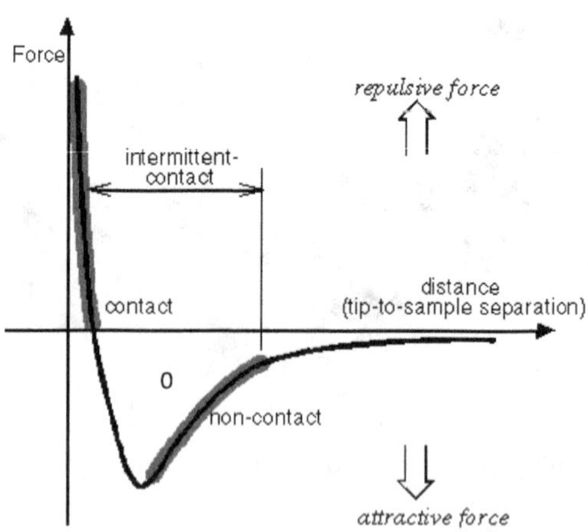

Figure 9. Variation of the van der Waals interaction with the distance between the tip and the surface (reprinted with permission from Ref. 1).

The van der Waals force is not the only force acting on the tip. When the tip is close to the surface, any residual trace of water on the surface will cause an attractive force on the tip due to capillary action.

3.1. *Contact AFM*

In the contact method, there is a sharp tip, a few microns in length, with the tip ground to a point of radius of the order of 100 Å. This tip is attached to a cantilever a few hundred micrometers long and fixed at one end. The cantilever is thin so that its stiffness is low. The tip is brought to make soft contact with the surface (i.e.) **within a distance of a few angstroms from the surface.**

The total repulsive force exerted on the tip by the van der Waals and capillary forces is of the order of 10^{-6}–10^{-8} Newtons. It causes the cantilever to bend upwards till the elastic force in the cantilever balances the van der Waals and capillary forces. For contact AFM, the deflection of the cantilever is measured by an optical device such as the one shown in Figure 10.

Light, from a laser diode, falls on the cantilever. The point of incidence of the light on the cantilever is close to the tip. The light gets

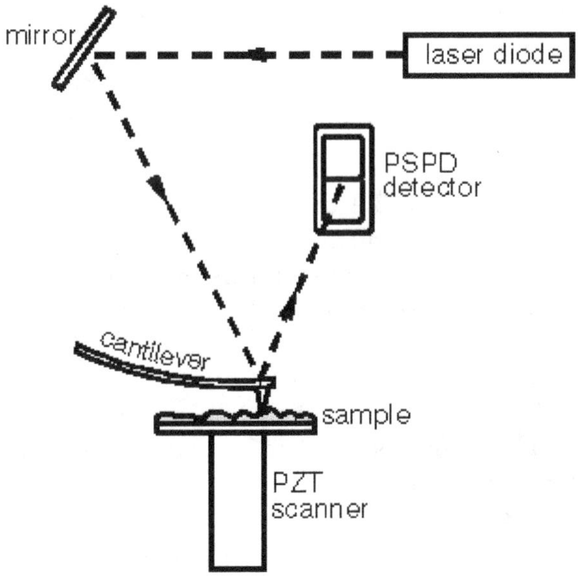

Figure 10. Optical method to detect the deflection of the cantilever (reprinted with permission from Ref. 1).

reflected to a position-sensitive detector. As the deflection of the cantilever changes, the position of the reflected beam falling on the PSD changes.

The surface topography is recorded by scanning the surface through a piezo scanner and recording the position-sensitive detector reading.

The advantages of contact AFM are as follows: (1) it produces a very high resolution of the order of a fraction of a nm; (2) the surface topography is accurately reflected even in the presence of an absorbed water layer.

The disadvantage is as follows: the contact AFM produces damage to the surface and alters the surface topography after a few scans.

3.2. *Non-contact AFM*

The tip is kept at a distance, of the order of 100 Å, from the surface. The attractive force is of the order of 10^{-12} Newtons. In non-contact AFM, the cantilever is made to vibrate near its resonance frequency (100–400 kHz), and the change in the resonance frequency due to the force on the tip is measured, keeping the amplitude constant at about a nanometer.

The advantages of non-contact AFM are as follows: (1) it can be used on soft materials; (2) it produces no surface damage.

The disadvantages are as follows: (1) its resolution is low, of the order of 100 nm; (2) it does not reproduce the true topography of the surface in the presence of a surface layer of water about a nanometer thick.

3.3. *Intermittent contact AFM*

In this method of using AFM, the tip is kept at about the same distance as in the non-contact method. But the amplitude of vibration of the tip is made large. This makes the tip come close to the surface, as in contact method, during part of its vibration. It gets over the problem of the surface layer of water. This mode is also called the tapping mode.

Non-contact AFM is the method used most often. It has been used in the study of biological materials also.

Figure 11 gives a photograph of an AFM setup. A comparison of this photograph with the photograph of STM in Figure 8 shows how less cumbersome the AFM setup is.

Figure 12, taken in the non-contact mode, shows dimer formation in the reconstruction of Si(100) surface. We have already discussed the 7×7 reconstruction of Si(111) surface. In the case of (100) surface of silicon, reconstruction takes place by dimer formation.

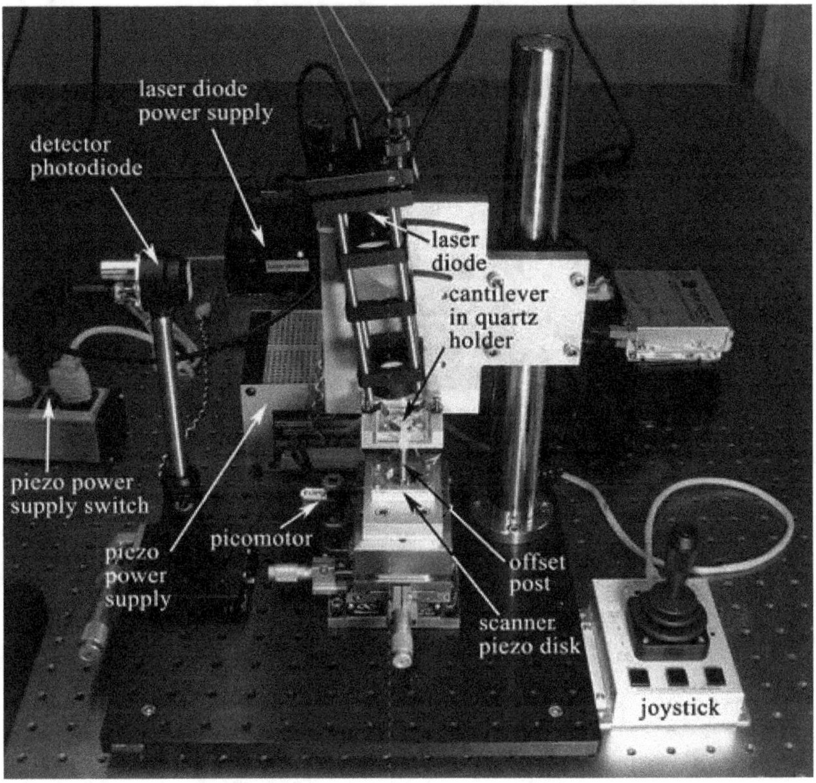

Figure 11. Photograph of an AFM setup.

Figure 13 shows an image of carbon nanotubes. The inset shows a single nanotube, and the graph shows the distribution in size of the nanotubes.

Figure 14 shows the surface roughness of a clean glass surface to a resolution of 0.8 nm.

We can also measure the lateral force, parallel to the surface, on the tip. This force may arise due to atoms on the surface of the sample or due to changes in slope of the surface topography. There are other modifications of the atomic force microscope.

Atomic force microscopy finds a wide variety of applications in the following: (a) imaging semiconductor surfaces on an atomic scale, (b) writing nanoscale electronic structures, (c) visualizing biological structures in various scales, and (d) imaging polymers.

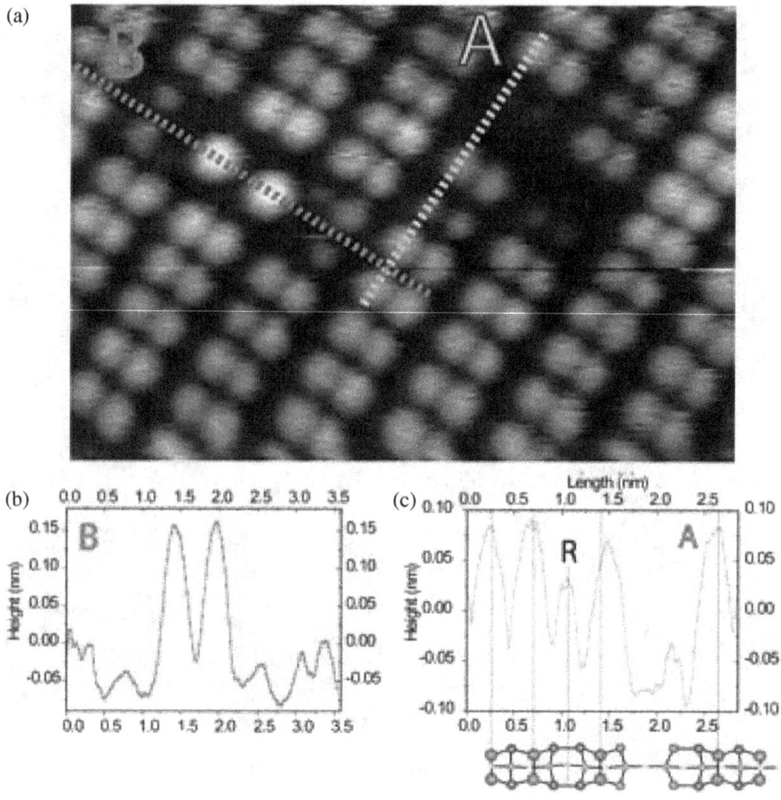

Figure 12. **Si(100) surface taken at 77 K with non-contact AFM (reprinted with permission from Ref. 4).**

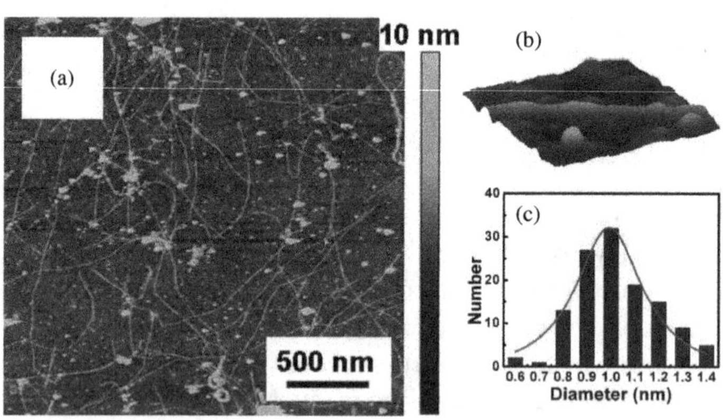

Figure 13. **AFM image of carbon nanotubes of different sizes. The size distribution of the nanotubes is shown in inset (c) (reprinted with permission from Ref. 2).**

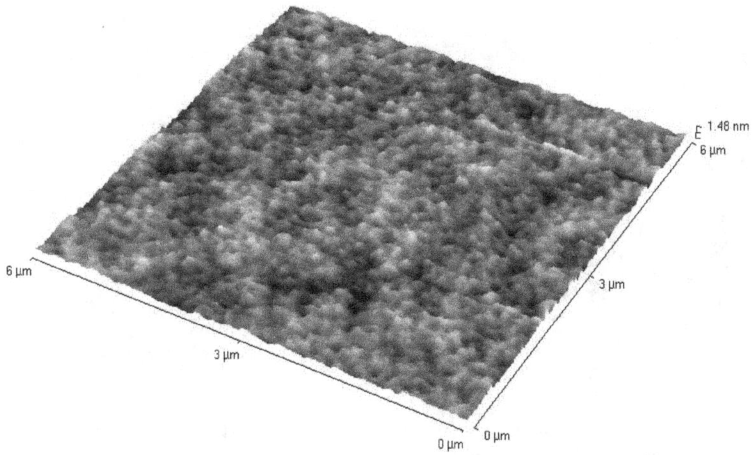

Figure 14. Surface roughness of a clean glass surface (reprinted with permission from Ref. 4).

4. Magnetic Force Microscopy

In this microscopy technique, the tip is coated with a thin film of a magnetic material. The microscope operates in the non-contact mode, and the change in frequency of the cantilever occurs due to the magnetic force between the tip and the surface of a magnetized material. Since the magnetic force decreases more slowly with distance than the van der Waals force, the distance between the tip and the surface can be increased beyond what is used in non-contact AFM (Figure 15). In this mode, as the distance of the tip from the surface is increased, one goes from a study of surface topography to the study of magnetic domains in a ferromagnetic material.

5. Media in Which Scanning Microscopes Operate

Ultrahigh vacuum: Since STMs are usually used to probe ultraclean surfaces, they operate in ultrahigh vacuum. In AFM, we have to position the tip and measure its deflection. This is difficult under ultrahigh vacuum conditions. Though UHV AFMs have been built, non-vacuum AFMs are more common.

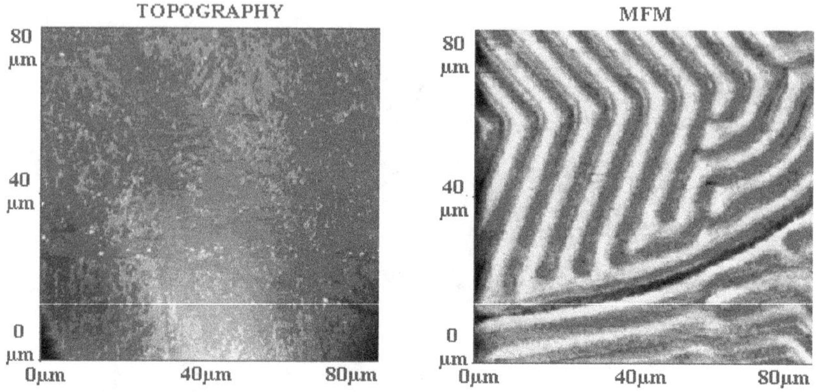

Figure 15. Simultaneous images of surface topography (left) by AFM and magnetic domains (right) by MFM of a Garnet film.

Liquid: One can study surfaces immersed in a liquid using AFM techniques. These are useful in studies on biological and geological materials.

6. Advantages of Scanning Probe Microscopes

1. Compared to optical microscopes (magnifying power 1000) and scanning electron microscopes (magnifying power 10,000), the magnifying power of SPMs is high (100,000). The SPMs can be used to study surfaces on a sub-nanometer scale.
2. Unlike electron microscopes which require UHV environment, AFMs operate in ambient environments.
3. Optical and electron microscopes only measure dimensions in the plane of the sample and not in a perpendicular direction. AFMs measure the z component also.
4. With phase imaging in AFM, one may identify contaminants, regions of low and high surface adhesion, and map electric and magnetic properties. With lateral force AFM, one can measure elastic properties of the material on the surface.

All AFMs need vibration isolation mounts.

7. Conclusion

Advances in surface probe techniques have happened rapidly since the discovery of STM. Scanning probes have found essential applications not only in surface physics but also in biology, pharmaceuticals, polymer science and microelectronics. It will be difficult to refer to all the advances in this chapter, which has the limited scope of providing an introduction to the subject. For in-depth study, books written by experts, such as Refs. 5 and 6, may be consulted.

References

1. Howland, R. and Benatar, L. (1996). *A Practical Guide to Surface Probe Microscopy*, https://gato,docs.its.txst.edu.>AFM>STM1.
2. Scanning Tunnelling Microscope Images- Purdue Physics, www.physics.purdue.edu>nanophys>stm.
3. Takayanagi, K., Tanishiro, Y., Takahashi, S., and Takahashi, M. (1985). *Surf. Sci.* **164**, 367.
4. Lecture 10: Basics of Atomic Force Microscope (AFM), https://my.eng.utah.edu > ~ | zang > images> Lecture 10 AFM.
5. Bhushan, B. and Fuchs, H. (eds.) (2006). Scanning Probe Microscopy Techniques: Chapter 11. In *Applied Scanning Probe Methods II*, Springer ebook.
6. Magonov, S. and Whangbo, M.-H. (1996). *Surface Analysis with STM and AFM: Experimental and Theoretical Aspects of Image Analysis* (1st Edition). Wiley-VCH.

Chapter 8

POSITRON ANNIHILATION SPECTROSCOPY AS A TOOL FOR THE STUDY OF DEFECTS IN SOLIDS

1. Introduction

In Chapter 6 on electron microscopy, we talked about point defects such as vacancies and interstitial atoms. Such defects can also be produced by irradiating a material with heavy ions and electrons of high energy, X-ray and gamma-ray photons and neutrons. Since the penetration of these different radiations is different, the damage produced by them extends from a few hundred nanometers to several millimeters.

Heavy ions and electrons interact with the charged particles in the material by Coulomb interaction. They knock out the electrons, and these in turn produce vacancies and interstitials. X-rays and gamma rays produce defects through the electrons ejected by photoelectric emission or through Compton scattering by the electrons in the material. Neutrons and heavy ions at the end of their range can directly knock out atoms through elastic and inelastic collisions. Fast neutrons can produce damage through nuclear reactions, which will result in the emission of beta particles or gamma rays that produce the damage. In addition, the transmuted atoms will form impurities. Slow neutrons can also induce fission in materials such as ^{235}U. The defects produced in the material in such reactions can result in helium or hydrogen bubble formation. Positron annihilation is the technique to be used to probe such defects.

There are two good articles on positron annihilation on the web. These are by Reinhard Krause-Rehberg (Ref. 1) and by Maciej Oskar Liedke (Ref. 2). Much of the material presented in this chapter is taken from these two references and from the work of Sundar (Ref. 3).

Positron is the antiparticle of an electron. It carries a positive charge equal in magnitude to the charge on an electron and has the same rest mass as the electron. When a positron and an electron come together, they annihilate each other and create two photons of equal frequency v. If the positron and electron are at rest, each of the two photons has an energy $hv_0 = 0.511$ MeV, which is the rest mass energy m_0c^2 of the positron (or the electron). The photons travel outward in opposite directions. Alternatively, if the positron is at rest and the electron has a momentum p_e, the two photon paths make an angle $(180 - \Delta\theta)$, as shown in Figure 1(b).

The laws of energy and momentum conservation, as given by Equations (1a) and (1b), should be satisfied:

$$2hv = 2hv_0 + p_e^2/2m_0 \tag{1a}$$

$$2(hv/c)\sin(\Delta\theta/2) = p_e \tag{1b}$$

Thus, the photon frequency v will differ from v_0 by a small shift, Δv. This shift Δv, called the Doppler shift of the photon frequency, and the

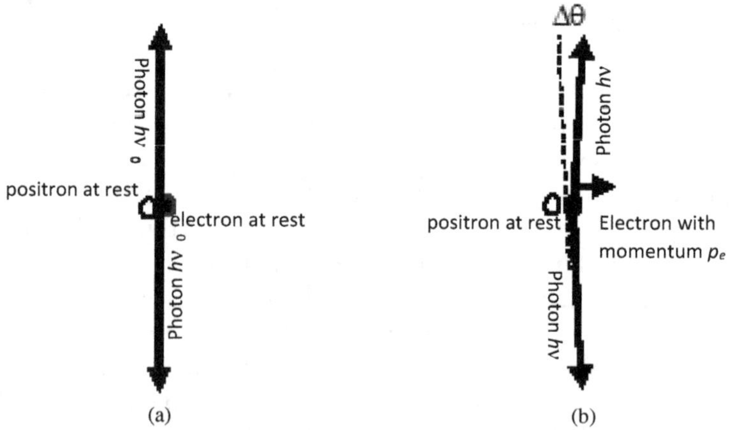

Figure 1. Annihilation of a positron by an electron, giving rise to two photons (a) when the electron and positron are at rest and (b) when the positron is at rest and electron moves with a momentum p_e.

angle $\Delta\theta$ are related by Equations (1a) and (1b). Since $\Delta\theta$ is small, we may use the approximation $\sin(\Delta\theta/2) = \Delta\theta/2$ in (1b). Since $\Delta\nu$ is small, we may use $h\nu = h\nu_0 = m_0c^2$ in Equation (1b) to get

$$p_e = m_0c\Delta\theta \tag{1c}$$

The theoretical treatment of the annihilation process reveals that the annihilation rate, λ, of the positron in the medium is proportional to the effective electron density n_e sampled by the positron, i.e., $\lambda = \pi r_c^2 c n_e$, where r_c stands for the classical electron radius and c is the speed of light in vacuum. The lifetime of the positron, τ, is the inverse of the annihilation rate, $\tau = \lambda^{-1}$. To get a rough idea of the magnitudes of the quantities λ, the Doppler shift ΔE $(= h\Delta\nu)$, and $\Delta\theta$, which characterize positron annihilation in condensed matter, one can insert realistic estimates of n_e and electron momentum p_e into the above expressions. Conduction electron densities in metals are typically of the order of 10^{29} m^{-3}. Core electron momenta in atoms may be taken roughly as $h/2r_a$, where the atomic size is characterized by Bohr's radius r_a and h is the Planck's constant. Then, the following estimates of λ, ΔE, and $\Delta\theta$ may be obtained: (i) positron lifetimes ~500 ps are expected in metals; (ii) Doppler shifts of annihilation photon energies amount to $\Delta E \sim 1$ keV; (iii) angular correlation curves should exhibit widths of a few millirads.

In using positron annihilation to investigate condensed media, we measure (1) the lifetime τ of the positron (the time interval between the creation of the positron and its annihilation), (2) the Doppler frequency shift $\Delta\nu$ of the gamma-ray photons, and (3) the angular correlation, i.e., what fraction of pairs of the gamma-ray photons come out with their paths making an angle $(180 - \Delta\theta)$ between them. This, in essence, is the basis of positron annihilation spectroscopy (PAS). The methodology of PAS can be explained with the help of the schematic diagram shown in Figure 2.

Positrons from radioactive sources, such as ^{22}NaCl, are injected into a solid, wherein they thermalize and annihilate with the electrons in the medium. The annihilation characteristics that are measured are the lifetime of the positron in the medium, the angular correlation of the two annihilation photons, and the Doppler-broadened line shape of the annihilation radiation.

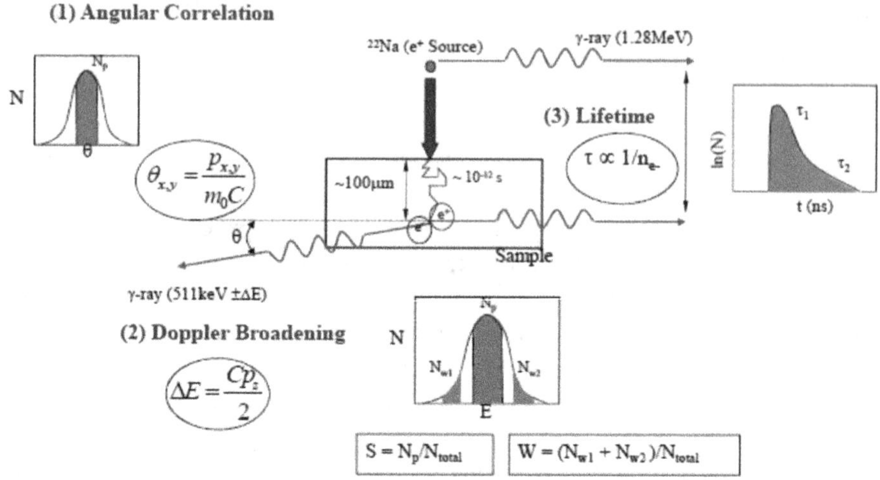

Figure 2. Schematic diagram of positron annihilation spectroscopy.

2. Positron Annihilation in a Solid

When energetic positrons from a radioactive source are injected into a solid, they rapidly lose their kinetic energy through Coulomb interaction with the electrons in the atoms of the material. When the energy becomes very low, they share their kinetic energy with the phonons in the material and reach thermal equilibrium with the phonons very quickly (in a time of about 10 ps). Phonons are quantized lattice vibration modes. In a perfect crystal, the positron moving in a periodic potential exists in a Bloch state, as shown schematically in Figure 3(a). Due to the strong Coulomb repulsion from the positive ion cores, the positron density distribution is at a maximum in the interstitial regions, and the positron mainly annihilates with the valence electrons with small contributions from the core electrons.

The positron behavior in crystalline materials is drastically affected by the presence of vacancy-type defects. At open-volume defects (monovacancies, larger vacancy clusters, dislocations, etc.), the potential sensed by the positron is lowered due to the reduction in the Coulomb repulsion. The transition from the delocalized state to the localized one is called positron trapping. As the local electron density at the defect site is lowered compared to that of the unperturbed regions, the lifetime τ_t of the trapped positrons is correspondingly longer than $\tau_b = \lambda_b^{-1}$, where λ_b is the annihilation rate of the trapped positron. Positron trapping is characterized by the trapping rate κ, which is proportional to defect concentration c_t in the sample: $\kappa = \mu c_t$. The trapping coefficient μ and the annihilation rate $\lambda_t = \tau_t^{-1}$ are specific for a given kind of defect. The positron localization in the presence of various defects, such as vacancies, vacancy clusters, and solute atom clusters, is shown schematically in Figure 3.

The association of experimentally measured lifetime with a specific defect site is made possible by the developments in the theoretical calculations of positron density distribution and its overlap with the electron density at the defect center, leading to the evaluation of annihilation characteristics. Briefly, the calculations proceed by numerically solving the Schrodinger equation for the positron with the positron potential given as a superposition of the Hartree electrostatic potential and the electron–positron correlation energy. The annihilation rate is then evaluated from

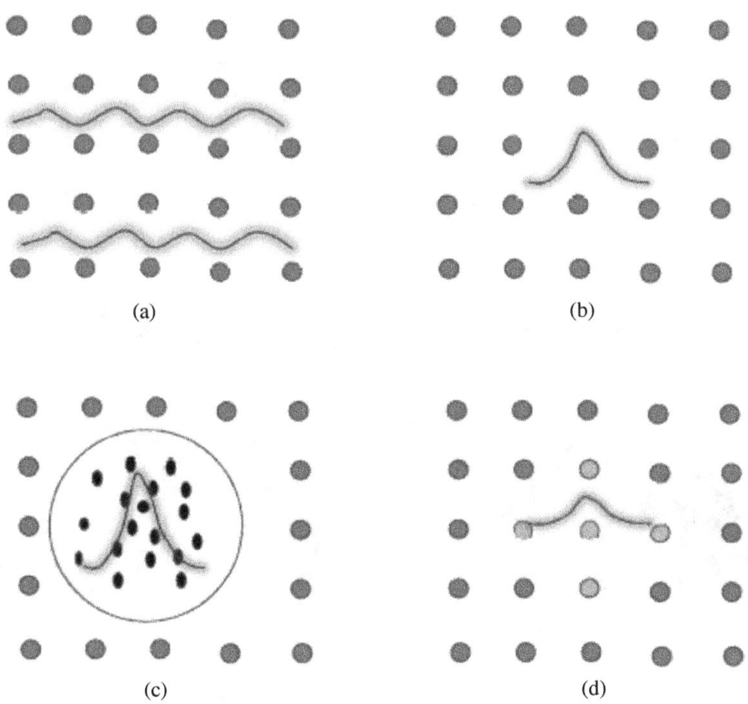

Figure 3. Positron distribution in (a) perfect crystalline solid, (b) vacancy, (c) a gas filled vacancy cluster, and (d) a solute cluster in a metallic matrix.

the overlap of the positron density with the enhanced electron density at the site of the positron. Such calculations have now been carried out for a variety of defects, such as monovacancies, vacancy clusters, helium-decorated vacancy clusters, and solute clusters. The annihilation characteristics of the positron trapped at defects is sensitively dependent on the size and even geometry of the defect clusters. For example, with the increase in the size of the vacancy cluster, the positron gets more localized at the defects and, consequently, overlaps less with the electron cloud around the defects. This results in an increase in the positron lifetime with the size of the defect cluster.

As an example, we provide the results of the calculation of the positron density distribution and lifetime in TiC (Figure 4). Such calculations, carried out by solving the Schrodinger equation, were necessitated by the attempt to understand the annihilation characteristics of TiC precipitates in steel, which were incorporated to improve its radiation resistance (Ref. 4).

The panel on the left shows the positron density distribution in a perfect crystal of TiC and indicates that the maximum of the positron density is in the interstitial region. In the presence of a C vacancy, the positron density can be seen to be localized at the vacancy. Using these positron density calculations, the positron lifetimes in perfect and defected TiC with carbon vacancies are calculated to be 108 and 133 ps, respectively. At a Ti vacancy, the lifetime is calculated to be 164 ps.

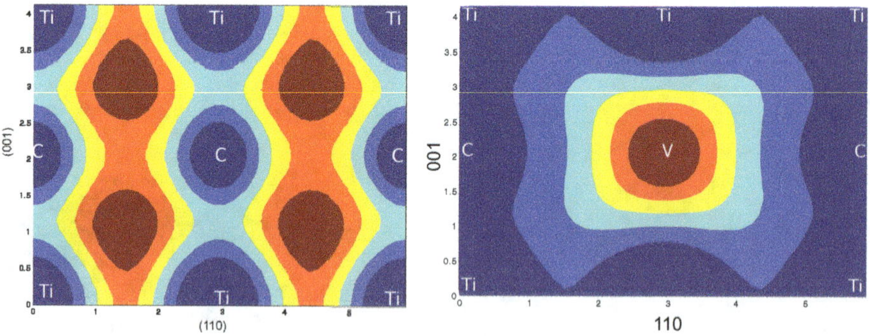

(a) (b)

Figure 4. **Positron density distribution in the (110) plane of TiC: (a) shows the maxima (indicated by dark red) of the positron density in the interstitial region that gets localized at a C vacancy, as shown in (b).**

The results of such calculations have been extremely useful in understanding the characteristics of the precipitates.

The sensitivity of positron lifetime to the size of small vacancy clusters is illustrated through the results of calculations in Ni, as shown in Figure 5. Also shown is the reduction in lifetime with He decoration of the vacancy cluster. The extreme sensitivity of the positron lifetime to the size of small vacancy clusters and the decoration with gas atoms is of great value in the studies on radiation damage in reactor materials, where He, formed by neutron-induced reactions, agglomerates to form He bubbles. The annihilation characteristics are also seen to be sensitive to the presence of small solute atom clusters in quenched alloys, and this has been used in the study of the early stages of phase separation in quenched alloys.

In insulating materials with low electron density, the positron can form a bound pair with an electron, in contrast to annihilation from a free state in a metallic sample. This bound pair is similar to an electron orbiting the proton in the hydrogen atom. This is called a positronium (Ps). Since the positron and electron have each a spin angular momentum of ½$\hbar$, the total spin angular momentum of the positronium can be 0 or $\hbar$. In the former case, the spin of the positron is opposite to the spin of the electron. Such a bound state is called para-positronium (p-Ps). This can

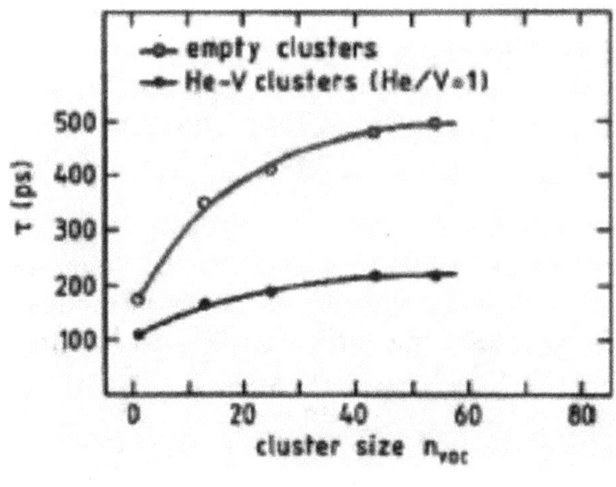

Figure 5. Positron lifetime versus cluster size for empty vacancy clusters and He-decorated clusters. The results of theoretical calculations, as shown above, are valuable for interpreting the experimental results (reprinted with permission from Ref. 5).

rapidly self-annihilate without interaction with host material. If the spins of the electron and positron in a positronium are parallel, it is called an ortho-positronium (o-Ps). In vacuum, o-Ps can self-annihilate but with a long lifetime of 142 ns. In a material, o-Ps can pick off an electron from its surroundings and can spin-flip, resulting in a faster decay, producing two gamma-ray photons. The annihilation of o-Ps via the pick-off process is characterized by lifetimes of a few nanoseconds. In a polymeric material, o-Ps can get trapped inside a free volume, and the corresponding pick-off lifetime, τ_{oPs}, can serve as a measure of the free-volume hole size. A simple model of Ps in a spherical potential well of radius R leads to a correlation between τ_{oPs} and R. This has formed the basis of numerous fruitful studies about polymers, such as the changes in free volume with temperature across the glass transition, with pressure, blending, etc. (Ref. 6).

3. Positron Annihilation Techniques

After providing a survey of the sensitivity of the positron annihilation process to various types of defects in solids, we provide details on the experimental methodology of positron annihilation techniques, as shown in Figure 2.

3.1. *Positron sources*

Positrons are emitted during the radioactive decay of certain nuclei. These nuclei serve as the source of positrons. Examples of some radioactive nuclei which are positron emitters are ^{22}Na, ^{64}Cu, ^{58}Co, and ^{68}Ge. The superscripts are the mass numbers. The most commonly used source is ^{22}NaCl, which has a half-life of 2.6 years, which is fairly long. The decay scheme of ^{22}Na is shown in Figure 6(a).

^{22}Na nucleus decays to an excited state of the ^{22}Ne nucleus by positron emission. The excited neon nucleus has a lifetime of 3.2 ps and quickly decays to the ground state by emitting a gamma-ray photon of energy 1.274 MeV. The sodium nucleus can also decay to the excited state of the neon nucleus by capturing an electron in its K shell. The fraction of sodium atoms decaying due to positron emission is 90.4%, while the fraction decaying due to electron capture is 10 times less.

The decay process by emission of a positron is a three-body process. Along with the positron emission, an electron neutrino is also emitted, and the nucleus of neon recoils. The neutrino carries away part of the

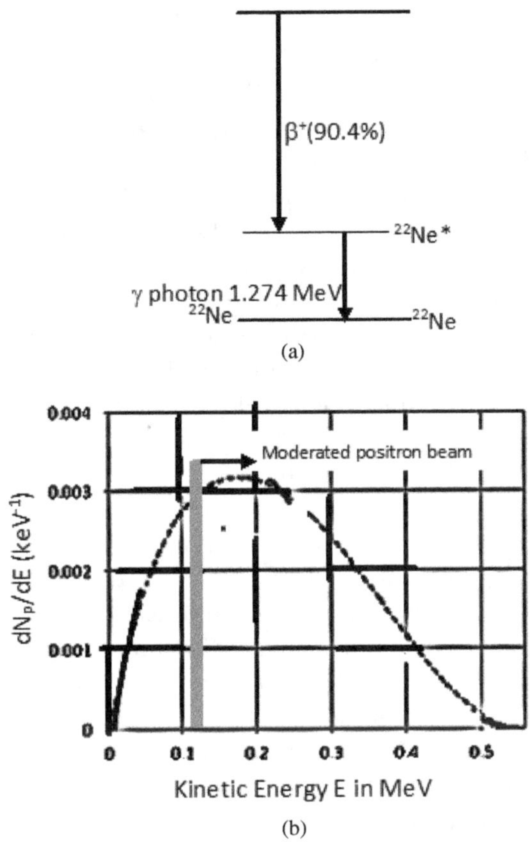

Figure 6. (a) Decay scheme of ^{22}Na: the asterisk on ^{22}Ne denotes excited state. (b) Kinetic energy distribution of the positrons.

energy and momentum during the transition. So, the positrons arising from the decay process have a distribution in energy as shown in Figure 6(b). The maximum number of positrons are emitted with a kinetic energy of around 180 keV. The vertical band corresponds to the moderated positron energy that is used in low energy positron beam experiments, discussed in section 6.

3.2. *Positron lifetime measurements*

The ^{22}Na radioactive source, deposited on a thin foil of Ni, is sandwiched between two samples, say, Al sheets, in which the positron lifetime is to be measured. To measure the positron lifetime in a medium, we need to measure the time delay between the 1.28 MeV gamma ray that signals the birth of the positron and the 0.511 MeV photon that signals the annihilation of the positron. The lifetime of the positron in a metallic system is typically ~150 ps. To measure such short lifetimes, an ultrafast timing spectrometer, derived from the methods developed in nuclear spectroscopy for the measurement of the lifetimes of excited states, is used.

Figure 7 shows a schematic diagram of a fast–fast coincidence spectrometer used for positron lifetime measurements. The main components of the timing circuit are: (1) a photomultiplier tube with plastic scintillator for converting the signal from the incident gamma rays to electrical pulses; (2) a constant fraction discriminator that provides the precise timing signal from the incoming electrical pulse; (3) a fast coincidence unit that selects, among the several gamma rays incident on the scintillators, the pair of START 1.28 gamma ray and STOP 0.5 MeV photon, corresponding to a particular annihilation event; (4) a time-to-pulse-height converter (TPHC) that converts the time difference between the START and STOP input signals into pulse height. The pulse height distribution, obtained as an output of the TPHC, is recorded in a multichannel analyzer (MCA).

The MCA records the pulse height distribution, i.e., a plot of coincidence counts as a function of channel number, that reflects the pulse height. As we have indicated, the TPHC converts the time difference into pulse height, and hence, the channel number reflects the time difference between the start (birth) and stop (annihilation) photons. To calibrate the channel number into time in ns, one introduces standard delay cables (RG 58 coaxial cable) in the stop channel of the spectrometer and measures the shift in the centroid of the pulse height spectrum recorded in the MCA.

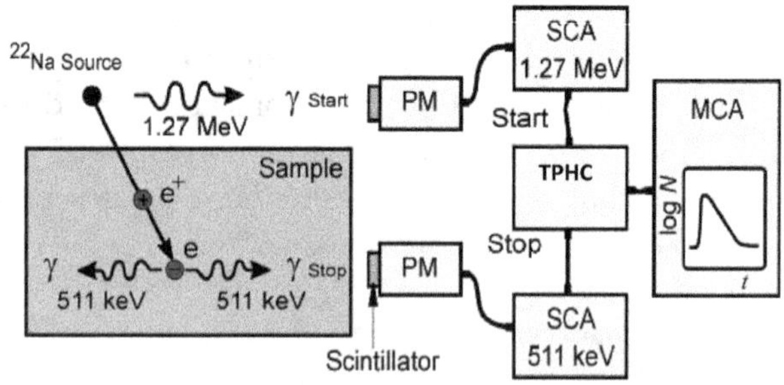

Figure 7. Schematic illustration of the measurement of positron lifetime with a ^{22}Na source: PM: photomultiplier, SCA: single-channel analyzer, MCA: multichannel analyzer, TPHC: time-to-pulse-height converter (reprinted with permission from Ref. 1).

Figure 8 shows the lifetime spectrum in as-grown and plastically deformed Si, taken from Ref. 1. In the case of as-grown Si, it is seen that the spectrum has a sharp rise on the left-hand side and decays exponentially on the right-hand side. The slope on the right-hand side is due to the delay in the arrival of the stop signal, which is a measure of the lifetime of the positron. In comparison to the as-grown Si, the lifetime spectrum in the case of plastically deformed silicon has a larger slope, and two distinct slopes can be seen. In general, the material may contain different types of defects or trapping sites j with different annihilation rates $1/\tau_j$. The number of counts as a function of time will follow the equation

$$N(t) = \Sigma_j I_j/\tau_j \exp(-t/\tau_j) \tag{2}$$

The measured spectrum is a convolution of the exponential decay due to the lifetime of the positron with the intrinsic time resolution of the spectrometer. The time resolution of the spectrometer is measured using a radioactive source, ^{60}Co, that emits simultaneously a cascade of gamma rays of energy 1.17 MeV and 1.33 MeV. For the simultaneous gamma-ray inputs to the two scintillators, the spectrometer records a Gaussian time profile, whose FWHM is a measure of the resolution of the spectrometer. A typical FWHM value for positron lifetime spectrometers is ~180 ps.

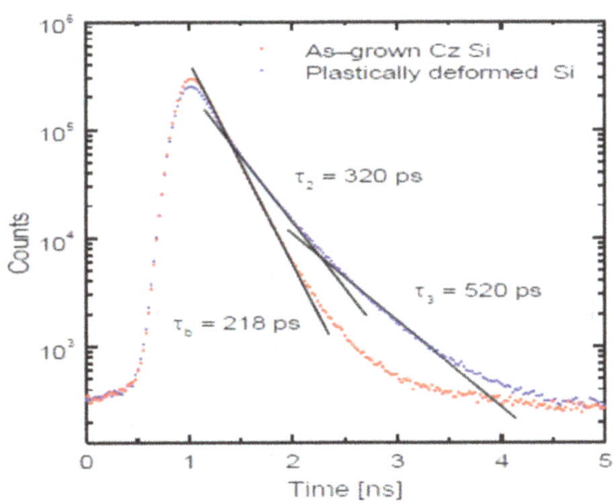

Figure 8. Positron lifetime spectra: As-grown Si (red) and plastically deformed Si (blue) (reprinted with permission from Ref. 1).

The lifetime spectrum, as seen in Figure 8, is analyzed using a computer program that fits the measured spectrum to the convolution of Equation (2) with a Gaussian. From such a fit, the positron lifetime components τ_j and the relative fractions I_j are extracted.

In as-grown silicon, a good fit is obtained by using a single exponential that yields the lifetime τ_1 of 218 ps. In plastically deformed Si, the fit requires three exponentials. The 218 ps lifetime corresponds to annihilation from the bulk, and the second with a lifetime of 320 ps corresponds to annihilation at the defect traps. The third lifetime component τ_3 of 520 ps arises due to annihilation in the source foil.

4. Select Examples of Defect Studies

4.1. *Vacancy formation energy in metals*

One of the significant early applications of PAS was in the determination of the vacancy formation energy in metals. It is well known that at any finite temperature, there exists an equilibrium of vacancies, the concentration of which is given by $C_v = n_0 e^{-(E_V/kT)}$, where E_V is the vacancy formation energy, k is the Boltzmann constant, and T is the absolute temperature. We have seen that in the presence of vacancy-type defects, positrons are trapped in them and annihilate with a characteristic lifetime that is higher than the bulk value. With an increasing temperature, as the concentration of vacancies increases, an increasing fraction of positrons get trapped at the vacancies. This results in the positron lifetime changing from a value characteristic of the bulk, τ_b, to that of vacancies, τ_v. The variation of positron lifetime with temperature in Ni is shown in Figure 9.

The sigmoidal variation of lifetime is analyzed in terms of the trapping model to extract the vacancy formation energy. The sigmoidal variation of positron lifetime with temperature is well described by the two-state trapping model. The central assumption of the simple trapping model is that positrons annihilate in a solid from a free or a trapped state. Escape from traps is neglected, and the rate of trapping κ is assumed to be proportional to the concentration of traps C_T. With characteristic lifetime values τ_b $(=\lambda_b^{-1})$ for the annihilation from the bulk and τ_v for the trapped state at a vacancy, the mean lifetime value for a two-state trapping model is given by

$$\tau = (\tau_b F_b + \tau_v F_v)/(F_b + F_v)$$

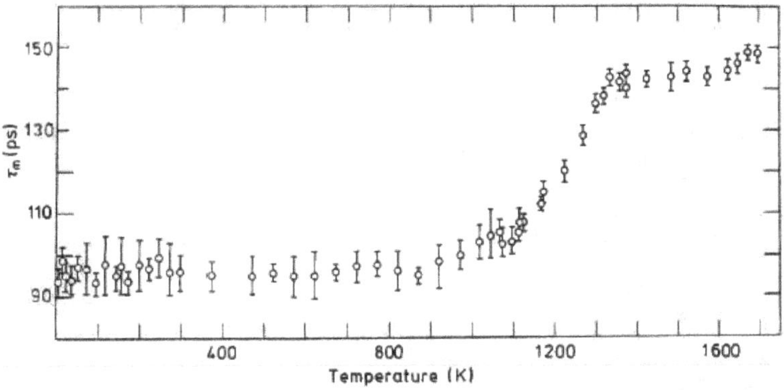

Figure 9. Positron lifetime vs. temperature in Ni (reprinted with permission from Ref. 7).

where F_b and F_v are the fractions of annihilations in the free and trapped states given by $F_b = \lambda_b/(\lambda_b + \kappa)$ and $F_v = \kappa/(\lambda_b + \kappa)$. It can be readily seen that the trapping rate $\kappa = (\tau - \tau_b)/[\tau_b(\tau_v - \tau)]$. From the measured value of $\tau(T)$, as shown in Figure 9, the trapping rate $\kappa(T)$ can be evaluated. Since the trapping rate $\kappa = \mu\, C_v$, with C_v varying as $e^{-(E_V/kT)}$, from an Arrhenius plot of ln $[(\tau - \tau_b)\ /(\tau_b(\tau_v - \tau))]$ vs. $1/T$, the vacancy formation energy can be extracted. Such an analysis of the results for Ni, shown in Figure 9, yields a value of the vacancy formation energy of 1.54 eV. The positron technique for the determination of the vacancy formation energy offers a distinct advantage over the other methods, such as dilatometry, in that it is possible to carry out the positron studies under thermal equilibrium conditions and at very low vacancy concentrations ($\sim 10^{-6}$), where only monovacancies dominate the whole equilibrium ensemble.

4.2. *Annealing behavior of defects*

An understanding of the evolution of nonequilibrium defects produced by quenching at high temperatures, plastic deformation, irradiation, etc., is of great importance in materials science. PAS, with its inherent sensitivity to vacancy-type defects, has played an important role in these studies. In the following, we provide some examples (Ref. 3) of studies on defect annealing behavior using positrons.

4.2.1. Vacancy clustering in metals

Cold-worked Ni was annealed at different temperatures for the same length of time (isochronal annealing), and positron lifetimes were

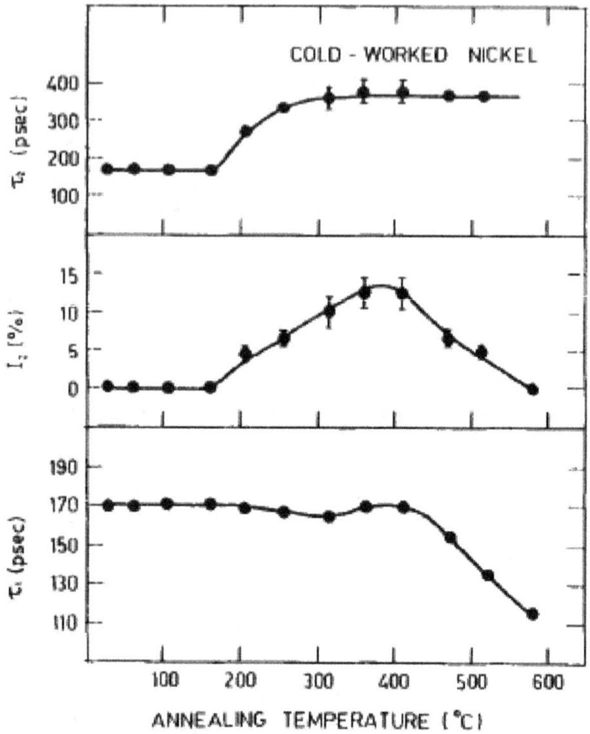

Figure 10. Variations in lifetime parameters τ_1, τ_2, and I_2 as functions of annealing temperature in cold-worked Ni (reprinted with permission from Ref. 3).

measured. The lifetimes were analyzed using Equation (2). Figure 10 shows the variation in the lifetimes τ_1 and τ_2 and the intensity parameter I_2 as functions of the annealing temperature.

Up to an annealing temperature of 160°C, only the component with a lifetime of 170 ps is seen. This is associated with annihilations at monovacancies and dislocation sites. Between 160°C and 400°C, τ_2 and I_2 grow, signaling an increase in the second component, which arises from the agglomeration of the mobile vacancies. By studying the variation of lifetime with isothermal annealing (annealing at a given temperature for different times), the dynamics of clustering has been studied using chemical kinetic rate equations.

4.2.2. Helium decoration of vacancies and helium bubble formation in α-irradiated Nickel

In studies on radiation damage in materials, helium bubble formation in materials irradiated with α particles and the growth of these bubbles causing embrittlement and weakening of the material were observed. The growth of helium bubbles proceeds in the following steps. In a material, there are vacancies and vacancy clusters. The alpha particles, which come to a stop in the material, absorb electrons to become neutral helium atoms. These atoms attach themselves to vacancies. This is called the helium decoration of vacancies. These decorated vacancies and decorated vacancy clusters are mobile. The clusters themselves come together and grow, leading to helium bubble formation. This process can be followed by positron lifetime studies.

Figure 11 shows the results of positron lifetime studies on α-irradiated Ni foil subjected to isochronal annealing at different temperatures.

There are two lifetimes, both dependent on the annealing temperature. The first lifetime τ_1 varies from 140 to 90 ps, and the second lifetime τ_2 varies from 240 to 380 ps, as the annealing temperature is changed. Theoretical calculations show that in the α-irradiated sample (at an annealing temperature of 300 K), the component with lifetime τ_1 of 140 ps and an intensity of 70% can be ascribed to one helium atom attached to a vacancy, while the component τ_2 of 240 ps with an intensity of 30% can be attributed to small vacancy clusters. When the annealing temperature increases from 500 to 750 K, the value of τ_1 decreases sharply to the bulk value, while the value of τ_2 decreases, exhibiting a minimum. The intensity of the second component increases sharply in this temperature range. This behavior can be attributed to the helium atoms migrating from monovacancies to decorate vacancy clusters. The rapid increase in τ_2 beyond 750 K can be attributed to the growth of these helium-decorated vacancy clusters into bubbles. From a theoretical model for positron annihilation at such bubbles, one may calculate the radius of the bubble and the concentration of the bubbles. Figure 12 shows how the radius and concentration of the bubble change with the annealing temperature.

4.3. *Phase transition studies*

Given that the positron lifetime is dependent on the local electron density, it can be expected that lifetime experiments will be sensitive to probing

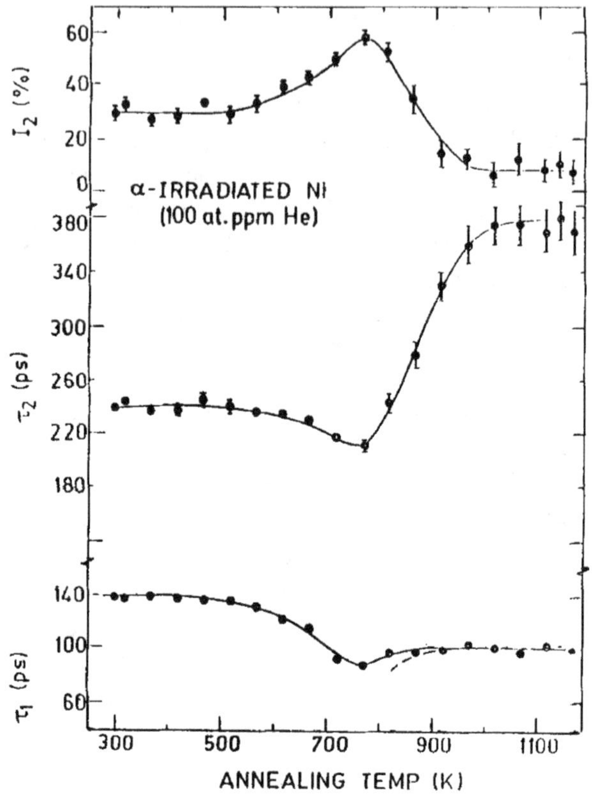

Figure 11. **Plot of lifetimes τ_1 and τ_2 and intensity parameter I_2 as functions of isochronal annealing in α-irradiated Nickel (reprinted with permission from Ref. 3).**

electronic phase transitions in materials. There are several positron studies on phase transitions, such as the metal–insulator transitions induced by changes in temperature or pressure.

Following the discovery of high-temperature superconductivity in cuprate superconductors, several experimental methods were used to unravel this phenomenon, and positron annihilation techniques were also used. It must be mentioned that the positron annihilation parameters do not change across the superconducting transition in conventional BCS superconductors. This is rationalized by the fact that the superconducting transition influences a small fraction of electrons close to the Fermi surface, whereas the positron annihilation parameters are determined by the whole Fermi sea and the core electrons. Thus, it came as a big surprise

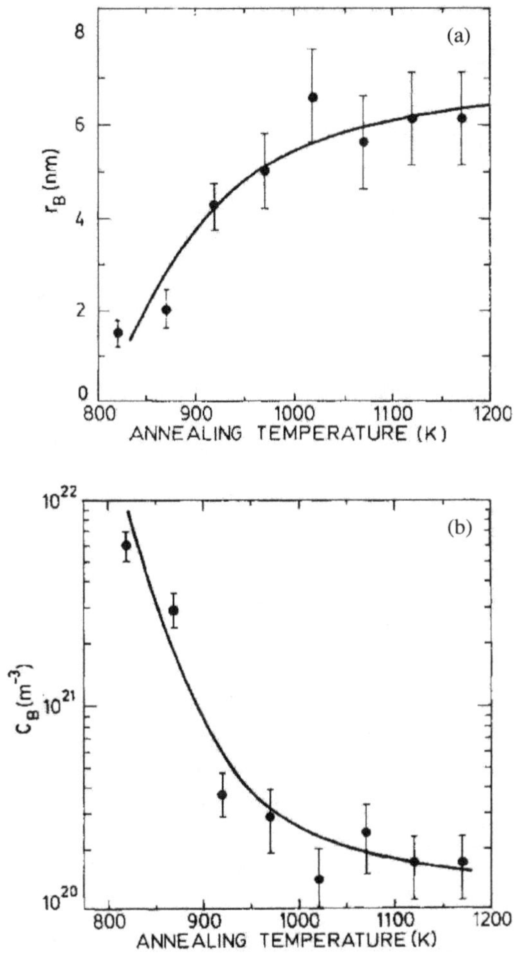

Figure 12. Change in radius and concentration of helium bubbles from a theoretical fit to the data in Figure 11 (reprinted with permission from Ref. 3).

that large changes in positron lifetime were observed across the superconducting transition in high-temperature superconductors.

Figure 13 shows the results of positron lifetime measurements in undoped $YBa_2Cu_3O_7$ and when samples were doped at the Cu site with different amounts of Zn to vary the transition temperature. We see that the positron lifetime in the undoped sample decreases as the sample becomes superconducting below T_C. But this trend is reversed in zinc-doped

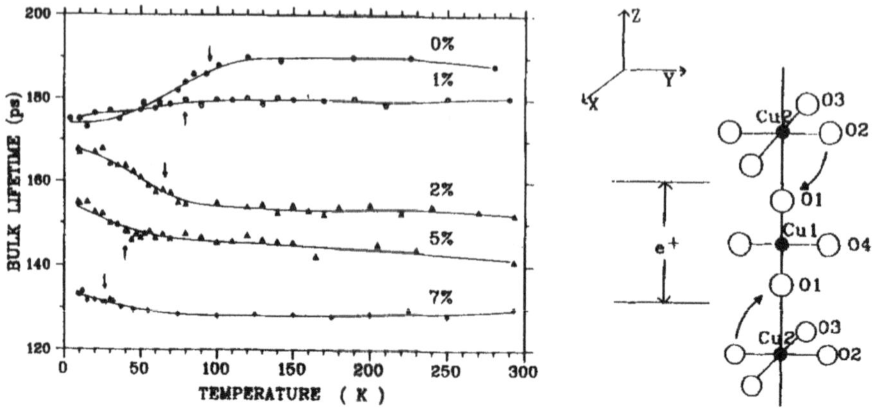

Figure 13. Temperature dependence of positron lifetime in undoped and in Zn-doped YBCO. The CuO planes and Cu–O chains in $YBa_2Cu_3O_7$ and the region sampled by the positron are shown in the right panel (reprinted with permission from Ref. 3).

samples. The high-temperature superconductor $YBa_2Cu_3O_{7-x}$ has two structural motifs: the CuO_2 planes responsible for superconductivity and the Cu–O chains that control the doping of the planes. Positron annihilation lifetime depends on the overlap of the positron density distribution (PDD) with the electron density distribution (EDD). Theoretical calculations of the PDDs along the CuO planes and the Cu–O chains show that in the undoped sample, the PDD is mainly along the Cu–O chain, and annihilation is dominated by the apical oxygen atoms. On the other hand, in zinc-doped samples, there is a significant overlap of the PDD with EDD in the CuO planes, and the oxygen in the planes also contributes to determining the lifetime. Theoretical calculations have explained the different temperature dependences in terms of the transfer of electron density from the planes to the chains. The superconducting properties of these oxide materials strongly depend on the presence of oxygen vacancies. Positron annihilation has proved to be a useful tool in unraveling the properties of the defects.

5. Doppler Broadening Spectrometry

As mentioned earlier, the random motion of electrons in a solid leads to a broadening of the annihilation radiation, centered at the rest mass value of 0.511 MeV. The halfwidth of the Doppler-broadened spectrum is given by $cp_x/2$ and is in the order of a few keV. This small broadening can be measured using a high-resolution Ge detector system, as shown in Figure 14.

In Figure 15, the line shape of the gamma-ray photon arising from the annihilation of a positron in GaAs is compared with the line shape of a gamma-ray photon of nearly the same energy arising from the radioactive decay of the isotope ^{85}Sr. The nucleus ^{85}Sr undergoes decay by electron capture to an excited state of ^{85}Rb. The ^{85}Rb comes to the ground state by emission of a γ-ray photon with energy 514 keV. The line width of the 514 keV gamma ray from ^{85}Sr gives the energy resolution of the

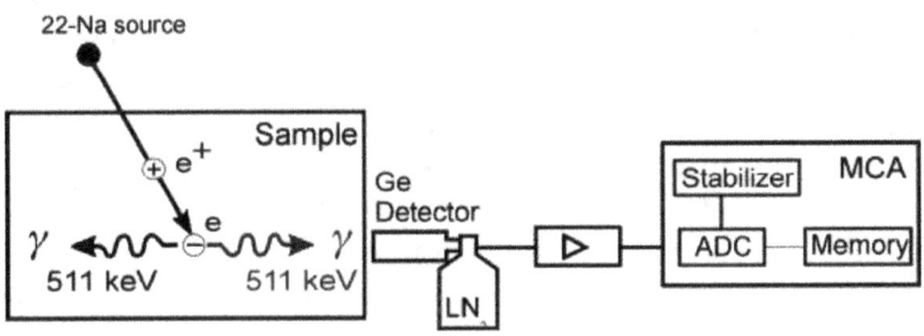

Figure 14. Measuring the doppler broadening of the positron annihilation γ ray (reprinted with permission from Ref. 1).

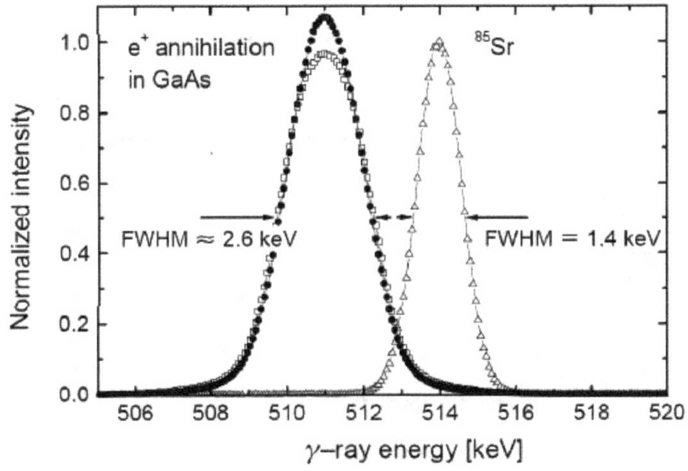

Figure 15. Doppler broadening of the positron annihilation γ ray in GaAs (reprinted with permission from Ref. 1).

germanium detector. Compare it with the line width of the annihilation photon in GaAs. The full width at half maximum for this line is nearly double that of the 514 keV line from ^{85}Sr. This increase in linewidth arises from the Doppler broadening of the positron annihilation gamma-ray photon due to the electron momentum distribution in GaAs.

For a positron trapped at a vacancy-type defect, the annihilation with the valence electrons increases as compared to the core electrons. This results in the narrowing of the Doppler-broadened spectrum for a positron trapped in a vacancy-type defect. This narrowing is quantified using the line-shape parameter *S*, which is defined as the ratio of counts in the central region to the total counts. This is indicated in Figure 16, where the red curve is the Doppler spectrum of the defective material and the blue curve is the Doppler spectrum of the defect-free material, and the ratio of A_S/A_{tot} is called the shape parameter *S*. It depends on the concentration of the defects. The wing regions of the Doppler-broadened spectrum that arise from the annihilation with core electrons are quantified using the *W* parameter, defined by the area ratio $(A_{W1}+A_{W2})/A_{tot}$.

In order to improve the signal-to-noise ratio in the tail region, one can employ the fact that there are two annihilation radiations at ~511 keV, which are emitted together in opposite directions. Through a coincidence measurement of these two radiations, using a pair of Ge detectors, the signal-to-noise ratio in the tail region can be dramatically improved, as

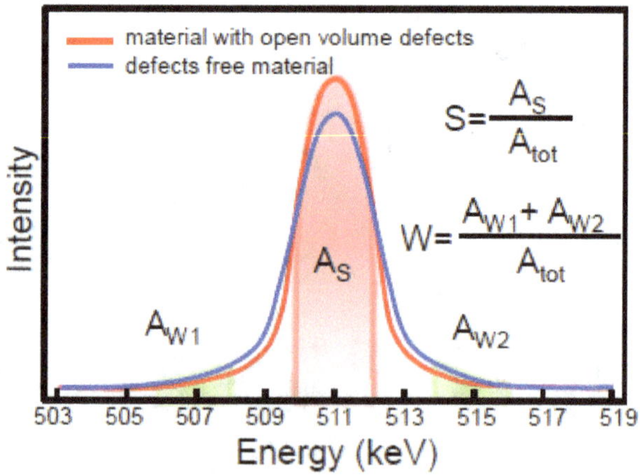

Figure 16. Doppler line shape for defect-free material (blue) and material with defects (red) (reprinted with permission from Ref. 2).

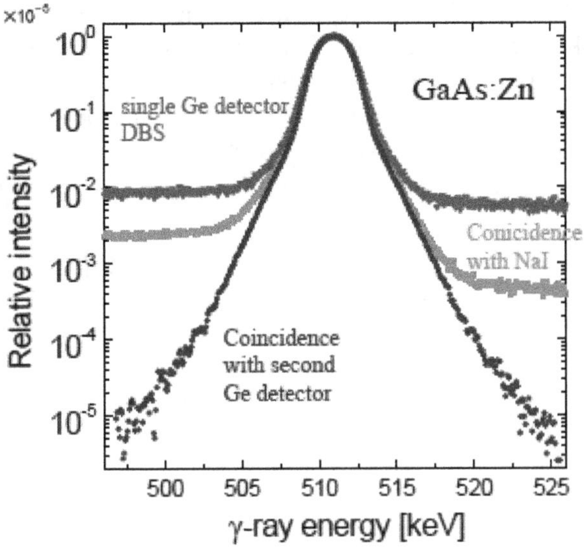

Figure 17. Coincidence Doppler broadening spectroscopy (reprinted with permission from Ref. 1).

shown in Figure 17. Also shown in the figure is the Doppler-broadened spectrum measured with a Ge detector in coincidence with a NaI scintillator. The technique of coincidence Doppler broadening spectrometry can be used to probe the annihilation from core electrons, which can provide information on the chemical nature of the annihilation site or defect.

In contrast to coincidence Doppler broadening spectrometry, the conventional Doppler broadening spectrometry involves the measurement of single photons (see Figure 14). This offers the advantage of faster data acquisition and is valuable for *in situ* studies of defect evolution at elevated temperatures. As an example of such an *in situ* Doppler broadening studies, we show the results on the investigation of dissolution of Ag clusters in quenched Al–Ag alloy. The study of the early stages of clustering of solute atoms in alloys resulting in the formation of precipitates is a topic of considerable interest in physical metallurgy and has been investigated by diffraction methods and electron microscopy. The sensitivity of positrons to the presence of small solute clusters accrues from the preferential affinity of positrons to one of the constituents of the binary alloy, e.g., Ag in the Al–Ag alloy. Thus, if the Ag atoms cluster together, then the attractive potential for the positron may become large enough to localize the positron at the Ag cluster (see Figure 3). Theoretical calculations show that the annihilation characteristics are very sensitive to the size, geometry, and composition of the solute cluster in the alloy.

The results of *in situ* Doppler broadening line-shape measurements in quenched Al – 1%Ag alloy are shown in Figure 18(a). In the as-quenched alloy, the line-shape parameter is close to that in Ag. This can

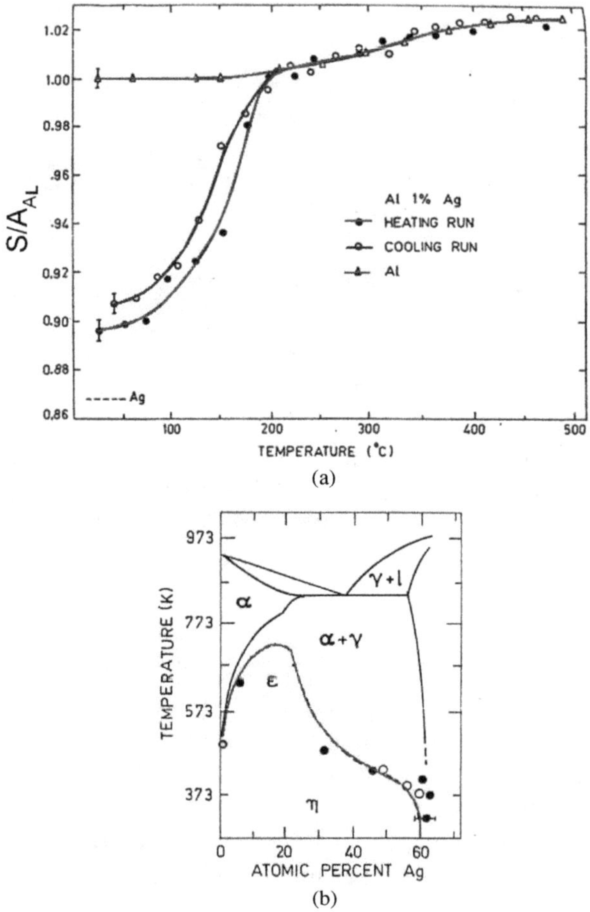

Figure 18. (a) Variation of *S* parameter, normalized to Al, as a function of temperature in quenched Al – 1%Ag alloy. The increase in the line-shape parameter from a Ag-like to Al-like value is associated with the dissolution of the quenched in Ag clusters, GP zones. The composition of the clusters, as estimated from the Doppler broadening measurements is plotted on the GP-zone solvus in the phase diagram shown in (b) (reprinted with permission from Ref. 8).

be understood in terms of positron trapping and annihilation from Ag-rich Guinier–Preston (GP) zones. GP zones are extremely fine-scaled (of the order of 3–10 nm in size) solute-enriched regions of the material. In the temperature range of 100°C–200°C, the line-shape parameter sharply increases and subsequently merges with that of pure Al. This sharp increase in the S parameter is associated with a decrease in the Ag content of the GP zones, rather than the coarsening of the precipitates, leading to a reduction in the number of precipitates. From the measured value of the S parameter, the composition of the GP zones at various temperatures can be estimated, and this forms a novel way of determining the GP zone solvus in the phase diagram of the Al–Ag alloy, as shown in Figure 18(b).

6. Low-Energy Positron Beam Spectrometry

In conventional PAS experiments, positrons from radioactive sources, such as ^{22}NaCl, are used. As mentioned earlier, this has a continuous beta spectrum with an end-point energy of ~300 keV and a range of ~25 μm. In recent times, positrons of controlled low energy have been used so that defects near the surface can be probed.

In a low-energy positron beam spectrometer, positrons from a radioactive source are implanted into a moderator that has a negative positron work function, such as tungsten. In these materials, a fraction of the implanted positrons, after thermalization in the medium, are reemitted from the surface. The thermal positrons emitted from the moderator are extracted and guided down a magnetic field, generated by the bent solenoid, and finally accelerated to the desired energy (0–20 keV) to impinge on the sample. The Doppler broadening annihilation radiation line shape is measured as a function of the positron beam energy. By tuning the energy of the positrons, the implantation depth in the medium can be controlled, and such experiments provide information about the depth distribution of near-surface defects and interfaces.

Figure 19 shows a schematic diagram of the Low Energy Positron Beam (LEPB) spectrometer at Kalpakkam. Here, a solenoidal magnetic field is used as an energy filter to separate out the slow positrons emitted by the moderator from the positrons streaming from the ^{22}Na source. In some other versions of the LEPB spectrometer, an $\boldsymbol{E} \times \boldsymbol{B}$ filter is used that provides a linear configuration of the spectrometer.

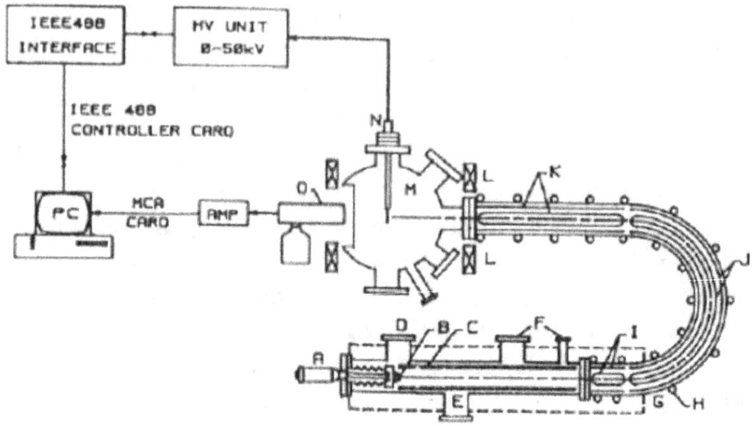

(a)

(b)

Figure 19. (a) Schematic diagram of a magnetically guided low-energy positron beam spectrometer, with the radioactive source, moderator foil, extraction electrostatic system, U-bend-shaped magnetic slow positron filtering and transport system, target chamber, and Ge detector indicated. (b) A photograph of the LEPB spectrometer at IGCAR, Kalpakkam. The source chamber is covered with lead shield.

As an example of positron beam studies, we provide the results of investigations on the Ni–Si interface, taken from Ref. 9. The metal silicides play an important role as interconnects in semiconductor technology. A controlled formation of stoichiometric metal silicides involves several processes, such as the diffusion of the overlayer metal and intermixing across the interface. Point defects play an important role in the processing of stoichiometric silicide phases, which are crucial for device performance. Positrons can be used to probe the vacancy defects at the interface.

Figure 20 shows the variation of the *S* parameter as a function of positron beam energy. The mean implantation depth of the positron and the location of the Ni–Si interface are shown in the top abscissa. As we proceed from the surface, with the increasing positron beam energy, the *S* parameter is seen to decrease from a value characteristic of the surface state to that of Ni (S_{Ni}) and then increase toward the Si value (S_{Si}). The *S* parameter in the interface region is seen to increase with the annealing temperature. Such an increase in the *S* parameter can arise due to the presence of vacancy defects at the interface or the formation of a silicide phase. In order to differentiate between the two, it is profitable to track the *S* and *W* parameters, with the latter representing the fraction of annihilations with high-momentum core electrons. The trajectory of the variation of *S* and *W* parameters as the Ni–Si sample is annealed is shown in Figure 20. The various silicide phases formed, as inferred from the glancing-angle X-ray diffraction, are indicated. In addition, positron beam studies have also been extensively used in the depth profiling of defects in the near-surface region.

While the ^{22}NaCl radioactive source, produced in a cyclotron, is the widely used positron source for laboratory experiments, other specialized positron sources are also employed that exploit access to a nuclear reactor or a high-energy particle accelerator. One such source is a neutron-induced positron source (NEPOMUC) that is available at Garching in Germany. Here, the high-energy γ radiation, produced by the nuclear reaction when ^{113}Cd absorbs a neutron to become ^{114}Cd, yields positrons through pair production. This, coupled to a moderator, produces a very high positron flux (5×10^8 positrons/s) of slow positrons. Another method, wherein photons arising from the acceleration or deceleration of electrons (called Bremmstrahlung), create monoenergetic positrons when they hit a tungsten target. Such sources of γ rays are used in positron production in Rosendorf, Germany, in Tsukuba, Japan, and in the Argonne National Laboratory in the US, among others. The ELBE Positron Source (EPOS) at Rosendorf produces pulsed positron beams with a pulse length of 250 ps and a pulse repetition rate of 1.625–13 MHz. The positrons can be accelerated, and their kinetic energy can be varied from 0.5 to 15 keV,

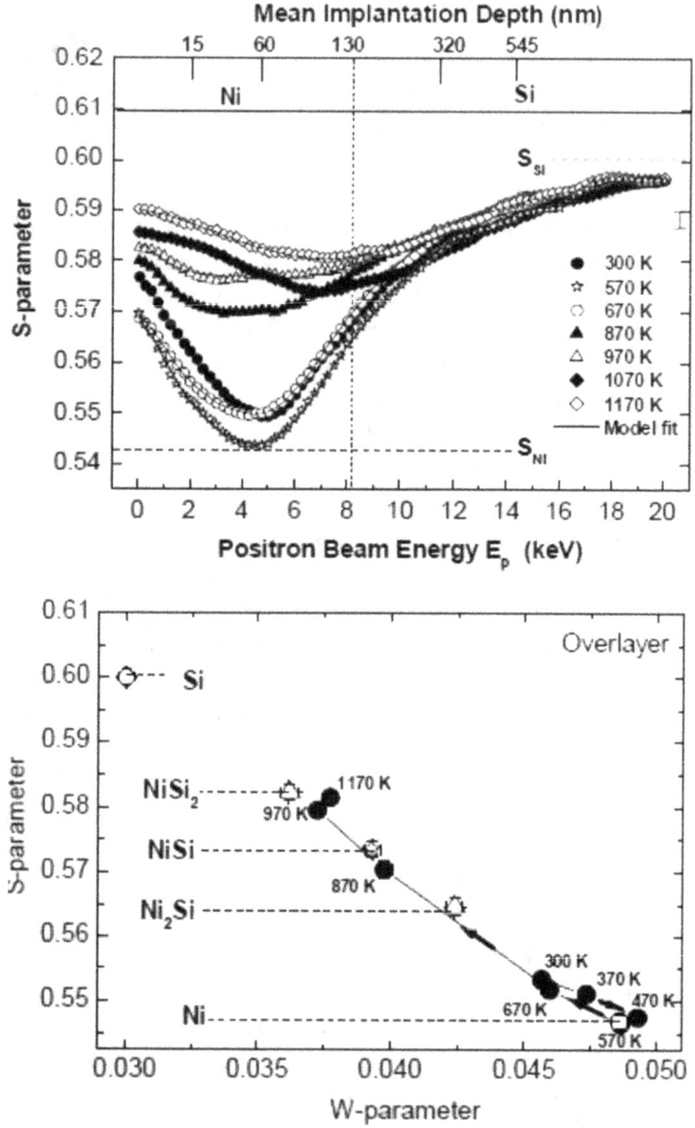

Figure 20. The variation of Doppler broadening line-shape parameter as a function of positron beam energy for a Ni–Si system subjected to annealing. The lower panel shows the correlated variation of the *S* and *W* parameters with annealing temperature. The formation of various silicide phases, as inferred from glancing-angle X-ray diffraction, are indicated (reprinted with permission from Ref. 9).

with a flux of 10^6 positrons/second. Because of the pulsed nature of the source, there is a time stamp of the injection of the positron in the medium, which, when coupled with the detection of annihilation radiation, can be used for positron lifetime measurements. Using this advanced pulsed positron beam facility, positron lifetime measurements can be carried out at various positron beam energies, and one can thus study the defect distribution in the material as a function of depth. It must be mentioned that in the low-energy positron beam spectrometer, as described in Figure 19, we have a continuous beam of positrons with no time stamp, and hence, only Doppler broadening measurements can be carried out.

7. Angular Correlation Positron Annihilation Spectroscopy

As indicated earlier, due to the momentum distribution of electrons in a medium, there is a spread in the angular deviation in the two gamma rays of about 180°. In an angular correlation experiment, one or two dimensional projections of the two-photon momentum density $\rho(\boldsymbol{p})$ is measured. The two-photon momentum density $\rho(\boldsymbol{p})$ can be expressed as the squared absolute value of the Fourier transform of the product wave functions of the positron, $\psi_p(\boldsymbol{r})$, with an electron $\varphi_{k,j}(\boldsymbol{r})$ in the kth state of the jth band, summed over all the k and j values (Ref. 10).

The 1D angular correlation curve, which represents the 1D projection of the electron momentum density $\rho(\boldsymbol{p})$, is given by

$$N(p_z) = \iint \rho(\boldsymbol{p})\, \mathrm{d}p_x\, \mathrm{d}p_y$$

In a 1D ACAR experiment, one measures the two-photon coincidence counts $N(p_z)$ as a function of the z component of the momentum, which is related to the angle between the two long-slit detectors via $\theta_z = p_z/mc$. The 1D ACAR apparatus consists of a positron source and a sample in the center, one fixed detector on one side, and a second movable detector on the other side of the sample. The two NaI scintillator detectors are collimated with long narrow slits in such a way that the active area is much smaller in one dimension than in the other. This results in the integration of the x and y components of momentum density. The two detectors are separated by a distance of ~3 m, and the slit width is set at ~3mm to achieve an angular resolution of a milliradian, which is required to trace the angular correlation curve in metals.

To understand how the ACAR experiment gives information on the Fermi surface, we consider the simple case of a nearly free-electron-type metal, characterized by a spherical Fermi surface. The measured 1D angular correlation curve has the appearance of an inverted parabola in the low-momentum region sitting on a Gaussian background in the high-momentum regions, as shown in Figure 21(a). The latter arises from annihilation with the core electrons. The parabolic behavior in the low-momentum region can be appreciated in the context of an isotropic Fermi sphere. The coincidence counts at each value of p_z are related to the area of the circular cross-section of the Fermi sphere at that value, as shown in Figure 21(b). This results in a parabolic variation of the angular correlation curve. The cut-off of the parabolic component represents Fermi momentum.

To investigate more complex Fermi surfaces, 2D angular correlation experiments are carried out using position-sensitive detectors, such as gamma cameras used for medical imaging or multi-wire proportional chambers used in particle physics experiments. In a 2D angular correlation experiment, the two 2D position-sensitive detector arrays are on either side of the sample, as shown in Figure 22. Coincidental counts are recorded for different $\Delta\theta$ values. $\Delta\theta$ has the following two components: (1) $\Delta\theta_x$, which is the angle between the x-axis and the projection of the photon path in the xz-plane; and (2) $\Delta\theta_y$, which is the angle between the y-axis and the projection of the photon path in the yz-plane.

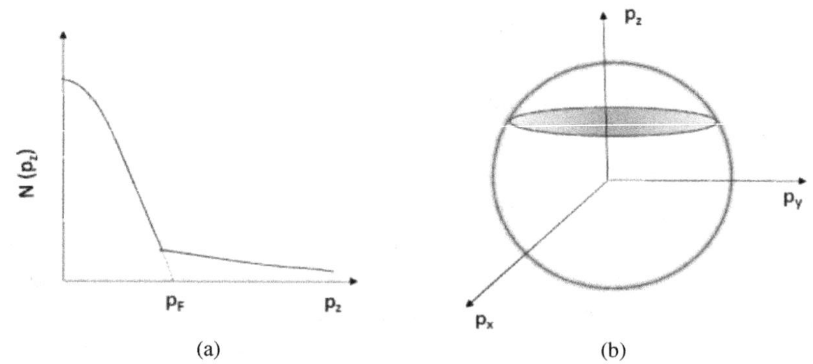

Figure 21. **(a) Schematic diagram of a 1D angular correlation curve from a free-electron-type metal showing the parabolic and Gaussian components. The parabolic variation in the low-momentum region can be rationalized in terms of the area of the circular cross-sections of the Fermi sphere, as shown in (b). The cut-off of the parabolic component, indicated by dotted line, is at the Fermi momentum p_F.**

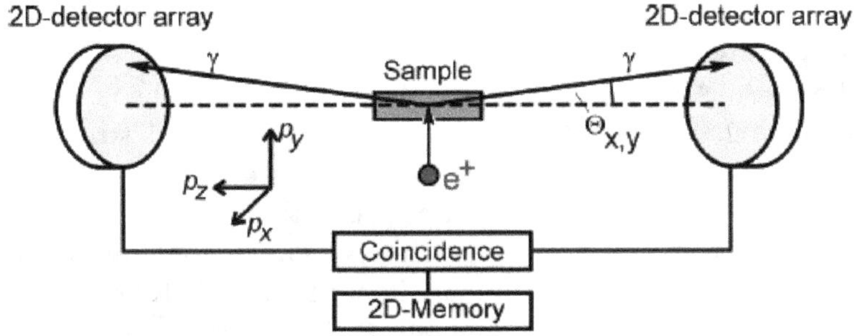

Figure 22. **Schematic illustration of 2D angular correlation spectroscopy (reprinted with permission from Ref. 1).**

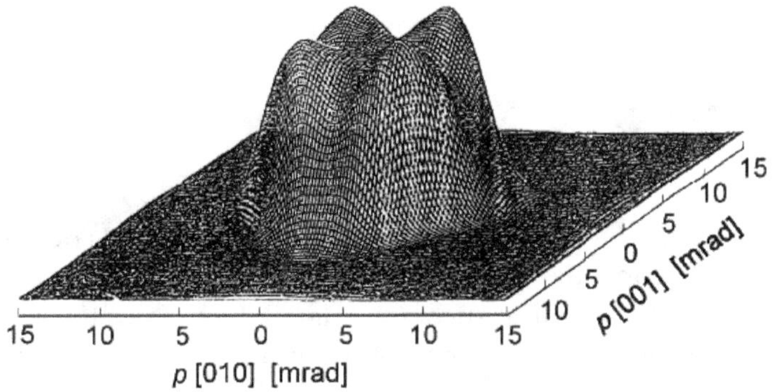

Figure 23. **2D angular correlation curve in defect-free GaAs (reprinted with permission from Ref. 1).**

In a 2D ACAR experiment, the 2D projection of $\rho(\boldsymbol{p})$ is measured, i.e.,

$$N(p_y, p_z) = \int \rho(\boldsymbol{p})\, dp_x$$

It is noted that, since the scintillator detectors do not have adequate energy resolution, to be sensitive to the Doppler broadening, the parallel component of $\boldsymbol{p}$ gets integrated.

Figure 23 displays the 2D angular correlation curve in GaAs. As ACAR measures projections of the TPMD, it is necessary to reconstruct $\rho(\boldsymbol{p})$ in order to recover the Fermi surface. From the measurements of the 2D ACAR for different orientations of the crystal and using the mathematical techniques derived from X-ray tomography, a complete reconstruction of the two-photon momentum density $\rho(\boldsymbol{p})$ is carried out. The discontinuities in $\rho(\boldsymbol{p})$ provide the Fermi surface features. Quantitative comparison with *ab initio* electronic structure calculations plays an important role in these studies.

The 2D ACAR has been extensively used to obtain the Fermi surface features of metals and alloys, including spin polarized structures. 2D ACAR played an important role in the understanding of electronic structure of oxide superconductors. Comparing with other techniques for the investigation of the electronic structure, such as angle-resolved photoemission (ARPES) and quantum oscillations, the ACAR has advantages in that it does not require low temperatures, high magnetic fields, or UHV conditions. However, it has disadvantages in that 2D ACAR is extremely sensitive to the presence of defects and hence requires high-quality single crystals. In recent times, the ACAR method has largely been replaced by ARPES, with the availability of synchrotron sources and the improvements in energy resolution.

8. Conclusion

In this chapter, we have seen how, by using the techniques derived from nuclear spectroscopy, the annihilation of a positron with an electron has been used to probe matter in such great detail. Even from this brief summary, one can see the wide range of applications of the techniques of

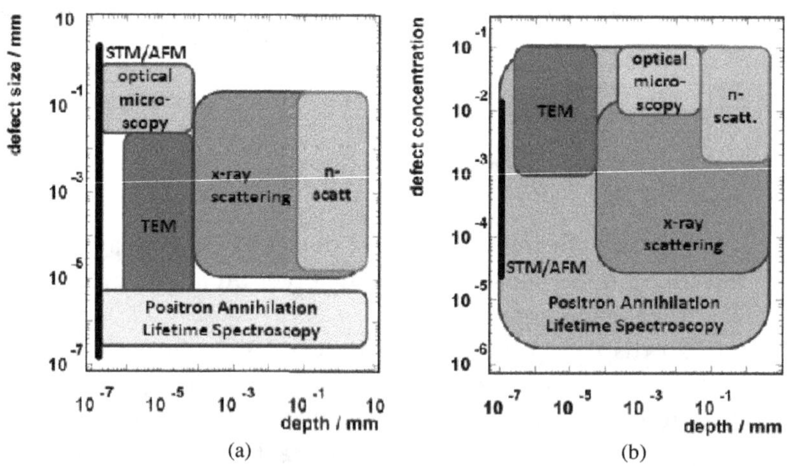

Figure 24. Comparing the range of usefulness of different techniques as a function of (a) size and depth of defects and (b) concentration and depth of defects (reprinted with permission from Ref. 2).

PAS in the study of defects in materials. Figures 24(a) and (b) show comparisons of various methods for studying defects of different sizes and for defects at various concentrations, respectively. From Figures 24(a) and (b), we see that positron annihilation is useful for defect sizes ranging from 2 $\times 10^{-7}$ mm (point defects) to about 5×10^{-6} mm (extended defects), occurring at depths of up to nearly 1 mm from the surface. Positron annihilation can be used over a wide range of concentration of defects.

References

1. Krause-Rehberg, R. Fundamentals in positron annihilation spectroscopy and its application to semiconductors, https//:positron.physik.uni-halle.de> talksICPA-15.
2. Liedke, M. O. Defects in solids, positron annihilation spectroscopy, apparatus for in-situ defect analysis, https://www.hzdr.de›CmsPDF.
3. Sundar, C. S. (1994). Positron annihilation spectroscopy in materials science. *Bull. Mater. Sci.* **17**, 1915.
4. Rajaraman, R., Amarendra, G., and Sundar, C. S. (2009). *Phys. Status Solidi C* **6**(11), 2285–2290.
5. Amarendra, G., Viswanathan, B., Bharathi, A., and Gopinathan, K. P. (1992). *Phys. Rev. B* **45**, 10231.
6. Jean, Y. C. (1995). In A. Dupasquier and A. P. Mills, Jr. (eds.) *Positron Spectroscopy of Solids*. IOS Press, Amsterdam, p. 563.
7. Lynn, K. G., Snead C. L. Jr., and Hurst, J. J. (1980). Positron lifetime studies in pure Ni from 4.2 to 1700 K. *J. Phys. F: Metal Phys.* **10**, 1753–61.
8. Bharathi, A. and Sundar, C. S. (1988). In L. Dorikens–Vanpraet and D. Segers (eds.) *Positron Annihilation*. World Scientific, Singapore, p. 479.
9. Abhaya, S., Amarendra, G., Panigrahi, B. K., and Nair, K. G. M. J. (2006). *Appl. Phys.* **99**, 033512.
10. West, R. N. (1973). *Adv. Phys.* **22**, 263.

Part III

Techniques for Measurement of Physical Properties

Chapter 9

ELASTIC PROPERTIES

1. Introduction

Consider a force d$\boldsymbol{F}$ acting on an area dA, as shown in Figure 1. The area d$\boldsymbol{A}$ is a vector in the direction of its normal. The force d$\boldsymbol{F}$ is a vector in some other arbitrary direction. d$\boldsymbol{F}$ and d$\boldsymbol{A}$ can be written in Cartesian coordinates as

$$\mathrm{d}\boldsymbol{F} = \mathrm{d}F_1\boldsymbol{e}_1 + \mathrm{d}F_2\,\boldsymbol{e}_2 + \mathrm{d}F_3\,\boldsymbol{e}_3 \tag{1}$$

and

$$\mathrm{d}\boldsymbol{A} = \mathrm{d}A_1\boldsymbol{e}_1 + \mathrm{d}A_2\,\boldsymbol{e}_2 + \mathrm{d}A_3\,\boldsymbol{e}_3 \tag{2}$$

Here, $\boldsymbol{e}_1$, $\boldsymbol{e}_2$, and $\boldsymbol{e}_3$ are unit vectors in the direction of x, y, and z axes respectively. dF_j and dA_j are the components of the vectors d$\boldsymbol{F}$ and dA, respectively, along $\boldsymbol{e}_j$. The quantities

$$\sigma_{ij} = \partial F_i/\partial A_j \tag{3}$$

are the components of a second-rank stress tensor.

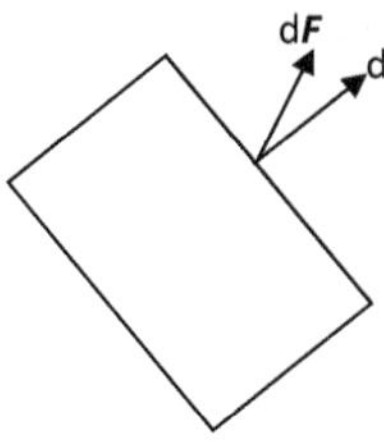

Figure 1. Force dF acting on an area dA.

If we rotate the coordinate system from x, y, z to x', y', z', then the components of the vectors d$\boldsymbol{F}$ and dA change. The components of the stress tensor σ_{ij} change to σ'_{ij}. σ'_{ij} are related to σ_{ij} by

$$\sigma'_{ij} = \Sigma_{mn}\alpha_{im}\alpha_{jn}\,\sigma_{mn} \tag{4}$$

Here, the sum over m and n runs from 1 to 3, and

$$\alpha_{im} = \partial x'_i / \partial x_m \tag{5}$$

The second-rank stress tensor has six independent components since it can be shown that $\sigma_{ij} = \sigma_{ji}$ $(i \neq j)$. σ_{ii} represents a tensile component of the stress, and σ_{ij} $(i \neq j)$ represents a shear component of the stress. They are measured in N/m^2.

Consider a body which is strained. Two points, O and A, are at a distance d$\boldsymbol{r}$ in the unstrained body. When the body is strained, the points move to O′ and A′. The vector distance O′A′ = d$\boldsymbol{r}'$. This is shown in Figure 2.

If dr'_j and dr_j are the components of d$\boldsymbol{r}'$ and d$\boldsymbol{r}$ along the unit vector e_j, respectively, we may write

$$\mathrm{d}r'_i = \Sigma_j\,(\partial r'_i / \partial r_j)\,\mathrm{d}r_j \tag{6}$$

The quantities

$$\varepsilon_{ij} = (\partial r'_i / \partial r_j) \tag{7}$$

are the components of the second-rank **strain** tensor. When the coordinates x, y, z, are rotated to x', y', z', the components of the strain tensor also change according to Equation (4) if we replace the components σ_{ij} of the stress tensor with the components ε_{ij} of the strain tensor.

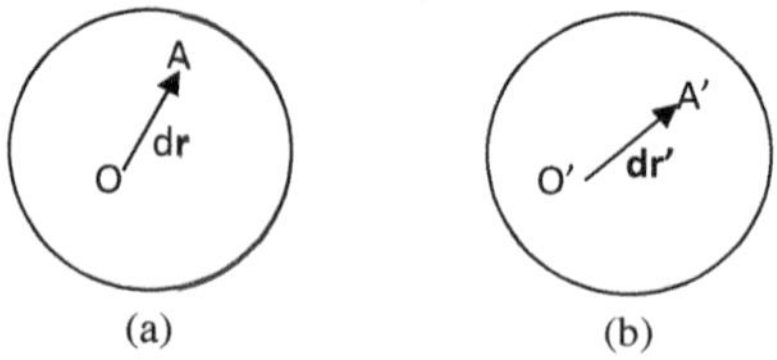

Figure 2. Two points O and A in (a) unstrained body and (b) strained body.

Again, it can be shown that the strain tensor has only six components because $\varepsilon_{ij} = \varepsilon_{ji}$ $(i \neq j)$. ε_{ii} represents a tensile strain along $\boldsymbol{e}_i$, while ε_{ij} $(i \neq j)$ represents a shear strain. The strain tensor components are dimensionless.

We often replace the pair of symbols *ij* by a single symbol *i*, as follows:

ij:	11,	22,	33,	23 = 32,	31 = 13,	12 = 21
i:	1	2	3	4	5	6

2. Stress–Strain Curve: Universal Testing Machine

The universal testing machine (UTM) is used by engineers to perform a variety of tests on engineering materials. One can perform a tensile or a compressive test. Figure 3 shows such a UTM.

The UTM has two parts. The setup on the left of the figure is used to hold the sample and subject it to tensile or compressive testing. The control unit is shown on the right-hand side of the figure.

The components of the UTM in Figure 3 are shown in Figure 4.

Figure 3. Universal testing machine.

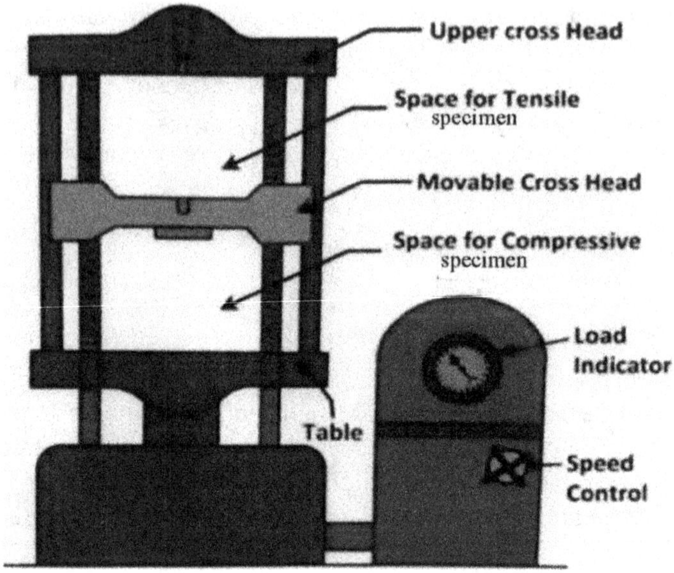

Figure 4. Components of the UTM.

There is a fixed upper crosshead and a movable crosshead. For tensile testing, the sample is held in grips fastened to the two crossheads. There is a fixed table below the movable crosshead. For performing compressive tests on blocks of materials, the block is laid on the table, and a force is applied to the movable crosshead to compress the block. The load applied to the specimen in kilonewton (kN) is indicated on the load indicator. There is a speed control to regulate the speed of the movable crosshead. There is an elongation scale to measure the movement of the movable crosshead.

The control unit has a hydraulic pump to produce a non-pulsating flow of oil into the main cylinder to apply load on the specimen smoothly. There is a pendulum which is moved by the piston of the oil pump. This pendulum is connected by a lever to produce a deflection of the pointer on the load indicator. The load indicator is calibrated in kN. One can choose the range of load applied by turning a knob on the control unit. The maximum load applied can be varied in stages from 100 to 1000 kN. The load can be applied or released using either electrical switches or hydraulic valves.

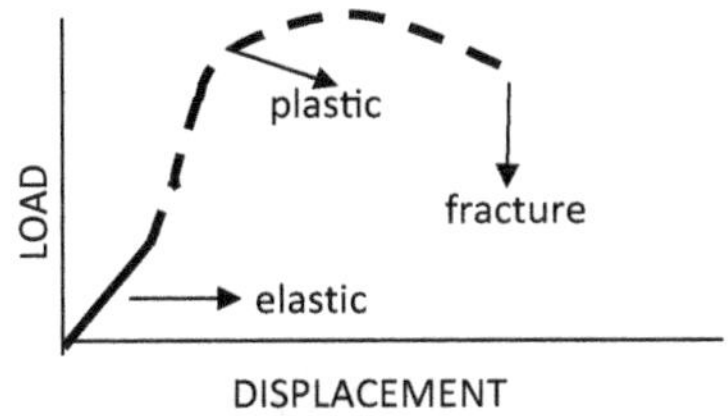

Figure 5. Typical load vs. displacement curve.

The force versus displacement curve of a specimen is shown in Figure 5. For small displacements (i.e., small strains), the load (i.e., proportional to stress) varies linearly with displacement. This is the elastic region described in Section 3. Then, the curve becomes nonlinear. At a certain load, the specimen elongates for very small changes in the load. This is the plastic region, the onset of which is indicated by "plastic" in Figure 5. The cross-section of the specimen decreases as it elongates, and at a certain displacement, the specimen ruptures.

Our interest in this chapter is in the elastic region of the curve to measure the elastic constants.

3. Hooke's Law and Elastic Constants

If the strain components are small, Hooke's law states that the components σ_{ij} of stress are linear functions of the components ε_{mn} of strain:

$$\sigma_{ij} = \Sigma_m \Sigma_n C_{ij,mn} \varepsilon_{mn} \tag{8a}$$

or, in the single-suffix notation,

$$\sigma_i = \Sigma_j C_{i,j} \varepsilon_j \tag{8b}$$

In Equation (8a), the summation over m and n go from 1 to 3. In Equation (8b), the summation over j goes from 1 to 6. $C_{ij,mn}$ form a fourth-rank tensor, transforming as

$$C'_{ij,mn} = \Sigma_{pqrs} \alpha_{ip} \alpha_{jq} \alpha_{mr} \alpha_{ns} C_{pq,rs} \tag{9}$$

Since $$C_{ij,mn} = C_{ji,mn} = C_{ij,nm} = C_{ji,nm} \tag{10a}$$

and

$$C_{i,j} = C_{j,i} \tag{10b}$$

the 81 components of $C_{ij,mn}$ reduce to 21 independent constants, namely six constants $C_{i,i}$ ($i = 1,2,\ldots,6$) and 15 independent constants $C_{i,j} = C_{j,i}$ ($i \neq j$; $i, j = 1,2,\ldots,6$). These are called the elastic constants, and they are properties characteristic of the material. They are expressed in N/m^2.

We may invert the Equations (8a) and (8b) and write the strain components ε_{ij} as linear functions of the stress components σ_{mn}:

$$\varepsilon_{ij} = \Sigma_m \Sigma_n s_{ij,mn} \sigma_{mn} \tag{11a}$$

or, using single-suffix notation,

$$\varepsilon_i = \Sigma_j s_{i,j} \sigma_j \tag{11b}$$

$s_{ij,mn}$ (or $s_{i,j}$ in single-suffix notation) are also fourth-rank tensors satisfying the transformation relation (9) for coordinate transformations, with C replaced by s. The relations (10a) and (10b) are also valid when C is replaced by s. Thus, $s_{ij,mn}$ (or $s_{i,j}$) have 21 independent constants for a material. These are called the elastic moduli of the material. They are in m^2/N.

The symmetry of a crystal will reduce the number of independent elastic constants. A cubic crystal has four three-fold axes of symmetry along the cube diagonals and three four-fold axes of symmetry along the cube edges. For cubic crystals, this symmetry imposes the conditions $C_{1,1} = C_{2,2} = C_{3,3}$, $C_{1,2} = C_{2,3} = C_{3,1}$, and $C_{4,4} = C_{5,5} = C_{6,6}$, and all the other coefficients $C_{i,j}$ are zero. Thus, there are only three independent coefficients.

If we have a polycrystalline cubic material in which the crystallites are oriented randomly, then the material will be isotropic. In such a material, every direction is equivalent to every other direction. For an isotropic material, there are only two independent elastic constants, which are usually termed the Young's modulus Y and the rigidity modulus n.

4. Elastic Constants and Sound Velocity

In a solid, sound will be propagated in any direction as three types of waves. One of these is the longitudinal wave, in which the particle displacement is in the direction of propagation. The other two waves are transverse waves, in which the particle displacement is along two mutually orthogonal directions in a plane perpendicular to the direction of propagation. We call the longitudinal sound wave L and the two transverse sound waves T1 and T2. The velocities of these waves, namely V_L, V_{T1}, and V_{T2}, vary with the direction of propagation relative to the crystallographic axes. For example, in a cubic crystal, for waves propagating along the (100) direction,

$$V_L^2 = C_{1,1}/\rho \tag{12a}$$

$$V_{T1}^2 = V_{T2}^2 = C_{4,4}/\rho \tag{12b}$$

The two transverse waves have the same velocity.

On the other hand, if the sound wave is propagating along the (110) direction, the velocities are given by

$$V_L^2 = (C_{1,1} + C_{1,2} + 2C_{4,4})/2\rho \tag{13a}$$

$$V_{T1}^2 = (C_{1,1} - C_{1,2})/2\rho \tag{13b}$$

$$V_{T2}^2 = C_{4,4}/\rho \tag{13c}$$

If the wave is propagating along the (1,1,1) direction, the sound velocities are given by

$$V_L^2 = (C_{1,1} + 2C_{1,2} + 4C_{4,4})/3\rho \tag{14a}$$

$$V_{T1}^2 = V_{T2}^2 = (C_{1,1} - C_{1,2} + C_{4,4})/3\rho \tag{14b}$$

For an isotropic medium,

$$(C_{1,1} - C_{1,2}) = 2C_{4,4} \tag{15}$$

If we substitute (15) in (13a, 13b, 13c) and (14a, 14b), then the velocities along the (100), (110), and (111) directions become equal and are given by (12a, 12b).

For an isotropic medium, we call $C_{1,1}$ as the Young's modulus Y and $C_{4,4}$ as the shear modulus n.

5. Static Methods for Measuring Elastic Constants

In the laboratory, elastic constants can be measured by static or dynamic methods.

In the static method, to measure the Young's modulus of a ductile material, a wire of the material is hung from a fixed support. The wire is loaded at the free end, and the displacement of the end of the wire is measured. Knowing the area of the cross-section and the original length of the wire, the Young's modulus is calculated.

One can use a tensile testing machine, attach a strain gauge to the sample, and measure the elastic modulus in the elastic region of the stress–strain curve.

If the material is available as a bar of sufficient length, it is mounted symmetrically on two knife edges, and a load is added at the center of the bar. This is shown in Figure 6.

Since the loading is symmetric about the center of the bar, we may consider it as two bars, each of length $(l/2)$ loaded with a weight $W/2$ at one end. Then, the bending moment at P, which is at a distance x from the knife edge, is $(W/2)[(l/2) - x]$. This should be equated to YAk^2/R, where R is the radius of curvature of the beam at the point P, A is the area of the cross-section of the beam, and k^2 is the square of the radius of gyration of the cross-section of the beam about a line in the middle of the cross-section. Ak^2 is called the geometric moment of inertia. For a beam of circular cross-section of radius r, Ak^2 is $\pi r^4/2$. For a beam of rectangular cross-section of breadth b and thickness d, the geometric moment of inertia is $bd^3/12$.

If y is the downward displacement of P, then $1/R$ can be approximated by d^2y/dx^2 when the displacement is small. Then,

$$YAk^2\, d^2y/dx^2 = (W/2)\,[(l/2) - x] \tag{16}$$

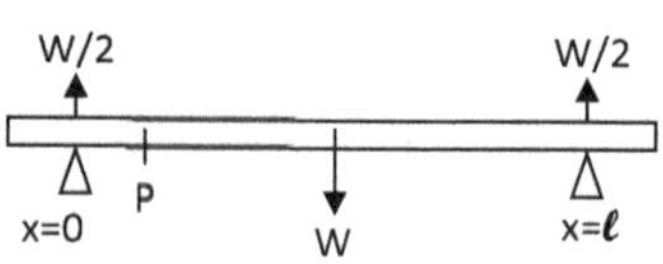

Figure 6. Bar supported on two knife edges and loaded in the middle.

Integrating, we get the displacement $y(x)$ for $0 < x < l/2$ as

$$y(x) = \{(W/(2YAk^2))\ [(lx^2/4) - x^3/6] \tag{17}$$

So, the displacement y at $x = l/2$ is given by

$$y(l/2) = ((W/(2YAk^2))\ (l^3/24) = Wl^3/(48YAk^2) \tag{18}$$

The displacement downward of the midpoint of the bar can be measured with a traveling microscope. Thus, one can determine the Young's modulus by taking a beam with a rectangular or circular cross-section.

To determine the shear modulus, one can fix a circular rod at one end and apply a torque τ at the other end. Then, by measuring the twist φ of the rod at the end where the torque is applied, we may find the shear modulus n from the relation

$$(\pi n r^4/4l)\ \varphi = \tau \tag{19}$$

In Eq.19, r is the radius of the rod, l its length, and n its shear modulus.

6. Dynamic Methods

We discuss the following two dynamic methods: (a) pulse echo method for measuring sound velocities in materials and (b) resonance method for measuring elastic constants.

6.1. *Pulse echo method for sound velocity measurement*

In this method, the solid is a slab with parallel end faces and a low attenuation.

In the "sing around" method, there are two transducers attached to the two ends of the slab. The pulse produced by the transmitter is communicated to the material by the first transducer. This pulse travels to the second transducer. When the pulse is received by the second transducer, it induces the transmitter to send a second pulse through the sample. The measurement of the pulse repetition rate (PRR) gives the reciprocal of the time of travel τ through the length l of the slab. The velocity of sound is given by

$$V = (l \times \text{PRR})/\ (1 - \text{e} \times \text{PRR}) \tag{20}$$

where the correction factor (1 – e × PRR) arises from the delay (i) in the transducer, (ii) in the coupling between the transducer and the specimen and (iii) in the electronic components.

The block diagram for the "sing around" technique is shown in Figure 7.

The trigger pulse goes through the sample and has a variable delay to reach the coincidence circuit. If the variable delay time is equal to the time τ of travel through the sample, the two parts of the trigger pulse are received in the coincidence circuit at the same time. When this happens, the next pulse is triggered to travel through the sample. The pulse repetition rate is then measured. The delay factor e can be estimated and corrected for by making measurements on homogeneous samples of different lengths. The sound velocity can be measured using this technique with moderate accuracy.

In materials of low attenuation, one can improve the accuracy by a modification of this technique used by Forgacs. The pulse travels to and fro in the sample. The travel lengths for the successive echoes will be l, 3l, 5l,…, as shown in Figure 8.

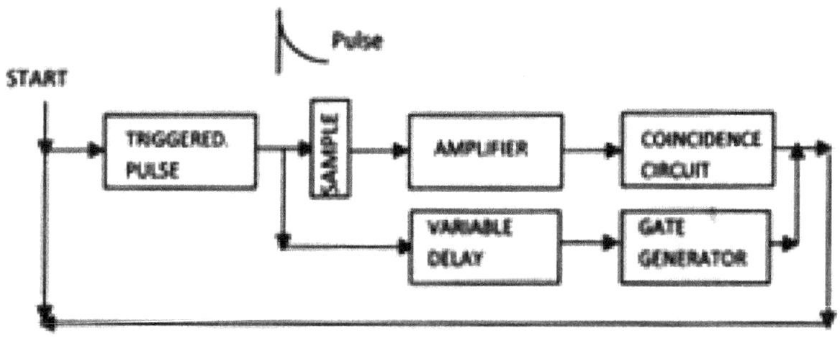

Figure 7. Block diagram for "sing around" technique.

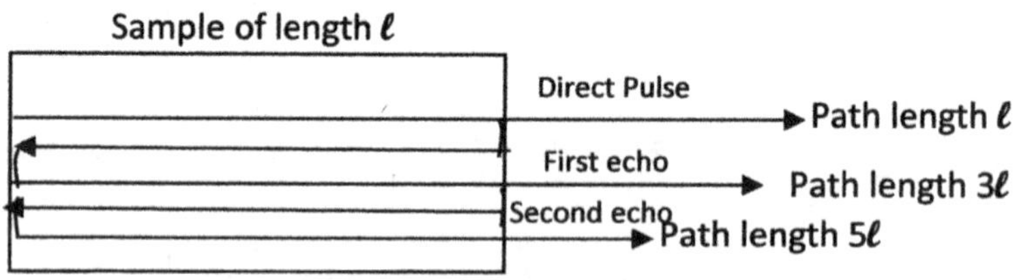

Figure 8. Successive echoes in a slab.

So, the nth echo would have a travel time τ_n approximately equal to $(2n + 1)l/V$. The transmitter is triggered after n echoes, not after the first echo. Thus, one can measure the travel time with greater accuracy.

The pulse superposition method is another variation of the pulse technique for measuring the velocity of sound.

As a practical example of the pulse superposition technique, we describe the experimental setup from Ref. 1 here. The sample holder is shown in Figure 9.

The sample was a rod of polycrystalline $Fe_{0.7}Al_{0.3}$ with a diameter of 6.0 mm and a length of 9.2 mm. The ultrasound transducers were fabricated from quartz, $BaTiO_3$, and $LiNbO_3$. They were cut into circular plates of 3.5 mm diameter and plated on both sides with a thin film of silver or gold. They were glued with conductive epoxy glue to the two ends of the sample rod. Depending on the crystal orientation of the transducer, the application of an AC voltage will produce a longitudinal or transverse displacement. Thus, one can generate longitudinal or shear waves

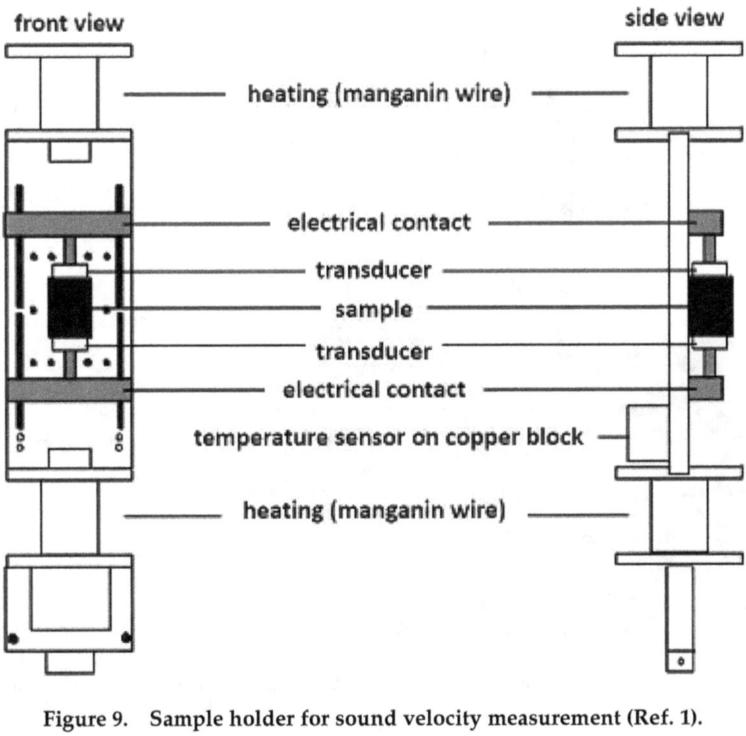

Figure 9. Sample holder for sound velocity measurement (Ref. 1).

in the sample rod. To enhance the amplitude of oscillations, the frequency of operation was chosen to be close to the resonant frequency of the transducer. This depends on the thickness of the transducer, its density, and its elastic constants. For a $BaTiO_3$ transducer of 0.2 mm thickness, for example, the resonance frequency is about 10 MHz. The sample holder was inserted into a stainless-steel (SS) tube, which could be evacuated and filled with helium exchange gas at low pressure. The SS tube was inserted into a tube made of fiber-reinforced glass and containing liquid nitrogen. This enabled experiments to be done at temperatures as low as 77 K. The entire arrangement was inserted between the pole pieces of an electromagnet to carry out measurements in different magnetic fields. At the top of the stainless-steel tube, there were vacuum feedthroughs for the high-frequency leads to the transducer, the leads to a platinum resistance thermometer, and leads to the heater on the sample holder. The tube was fitted with valves to evacuate it and fill it with exchange helium gas. The setup is shown in Figure 10.

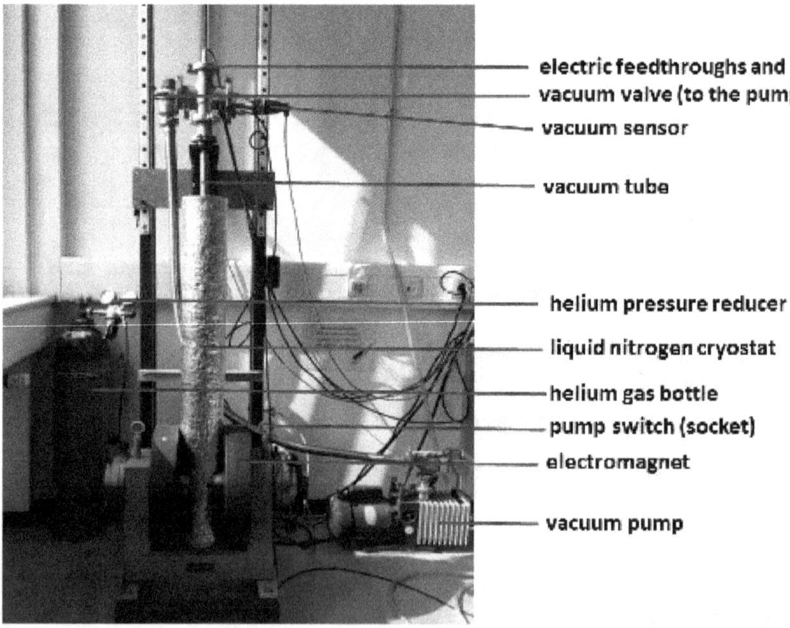

Figure 10. Experimental arrangement for sound velocity measurement in $Fe_{0.7}Al_{0.3}$ alloy (Ref. 1).

Figure 11 shows a schematic diagram of the electronic setup. A pulse generator generates square pulses periodically at a rate of 5–500 kHz. This is connected to a pulse divider. Depending on its setting, this pulse divider suppresses $n + 1$ pulses and allows pulse 1 and pulses $(n + 2)$, $(2n + 3)$,... to go through. If the selector is set at $n = 1$, the divider allows pulses 1, 3, 5,... to reach the pulse modulator. If n is chosen as 2, pulses 1, 4, 7,... reaches the pulse modulator. The pulse reaching the modulator triggers it to generate a pulse of sinusoidal oscillations of voltage. The pulse width can be adjusted to contain three to five peaks. The output of the modulator is amplified by the high-frequency amplifier and applied to the transducer T_1 pasted at the top of the sample.

The pulse and the sequence of echoes generated by it are picked up by the transducer T_2 at the bottom. They are amplified and fed to one channel of an oscilloscope.

If we set $n = 1$, then we see on the oscilloscope the direct pulse reaching the transducer T_2 and the first echo, and this is repeated on the oscilloscope. If we set $n = 2$, we see the direct pulse as well as the first and

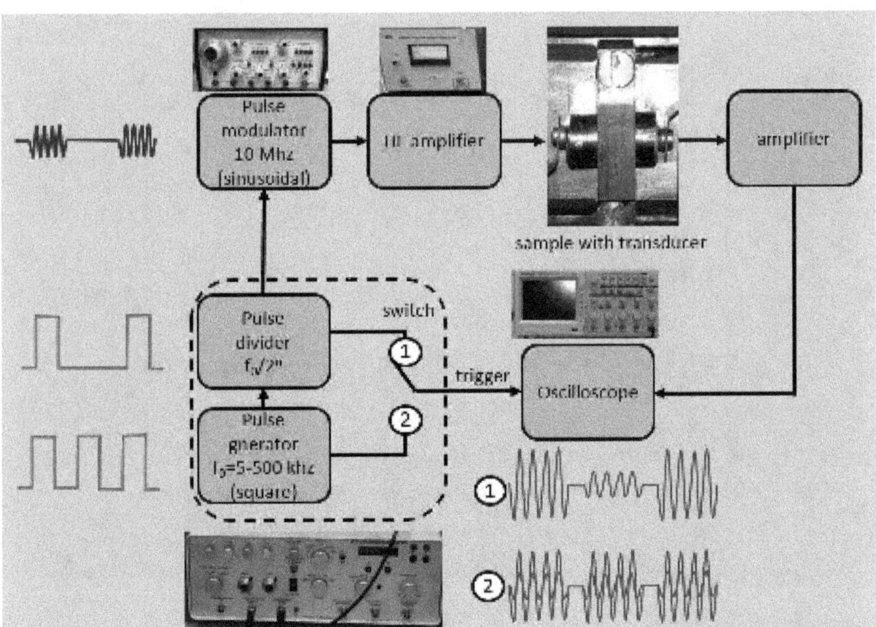

Figure 11. Schematic diagram of the electronic setup (Ref. 1).

second echoes, and this is repeated again and again. This is shown in Figure 12.

If we put $n = 1$, the first pulse arrives at T_2 after a time $\tau/2$, where $\tau = 2l/V$ is the roundtrip time of travel for sound. The echo arising from the reflected pulse arrives at T_1 at time τ . It gets reflected, and the echo arrives at T_2 at a time $3\tau/2$. When the pulse echo reaches T_1 again after a time 2τ, the modulator feeds a second pulse to T_1. If we look at T_2, it receives the first direct pulse at time $\tau/2$, the echo at time $3\tau/2$, and the second pulse at time $5\tau/2$. So, we see on the oscilloscope the big signals at $\tau/2$ and $5\tau/2$ and the smaller echo signal at $3\tau/2$. This is shown in the upper part of the figure on its left-hand side. If we put $n = 2$, then the second pulse is triggered after a time 3τ after the first signal. In this time, the echo makes two roundtrips between T_1 and T_2. So, we see at T_2 the first pulse arriving after a time $\tau/2$, two successive echo pulses arriving after a time $3\tau/2$ and $5\tau/2$, and the second pulse at $7\tau/2$. This is shown on the right-hand side of Figure 12. If the switch in Figure 12 is put in position 2, the oscilloscope trace also shows pulses at an interval of τ from the pulse modulator. We see that the echo pulses in Figure 12 exactly superpose on the successive pulses of the oscillator shown on the oscilloscope when the

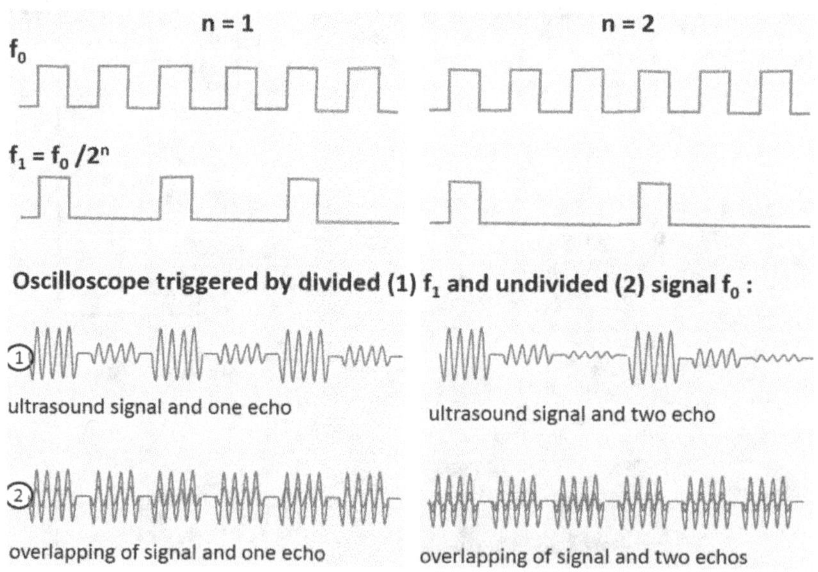

Figure 12. Output of transducer T_2 for $n = 1$ and $n = 2$ (from Ref. 1).

switch is in position 2. This superposition will not occur if the time between two pulses does not match the roundtrip travel time τ. Thus, the pulse superposition technique offers greater accuracy in the determination of the roundtrip travel time.

The second advantage is that it allows the measurement of the amplitude of successive echoes. The ratio of successive amplitudes should be $\exp(-\alpha \times 2l)$, where α is the attenuation coefficient of sound in the material. The attenuation of sound occurs because of an exchange of energy between the sound wave and the phonons, electrons, and magnetic moments of the atoms in the material. The attenuation will change at a phase transition. Thus, one obtains important additional information about the material by studying the attenuation.

6.2. *Resonance methods for measuring elastic constants*

Resonance methods for measuring elastic constants are reviewed in the book by Shreiber, Anderson, and Soge (Ref. 2).

In the resonance methods, one adjusts the frequency of an oscillator to coincide with one of the natural resonance frequencies of the material under certain boundary conditions. To understand this method, we consider the case of a rectangular bar of length l clamped at one end and free at the other. The equation of motion for transverse displacements perpendicular to the length of the bar is (from Ref. 3)

$$\rho A\partial^2 y/\partial t^2 + YAk^2\partial^4 y/\partial x^4 = 0 \tag{21}$$

The solution to this equation subject to the boundary conditions $y = 0$ and $\partial y/\partial x = 0$ at $x = 0$ is

$$y(x) = \{A[\cosh(\alpha x) - \cos(\alpha x)] + \mathrm{B}\,[\sinh(\alpha x) - \sin(\alpha x)]\}\exp(i\omega t)] \tag{22}$$

where

$$\alpha^4 = \rho\omega^2/Yk^2 \tag{23}$$

At the free end, $x = 1$, the boundary conditions are (i) $\partial^2 y/\partial x^2 = 0$ and (ii) $\partial^3 y/\partial x^3 = 0$, corresponding to the absence of bending moment and force. These two boundary conditions can only be satisfied for values of α satisfying the condition

$$1 + \cos(\alpha l)\cosh(\alpha l) = 0 \tag{24}$$

The values of αl satisfying this equation are at 1.875, 4.694,.... Therefore, the natural frequencies of vibration are

$$\omega_1 = [Yk^2/\rho]^{1/2}(1.875/l)^2 \quad (25a)$$

$$\omega_2 = [Yk^2/\rho]^{1/2}(4.694/l)^2 \quad (25b)$$

At the free end of the bar, one can attach a lightweight magnet and excite it with a coil to which an AC voltage is applied. Then, the amplitude reaches a maximum if the AC frequency matches the resonance frequency. For a rectangular bar, $k^2 = d^2/12$, where d is the thickness of the bar. Thus, knowing the density of the material of the bar, its length l, and its thickness d, the measurement of the resonant frequency enables one to get the value of the Young's modulus of the bar. Such a student experiment is described in Ref. 3.

This method is especially useful in the study of materials which are obtained in short lengths with small thicknesses and is called the "vibrating reed technique" for measuring elastic constants. One can refer to the Internet for various publications about the vibrating reed technique. We discuss in some detail one such setup described by Gamboa *et al.* (Ref. 4) shown in Figure 13.

The specimen was clamped between a flexible steel strip and a piezoelectric transducer mounted on the vertical pillar of an aluminum frame. A rubber band between the strip and the sample pressed the strip to the transducer to improve the contact. The sample was a 100 nm thick gold

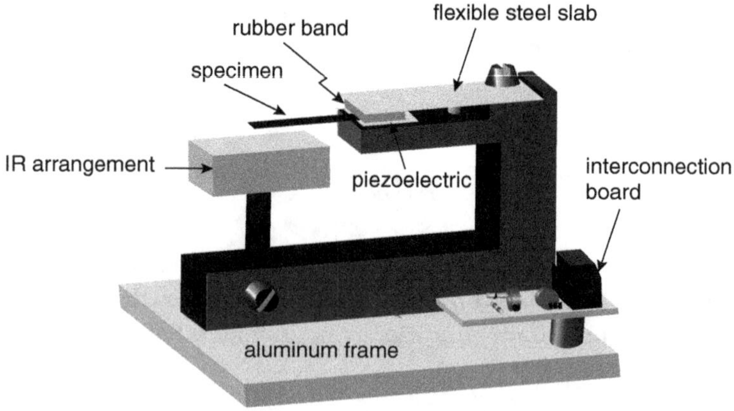

Figure 13. Vibrating reed setup by Gamboa *et al.* (redrawn from Ref. 4).

film deposited on a strip of polysulfone substrate, which was 5 mm wide, 30 mm long, and 130 μm thick. A length of 5 mm of the sample at one end was clamped. The piezoelectric transducer produced oscillations of the sample at the frequency of excitation. At the free end of the sample, a thin film of a few nm thickness of gold was deposited to a length of 5 mm to provide a good reflecting surface. Light from an IR laser was incident at an angle of 30° on this reflecting surface. The reflected light fell on a phototransistor. The intensity of the light beam produced a signal from the phototransistor, which peaked when the amplitude of the end of the substrate strip was at its maximum. The frequency of the transducer was varied in steps of 0.1 Hz to determine the frequency at which this peak in amplitude is seen. The frequency at peak amplitude is the resonant frequency of the strip.

First, the resonant frequency of the substrate strip without the 100 nm gold coating on the top face was measured repeatedly. The average of the resonant frequency thus obtained was 45.3 ± 0.2 Hz, which agreed with the value of 45.6 Hz calculated from the values of Young's modulus, the density of the substrate material, and its thickness specified by the supplier.

Then, the substrate was coated with 100 nm of gold, and the experiment was repeated. One observes a shift in the resonant frequency from the value for the gold-film-free substrate, as shown in Figure 14.

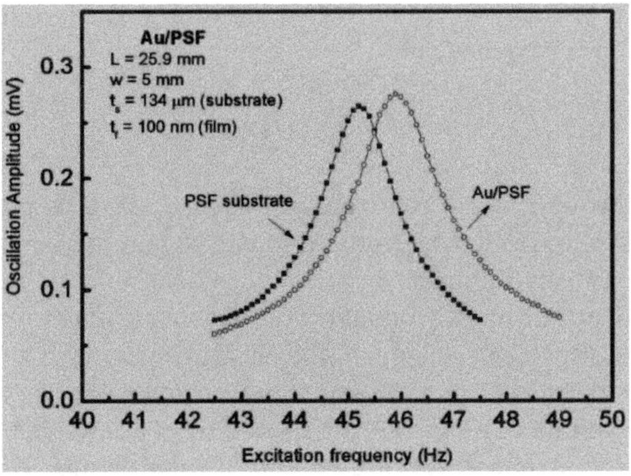

Figure 14. Shift in resonant frequency of the gold-coated substrate (reprinted with permission from Ref. 4).

This shift was 0.69 Hz. Using the theory of elastic vibrations of a bilayer material, the Young's modulus of gold was calculated as 62.7 ± 3.2 GPA, which is in good agreement with the value of Young's modulus for films of gold reported in the literature.

In vibrating reed experiments, often the amplitude of oscillation is measured using a capacitive technique. The reed with a metallic coating vibrates against a fixed electrode. The capacitance between the reed and the electrode varies. A DC voltage is applied across the reed and the electrode. Due to the change in capacitance, this produces a varying current in the circuit, which includes a resistance. The voltage across the resistance is detected with a lock-in amplifier.

A student experiment for measuring the shear modulus of the material of a wire using the resonant method is described in Ref. 5. Here, a wire is clamped taut at both ends. At its center, it carries a brass disk with two prongs along its diameter carrying lightweight magnets. I is the moment of inertia of the disk and the magnets about the axis of the wire. Using a pair of coils, the wire can be set into torsional oscillations. The equation of motion is

$$I\mathrm{d}^2\theta/\mathrm{d}t^2 + 2(\pi n r^4/(4(\ell/2)))\theta = 0 \tag{26}$$

Here, θ is the angle of twist of the brass disk, r is the radius of the wire, and ℓ is the length of the wire between two clamps. The torque exerted at the center of the wire by each half of the wire is $(\pi n r^4/(4(\ell/2)))$, and the factor of 2 takes care of the torque due to the two halves of the wire. The natural frequency is

$$\omega^2 = (\pi n r^4/\ell)/I \tag{27}$$

At this frequency of the current in the coils, the disk oscillates with maximum amplitude. Thus, a measurement of this frequency leads to a value for the rigidity modulus n.

The resonant ultrasound method of measuring elastic constants has now been developed to measure all the elastic constants of a material in a single-shot experiment by measuring the normal modes of a cubic or spherical crystalline particle varying in size from a few microns to a few centimeters and in mass varying from 100 micrograms to a kilogram. Maynard (Ref. 6) has given a brief review of this technique. In this technique, digital analysis of the data to extract all the elastic constants from the measured normal mode frequencies plays a very important role. This digital analysis was possible due to the development of computers of high speed and the development of complicated algorithms. The equations for displacement are given by

$$\rho\partial^2\psi_i/\partial t^2 = \Sigma_j\, \partial\sigma_{ij}/\partial x_j \tag{28}$$

Here, ψ_i ($i = 1$ to 3) are the Cartesian components of the displacement, and the summation over j goes from 1 to 3. The strain ε_{ij}, in terms of ψ_i, is given by

$$\varepsilon_{ij} = (1/2)\, [\partial\psi_i/\partial x_j + \partial\psi_j/\partial x_i) \tag{29}$$

Hooke's law (Equation (8a)) relates the stress and strain components.

The boundary condition is that all faces of the sample are stress-free. This gives rise to the condition

$$\Sigma_j\, \sigma_{ij} n_j = 0 \tag{30}$$

Here, n_j is the jth component of the normal to a bounding face of the sample, and Equation (30) must hold over all bounding faces.

Solutions to Equation (28), subject to the boundary conditions (30), give rise to several normal modes with different frequencies. The patterns of displacement in some of the normal modes for a rectangular parallelopiped are shown in Figure 15, taken from Maynard (Ref. 6). Each of these modes will have a different frequency. Experimentally, one measures the frequencies of several of these normal modes by varying the oscillator frequency in a single experiment. Using data analysis algorithms, we try to find the values of all the elastic constants which give a good fit to the experimentally measured normal mode frequencies. This inverse problem is difficult for a specimen of arbitrary shape. However, for specimens of rectangular or cubic shape, the inverse problem can be solved. Since the number of measured frequencies can be greater than the number of elastic constants, one starts with a guestimate of the values of the elastic constants and then refines the values by the least squares method to give the best fit to the measured frequencies.

The sample must be excited by one piezotransducer and the amplitude measured by another so that the resonant frequencies can be determined by continuously varying the frequency of the driver oscillator. The transducer must be attached so that the loading of the sample is

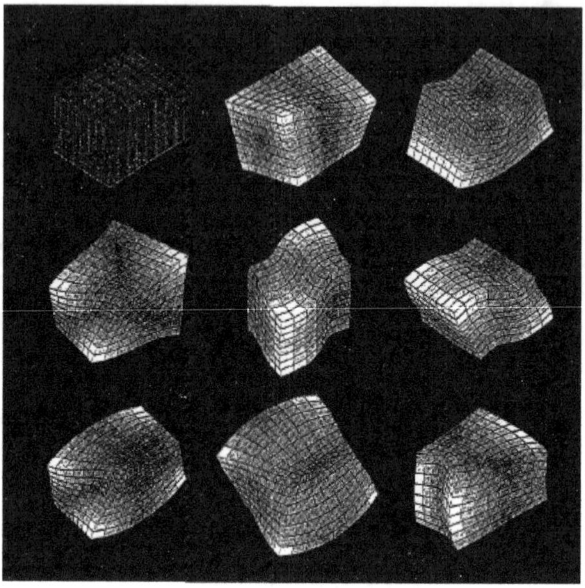

Figure 15. Pattern of displacement of some normal modes for a rectangular specimen (Reproduced with permission from J. Maynard; Resonant Ultrasound Spectroscopy. *Physics Today* 1 January 1996; 49(1): 26–31, with the permission of the American Institute of Physics).

minimized, and the stress-free condition is not violated. The sample is held lightly at two corners by the transmitter and receiver transducers. The corner is never a node for any standing wave normal mode. In the pulse method of measuring the velocity of sound, one needs a tight coupling between the transducers and the sample to facilitate maximum transfer of energy. In a resonant method, such tight coupling is unnecessary. At resonance, even a weak force will produce a large amplitude. The transducers are PVDF strips of 500 μm width. They are metallized partially on both sides so that the metal coatings overlap over a small region in the center, forming a capacitance. The sample is held lightly by the corners between the two PVDF films. The top PVDF film is the transmitter. The electrical leads from the metal coatings on the two sides are connected to an oscillator. The bottom PVDF film is the receiver. This set up is shown in Figure 16.

For the data analysis to converge, the sample frequencies must be correctly identified with the corresponding normal modes of the sample.

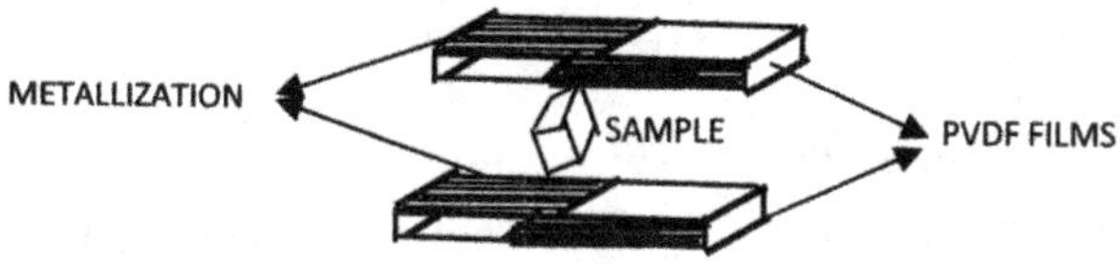

Figure 16. **Schematic diagram of PVDF films and sample.**

If we know the approximate values of the elastic constants of the sample from theoretical estimates, then the approximate normal mode frequencies of the sample can be calculated, and this helps in identifying the sample frequencies with the correct normal modes. Otherwise, the sample dimensions may be varied, and one measures the changes in the sample frequencies. There are other techniques for identifying the normal modes. If one wants to measure the elastic constants at high temperatures, one uses alumina buffer rods. The sample is held lightly between two alumina buffer rods in a high-temperature furnace. The transducers are bonded to the alumina rods at the other ends, which are at room temperature.

The resonant ultrasonic method has been invaluable in the study of the elastic constants of minerals at high temperatures, which is important in the study of geology. Phase transitions at high temperatures can be studied using changes in elastic constants. For further study, the references in Ref. 6 may be consulted.

7. Conclusion

In this chapter, we have given a description of the most commonly used methods for measuring the elastic constants of materials. There are also other methods, such as Brillouin scattering and diffuse X-ray scattering, but these are only of academic interest. It is possible to get third- and fourth-order elastic constants from the measurement of the velocity of sound in crystals subjected to high pressure. These will describe the nonlinear region of stress–strain curves. Knowledge about elastic constants will enable us to calculate the Debye temperature of a material and to predict roughly the behavior of the specific heat of that material as a function of temperature. Knowledge about the elastic behavior of materials is also of interest to geologists.

References

1. Pulse overlap method for measuring ultrasonic velocity in Fe0.7Al0.3 alloy Ultrasound-RWTH Aachen, https://institut-2A.physik.rwth-achen.de>praktikum.
2. Schreiber, E., Anderson, O. L., and Soga, N. (1974). *Elastic Constants and Their Measurements*. McGraw-Hill Book Co.
3. Srinivasan, R., Priolkar, K. R., and Ramesh, T. G. (2018). *A Manual on Experimental Physics*. Indian Academy of Sciences, p. 31.
4. Gamboa, F. *et al.* (2016). A simple vibrating reed apparatus for determination of thin film elastic modulus. In *1st International Congress in Instrumentation and Applied Sciences*. https://www.researchgate.net/publication/265975755.
5. Srinivasan, R., Priolkar, K. R., and Ramesh, T. G. (2018). *A Manual on Experimental Physics*. Indian Academy of Sciences, p. 27.
6. Maynard, J. (1996). Resonant ultrasound spectroscopy. *Phys. Today* **49**, 26.

Part III.1

Thermal Properties

Chapter 10

SPECIFIC HEAT

1. Introduction

The specific heat of a material is the amount of heat required to raise the temperature of unit mass of the material by one degree Kelvin. This is an important property of the material. It depends on the temperature. The specific heat of diamond as a function of temperature (taken from the NSM archive) is shown in Figure 1.

There are two noteworthy features in Figure 1:

1. There are two curves marked C_V and C_p. C_V is the specific heat at constant volume, and C_p is the specific heat at constant pressure. The difference between C_p and C_V is given by

$$C_p - C_V = TV\beta^2/\chi \tag{1}$$

   Here, T is the absolute temperature, V the specific volume, β the volume expansion coefficient, and χ the isothermal compressibility of the material. Since all the quantities are positive on the right-hand side of the equation, C_p for any material is **always** larger than C_V. In solids, however, the difference is small because the volume expansion coefficient of a solid is small. In an experiment, we measure C_p. However, from a theoretical point of view, C_V is important.
2. We see that the specific heat of a solid decreases to zero as one lowers the temperature. According to the Debye theory of specific heat of solids,

$$C_V = 3Nk\, D(\theta_D/T) \tag{2}$$

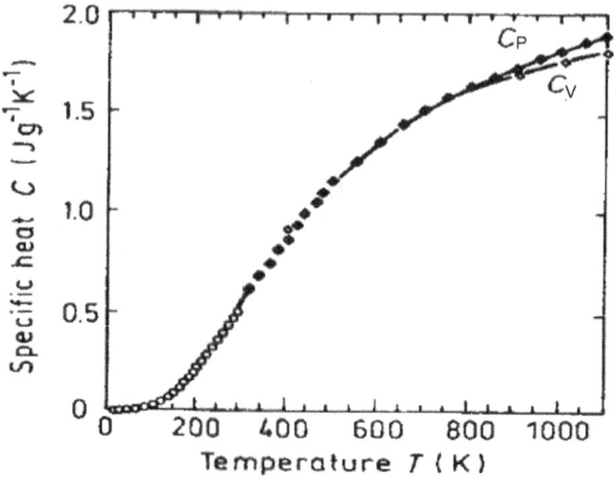

Figure 1. Temperature dependence of specific heat of diamond.

Here, N is the number of atoms in unit mass, k is the Boltzmann constant, and

$$D(X_D) = (3/X_{D^3}) \int_0^{X_D} [x^4 e^x/(e^x - 1)^2] dx \tag{3}$$

Here,

$$X_D = \theta_D/T \tag{4}$$

Thus, if C_V is plotted against T/θ_D, the specific heat of all solids should follow a universal curve. θ_D is called the Debye temperature and is characteristic of the material.

In this theory, the specific heat per atom (i.e., C_V/N) should tend to the value $3k$ when $T >> \theta_D$. When $T << \theta_D$, the specific heat of all solids should vary as T^3. If we have two materials A and B, the specific heat per atom of material A will be lower than the specific heat per atom of material B at any temperature T **if the Debye temperature of A is higher than the Debye temperature of B.**

The Debye theory is based on certain simple assumptions about the phonons, i.e., vibrations, of atoms in a solid. For an actual solid, the experimental data will not follow the Debye formula with a constant

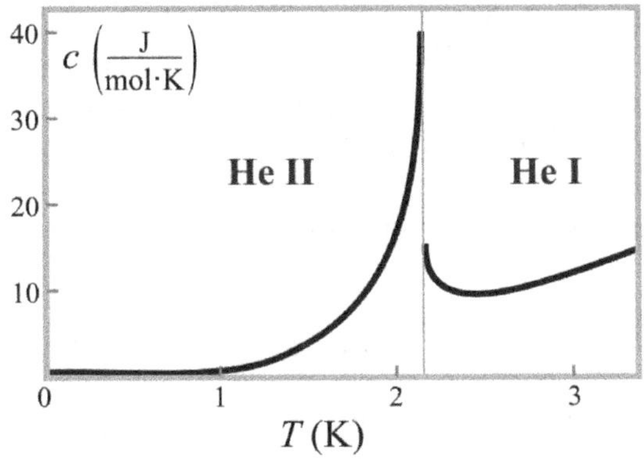

Figure 2. Specific heat behavior at the λ transition in liquid helium.

Debye temperature. One may fit the measured specific heat at each temperature to the expressions (2) and (3) and get a temperature-dependent Debye temperature for the solid. A plot of $\theta_D(T)$ vs. T will then be an equivalent representation of the variation of specific heat with temperature.

In a metal, the free electrons also contribute to the specific heat. This contribution varies as

$$C_{\text{Vel}} = \gamma T \tag{5}$$

Here, γ is a constant characteristic of a given metal. This contribution is very small compared to the phonon contribution at high temperatures. At very low temperatures, the specific heat of a metal varies as

$$C_{\text{V}} = \alpha T^3 + \gamma T \tag{6}$$

The T^3 term is due to the contribution of phonons, and the T term is due to that of free electrons.

In a second-order phase transition, the specific heat shows a discontinuity. This is seen in the transition from liquid helium I to liquid helium II at 2.37 K, as shown in Figure 2.

2. Procedure for Measuring Specific Heat

A good review of the techniques for measuring the specific heat of solids is given in Ref. 1. A more recent review is found in Ref. 2.

A calorimeter is a container which holds the substance whose specific heat is to be measured. The calorimeter contains a heater of resistance R. The calorimeter has a mass m and specific heat c. The material of mass M is held in the calorimeter. The calorimeter is surrounded by an enclosure at a temperature of T_0. A quantity of heat Q, in Joules, is given to the calorimeter with the material by passing a current I through the resistance R for a time τ:

$$Q = I^2 R\tau \tag{7}$$

The temperature, T, of the calorimeter and its contents will rise as a function of time t. If all the heat supplied goes only to raise the temperature of the calorimeter and its contents, then the final maximum temperature T_{final} reached by the calorimeter and its contents is given by

$$Q = (mc + MC)\,(T_{\text{final}} - T_{\text{init}}) \tag{8}$$

In Equation (8), m is the mass of the calorimeter with the heater and temperature sensor, mc is the thermal capacity of the empty calorimeter, C is the specific heat of the material, and T_{init} is the initial temperature of the calorimeter and its contents. From Equations (7) and (8), one can find $(mc + MC)$. A run with an empty calorimeter will give mc. Subtracting the value of mc, derived from the empty run, from the value of $(mc + MC)$ obtained from the run with the material, one may find C.

This is the principle of the method. Note that the material of the calorimeter must be chosen so that mc is a small fraction of $(mc + MC)$ to get an accurate value of the specific heat C of the material. So, the calorimeter must have a low mass m, and its specific heat c must be small. Since the specimen is inside the calorimeter and we measure the temperature of the calorimeter, the material of the calorimeter must have high thermal conductivity so that temperature gradients in the calorimeter are minimized. Sapphire and diamond are suitable materials for making the thin plate of the calorimeter. Both of these materials have a low density, a low specific heat, and a large thermal conductivity. A thin film of nichrome can be deposited as the heater on these plates.

The temperature can be conveniently measured by a suitable thermocouple or a resistance thermometer. These thermometers have a very low mass and are available commercially.

Heat will be lost by the calorimeter and sample to the environment through conduction, convection, and radiation. By enclosing the calorimeter in a vacuum, we may reduce the loss of heat due to convection. But heat will be lost through (a) radiation from the calorimeter and its contents to the environment and (b) conduction in the leads of the temperature sensor and heater. If the difference between $T(t)$ and T_0 is small compared to T_0, we may write this heat loss as

$$P_{\text{Loss}} = \alpha\theta(t) \tag{9}$$

Here,

$$\theta(t) = T(t) - T_0 \tag{10}$$

If we want to make this heat loss exactly zero, then we should make the temperature of the environment, T_0, equal to $T(t)$ at all times. This means that the environment must also be heated, and we must ensure that the environment temperature follows the rise in the sample temperature. If we do so, then the process of measuring specific heat is called **adiabatic calorimetry**. For adiabatic calorimetry, we have to enclose the calorimeter in a shield over which another heater is wound and adjust the current supplied to this heater so that the temperature of the shield faithfully follows the temperature of the calorimeter and its contents. This is difficult to achieve experimentally. So, we adopt a quasi-adiabatic method.

We can maintain the temperature T_0 constant such that $T(t) - T_0$ is only a few degrees at all time t. Then, some heat will be lost to the surroundings, and we will have to correct our temperature readings to take care of this. In this situation, taking T_{init} to be T_0, the rate of change of temperature $T(t)$ will be given by the equation

$$(mc + MC)\, \mathrm{d}\theta/\mathrm{d}t + \alpha\theta = I^2R \ (0 < t < \tau) \tag{11a}$$

$$(mc + MC)\, \mathrm{d}\theta/\mathrm{d}t + \alpha\theta = 0 \ (t > \tau) \tag{11b}$$

The solutions to these two equations are

$$\theta(t) = (I^2R/\alpha)\,(1 - \exp(-\beta t))\ (0 < t < \tau) \tag{12a}$$

and

$$\theta(t) = \theta(\tau)\exp(-\beta(t - \tau))\ (t > \tau) \tag{12b}$$

In the above, $\theta(0)$ is zero and

$$\beta = \alpha/(mc + MC) \tag{13}$$

In quasi-adiabatic calorimetry, we make β very small, i.e., $\tau_{Rel} = 1/\beta$ very large, and **heat the specimen for a time $\boldsymbol{\tau}$ that is smaller than $\boldsymbol{\tau_{Rel}}$**. If βt is small compared to 1, we may write

$$\exp(-\beta t) = 1 - \beta t + (\beta t)^2/2$$

and

$$\theta(\tau) + \Delta/2 = I^2R\tau/\ (mc + MC) \tag{14}$$

Here,

$$\Delta = \theta(\tau)\ \beta\tau = \theta(\tau) - \theta(2\tau) \tag{15}$$

After switching off the heater at time τ, we continue the measurement of temperature $T(t)$ to a time greater than 2τ. We plot the time–temperature graph, as shown in Figure 3.

In this graph, $I^2R = 0.2$ W, $(mc + MC) = 1$ J/K, $\tau = 10$ s, and $\beta = 0.005$ s^{-1}. So, $\tau_{Rel} = 200$ s. If there is no heat loss (i.e., $\beta = 0$), the final temperature reached when the heating current is switched off at time τ will be two degrees above the initial temperature (i.e., θ_{final} will be 2). With the heat loss, it is only 1.950°. After the heating current is switched off, the temperature falls from 1.950° at τ to 1.856° at 2τ. The difference Δ is 0.094°. $\Delta/2$ added to 1.950° gives a temperature of 1.997°, which differs from the

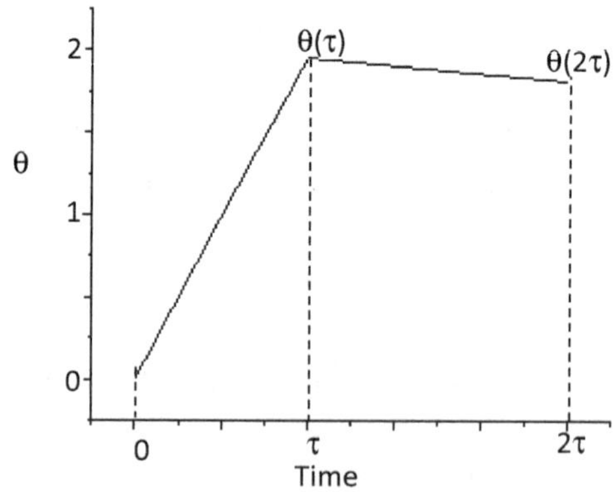

Figure 3. Temperature vs. time in quasi-adiabatic calorimetry.

value of 2° by 0.003°, i.e., by ≈0.1%. Thus, this is a good method of correcting for the heat loss.

3. Schematic Diagram of a Quasi-Adiabatic Calorimeter

A schematic diagram for such a calorimeter is shown in Figure 4. A denotes a stainless-steel enclosure, which can be evacuated through tube T_1 attached to a flange F_1. The vacuum seal at the flange must be made of a material suitable for the temperature of operation. Inside A hangs a tube T_2 with a hole. At its top end, the tube is brazed to flange F_1. At its bottom end, it has a copper flange F_2. From this flange hangs a thin-walled copper cup B. This is the shield of the calorimeter. On this, a heater is wound to raise the temperature of the shield to whatever value one wants. From the bottom of flange F_2 hangs a thin plate of sapphire C. On the bottom face of the sapphire plate, a nichrome heater is deposited as a thin film. A suitable resistance thermometer is also fixed. The leads of the nichrome heater and the resistance thermometer come out through the hole in T_2.

Along with the leads of the heater and thermometer on the shield B, the leads then come out through the electric feed through FT, which is sealed to the flange F_1 to make it vacuum tight. The sample S is in the form of granules or a pellet and is pressed with a thin layer of silicone grease to the top surface of the sapphire plate C.

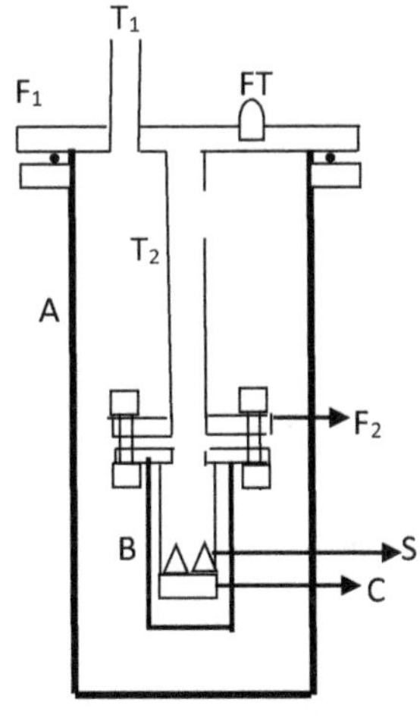

Figure 4. Schematic diagram of a calorimeter.

First, the vacuum space is evacuated and filled with helium gas to a pressure of about 1 mbar. The heater on shield B is switched on so that the shield reaches the desired temperature T_0. Through convection through the helium gas, the sample and the sapphire plate will reach the temperature T_0 of the shield after waiting for a suitable time. The helium gas is then pumped out to reach a vacuum of about 10^{-6} mbar. A pulse of current I is passed through the nichrome heater on the sapphire plate for a time τ. The temperature of the sample will rise. The temperature values are collected using a computer at equal intervals of time for a time of about 3τ. From a plot of temperature vs. time, the final corrected temperature T_{final} is estimated using Equation (15). Knowing $I^2R\tau$ and the corrected value of $(T_{final} - T_{init})$, one may get the value of $(mc + MC)$. A similar run without the sample gives the value of mc. By subtracting this from the value of $(mc + MC)$ determined above and knowing the mass M of the sample, the specific heat, C, of the sample can be found.

For measurements at 4.2 K and above, this arrangement can be suspended in a bath of liquid helium. The specific heat of $YBa_2Cu_3O_7$ (YBCO) was measured from 4.2 to 60 K by Shankar *et al.* (Ref. 3) using such a calorimeter. Two samples were used, weighing 1.033 and 0.6643 g. A germanium resistance thermometer was used to measure temperature. The specific heat values for the two samples agreed to within 5% (Figure 5).

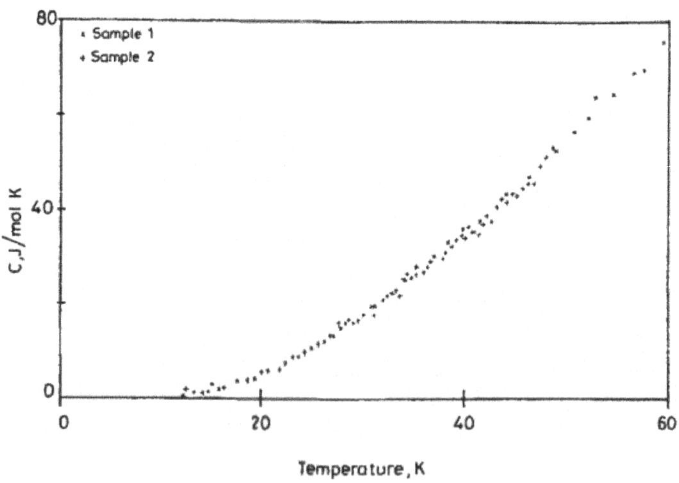

Figure 5. Variation of specific heat of two samples of YBCO with their temperatures in the range 4.3–60 K (reprinted with permission from Ref. 3).

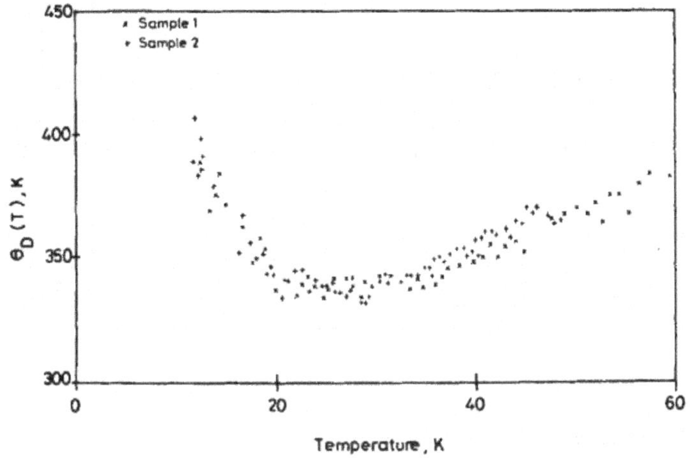

Figure 6. Variation of Debye temperature with the temperature of YBCO samples (reprinted with permission from Ref. 3).

The Debye temperatures were derived from the measured specific heat. The variation of Debye temperature, θ_D, with the temperature of the YBCO sample is shown in Figure 6.

4. Relaxation Method

In quasi-adiabatic calorimetry, we make the relaxation time τ_{Rel} as large as possible by reducing the heat loss through the leads and through radiation. This takes a lot of effort. **In the relaxation method, the heat loss and, hence, the value of β can be 10–100 times larger than in quasi-adiabatic calorimetry.**

One can employ the relaxation method in two ways. Let β be a large value, i.e., τ_{Rel} is small. Then, heat the sample by supplying power input P for a time $\tau > 8\tau_{Rel}$. Then, the specimen reaches a constant final value of temperature given by

$$T_{final} - T_{init} = P/\alpha \tag{16}$$

Then, switch off the power, and take temperature readings for a further time interval τ', which is a few times τ_{Rel}. This is called the 1-τ_{Rel}

method. The time–temperature behavior for such a situation is shown in Figure 7, taking $P = 1$ W, $mc + MC = 1$ J/K, $\beta = 0.2$ s^{-1} ($\tau_{\text{Rel}} = 5$ s), and $\tau = 40$ s.

We see that at the end of the heating period, τ, the sample reaches the steady-state temperature value given by Equation (16). For the condition where the power is switched off, we fit the curve to the equation

$$\theta(t) = \theta_{\text{Final}} \exp(-\beta(t - \tau))$$

and get the value of β. Then,

$$mc + MC = P/(\beta\theta_{\text{Final}}) \tag{17}$$

Thus, the specific heat is found.

In a modification of this method, called the 2-τ_{Rel} method, the heating period τ is taken to be a fraction of τ_{Rel}. The variation of time with temperature for $P = 1$ W, $(mc + MC) = 1$ J/K, $\beta = 0.1$ s^{-1} ($\tau_{\text{Rel}} = 10$ s), and $\tau = 5$ s is shown in Figure 8.

θ_{MAX}, the maximum temperature reached in 2-τ_{Rel} method, is much less than θ_{Final}, the steady-state temperature, which would have been reached if the power had been on for more than $8\tau_{\text{Rel}}$. In this case, $\theta_{\text{Final}} = (P/((mc + MC)\beta) = 10°$, while θ_{MAX} is only 3.93°.

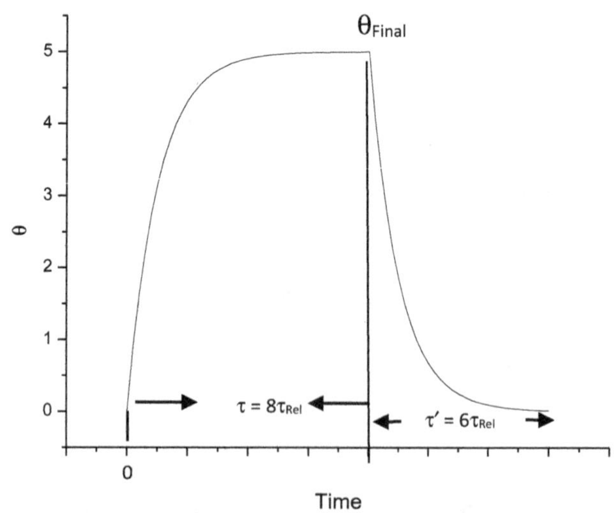

Figure 7. 1-τ_{Rel} method: time–temperature graph.

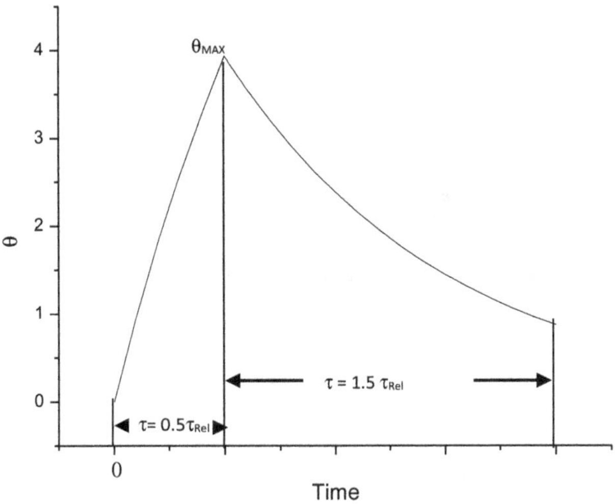

Figure 8. 2-τ_{Rel} method: time–temperature graph.

θ_{MAX} is related to θ_{Final} by

$$\theta_{\text{MAX}} = \theta_{\text{Final}}\,(1 - \exp(-\beta\tau)) \tag{18}$$

An analysis of the fall in temperature when the power is switched off gives the value of β. So, by measuring θ_{MAX} and β, one may find θ_{Final} and hence the value of $(mc + MC)$. Codes are available for carrying out this procedure.

Newsome Jr. and Andrei (Ref. 4) describe a 2-τ_{Rel} calorimeter for measuring the specific heat of polymer films at low temperatures. To test how well the calorimeter works, they measured the specific heat of a copper sheet weighing only 22.8 mg in the temperature range 3.5–8.5 K. They applied a succession of pulses of current and chose the time τ of the heating pulse to be either τ_{Rel} or $\tau_{\text{Rel}}/2$. What they denote as ΔW in their paper corresponds to our τ, and what they denote as τ in their paper corresponds to our τ_{Rel}. They analyzed two sets of data, one when the starting time t_0 of the pulse was 6 s and $\tau = \tau_{\text{Rel}}$ (in our notation), and the second when t_0 was 1 s and $\tau = \tau_{\text{Rel}}/2$ in our notation. Their results for the specific heat of copper are shown in Figure 9.

The agreement of their results with the measurements by Holste *et al.* is also shown in the figure.

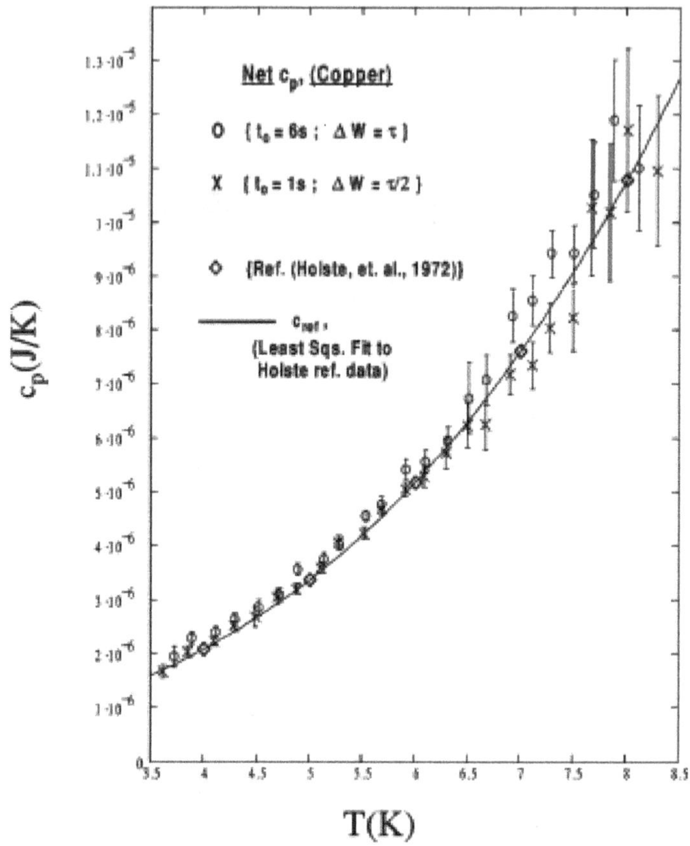

Figure 9. Specific heat of copper in the range 3.5–8.5 K determined by the 2-τ_{Rel} method (reprinted with permission from Ref. 4).

The advantages of the 2-τ_{Rel} method are the shorter time it takes for measurement and the smaller weight of the samples that can be used.

5. AC Calorimetry

In this method, we employ a heater. The resistance $R(T)$ of the heater at a temperature T, around a base temperature T_0, is given by

$$R(T) = R(T_0)\,[1 + \varepsilon(T - T_0)] \tag{19}$$

Here, ε is the temperature coefficient of resistance of the heater.

An AC current

$$I = I_0 \sin(\omega t) \tag{20}$$

is passed through the heater. Then, the power generated is

$$P(t) = I^2(t)\, R(T(t)) = P_0 + P_1\cos(\omega t) - P_2 \cos(2\omega t) + P_3\cos(3\omega t) \tag{21}$$

The steady heat input P_0 and the second harmonic P_2 arise from the fact that $\sin^2\omega t = [1 - \cos(2\omega t)]/2$. P_1 and P_3 arise from the temperature coefficient of resistance. If $\varepsilon = 0$, the components P_1 and P_3 of power will be zero.

The temperature will correspondingly have the components

$$T(t) = T_{\text{Steady}} + T_1 \cos(\omega t - \varphi_1) + T_2 \cos(2\omega t - \varphi_2) + T_3\sin(3\omega t - \varphi_3) + \cdots \tag{22}$$

$$T_{\text{Steady}} = [I_0{}^2R(T_0)/2]/\ [(mc + MC)\,\beta] \tag{23}$$

$$T_2 = [I_0^2R(T_0)/2]/\{[(mc + MC)\,\beta]\ [1 + 4\omega^2\tau_{\text{Rel}}{}^2]^{1/2}\} \tag{24}$$

$$\varphi_2 = 2\omega\tau_{\text{Rel}} \tag{25}$$

In the above,

$$\tau_{\text{Rel}} = 1/\beta \tag{26}$$

The ω and 3ω components in the temperature of the sample arise due to the temperature variation in the resistance of the heater. These will not be discussed here.

Obviously, the amplitude T_2 is less than the value of T_{Steady} by the factor $1/[1 + 4\omega^2\tau^2_{\text{Rel}}]$. T_1 and T_3 are even smaller than T_2.

However, we can pick up the amplitude of T_2 in the presence of noise using a lock-in amplifier. The lock-in amplifier is capable of measuring AC signals of a few hundred nanovolts in the presence of noise.

If we chose the frequency such that $\omega\tau_{\text{Rel}} >> 1$, then from Equations (24) and (26),

$$T_2 = [I_0^2R(T_0)/4\omega]/\ [(mc + MC)] \tag{27}$$

We see that the amplitude of the second harmonic temperature excursion does not depend on β and varies as $1/\omega$. Whatever the magnitude of the heat loss coefficient, the amplitude of the second harmonic coefficient will be the same. This is the advantage of AC calorimetry.

The frequency chosen must be one for which $4\omega^2\tau_{\text{Rel}}^2 >> 1$. At the same time, the amplitude of the second harmonic signal is inversely proportional to the frequency ω. So, the frequency of heating must have a value consistent with the condition $4\omega^2\tau_{\text{Rel}}^2 = 100$ to get an accuracy of 1% of the value of T_2. If the relaxation time is 1 s, then the frequency of the heating current should be greater than 1 Hz.

As in all calorimetry experiments, we must ensure very good thermal contact between the temperature sensor and the sample. If the sample is a metal, we may pass the AC current directly through the sample so that the heat is generated within the sample. If the heater is external to the sample, then we will encounter a more complicated theory, taking into account the thermal conductance between the sample and the heater.

Thiruvikraman (Ref. 5) used this technique for the measurement of the specific heat of nickel. The sample was in the form of a thin wire, 0.14 mm in diameter. A chromel–alumel thermocouple made of 0.01 mm thin wire was spot-welded at the center of the sample. Square pulses of constant current at a frequency of 3 Hz or more were sent through the specimen. The current was on for half the period and off for the other half. Such a square pulse has no second harmonic of the current. The sample was mounted on a furnace, which could be heated. The temperature of the furnace could be controlled to within 0.1°C. The amplitude of the current pulse was adjusted to keep the power dissipation constant throughout the experiment. The amplitude of the second harmonic temperature fluctuation was measured with a lock-in amplifier. Using the four-probe method, the sample resistance was measured to calculate the power.

Figure 10 shows that the amplitude of the second harmonic varies linearly with $1/\omega$, where ω is the angular frequency of the current pulse.

This behavior shows that at a frequency above 3 Hz, the condition $4\omega^2\tau^2 >> 1$ is well satisfied.

Figure 11 shows the variation of the specific heat of nickel as a function of temperature. Nickel undergoes a ferromagnetic-to-paramagnetic transition at a temperature of 354°C. We see the specific heat jump near the transition temperature.

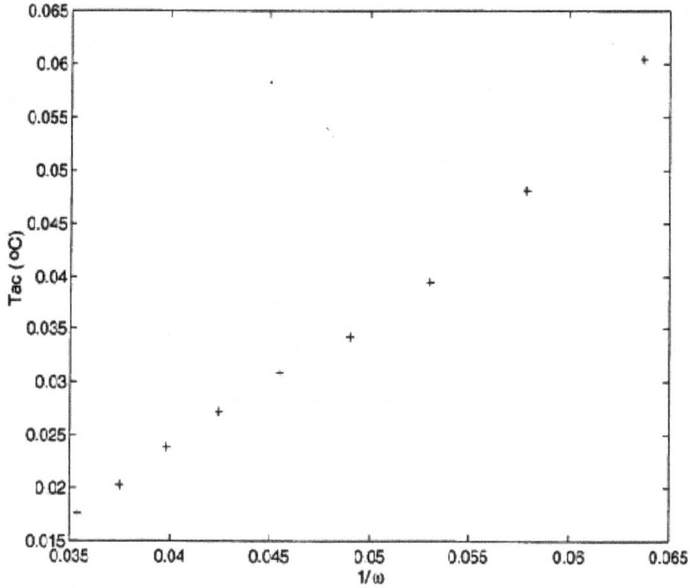

Figure 10. Variation of the second harmonic temperature amplitude as a function of the frequency of the current pulse (reprinted with permission from Ref. 5).

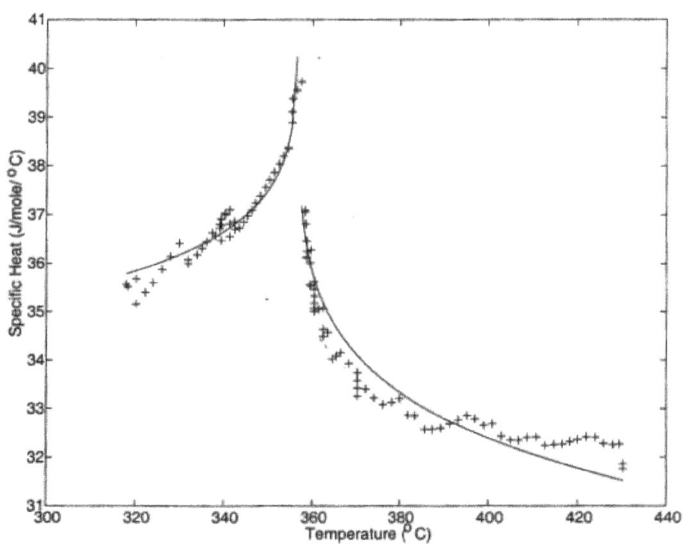

Figure 11. Variation of specific heat of Nickel as a function of temperature (reprinted with permission from Ref. 5).

The 2ω-AC calorimetric technique is useful for measuring the specific heat of materials in the form of thin metallic films. It has also been used with polymer films, but with an external heater.

6. Conclusion

We have given a brief introduction to the more commonly used techniques for the measurement of specific heat. The literature on specific heat is extensive. For further reading, we list a few references in the following.

References

1. Lakshmikumar, S. T. and Rajagopal, E. S. (1981). Heat capacity measurements — Progress in experimental techniques. *J. Indian Inst. Sci.* **63A**, 277–329.
2. Ventura, G. and Risegari, L. (2008). *The Art of Cryogenics*. Elsevier, Part V: Chapter 12, pp. 267–287.
3. Sankar, N., Sankaranarayanan, V., Srinivasan, R., Rangarajan, G. and Subba Rao, G. V. (1988). *Pramana J. Phys.* **30**, 199.
4. Newsome, Jr., R. W. and Andrei, E. Y. (2004). *Rev. Sci. Instrum.* **75**, 104.
5. Thiruvikraman, P. K. *Study of Electronic Phase Transitions at High Pressures*. Ph.D Thesis, Raman Research Institute, http://hdl.handle.net/2289/3539.

Chapter 11

THERMAL EXPANSION OF SOLIDS

1. Introduction

When a solid material is heated, its linear dimensions change. If $\ell(T_0)$ is the length of the material at temperature T_0, its length at a temperature T, slightly different from T_0, is given by

$$\ell(T) = \ell(T_0)\,[1 + \alpha(T - T_0)] \tag{1}$$

The coefficient $\alpha = (1/\ell(T_0))\, \mathrm{d}\ell/\mathrm{d}T$ is called the coefficient of linear expansion of the material. It is generally positive. In an isotropic material or a cubic crystal, α is independent of the direction of measurement. In an anisotropic material, the linear expansion coefficient, α, will depend on the direction of measurement. In such materials, a plot of α as a function of direction in three-dimensional space will be an ellipsoid with three principal axes, labeled 1, 2, and 3. The linear expansion coefficients along the three axes, called the principal expansion coefficients, are labeled α_1, α_2, and α_3, respectively. If the material has axes of rotational symmetry, the principal axes of the ellipsoid of the expansion coefficient will lie along these axes of symmetry. For an isotropic material or a cubic crystal, α_1, α_2, and, α_3 will be equal.

The change in the linear dimensions will be accompanied by a change in the volume V of the material. The volume expansion coefficient

$$\beta = (1/V)\,(\mathrm{d}V/\mathrm{d}T) = \alpha_1 + \alpha_2 + \alpha_3 \tag{2}$$

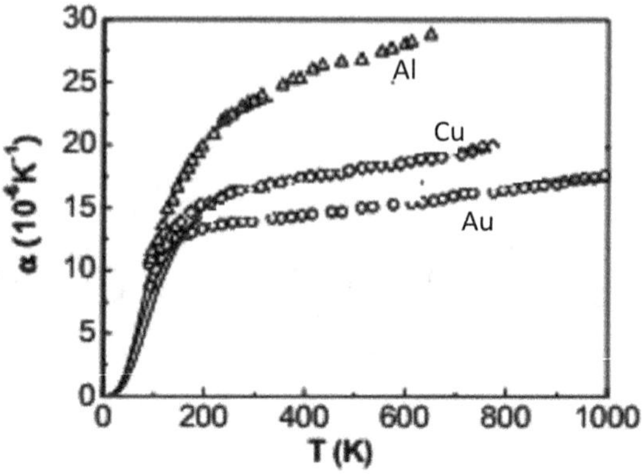

Figure 1. Temperature variation of the linear expansion coefficient α for aluminum, copper, and gold.

The linear expansion coefficients α_j and the volume expansion coefficient β vary with temperature. The temperature variation of the linear expansion coefficient for some metals is shown in Figure 1.

The linear expansion coefficient decreases as the temperature decreases and tends to zero as the temperature tends to zero. Its curve is similar to that of the temperature variation of specific heat.

Thermal expansion arises from the anharmonic nature of the interatomic vibrations in the solid material. A consequence of this anharmonic nature is the dependence of the frequency of the vibration, ω, on the volume V of the material. We may define a parameter γ as follows:

$$\gamma = -\mathrm{d}(\log(\omega))/\mathrm{d}(\log(V)) \tag{3}$$

which is called the Gruneisen parameter, named after Eduard Gruneisen. We may then relate the volume expansion coefficient to the contribution of the specific heat due to lattice vibrations by the equation

$$\beta V = \chi \sum_{j=1}^{3N} \gamma_j c_V(\omega_j) \tag{4}$$

In this equation, χ is the isothermal compressibility of the material, N is the number of atoms in volume V, $c_V(\omega_j)$ is the contribution to the specific heat at constant volume V of the material due to the vibration of frequency ω_j, and γ_j is the Gruneisen parameter of mode j. If all the $3N$ lattice modes in the solid volume V have the same Gruneisen parameter, γ, then β will be proportional to C_V given by

$$C_V = \sum_{j=1}^{3N} C_V(\omega_j) \tag{5}$$

C_V is the total specific heat of the material of volume V. This accounts for the similarity in behavior of β and C_V as the temperature is varied. At high temperatures, C_V becomes independent of temperature based on Debye's theory. So, β should also become constant. At low temperatures, C_V varies as T^3 for an insulator and as $\gamma_{el}T + \delta T^3$ for a metal. The volume expansion coefficient must also show a similar dependence on temperature. The theory of thermal expansion and some of the techniques used to measure thermal expansion are discussed in Ref. 1.

The value of α for most materials is of the order of 10^{-5}/K at room temperature, and it decreases as the temperature falls. So, the fractional change in the linear dimensions of the specimen for a change in temperature of 10 K will be of the order of 10^{-3} or less.

2. Procedure for the Measurement of α

We take a specimen of the material of length ℓ at the initial temperature T_{init} and measure the change in length, $\Delta\ell$, as the temperature is increased. We draw the graph of the length of the material as a function of temperature and determine the slope at a temperature T. From the slope, we can calculate the linear expansion coefficient.

We describe the following four methods for determining α over a wide temperature change, which use (1) the interference of light, (2) the change in capacitance when the distance between two plates is changed, (3) the linear voltage differential transformer, and (4) X-ray diffraction.

3. Interferometric Method

There are many variations of the interferometer technique used for the measurement of the thermal expansion of materials. Here, we describe a simple setup that can be built easily in the laboratory. This setup is shown in Figure 2.

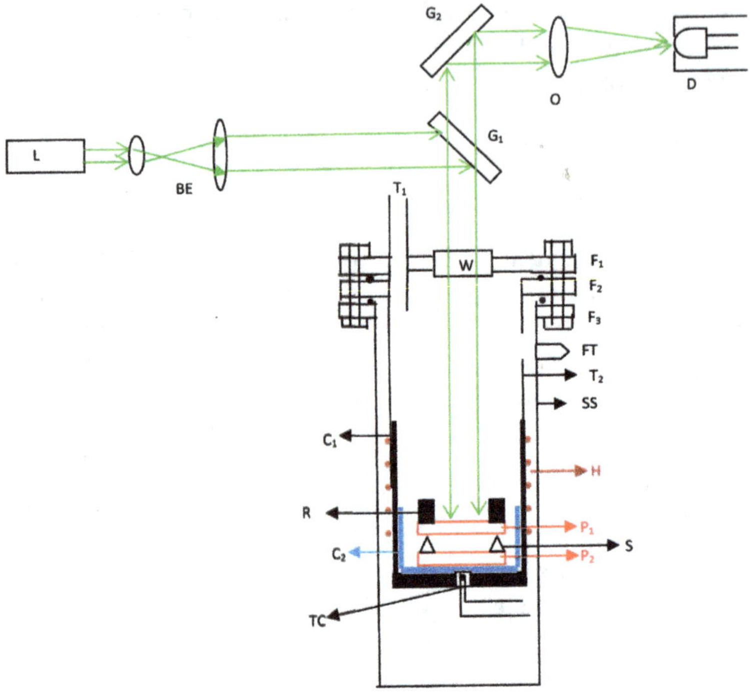

Figure 2. Schematic diagram of the apparatus for an interferometric measurement of thermal expansion.

The flange F_1, made of stainless steel, has a tube T_1 brazed to it for evacuation. At the center of the flange is an optical window, W, made of a flat transparent plate of fused silica and fixed to F_1 with a vacuum-tight seal. A thin-walled tube T_2, with a hole in it, is suspended from flange F_2. At the lower end of T_2, a thick-walled copper cup C_1 closed at the bottom is brazed. A heater H, made of nichrome wire, is wound on the outer surface of C_1.

One junction of a chromel–alumel thermocouple, TC, is welded to the midpoint of the bottom face of C_1. To the flange F_3 is attached a thin-walled stainless-steel tube, SS, closed at the bottom. The leads of the heater and the thermocouple come out of the feed-through, FT, located on the wall of SS just below the flange F_3.

The interferometer assembly is placed in another copper cup C_2, which fits into cup C_1. Two transparent optical flats, P_1 and P_2, separated by three pyramids of the sample, S, form the interferometer. First, the copper cup C_2 is removed from C_1, then the samples S are placed on plate P_2, and the plate P_1 is placed on the conical tips of the pyramids. An iron ring, R, is placed on the plate P_1 to add stability to the system. The optical flats P_1 and P_2, made of fused silica, have a diameter of about 2.5 cm and a thickness of about 3–5 mm. The bottom face of P_2 is ground, while its top face is polished. The plate P_1 is actually wedge-shaped, with a wedge angle of about half a degree between the two faces, with both faces being well polished. Due to this wedge shape, any ray of light incident vertically on the top face of P_1 is reflected away from the vertical direction. The heights of the three pyramids may differ by a few hundredths of a millimeter. Hence, a wedge with a very small wedge angle will be formed between the bottom face of P_1 and the top face of P_2, and the light rays reflected by these two surfaces will interfere, forming fringes, which may be recorded by the detector. The light rays reflected from the top face of P_1 do not contribute to the interference since they are deflected away from the vertical direction. Rays are scattered diffusely by the rough bottom face of P_2 and do not contribute to the interference. The interference will result in nearly straight-line fringes with equal spacing. The fringes are not formed at infinity. After assembling the interferometer outside and ensuring that fringes are formed, the cup C_2 is lowered carefully into the cup C_1 with the aid of tongs.

The optical arrangement is also shown in Figure 2. L is a diode laser that gives green light. The laser beam is expanded using a beam expander BE. This beam is partially reflected downward into the interferometer by the transparent glass plate G_1. The reflected beams from the interferometer plates pass through G_1 and fall on another glass plate G_2, whose front face is fully silvered; G_2 functions as a mirror. The lens O focuses the rays on a pinhole in a screen, forming an image of the fringes at the pinhole. A photodetector D is placed behind the pinhole for recording the interference fringes.

The specimens are in the form of three pyramids with conical tips. The height of each pyramid is about 5 mm. The difference in the height of the pyramids should not exceed 1/100 mm, as measured on a dial gauge. The base of the pyramid may be a square of side 5 mm. In an anisotropic material, the pyramids must all be oriented in the same direction relative to the crystallographic axes.

The condition for constructive interference is

$$2\mu\ell \cos(\theta) = (n + ½)\lambda \tag{6}$$

Here, ℓ is the height of the specimen (i.e., the pyramids), θ the angle of incidence, μ the refractive index of the medium between the plates P_1 and P_2, λ is the wavelength of light, and n is an integer. When light rays, traveling downward from G_1, undergo reflection from the lower face of P_1, they do not suffer a sudden phase change due to the reflection process. However, light rays undergoing reflection from the top face of P_2 will suffer a phase change of π since these rays, traveling in vacuum, are incident on glass, which has a higher refractive index. This is the reason for the factor ½ on the right-hand side of Equation (6).

Since we evacuate the interferometer through tube T_1 during the measurement, μ is unity. From Equation (6), it follows that n takes the maximum value when the angle of incidence is zero. For higher values of θ, n takes lower integer values. The incident angle will be zero for the maximum value of n and will increase as n decreases.

When the temperature increases, the height ℓ increases, and the fringes will move across the photodetector. The output of the photodetector will change from maximum to minimum and back again to maximum, as one fringe moves across the photodetector and the next fringe replaces it. The output of the photodetector D and the thermo-emf of the thermocouple TC are recorded on a computer. The heating rate is kept small (of the order of 1°C/min), and the heating rate is kept constant by adjusting the heating power continuously.

For measurement at 4.2 K and above, the outer tube SS is immersed in a liquid helium bath. The bath is filled with liquid helium after removing the air from the container SS. The tube SS is then filled with helium exchange gas at low pressure. The copper cup C_1 and its contents are slowly cooled to nearly 4.2 K by heat exchange through the helium gas. The helium gas is then pumped out, and the heating current is switched on. At such low temperatures, the sample temperature is measured by a germanium resistance thermometer attached to the bottom of cup C_1.

For one fringe shift, i.e., when the photodetector output changes from one peak value to the next, the increase in length $\Delta\ell = \lambda/2$. So, by counting the number of fringe shifts, we may plot the increase in length $\Delta\ell$ against the temperature T in K.

Knowing the initial length of the sample pyramids and the slope of this plot, the linear expansion coefficient can be calculated. Figure 3 shows the variation in the linear expansion coefficient of NaCl from −120°C to 300°C determined using this method by Srinivasan (Ref. 2).

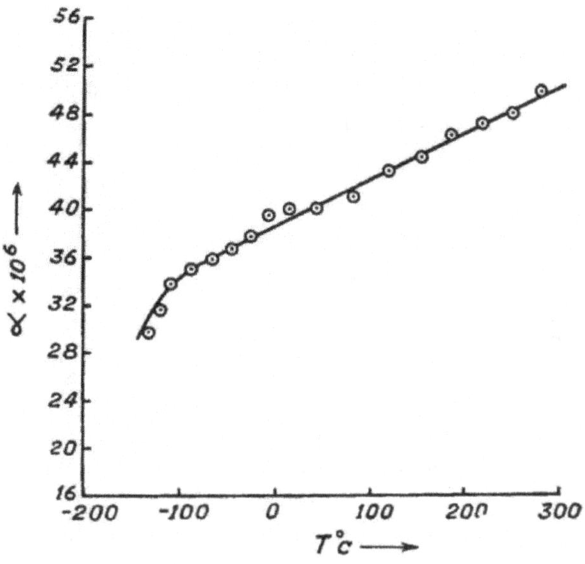

Figure 3. Linear expansion coefficient of NaCl from 90 to 300K (reprinted with permission from Ref. 2).

Other interferometric methods based on Michelson and Fabry–Perot interferometers are described in the literature. The setup described above is the simplest to use.

4. Three-Terminal Capacitance Technique

In this technique, one uses a three-terminal capacitor shown in Figure 4(a).

In this arrangement, the plates P_1 and P_2 form a parallel-plate capacitor with the capacitance C_{12}. The plates are surrounded by a metal guard G (i.e., a metal box), which is earthed as well as insulated from the plates P_1 and P_2. The capacitance C_{12} of the parallel-plate capacitor is

$$C_{12} = \varepsilon_0 A/d \tag{7}$$

It is assumed that the medium between the capacitance plates is a vacuum. Here, ε_0 is the permittivity of free space, which has a value of 8.854×10^{-12} F/m, A is the area of the plate in m^2, and d is the distance between the plates in m. C_{13} and C_{23} are the capacitances between G and the plates P_1 and P_2, respectively. H and L are the high- and low-voltage

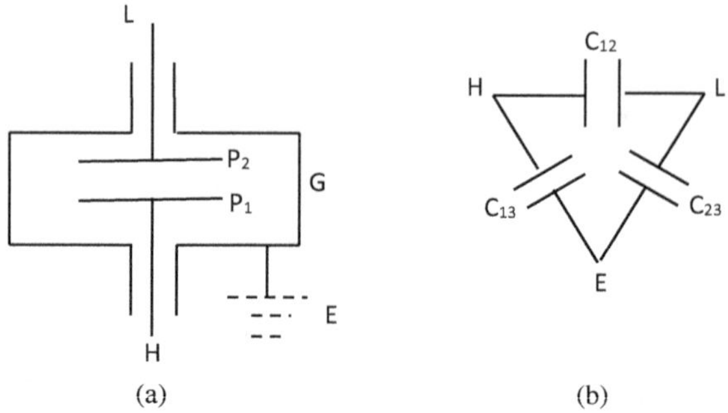

Figure 4. Three-terminal capacitance (a) schematic and (b) equivalent circuits.

terminals of the parallel-plate capacitor. If the low-voltage terminal L of the parallel-plate capacitor is earthed, then the plates P_1 and P_2 will pick up other stray capacitances due to metal objects in the surroundings. The stray capacitance C_0 acts in parallel with C_{12} and can vary depending on the position of the surrounding objects. By insulating the low-voltage terminal L and connecting the guard G to earth, we ensure that the stray capacitance due to any external object is eliminated. Thus, we may consider the arrangement to be the capacitance C_{12} in parallel with a fixed stray capacitance C_0 due to C_{13} and C_{23}.

The plate P_2 is fixed, and the plate P_1 is attached to the solid sample. When the sample is heated, the plate P_1 moves due to the expansion of the specimen. The resulting change in the gap d leads to a change in the capacitance C_{12}, while C_0 (i.e., C_{13} and C_{23}) remains unchanged. If ΔC_{12} is the change in capacitance C_{12} due to a change, Δd, in the distance between the plates P_1 and P_2, we can show that

$$\Delta d = -\varepsilon_0 A[\Delta C_{12}/C_{12}^2] \tag{8}$$

The advantages of the capacitance method are as follows:

(1) One can make a very accurate measurement of capacitance. A resolution of one part per million is achievable in the capacitance measurement. Hence, Δd can be determined with high accuracy.
(2) The size of the sample can be small, of the order of 0.5–1 cm.
(3) The entire size of the capacitance cell is small, facilitating low-temperature measurements.

To get accurate results, one must ensure that the plates P_1 and P_2 are parallel within a few minutes of an arc.

The cell itself will expand on heating, changing the capacitance between the plates. So, one should first make measurements with a standard sample to find the expansion coefficient of the material of the cell itself. We must subtract this value from the value determined for the sample being studied.

We describe briefly one such cell constructed by Subrahmanyam and Subrahmanyam (Ref. 3). A schematic of the capacitance cell is shown in Figure 5.

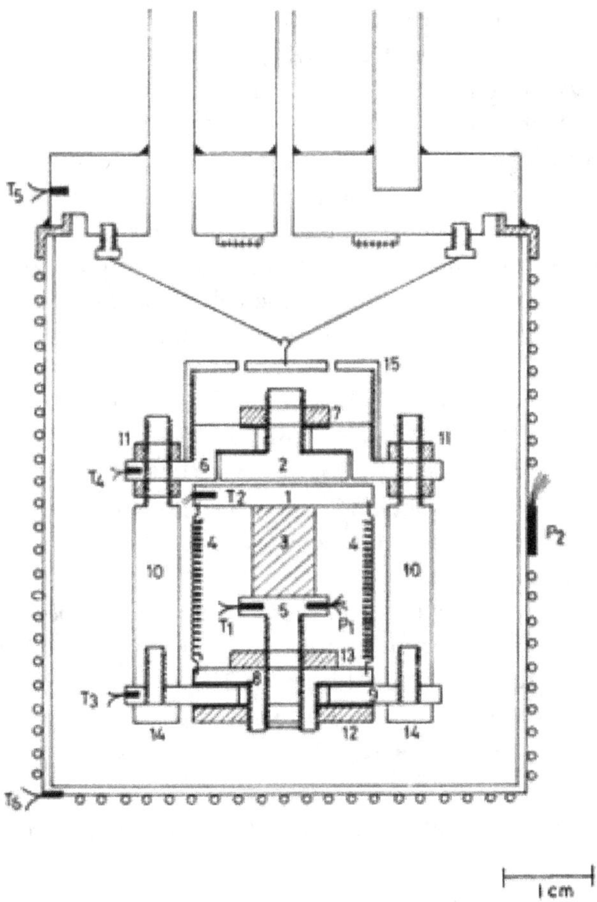

Figure 5. Three-terminal capacitance cell for thermal expansion measurement (reprinted with permission from Ref. 3).

The cell is made of stress-relieved, high-conductivity copper. The cell is cylindrical in shape, with a diameter of 3.5 cm and a length of 4 cm. The high-voltage plate 1 is held pressed to the sample 3 by three stainless-steel springs 4. Plate 1 has a diameter of 2 cm and a thickness of 0.25 cm. The two faces of plate 1 are polished so that the faces are parallel to within 1′ of arc. The low-voltage plate 2 has a diameter of 1.5 cm and is firmly fixed in a guard ring 6, which is fixed to the cell by nut 7. The plate is insulated from the guard ring by thin mica sheets (shown by a thick black line in Figure 5). The face of plate 2 and the guard ring are lapped to a few millionths of an inch. A copper cup, 15, surrounds the guard ring, and the two are electrically connected and earthed. The sample platform base 8 is fixed to the base of the capacitance cell 9 by the nut 12. The sample platform can be raised or lowered to the desired position and fixed by the nut 13. This allows samples of different lengths to be accommodated in the cell. The sample length can vary between 0.5 and 1.5 cm. The upper and lower portions of the cell are fixed to three posts 10 by nuts 11 and 14. The parallelism of the plates can be adjusted with nut 11. The entire assembly is suspended inside a metal chamber by Teflon threads.

A platinum resistance thermometer is used to measure the sample temperature and the temperature of the cell. A heater of 25 ohms (indicated by open circles) around the external metal container serves to heat the sample.

A six-decade ratio transformer bridge, using a lock-in amplifier as a null detector, was employed to measure the capacitance of the cell. This bridge is described in detail by Krishnapur (Ref. 4). An imported 10 pf capacitor with 0.1% accuracy and 1 ppm stability was used as the reference. The plates of the capacitor were connected to the bridge through stainless-steel coaxial cables.

Figure 6 shows the expansion coefficient of the copper cell as a function of temperature using aluminum and germanium as two standard materials. The accuracy obtained is about 3%.

This cell was used to measure the expansion coefficients of γ-irradiated polymers. For further details, Ref. 3 may be consulted.

5. Linear Voltage Differential Transformer Method

The linear voltage differential transformer (LVDT) is a transducer which converts a mechanical displacement into a proportionate electrical signal. The components of an LVDT are shown in Figure 7 (Ref. 5).

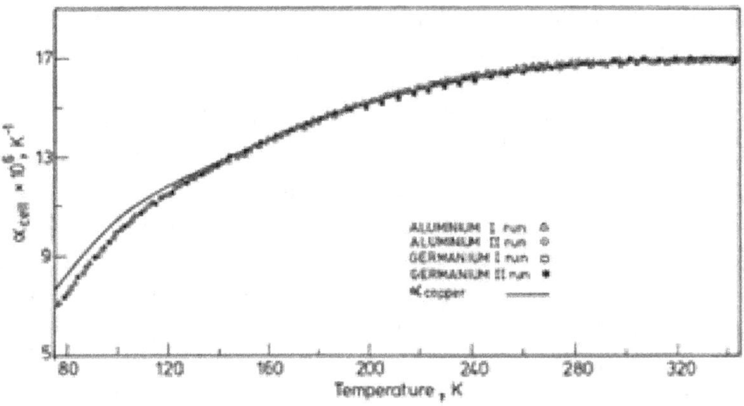

Figure 6. Expansion of the copper cell using Al and Ge as standard materials.

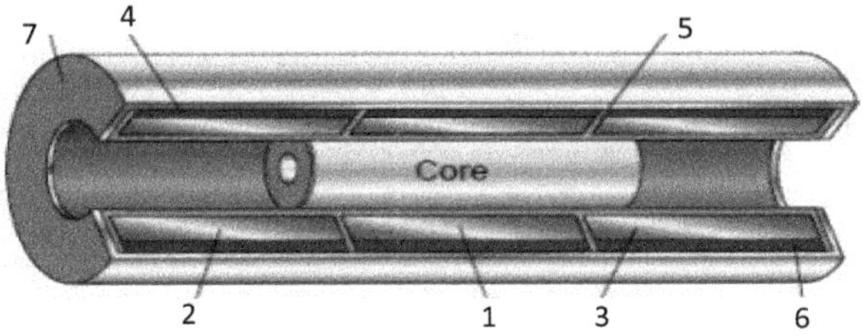

Figure 7. LVDT transducer (reprinted with permission from Ref. 5).

An LVDT consists of a primary coil 1, and two identical secondary coils 2 and 3, wound on a high-density glass-filled polymer former 4. The coils are potted in epoxy 5. This coil assembly is surrounded by a high-permeability magnetic shield 6. The whole assembly is enclosed in a stainless-steel housing with end caps 7. The core is made of a high-permeability magnetic material. It is a threaded cylinder and can move axially within the coil assembly. There is a large clearance between the core and the coil assembly, enabling the core to move without friction along the axis of the coil.

When the primary coil is excited by an AC current, voltages v_2 and v_3 are induced in the secondary. The voltages v_2 and v_3 are combined in opposition, leading to a net output AC voltage from the secondary coils, which is given by

$$v = v_2 - v_3 \tag{9}$$

This output voltage will depend on the position of the core relative to the secondary coils. When a large portion of the core is within coil 2, v_2 is greater than v_3. On the other hand, if a large portion of the core is within coil 3, then v_3 is more than v_2. So, as the core moves from an extreme end on the left to an extreme end on the right, the amplitude of the output voltage changes both in magnitude and sign.

This output AC signal is further processed to give a large DC output voltage, which varies with the position of the core, as shown in Figure 8.

We see that the variation of the output DC voltage with the position of the core, as we move the core from one end to the opposite end, is linear, except near the two ends.

The advantages of LVDT for measuring displacement are as follows:

1. A high resolution in the measurement of small displacements of the order of 10^{-7}–10^{-8} m is achievable.
2. The null point of the LVDT is very stable and repeatable.
3. It has a fast dynamic response due to the frictionless movement of the core.
4. It has a small size, and an LVDT of 20 mm overall size can measure displacements in a range of ±1 mm.

A common arrangement for a dilatometer using the LVDT is shown in Figure 9 (from Ref. 6). There are two push rods. One of them is in contact with the top end of the sample. The other push rod is in contact with the base on which the sample is mounted. The core of the LVDT is

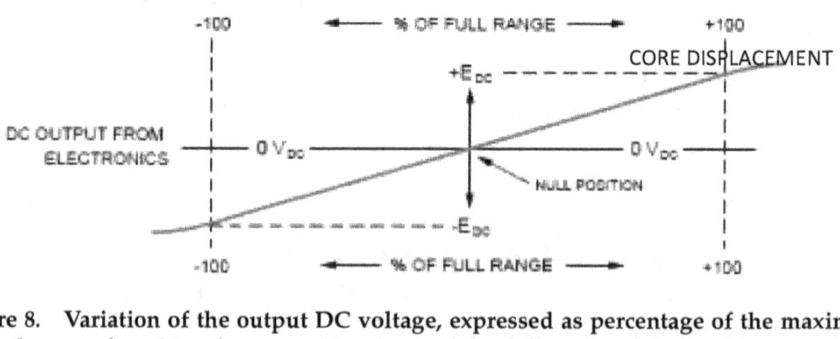

Figure 8. Variation of the output DC voltage, expressed as percentage of the maximum DC value as a function of core position in % of the full range of the position (reprinted with permission from Ref. 5).

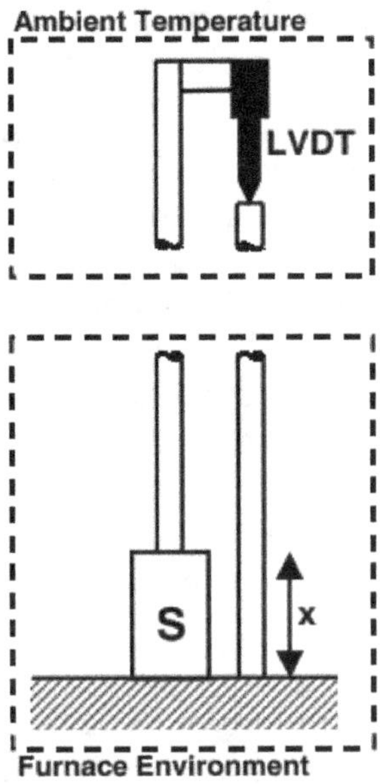

Figure 9. Common arrangement of a dilatometer using LVDT (reprinted with permission from Ref. 6).

connected to the push rod connected to the bottom face of the sample, and the outer former on which coils are wound is connected to the push rod residing on the top face of the sample, as shown in the figure. The sample and parts of the push rods are in a furnace. The LVDT is at ambient temperature and measures the difference in displacement of the two push rods.

If the two push rods are of identical material, the LVDT measures the difference in expansion between the sample S and the portion of the push rod of the same length as the sample. To reduce this correction in the expansion of the sample, the push rods are made of fused silica with a very low expansion coefficient (of the order of 0.5×10^{-6} /K).

The LVDT can also be used to measure magnetostriction, which involves a change in the linear dimensions of a solid due to the application of a magnetic field. There are several publications on the determination of thermal expansion and magnetostriction of rare-earth magnetic materials.

Dilatometers for measuring the expansion of samples up to a temperature of 1000 K using LVDT are commercially available.

6. X-ray Technique

When X-rays of wavelength, λ, are incident at a glancing angle, θ, on a set of lattice planes with Miller indices (h, k, l) with a spacing d_{hkl}, they will be Bragg-reflected if the following condition is satisfied:

$$2d_{hkl}\sin(\theta) = n\lambda \tag{10}$$

If the temperature of the specimen is changed, the spacing d_{hkl} changes by Δd. This changes the glancing angle for reflection by $\Delta\theta$ for a fixed value of n. By measuring the change in the glancing angle at different temperatures, one can determine the linear expansion coefficient in the direction normal to the h, k, l planes.

The advantages of the X-ray diffraction method are as follows:

1. It gives the intrinsic expansion coefficient, which is obtained from the diffraction pattern of the crystalline material. On the other hand, the methods described earlier yield expansion coefficients, which depend on the microstructure of the material since large samples are used.
2. In an anisotropic material, X-ray diffraction allows the measurement of the linear expansion coefficient in different directions.

The fractional change in length of the lattice planes, $\Delta d/d$, is related to the change in glancing angle $\Delta\theta$ by

$$\Delta d/d = -\cot(\theta)\Delta\theta \tag{11}$$

So, a given change $\Delta d/d$ will cause a large change in $\Delta\theta$ if the glancing angle θ is close to 90° (i.e., the X-rays are back-reflected).

An X-ray camera in the symmetric back-reflection geometry is shown in Figure 10.

From the target of the X-ray tube, X-rays enter through a slit and fall on a powder sample of the material being studied. The diffracted X-rays fall on a photographic film in a circular holder. The slit and the material are on the same circle. The reflections from the (h, k, l) planes in the material fall symmetrically on either side of the normal to the material.

If R is the radius of the film holder, the spacing between these two symmetrical diffraction spots is separated by a distance $4(\pi - 2\theta)R$, where

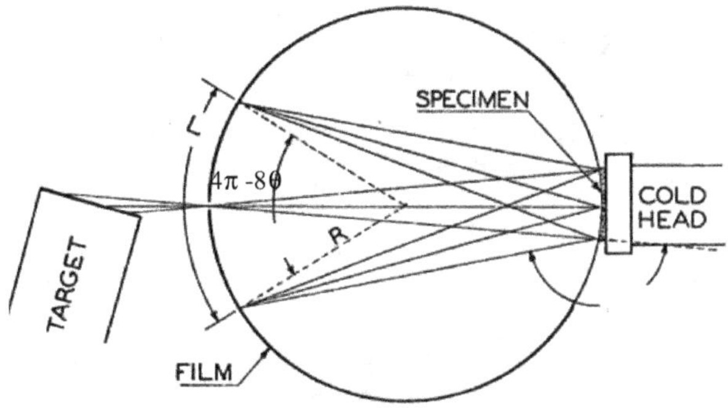

Figure 10. Symmetric back-reflection focusing X-ray diffraction camera for thermal expansion studies (reprinted with permission from Ref. 7).

θ is the glancing angle of reflection in radian. For high sensitivity, the plane (h, k, l) in the material is chosen such that θ lies between 75° and 90°. A filter must be used to suppress the K_β radiation from the X-ray target. A narrow slit will enhance the accuracy of the measurement of the separation between the symmetrically spaced diffraction peaks. One can take multiple pictures of the diffraction peaks by moving the film holder along the axis of the camera. The specimen can be mounted on a backing plate made of silver, which can be heated by an electric heater. A suitable resistance thermometer is pasted onto the silver plate to measure the temperature. For low-temperature measurements, the silver backing plate, on which the specimen is mounted, is placed in contact with the cold head of the cryocooler (Figure 11). One should ensure that the material is not displaced from the circle of radius R; and the photographic film does not buckle when its temperature is changed. The camera and the specimen are kept within the tight vacuum space of the cryocooler.

Figure 12 shows six different exposures at different temperatures using Cu K_α radiation. The separation between the $K_{\alpha 1}$ and $K_{\alpha 2}$ lines is seen to be large compared to the shift in the lines due to the change in temperature.

The linear expansion coefficient of aluminum measured with this arrangement is plotted against temperature in Figure 13. The data obtained by Woodard (Ref. 7) using this camera are compared with the X-ray expansion data from Figgins *et al.* and the interferometric measurements by Gibbons.

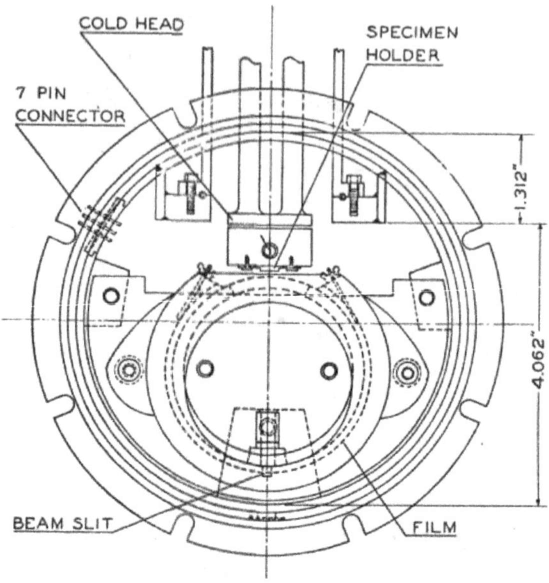

Figure 11. Top view of the camera showing the specimen holder attached to the cold head, the entrance slit for X-rays, and the film (reprinted with permission from Ref. 7).

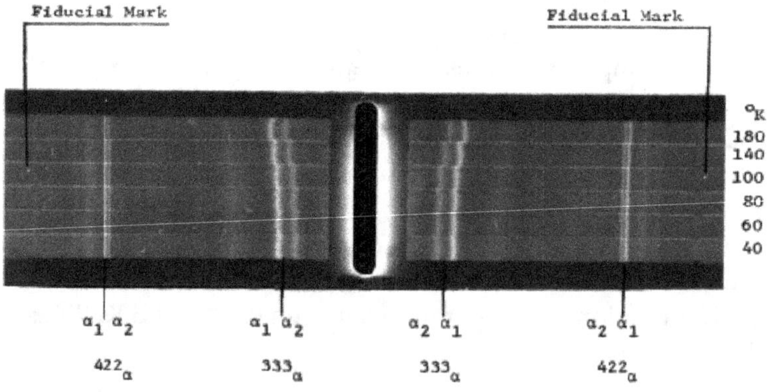

Figure 12. Exposures at six different temperatures on the same film for an aluminum specimen (reprinted with permission from Ref. 7).

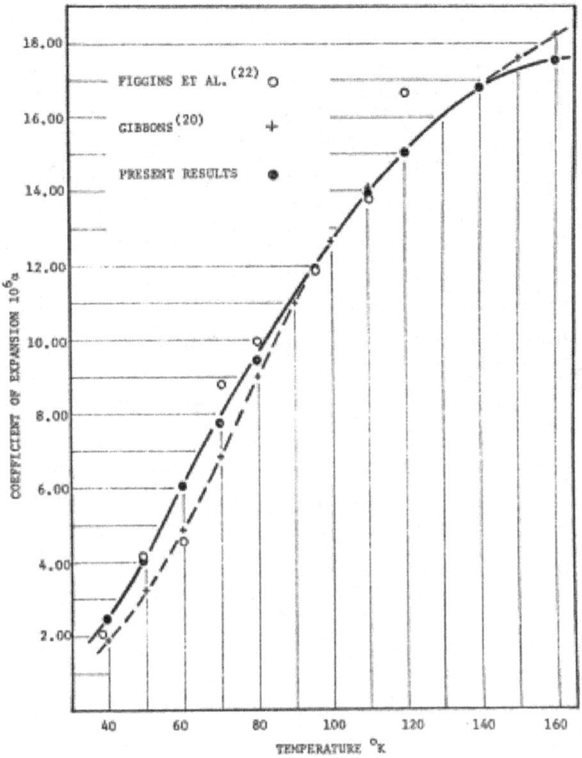

Figure 13. Linear expansion coefficient of aluminum as a function of temperature (reprinted with permission from Ref. 7).

There is good agreement between the two X-ray diffraction measurements. But the interferometric measurement by Gibbons yielded values lower than those yielded by X-ray diffraction at low temperatures.

Commercial X-ray powder diffraction cameras for thermal expansion measurements are available with scintillation detectors and automatic data acquisition systems.

7. Conclusion

In this chapter, different methods of measurement of the thermal expansion coefficient of materials were briefly presented. For greater details, readers may consult the review papers available on the web.

References

1. Krishnan, R. S., Srinivasan, R., and Devanarayanan, S. (1979). *Thermal Expansion of Crystals*. Pergamon Press, Oxford.
2. Srinivasan, R. (1955). *J. Indian Inst. Sci.* **37**, 232.
3. Subrahmanyam, H. N., and Subrahmanyam, S. V. (1986). *Pramana- J. Phys.* **27**, 647.
4. Krishnapur, P. P. (1983). Development of three-terminal capacitance bridge system for thermal expansion studies between 77 K and 350 K. Ph.D. thesis, Indian Institute of Science, Bangalore, India.
5. LVDT Tutorial | LVDT Basics — What is an LVDT | TE Connectivity, www.te.com › Sensor Solutions › Insights.
6. James, J. D., Spittle, J. A., Brown, S. G. R., and Evans, R. W. (2001). *Meas. Sci. Technol.* **12**, R1–R15.
7. Woodard, C. L. (1969). X-ray determination of lattice parameters and thermal expansion coefficients of aluminium, silver and molybdenum at cryogenic temperatures, https://scholarsmine.mst.edu/doctoral_dissertations/2322.

Chapter 12

THERMAL CONDUCTIVITY AND DIFFUSIVITY

1. Introduction

Heat applied to one side of a material is conducted to the other sides. In this process, a temperature gradient is set up in the material. If we take a unit surface area in the material at point P, normal to the gradient of temperature, the amount of heat flux, j_h, per second through the unit area is given by

$$j_h = -K\mathbf{grad}(T) \tag{1}$$

where K is called the thermal conductivity of the material. The direction of heat flux is opposite to the direction of the temperature gradient, i.e., heat flows from the high-temperature side to the low-temperature side. Under steady-state conditions, j_h and **grad** (T) are independent of time. If we consider the gradient to be in the Z direction, then the heat flowing per second across an area A, normal to the Z direction, is

$$Q = -KA\mathrm{d}T/\mathrm{d}z \tag{2}$$

For an isotropic material, the thermal conductivity is independent of the direction of measurement. For an anisotropic material, such as a single crystal belonging to the non-cubic system, the value of thermal conductivity depends on the direction of measurement. If a polar diagram of the thermal conductivity values is plotted, it will be an ellipsoid in general. The values of K along the three principal axes of the ellipsoid are denoted by K_1, K_2, and K_3. If the crystal has axes of symmetry, the principal axes of the thermal conductivity ellipsoid will lie along the axes of symmetry. For an isotropic material, $K_1 = K_2 = K_3$.

In a metal both electrons and phonons will contribute to thermal conductivity.

$$K = K_{\text{el}} + K_{\text{ph}} \tag{3}$$

In an insulating material, there are no free electrons, and the thermal conductivity arises **only** from phonons.

$$K = (1/3)v\Lambda c_V \tag{4}$$

Here, v is the velocity of the particle, c_V is the contribution of the particle to the specific heat per unit volume, and Λ is the mean free path of the particle. This expression is derived from the kinetic theory of gases. The mean free path is the average distance a particle travels between two successive collisions.

There can be many different processes which scatter a particle. For example, the electron may be scattered by the ubiquitous phonons, defects in the metal, and other electrons. Each of these scattering processes will have a mean free path Λ_j. **However, only those processes in which the heat current j_h is changed will matter for thermal conductivity**. Considering the scattering by all such processes, the effective mean free path for the particle will be

$$1/\Lambda = \Sigma_j(1/\Lambda_j) \tag{5}$$

However, in phonon — phonon scattering, there are two types of phonon–phonon processes. This distinction arises because the wavelength of a phonon in a crystal lattice is not uniquely determined. If we take the wave vector k of a phonon of frequency ω, the mode is unchanged if k is replaced by $k' = k + G_h$, where G_h is 2π times a reciprocal lattice vector. This will be discussed in detail in Chapter 13. To prevent counting a normal mode again and again, we confine the k values of all normal modes of vibration to the first Brillouin zone. A wave vector k' lying outside the first Brillouin zone is brought into the first Brillouin zone to describe the wave vector k of the mode by adding an appropriate reciprocal lattice vector G_h.

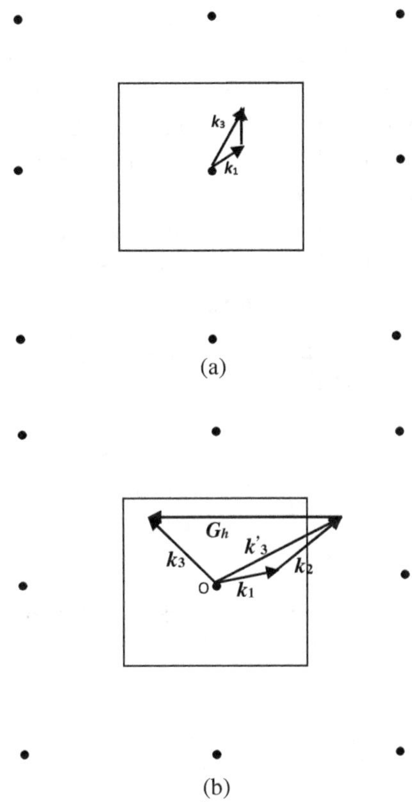

Figure 1. Three-phonon scattering processes in a crystal: (a) normal process; (b) umklapp process. The first Brillouin zone is shown as the square.

Let us consider the two three-phonon scattering processes shown in Figure 1. The reciprocal lattice (expanded by a factor of 2π) for a square lattice in two dimensions is shown. A vector, G_h, connecting two points in the figure is 2π times a reciprocal lattice vector.

Phonon 1 has a wave vector $\boldsymbol{k}_1$ and a frequency $\omega(\boldsymbol{k}_1)$. Phonon 2 has a wave vector $\boldsymbol{k}_2$ and a frequency $\omega(\boldsymbol{k}_2)$. Due to the anharmonicity of the lattice vibrations, these two phonons interact to give a third phonon with wave vector $\boldsymbol{k}_3$ and frequency $\omega(\boldsymbol{k}_3)$. In such a three-phonon scattering, all $\boldsymbol{k}$ vectors must lie in the first Brillouin zone. The conservation of energy gives

$$\hbar\omega(\boldsymbol{k}_1) + \hbar\omega(\boldsymbol{k}_2) = \hbar\omega(\boldsymbol{k}_3) \tag{6}$$

$$\boldsymbol{k}_1 + \boldsymbol{k}_2 = \boldsymbol{k}_3' = \boldsymbol{k}_3 + \boldsymbol{G}_h \tag{7}$$

There are two types of scattering processes, namely (a) those in which k_3' lies within the first Brillouin zone and (b) those in which k_3' lies outside the first Brillouin zone. Processes of type (a), called normal processes, are shown in Figure 1(a). In these processes, $G_h = 0$ and $k_3' = k_3$ since k_3' lies in the first Brillouin zone. In the second type of process shown in Figure 1(b), k_3' lies outside the first Brillouin zone. Then, one should subtract an appropriate reciprocal lattice vector G_h so that the wave vector k_3 lies within the first Brillouin zone. The direction of energy flow after scattering is along k_3.

In the normal process, the direction of energy flow of the incoming phonons is along $\boldsymbol{k}_1$ due to the first phonon and along $\boldsymbol{k}_2$ due to the second phonon. After scattering, the same total energy is flowing in the direction $\boldsymbol{k}_3$, which is the wave vector in the first Brillouin zone of the resulting phonon. So, the incoming heat current through any area $\boldsymbol{A}$ is $\hbar v[(\boldsymbol{k}_1 + \boldsymbol{k}_2)\boldsymbol{A}]$ since in the simple Debye model $\omega(k) = vk$, where v is the velocity of the lattice wave. After scattering, the energy flow across $\boldsymbol{A}$ is $\hbar v[\boldsymbol{k}_3 \boldsymbol{A}]$.

In the normal process, there is no change in the heat current across A due to scattering since k_3 is equal to $(k_1 + k_2)$.

In the umklapp (U) process, the heat current changes direction because $\boldsymbol{k}_3$ is not equal to $\boldsymbol{k}_1 + \boldsymbol{k}_2$. **We may say that the original heat currents carried by phonons 1 and 2 are flipped by the reciprocal lattice vector G_h. This is called umklapp scattering because *umklappen* in German means flip. So, in a crystal, only umklapp processes can give rise to thermal resistance due to phonon–phonon scattering. In calculating the effective mean free path for phonon contribution to thermal conductivity, we have to consider only U processes in three-phonon scattering. It is obvious that for U processes, at least one of the phonons must have a wave vector larger than half the side of the Brillouin zone. In the simple Debye model, this implies that at least one of the phonons that is scattered should have a frequency greater than $\omega_D/2$, where ω_D is the Debye frequency. It is important to remember this.**

In a metal, the free electrons have a velocity v_F, the Fermi velocity. This is of the order of 10^6 m/s. The electronic specific heat per unit volume, $c_{v,el}$, is γT, which is small compared to the lattice specific heat $c_{v,lat}$, except at very low temperatures. **In a pure metal, the mean free path at high temperatures is dominated by electron–phonon interaction.** At high temperatures, the number of phonons, $N(\omega)$, of frequency ω in unit volume is $kT/\hbar\omega$. So, the mean free path due to electron–phonon interaction varies as $1/T$. Hence, the contribution to K_{el} is independent of temperature. As the temperature is reduced, the phonon density, $N(\omega)$, of high-frequency phonons decreases as

$$N(\omega) = 1/[\exp(\hbar\omega/kT) - 1] \tag{8}$$

The number of phonons decreases faster than T. So, the mean free path of electrons due to electron–phonon interaction increases faster than $1/T$. The thermal conductivity K_{el} increases as the temperature is lowered. This increase persists until the mean free path due to electron–phonon scattering, $\Lambda_{el\text{-}ph}$, becomes comparable to the mean free path, $\Lambda_{el\text{-}def}$, due to the scattering of electrons by defects. Then, the effective mean free path, Λ_{el}, reaches a maximum at a certain temperature. A further fall in temperature makes $\Lambda_{el\text{-}ph}$ become much larger than $\Lambda_{el\text{-}def}$. $\Lambda_{el\text{-}def}$ is independent of temperature and depends only on the average distance between defects (i.e., on the number of defects per unit volume). Impurities and isotopes are included in defects. So, the more impure the metal, the shorter the $\Lambda_{el\text{-}def}$. At very low temperatures, defect scattering dominates in determining the effective mean free path for electrons. This effective mean free path becomes independent of temperature, while the specific heat decreases with temperature as γT. The Fermi velocity of the electrons depends weakly on temperature. So, K_{el} decreases to zero as the temperature is reduced to absolute zero. The more impure the metal, the higher the temperature at which the conductivity becomes a maximum, and the lower the maximum value of conductivity. For a pure metal, the electronic contribution to thermal conductivity, K_{el}, dominates the phonon contribution, K_{lat}.

Figure 2 shows the thermal conductivity data for electrolytic tough pitch copper of different purities, as indicated by the residual resistivity ratio (RRR). The higher the value of RRR, the purer the copper. This figure bears out the theoretical analysis presented above.

The electrical conductivity, σ, of a pure metal at high temperature is also determined by the electron mean free path due to electron–phonon interaction. Hence, we should expect a correlation between the thermal and electrical conductivities of pure metals at high temperatures. Free electron theory predicts that $K/\sigma T$ of any pure metal should be a universal constant given by

$$K/\sigma T = (\pi^2/3)\,(k/e)^2 \tag{9}$$

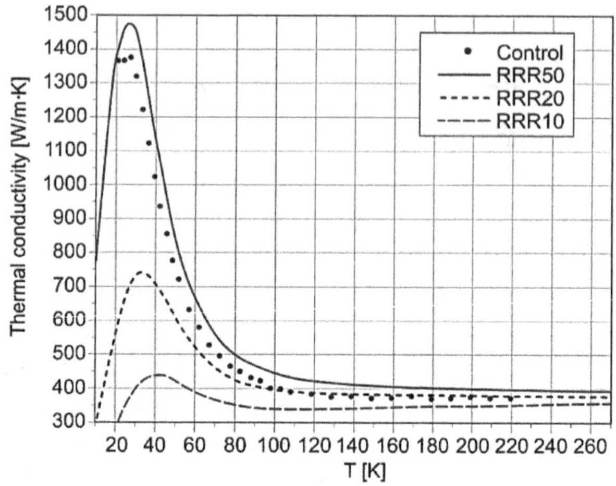

Figure 2. Thermal conductivity of Copper ETP with different residual.

Here, k is the Boltzmann constant and e is the electronic charge. This is called the Wiedemann–Franz law.

In an alloy, such as brass, the electron–defect scattering reduces the effective mean free path Λ_{el}. So, the thermal conductivity of an alloy is less than the thermal conductivity of a pure metal. In the case of an alloy, the contribution of phonons to thermal conductivity becomes comparable to the contribution due to electrons.

In an insulator, there are no free electrons. So, the only contribution to thermal conductivity comes from phonons. The velocity of the phonons is in the 10^4 m/s range. If the material is very pure and the temperature is high compared to the Debye temperature of the metal, the effective phonon mean free path, Λ_{ph}, is determined only by phonon–phonon scattering. This mean free path is determined by the U processes. When $T >> \theta_D$, we see from Equation (8) that the number of phonons causing U scattering varies as T, and so, the mean free path $\Lambda_{ph\text{-}ph}$ varies as $1/T$. The lattice contribution to the specific heat $c_{v,lat}$ is independent of temperature. So, the thermal conductivity of an insulator at high temperatures should vary as $1/T$, as seen from Equation (4).

The U processes require at least one of the participating phonons to have an energy greater than $k\theta_D/2$. The number of such phonons will decrease rapidly as the temperature becomes less than θ_D. So, $\Lambda_{ph\text{-}ph}$ will increase rapidly as temperature is reduced, roughly as $\exp(\theta_D/\beta T)$, where

β is a constant close to the value 2. The specific heat will decrease slowly as the temperature is lowered. So, the thermal conductivity K_{ph} will increase rapidly as $T^{\alpha}\exp(\theta_D/\beta T)$. When the mean free path due to phonon–phonon scattering, $\Lambda_{ph\text{-}ph}$, becomes comparable to the mean free path due to phonon–defect scattering, $\Lambda_{ph\text{-}def}$, then the increase in the effective mean free path, Λ_{ph}, will become progressively slower as the temperature is decreased. On the other hand, the specific heat decreases fast as the temperature is lowered. So, the thermal conductivity reaches a maximum value at a temperature T_{max}. When $T << \theta_D$, the effective mean free path, Λ_{ph}, becomes temperature independent, while the specific heat $c_{v,lat}$ will vary as T^3. So, the thermal conductivity will decrease as T^3. At very low temperatures and in very pure materials, the phonon scattering by the boundaries of the material becomes dominant. Then, the thermal conductivity, in addition to showing T^3 temperature dependence, will also depend on the dimensions of the material.

Figure 3 shows the thermal conductivity data for LiF from Berman and Brock (Ref. 1). Li has two isotopes with mass numbers 7 and 6. ^{7}Li is the most abundant isotope. We replace ^{7}Li atoms with ^{6}Li atoms up to a concentration of 50.1%. Note that the peak thermal conductivity value for pure ^{7}LiF is close to 10,000 W/mK, which is much higher than the value of pure copper shown in Figure 2. We see that the peak value of thermal conductivity decreases as the concentration of ^{6}Li increases. Atoms of ^{6}Li act as defects in the material. At high temperatures, the thermal conductivity varies as $1/T$, and at low temperatures, it varies as T^3, as mentioned in the earlier discussion.

2. Steady-State Method for Measuring Thermal Conductivity

For a brief review of the various methods of measuring thermal conductivity, one may consult Ref. 2.

In the steady-state method, we deal with one-dimensional heat flow, which is described by Equation (2). We use different configurations of the sample depending on whether it is a good conductor of heat, such as metals, or a poor conductor of heat, such as plastics.

For determining the thermal conductivity of a good conductor, we use either a rod or a tube of the material with a small cross-sectional area. An arrangement for measuring the thermal conductivity of such a material is shown in Figure 4.

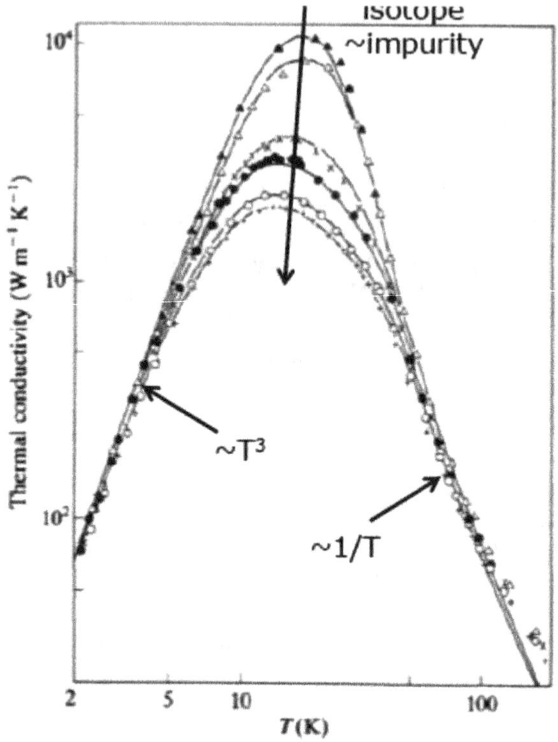

Figure 3. Thermal conductivity of ^{7}LiF, with ^{6}Li concentration varying from 0.01% to 50.1% (Ref. 1). ^{6}Li isotope concentration is increasing in the direction of the arrow.

A vacuum can C made of stainless steel (SS) hangs from a flange F_2 which can be fitted to an outer flange F1 on the inside through screws. An O-ring between the flanges makes for a vacuum-tight seal. The can is evacuated through tube T. From the flange F1, a thick copper plate CP hangs through thin-walled SS tubes SST. On the bottom of CP, a heater H1 is pasted. The leads of H1 are connected to pins P1 on the top of CP but electrically insulated from CP. The pins can be connected through copper wires to one of the feedthroughs, FT, on the top flange F_1. There is a thick-walled copper tube CT. On CT, a heater H2 of 1–2 W capacity is pasted. CT is fixed at the top to a fiberglass plate D of low thermal conductivity with a suitable glue. The leads from heater H2 come to the pins P3 on the top of the fiberglass plate. The leads are led out from P3 through one of the electrical feedthroughs, FT, on flange F_1.

The sample is a tube or rod made of a good thermal conductor. At the top end, it can be fixed tightly to the tube CT. If the sample is a metal rod,

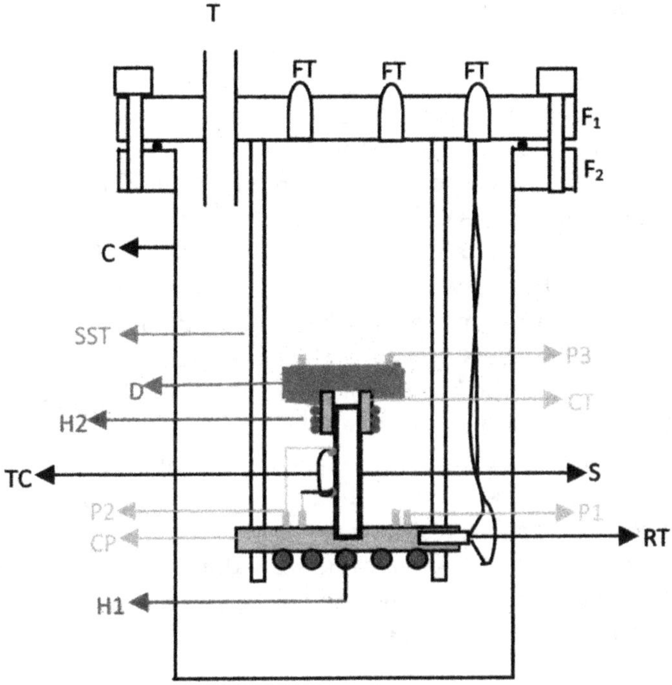

Figure 4. Schematic of a setup to measure the thermal conductivity of a good conductor.

it can be threaded and screwed into the copper tube CT. At the bottom end, it is fixed tightly, either by screwing or by other means, to a central hole in the copper plate CP. Thus, at one end, the sample is in good thermal contact with the heater H2 and, at the other end, with the copper plate CP. On the sample, we fix two junctions of a differential thermocouple with an epoxy of high thermal conductivity, such as Deltabond or Stycast. For experiments at temperatures above 40 K, one may use a copper–constantan thermocouple, with the constantan wire between the junctions and the copper wires as leads. If we want to conduct experiments at temperatures below 40 K, we may use gold (doped with a few atomic percent of Fe) and chromel as the thermocouple junction. Here, the chromel wire will be between the two junctions, and the gold wires will form the leads. The thermocouple must be made of very thin wires. The two leads of the thermocouple are connected to the pins P2 on the top of the copper plate CP but are electrically insulated from the plate.

These pins are connected to one of the feedthroughs on the flange F_1. These connections are not shown in the figure. A resistance thermometer RT measures the temperature of the bottom plate CP. For temperatures above 20 K, the thermometer used is a platinum resistance thermometer (PRT). For temperatures below 20 K, it will be a germanium resistance thermometer (GRT). The leads of the thermometer are connected to one of the electrical feedthroughs on the flange F_1.

To assemble the system, we first take the sample with thermocouples attached and fit the two ends of the sample tightly into the copper tube CT and the copper plate CP. Then, the leads of the thermocouple are attached to the pins P2 through soldering. The leads from the feedthroughs are attached to the pins P1, P2, and P3. The can C is slid up, and the flange F_2 is fixed to F_1 with screws.

After assembly, the can C is evacuated to a vacuum of 10^{-6} mbar pressure. A current is passed through the heater H1. Through a temperature controller connected to the resistance thermometer RT, the temperature of CP is maintained at the desired value to within 1 mK. Then, a suitable current, I, is sent through H2, and the voltage, V, across the heater H2 is measured. The heat power Q is then

$$Q = IV \tag{10}$$

Most of the heat flows through the sample, which is a good conductor. Some heat is lost through the fiberglass disk D. Some heat is lost by radiation from the surface of the sample to the can. By polishing the inside of C, we can reduce the radiation loss. The low thermal conductivity of plate D reduces the heat loss through it. We wait for a duration of about 30 min for the steady state to be attained. Then, we measure the thermo-emf of the thermocouple on a nanovoltmeter. Knowing the thermo-power of the thermocouple, we can convert the thermo-emf to a difference in temperature, ΔT, between the two junctions. We know the area of the cross-section A of the sample and the distance ℓ between the two thermocouple junctions. The thermal conductivity is calculated from

$$K = Q\ell / A\Delta T \tag{11}$$

We can change the current through H2 and repeat the measurement. This will give the thermal conductivity of the material at a mean temperature of $\approx T + (\Delta T/2)$, where T is the temperature of the copper plate, CP, measured by RT. With this apparatus, one may get the value of thermal conductivity to an accuracy of 5% or better.

For measurements at liquid helium temperature, the construction of the setup is more complicated. One may consult Ref. 1 and the references cited therein for such measurements.

If the sample is a poor conductor of heat, we need to reduce the heat flux as well as the temperature gradient in the sample for a given heat input. This is accomplished by using a large, thin disk of the material. The apparatus is designed such that heat loss by radiation occurs only through the thin edge of the disk, which is, therefore, very small.

The sample of a poor thermal conductor is prepared as a circular disk of a few centimeters in radius and a few millimeters in thickness. The thermal conductivity can be measured using the setup shown in Figure 5.

As in Figure 4, there is an SS can, C, polished on the inner surface and fitted with a flange F_2. The flange F_1 is fixed to F_2 with an O-ring seal

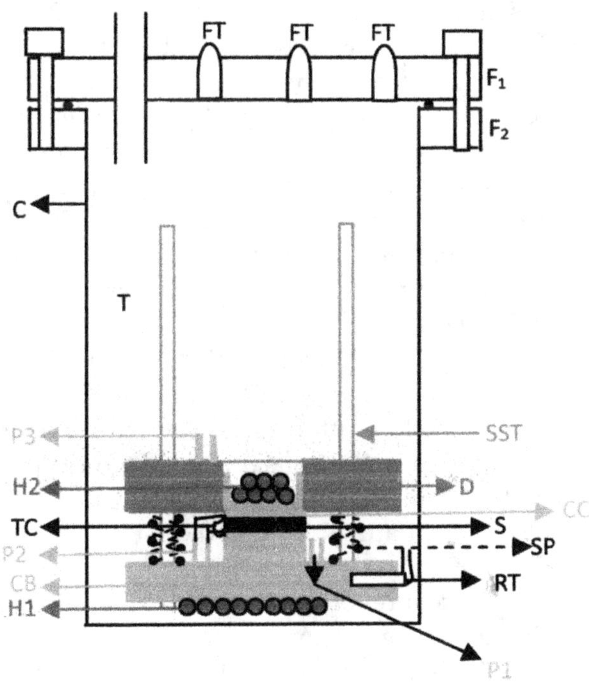

Figure 5. Schematic diagram of the setup for measuring the thermal conductivity of a poor conductor.

which is vacuum-tight. Then, C can be evacuated by connecting a vacuum pump to tube T. Electrical feedthroughs, FT, are fixed to the flange F_1 to bring out the electrical leads. A copper base plate CB is hung from F_1 by two thin-walled SS tubes. It has a centrally raised platform about 3 cm in diameter and 5 mm in height. A heater H_1 is pasted to the copper base plate. The leads of the heater come to the pins P1 fixed to, but insulated from, the base plate. From P1, the leads can be connected to one of the electrical feedthroughs, FT, on flange F1. A resistance thermometer RT (either PTR for temperatures above 20 K or GTR for temperatures below 20 K) is fixed to the copper base plate to monitor its temperature. Its leads are brought out through one of the electrical feedthroughs fixed to the flange F_1. There is a copper cup, CC, 3 cm in diameter with a base 3 mm in thickness. A second heater H2 of 1–2 W capacity is pasted inside this copper cup. The copper cup fits into a fiberglass plate D. This plate slides freely over the SS tubes. The leads of heater H2 are connected to the pins P3 fixed to the top of D. The leads from P3 can be connected to one of the feedthroughs on flange F_1.

The sample is in the form of a circular plate that is 3 cm in diameter and 1–2 mm thick. The sample fits between the base of the copper cup CC and the raised central platform on the base plate CB. Two SS springs connected between CB and D hold the sample tight between the copper cup and the raised platform of the base plate. One junction of a differential thermocouple, TC (copper–constantan above 40 K and gold (Fe)–chromel below 40 K), is fixed to the copper cup close to its base. The other junction of TC is fixed to the platform of CB close to its top face. The leads of TC are fixed to the pins P2. From P2, copper leads can be connected to one of the electrical feedthroughs FT.

For assembly, cup C is removed. A thin film of silicone grease is applied to the two faces of the sample. Then, the plate D is pulled up, and the sample S is placed on the platform. The plate D is then released, and it is checked to see that the sample S sits properly between the bottom face of CC and the top face of the raised platform on BP. The pins P1, P2, and P3 as well as the leads of RT are soldered to the corresponding pins on the feedthroughs.

The rest of the procedure for measuring thermal conductivity is the same as that in the case of a good conductor.

Ravindran *et al.* (Ref. 3) determined the thermal conductivity of yttrium barium copper oxide (YBCO) from 7 to 260 K using the plate

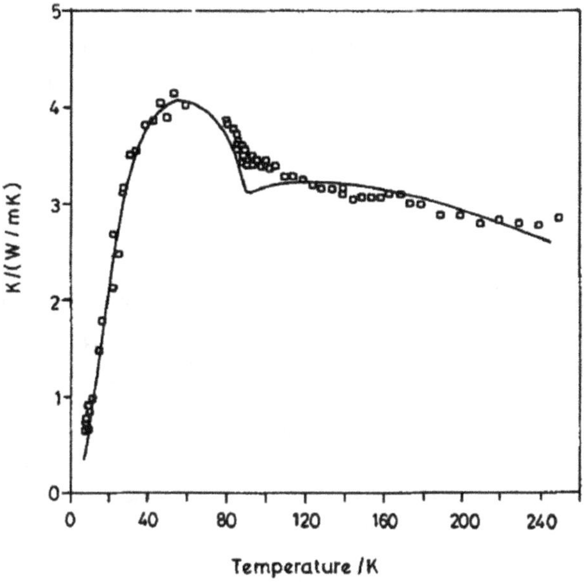

Figure 6. Thermal conductivity of YBCO determined by the plate method (reprinted with permission Ref. 3).

method and a continuous flow cryostat to cool the sample. For details of the setup, the reference must be consulted.

Figure 6 shows the thermal conductivity of a sample of YBCO determined by Ravindran *et al.* (Ref. 3). The continuous curve is fit to theoretical values.

3. Non-Steady-State Methods to Measure Thermal Diffusivity

When the heat supplied to the material varies with time, heat transfer in a material is characterized by the parameter called the thermal diffusivity of the material. The thermal diffusivity D is related to the thermal conductivity K by

$$D = K/\rho c_m \tag{12}$$

Here, ρ is the density of the material and c_m is its specific heat per unit mass. We note that ρc_m is the specific heat, c_V, per unit volume of the material used in Equation (4).

Consider an infinite cylindrical rod of a material extending from $X = 0$ to $X = \infty$. The rod has a cross-sectional area A and a circumference P. At the end $X = 0$, we supply heat power $Q(t)$, which varies sinusoidally with time at frequency f as follows:

$$Q(t) = Pe^{i\omega t} \tag{13}$$

where ω is $2\pi f$. Take an element of the rod between X and $X + \mathrm{d}X$. The temperature at that point is $T(X, t)$. The temperature at the far end $X = \infty$ is T_0, the room temperature.

We represent the temperature by a new variable $\theta(X, t)$ defined as

$$\theta(X, t) = T(X, t) - T_0 \tag{14}$$

The heat flowing per unit time into the differential element at the point X of the rod due to thermal conduction is $-KA(\partial\theta/\partial X)_X$. The heat flowing out per unit time at $X + \mathrm{d}X$ is $-KA(\partial\theta/\partial X)_{X+\mathrm{d}X}$. As a good approximation, we may write

$$(\partial\theta/\partial X)_{X+\mathrm{d}X} = (\partial\theta/\partial X)_X + (\partial^2\theta/\partial X^2)_X\mathrm{d}X \tag{15}$$

So, the heat lost by an element of length $\mathrm{d}X$ in unit time due to thermal conduction is $-KA(\partial^2\theta/\partial X^2)_X\mathrm{d}X$.

The element loses heat to its surroundings. If $\theta(X, t)$ is small compared to T_0, the heat loss per unit time is governed by Newton's law of cooling. So, heat lost to the surroundings in unit time is $E(P\mathrm{d}X)\theta$, where P is the perimeter of the rod, $P\mathrm{d}X$ is the surface area of the element from which heat is lost to the surroundings, and E is a constant.

The temperature of the element changes with time. So, the heat used up by the element $\mathrm{d}X$ in changing its temperature in unit time is $(\rho A\mathrm{d}X)$ $c_m\partial\theta/\partial t$. Here, $(\rho A\mathrm{d}x)$ is the mass of the element. There is no heat supply to the rod at any point X except at $X = 0$. So, by adding the total heat partly lost by transmission through the rod and by radiation from its surface and partly used up in raising the temperature of the element $\mathrm{d}X$ and dividing by $\rho A\mathrm{d}X$, we get the partial differential equation satisfied by $\theta(X, t)$ given by

$$\partial\theta/\partial t - \mathrm{D}\,(\partial^2\theta/\partial X^2) + \varepsilon\theta = 0 \tag{16}$$

Here, $$\varepsilon = EP/\rho A \tag{17}$$

The temperature at every point will also vary sinusoidally at the same frequency f. So, writing

$$\theta(X, t) = \Theta(X)e^{i\omega t} \tag{18}$$

and substituting this in Equation (16), we get the differential equation for $\Theta(X)$ as

$$d^2\Theta/dX^2 - \eta^2\Theta = 0 \tag{19}$$

Here, $$\eta^2 = (\varepsilon + i\omega)/D \tag{20}$$

Writing $$\eta = \alpha + i\beta, \tag{21}$$

we obtain

$$\alpha^2 - \beta^2 = \varepsilon/D \tag{22}$$

and $$2\alpha\beta = \omega/D \tag{23}$$

Equation (18) has to be solved subject to the boundary condition

(1) at $X = 0$ $$-KA(d\Theta/dX)_{X=0} = P \tag{24}$$

(2) at $X = \infty$ $$\Theta = 0 \tag{25}$$

In the above, P is the heat power supplied at the end $X = 0$.

The general solution to Equation (18) for $\Theta(X)$ satisfying the condition (25) is

$$\Theta(X) = A\exp - [(\alpha + i\beta)X] \tag{26}$$

Here, A is to be fixed using Equation (24).

So, the solution to $\theta(X, t)$ may be written as

$$\theta(X, t) = A\exp - (\alpha X)\exp[i(\omega t - \beta X)] \tag{27}$$

The above expression indicates that the temperature propagates along the rod like a wave of frequency f with an amplitude that decays exponentially with distance X as follows:

$$A(X) = A\exp(-\alpha X) \tag{28}$$

This implies that

$$\alpha = \ln[A(X_1)/A(X_2)]/[X_2 - X_1] \tag{29}$$

The phase, $\varphi(X)$, of the wave at a given time, increases linearly with X as

$$\varphi(X) = \beta X \tag{30}$$

or

$$\beta = [\varphi(X_2) - \varphi(X_1)]/[X_2 - X_1] \tag{31}$$

Figure 7 illustrates such a damped temperature wave.

From a measurement of α and β of the temperature wave, we can find the thermal diffusivity from Equation (23), knowing the frequency of the wave.

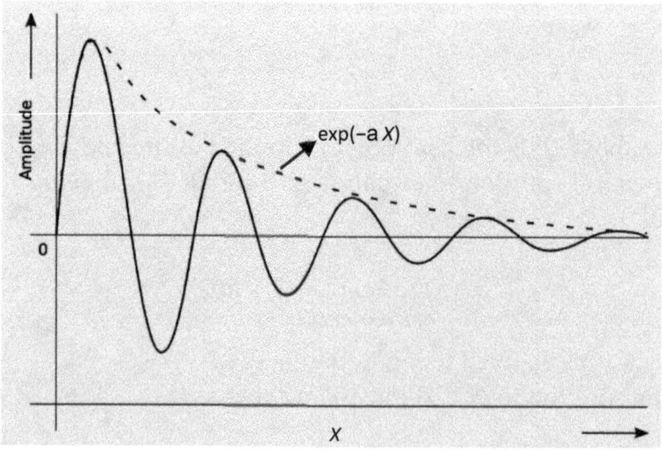

Figure 7. Damped temperature wave in a material for a periodic heat input.

Any periodic heat power $Q(t)$ can be expanded as a Fourier series of sine waves of varying amplitudes and frequencies $n\omega$ ($n = ..., -2, -1, 0, 1, 2,...$). Here, $\omega = 2\pi/T$ is the fundamental frequency, where T is the period of the heat pulses. Thus, we may write

$$Q(t) = Q_0 + \Sigma_n Q_n \exp(in\omega t) \tag{32}$$

Since Q is real,

$$Q_{-n} = Q_n^* \tag{33}$$

where Q_n is a complex number and Q_n^* is the complex conjugate of Q_n.

The temperature at any point is made of a superposition of harmonic thermal wave components

$$T(t) = T_0 + \Sigma_n [T_n(X) \exp(in\omega t + \varphi_n(X))] \tag{34}$$

with

$$T_{-n} = T_n^*$$

$$\varphi_{-n}(X) = -\varphi_n(X) \tag{35}$$

The amplitude of the nth temperature component T_n decays with increasing X as $\exp(-\alpha_n X)$ and the phase $\varphi_n(X)$ varies as $\beta_n X$, with α_n and β_n satisfying the equation

$$2\alpha_n\beta_n = n\omega/D \tag{36}$$

Let us consider a good conductor, such as copper. Its thermal diffusivity is about 1 cm^2/s. We keep the frequency of sinusoidal heating at 1 Hz. Then, the amplitude decay coefficient α will be 1.7 /cm or less. If we take a length of the rod larger than $5/1.7 \approx 3$ cm, the thermal wave will decay to less than 1% of its maximum value at the heated end. The value of α will be larger (a) for higher frequencies of heating and (b) for lower diffusivities of the material. So, thermal diffusivity measurements are possible with thin films of the material. Such measurements take much less time than the steady-state methods for measuring thermal conductivity. These are the advantages of measuring thermal diffusivity.

4. Methods for Measuring Thermal Diffusivity

Some of the methods for measuring thermal diffusivity are reviewed by Cemushi *et al.* (Ref. 4).

Here, we give a brief description of two commonly used techniques:

4.1. *Thermal wave method*

In this method, we apply a sinusoidal heat input of a low frequency (few millihertz) at one end of a long rod a few centimeter in length. We measure the amplitude and phase of the temperature wave at two points, X_1 and X_2, along the length of the rod using a lock-in amplifier.

A student experiment for measuring the thermal diffusivity of brass using a long rod is described by Srinivasan *et al.* (Ref. 5). A schematic diagram of the setup is shown in Figure 8.

Here, the brass rod is 30 cm long, and heat is applied by an electric heater at the center of the rod. The current through the heater is on for 5 min and off for 5 min periodically. The thermo-emfs at two points, separated by 3 cm, are measured for one full period at intervals of 30 s. A Fourier analysis of the thermo-emf gives the amplitude and phase of the temperature wave at the fundamental frequency, which is 0.00167 Hz. From the amplitudes at the two points, the decay constant α is determined. From the phases at the two points, the constant β is determined. The thermal diffusivity is calculated from Equation (23).

A thermal wave setup suitable for research is shown in Figure 9 (from Ref. 4). In this figure, the sample is a stainless-steel strip of length 250 mm, width 40 mm, and thickness 1.5 mm. The thermal wave is generated by a thermoelectric device through which a periodic current is passed, the

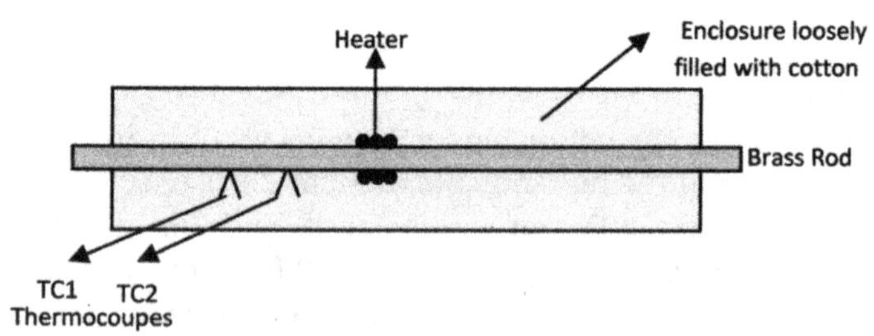

Figure 8. Schematic diagram of a setup for measuring thermal diffusivity (reprinted with permission from Ref. 5).

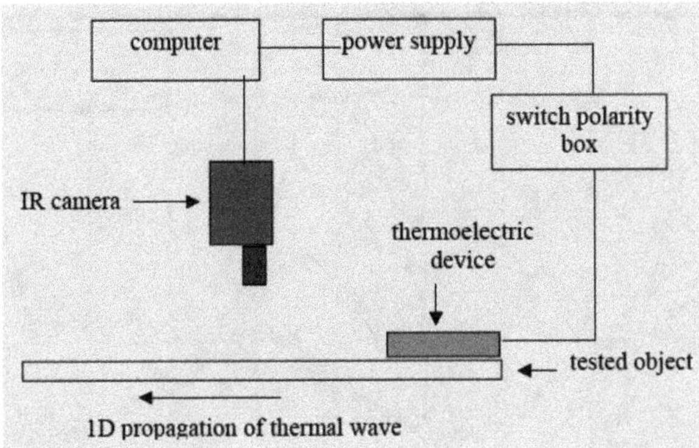

Figure 9. Schematic of the setup for thermal diffusivity measurements using a thermal wave (reprinted with permission from Ref. 4).

current being controlled by a computer. The temperature at any point is measured by an IR camera for several cycles. The camera is moved to measure the temperature wave at different points along the slab.

Figure 10 shows the temperatures recorded at three points along the rod as a function of time. Note that the period of the heater current ranges from 250 to 300 s.

The amplitude and phase of the sine wave are found from the above graph at different points. From this, the constants α and β are found, and the diffusivity is evaluated.

4.2. *Laser flash method*

In this method, the top surface of a film or plate with surface area A and thickness d is uniformly irradiated with a delta-function laser pulse of energy Q. The temperature on the bottom surface is measured as a function of time. When no heat is lost to the surroundings, the temperature rises after a long time to θ_∞ given by

$$\theta_\infty = Q/[(\rho A d)c_m] \tag{37}$$

One measures the time $t_{1/2}$ required for the bottom surface temperature to increase by $\theta_\infty/2$. The thermal diffusivity can then be evaluated from

$$D = 0.1388 d^2/t_{1/2} \tag{38}$$

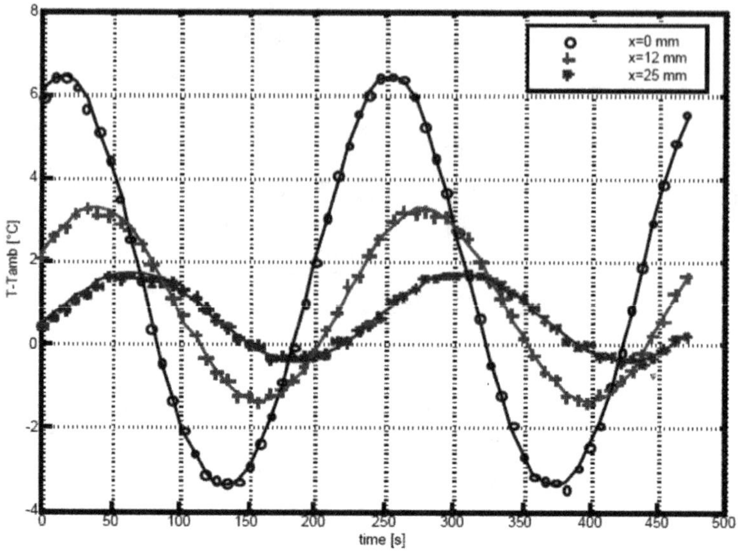

Figure 10. Temperature vs. time at three points of the SS strip in the thermal wave method (reprinted with permission from Ref. 4).

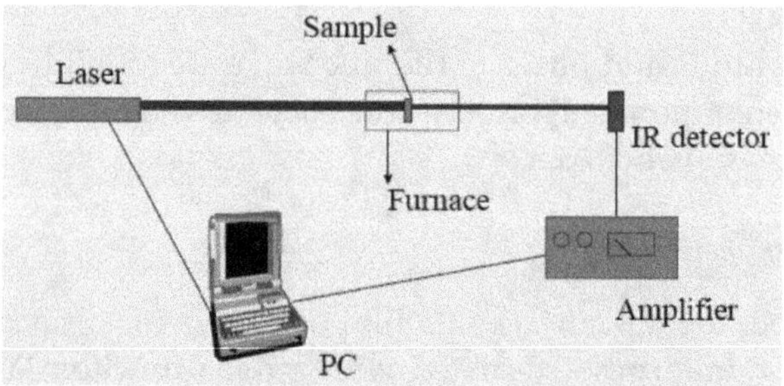

Figure 11. A schematic diagram of the experimental arrangement for laser flash method of measuring thermal diffusivity (reprinted with permission from Ref. 4).

An experimental arrangement for measuring the thermal diffusivity of a SS disk at different temperatures is described in Ref. 4. A schematic diagram of the experimental arrangement is shown in Figure 11.

The sample, S, of stainless steel is a disk with a diameter of 10 mm and a thickness of 1.474 mm. It is in a molybdenum furnace, which can go up to 1500°C. The furnace and sample are kept in a vacuum chamber.

The sample is heated by a pulse of light at 1.06 μm wavelength from a Nd-YAG laser. The light spot is circular with a uniform intensity over the cross-section of the circle. One can vary the energy of the pulse from 8 to 50 J. The temperature of the rear surface of the SS disk is sensed by an IR camera. The temperature is measured as a function of time, and $t_{1/2}$ is estimated.

There are other methods described in Ref. 4. However, the theory involved in these methods is too complicated and beyond the scope of this book.

5. Conclusion

The literature on thermal conductivity and thermal diffusivity is extensive. We have discussed some of the common methods for measuring thermal conductivity and thermal diffusivity. The interested reader will have to refer to the extensive literature available on the World Wide Web.

References

1. Berman, R. and Brock, J. C. F. (1965). *Proc. R. Soc. Lond. Ser. A Math. Phys. Sci.* **289**, 46, Downloaded from http://rspa.royalsocietypublishing.org/ on January 26, 2018.
2. Llavona, M. A., Zapico, R., Blanco, F., Verdeja, L. F., and Sancho, J. P. (1991). Methods of measuring thermal conductivity. *Rev. Minas*. 89.
3. Ravindran, T. R., Sankaranarayanan, V., and Srinivasan, R. (1992). *Pramana J. Phys.* **39**, 109.
4. Cemushi, F. Comparison of thermal diffusivity measurement techniques, http://www.qirt.org › archives › qirt2002 › papers.
5. Srinivasan, R., Ramesh, T. G., and Priolkar, K. R. (2018). *Experiments in Physics — A Laboratory Manual*. Indian Academy of Sciences, Bangalore.

Part III.2

Electrical Transport Properties

Chapter 13

ELECTRICAL CONDUCTIVITY OF METALS AND SEMICONDUCTORS

1. Introduction

Electrical conductivity has the unique distinction among the various physical properties of a material in that it exhibits the widest range in its magnitude, from nearly $10^6\Omega^{-1}\text{cm}^{-1}$, which is typical of a metal, to values smaller than $10^{-15}\Omega^{-1}\text{cm}^{-1}$, appropriate to an insulator. Semiconductors have electrical conductivity values in between these two extremes, in the range 10^{-5}–10^3 $\Omega^{-1}\text{cm}^{-1}$. The theoretical explanation seeking answers to questions like why some materials crystallize as metals and others as insulators or semiconductors is at the heart of the band theory of solids. In what follows, we present the simple yet highly successful Drude theory of the electrical conductivity of metals, which was developed prior to the advent of quantum mechanics. The subsequent developments, namely the Sommerfeld model and the band theory of solids, are discussed in some detail. Important concepts, such as "crystal momentum," the velocity of a Bloch wave, and the effective mass of a Bloch electron, are discussed with reference to their roles in determining the electrical conductivity of metals and semiconductors. The energy band structures of both intrinsic and extrinsic semiconductors are discussed, emphasizing the role of the chemical potential. The techniques of measuring the bulk resistivity of metallic samples in different geometries, namely infinite or bulk samples, essentially two-dimensional thin metallic sheets, and close to one-dimensional geometry applicable to samples in wire form, are discussed. The van der Pauw technique of measuring the electrical resistivity of especially irregular samples of either semiconductors or metallic sheets is discussed, and some typical data for a stainless-steel sheet cut in different shapes are presented. High-resolution resistivity measurement as a function of temperature near a critical point is of great interest. An AC resistivity technique based on phase-sensitive detection is discussed, and some typical results on the ferromagnetic–paramagnetic transition in nickel are presented. Comprehensive treatments of the theoretical framework discussed in this chapter can be found in Refs. 1–4.

2. Drude Theory of Electrical Conductivity

Metals exhibit striking properties such as high electrical and thermal conductivity and also high reflectivity in the optical region of the spectrum. Soon after the discovery of the electron by J. J. Thomson, Drude developed a remarkable theory of electrical and thermal conduction in metals. The theory is simple and agrees well with many experimental results. It equally fails to explain other observations. Lorentz carried Drude's postulates to their logical conclusion in an extensive treatment of the problem. In particular, he developed a model to study the response of the electrons to electromagnetic waves and treated thermoelectric effects in great detail.

Drude reasoned that the close proximity of atoms in a metal would lead to electrons getting detached from their parent atom, and these electrons are free to move in the solid. We call such electrons as mobile or itinerant electrons, and these are responsible for the high electrical conductivity of metals. Consider a typical alkali metal, such as sodium. The sodium atom has the electronic configuration $1s^2 2s^2 2p^6 3s^1$. The outermost 3s electron is loosely bound to the atom and gets detached from the parent atom in the solid state. This itinerant electron roams freely against the background of positively charged Na^+ ion cores. Thus, a monovalent metal containing n atoms will have n free or conduction electrons and n positive ion cores.

Drude treated the free electrons as a classical ideal gas obeying the Maxwell–Boltzmann statistics. Furthermore, he made the important assumption that electrons should collide with the stationary ions which form the background and not with each other. Electron–electron collision conserves momentum and would not contribute to the electrical resistivity of a metal.

We now develop the theory of conduction in some detail, which shows the essential features of a more sophisticated theory.

The electrical conductivity σ is given by the relation

$$J = \sigma E \quad (1)$$

where J is the electric current density and E is the applied electric field. In the absence of an electric field, the free electrons are moving in random directions, and there is no current flow. The velocity distribution is essentially isotropic. If we apply an electric field in the x-direction, the electrons acquire an average velocity $<v_x>$ in the x-direction, where the average is taken over the electrons per unit volume. $<v_x>$ is also known as the drift velocity v_d. From a microscopic viewpoint, electric current is ascribed to the flow of electrons. Consider a conductor of unit area of cross-section normal to the x-direction, and subject it to an electric field in the x-direction. The current density J_x is the flow of electrons contained in a volume element of length v_d and unit area of cross-section. If n is the number of conduction electrons per unit volume, the current density turns out to be

$$J_x = -nev_d \quad (2)$$

where $-e$ is the electronic charge. The drift velocity is related to the applied electric field, and we now derive this relation based on the physical processes visualized in the Drude model. A free electron subjected to an electric field E would acquire an acceleration, which is given by the expression

$$a = -eE/m \quad (3)$$

Since E is a constant, the drift velocity (as well as the current density) continuously increases with time without any limit, and one cannot achieve a steady state as required by Ohm's law. We have to invoke some kind of "frictional" force, which, together with the external electric field, would lead to a finite drift velocity. The origin of this frictional force in the Drude model is the collision of an electron with the positive ions that form the background. It is assumed that in this collision process, the electron loses all the energy it has acquired due to the electric field, and its velocity after the collision is completely random. In simple terms, this implies that the electron after the collision has no memory of the situation before the collision. The rate of change of the average or drift velocity due to electric field alone is given by the relation

$$[\partial v_d/\partial t]_{\text{field}} = -eE/m \tag{4}$$

The rate of change of the average velocity due to the collision process alone is given by the expression

$$[\partial v_d/\partial t]_{\text{collision}} = -v_d\ /\tau \tag{5}$$

where $1/\tau$ is the probability of a collision per second, and it is assumed that after the collision, the velocity is random. Thus,

$$[\partial v_d/\partial t]_{\text{total}} = [\partial v_d/\partial t]_{\text{field}} + [\partial v_d/\partial t]_{\text{collision}} = -eE/m - v_d/\tau \tag{6}$$

In the steady state, $[\partial v_d/\partial t]_{\text{total}} = 0$. This leads to the following expression for the drift velocity:

$$v_d = -eE\tau/m \tag{7}$$

Hence, the expression for electrical conductivity σ turns out to be

$$\sigma = ne^2\tau/m \tag{8}$$

This is the celebrated Drude formula for the electrical conductivity of a metal.

Let us explore the physical significance of the quantity τ in the above expression. After attaining steady state, suppose we switch off the electric field at, say, time $t = 0$. The average drift velocity corresponding to the steady-state value should gradually approach zero due to collisions, and the expression for the average drift velocity as a function of time is given by

$$v_d(t) = v_d(0)\exp(-t/\tau) \tag{9}$$

where $v_d(0)$ is the average drift velocity at $t = 0$. Because of the exponential form, the quantity τ is denoted as the relaxation time. It must be mentioned that τ is a phenomenological parameter in this theory, and there was no way to calculate it through first principles. However, we can estimate the order of magnitude of τ by invoking the concept of **mean free**

path, ℓ_0, which is the characteristic distance traveled by an electron between two collisions. Clearly,

$$\ell_0 = v\tau \tag{10}$$

and v can be estimated using the equipartition theorem:

$$(1/2)mv^2 = (3/2)kT \tag{11}$$

Using the known electron mass and the Boltzmann constant, $v \approx 10^7$ cm/s at room temperature. Drude visualized the collision process as one in which the electron bumps into stationary positive ions in the background, and we can assume ℓ_0 to be in the range of 1–10 Å. The estimate of τ using these values turns out to be in the range of 10^{-14}–10^{-15} s. We can also estimate τ by measuring the conductivity of a monovalent metal and noting that the conduction electron density is in the range of 10^{22} per cm^3. This value is in the same range as 10^{-14}–10^{-15} s. However, serious discrepancies occur at low temperatures, especially in pure crystals, and the mean free path is orders of magnitude higher, in the range of 2000–5000 Å. Such large values of mean free path cannot be comprehended in the Drude model.

Let us now turn our attention to the prediction of the Drude model regarding the temperature dependence of the electrical conductivity. The mean free path is not temperature dependent; therefore, the temperature dependence of τ comes from the temperature dependence of the average drift velocity acquired by the electron due to an applied electric field. It is clear from the equipartition theorem that v would vary as $\sqrt{T}$. Thus, σ, the electrical conductivity, would vary inversely as $\sqrt{T}$ or the electrical resistivity, ρ, would vary directly as $\sqrt{T}$. Typical behavior of the electrical resistivity of a pure metal over a wide temperature range shows that the electrical resistivity varies as T in the high-temperature region and varies roughly as T^5 in the low-temperature region. **The Drude model thus predicts a wrong temperature dependence for the electrical resistivity.**

3. Sommerfeld Model of Free Electron Gas

Arnold Sommerfeld developed a successful model of a metal, incorporating the idea that the electrons in a metal obey the laws of quantum mechanics and, importantly, obey the Pauli exclusion principle. Although electrons are strongly interacting among themselves and with the Coulomb attraction due to the positive ions, the free electron gas model, wherein we neglect these interactions, works well. In quantum mechanics, the effect of one electron on all the others can, to a large extent, be *averaged*. Thus, an electron can be treated as if it were moving in the field of the positive ions and in the average field of all the other electrons. This average field is obtained by the "self-consistent" field method. We may note that when an electron passes near a positive ion, its velocity is large, and thus, it spends only a fraction of its time in the region of the strong Coulomb field of the ion. The requirement that the wave function of the free electron be orthogonal to all the core wave functions of the ion leads to a strong gradient of the wave function. This leads to the concept of conduction electrons interacting with the lattice essentially through a weaker *pseudopotental*. This is particularly true in the case of alkali metals. Let us now turn our attention to electron–electron interactions. There are primarily two reasons for the weakness of this interaction. First, the Pauli exclusion principle forces electrons with the same spin to stay away from each other. Second, even electrons of opposite spin tend to be apart to minimize the strong Coulomb repulsion energy. Thus, each electron is surrounded by a spherical region devoid of other electrons. This hole (called the exchange and correlation hole) is roughly a spherical region with a radius of 1 Å for typical electron densities in a metal. If we consider the interaction between two electrons, the other electrons redistribute themselves in such a manner that their interaction is screened. These arguments justify the *free electron gas* model.

3.1. *One-dimensional metal*

Let us consider, for simplicity, a one-dimensional metal in which the free electrons move along, say, the x-axis. The electrons move in a potential well in which the potential energy φ is zero within the metal and rises to some finite value at the boundaries on either side. The electron, although a particle with well-defined charge and mass, exhibits wave-like properties in the quantum regime and is described by a wave function $\psi(x)$, satisfying the Schrödinger wave equation:

$$-(\hbar^2/2m)\, d^2\psi/dx^2 + \varphi(x)\, \psi = E\psi \tag{12a}$$

$$\varphi(x) = 0 \quad 0 < x < L \tag{12b}$$

$$\varphi(x) = \infty \quad x > L \quad \text{and} \quad < 0 \tag{12c}$$

The wave function $\psi(x)$ must vanish at $x = 0$ and at $x = L$. This is satisfied if the wavefunction is sine-wave-like, and the length L is an integral multiple of half wavelengths. Thus,

$$\psi_n = A\sin\left(\frac{2\pi}{\lambda_n}x\right) \quad \text{and} \quad L = \frac{n\lambda_n}{2} \tag{13}$$

Solving the Schrödinger wave equation, the allowed energies E_n are given by

$$E_n = \frac{\hbar^2}{2m}\left(\frac{n\pi}{L}\right)^2 \tag{14}$$

The wave functions for $n = 1$, $n = 2$, and $n = 3$ are shown in Figure 1.

The greatest advance made by Sommerfeld was the application of the Pauli exclusion principle to filling up the allowed energy states for the

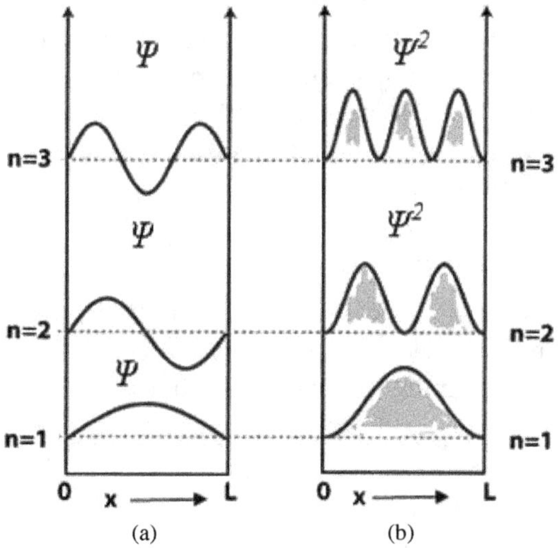

Figure 1. (a) Allowed energy levels and corresponding wave functions for an infinitely deep potential well. (b) Corresponding probability densities for $n = 1$, $n = 2$, and $n = 3$.

electron. According to this principle, not more than two electrons (one with each spin direction) can have the same wave function. The symmetry of the wave function with respect to the exchange of identical (indistinguishable) particles allows only two possibilities, namely symmetric or antisymmetric. Quantum particles, such as electrons with half-integral spins are characterized by an antisymmetric wave function, while photons and phonons are characterized by a symmetric wave function. Thus, not more than two electrons (one with each spin direction) can ever be in the same state. That is, not more than two electrons can have the same spatial part of a wave function. Applying these ideas to our one-dimensional metal, we note that the lowest energy state corresponding to $n = 1$ will have two electrons, the next state will have two electrons, and so on. Assume that our one-dimensional metal has N atoms (an even number), and each atom contributes one free electron. These N electrons are distributed over the states $n = 1, 2, \ldots, N/2$, with an occupancy of two electrons in each state. The condition $2n_F = N$ thus determines the energy of the highest occupied state, which we call the **Fermi energy E_F**. We therefore have for E_F the relation

$$E_F = \frac{\hbar^2}{2m}\left[\frac{N\pi}{2L}\right]^2 \tag{15}$$

It is clear that E_F corresponds to an electron having a wavelength of the order of atomic dimensions; therefore, its kinetic energy is in the range of several electron volts. **The most important conclusion we draw is that the electrons in a one-dimensional metal have energies lying between zero and a certain maximum E_F.**

3.2. *Three-dimensional metal*

The earlier treatment can easily be extended to the case of a three-dimensional metal. Consider a piece of metal in the form of a cube of side L. The wave function has to satisfy the boundary conditions that ψ must vanish at $x = 0$, $x = L$; $y = 0$, $y = L$; $z = 0$, and $z = L$. In three dimensions, the energy for a wave vector $\boldsymbol{k}$ is

$$E_k = \frac{\hbar^2}{2m}(k_x^2 + k_y^2 + k_z^2) \tag{16}$$

Since the boundary conditions at the edges of the cube of length L demand that $k_x = n_1\pi/L$, $k_y = n_2\pi/L$, and $k_z = n_3\pi/L$, with n_1, n_2, and n_3 being positive integers, the corresponding energy values are

$$E = \frac{\hbar^2}{2m}\frac{\pi^2}{L^2}(n_1^2 + n_2^2 + n_3^2) \tag{17}$$

The components of the wave vector $\boldsymbol{k}$, namely $(\pi/L)n_1$, $(\pi/L)n_2$, and $(\pi/L)n_3$, are the quantum numbers of the electronic state along with the spin quantum number $\boldsymbol{m}_s$.

The Fermi energy corresponding to a sphere of radius k_F is given by the relation

$$E_F = \frac{\hbar^2}{2m}k_F^2 \tag{18}$$

We note that there is one allowed wave vector for a distinct triplet of quantum numbers k_x, k_y, and k_z for the volume element $(\frac{\pi}{L})^3$, with an occupancy of two, corresponding to the two orientations of spin. k_x, k_y, and k_z must be positive and lie in the positive octant of a sphere of radius k_F, occupying a volume of $(1/8)(4\pi k_F^3/3)$. The total number of states in a sphere of this volume is given by

$$2\left(\frac{1}{8}\right)\left[\frac{4\pi k_F^3}{3}\right] \Big/ \left[\frac{\pi}{L}\right]^3 = \frac{V}{3\pi^2}k_F^3 = N \tag{19}$$

The Fermi wave vector k_F, which depends only on the electron concentration, is thus given by

$$k_F = \left[\frac{3\pi^2 N}{V}\right]^{1/3} \tag{20}$$

The Fermi energy is thus given by the relation

$$E_F = \frac{\hbar^2}{2m}\left[\frac{3\pi^2 N}{V}\right]^{2/3} \tag{21}$$

For a typical value of carrier concentration of the order of $10^{22}/\text{cm}^3$, $E_F \approx 5$ eV.

From the above expression of the Fermi energy, we can also derive an expression for the density of states $N(E)$, i.e., the number of electronic states per unit energy interval. Noting that the relation holds good for any energy value less than E_F, differentiating that expression with respect to energy yields

$$N(E)\mathrm{d}E = \frac{V}{4\pi^2}\left[\frac{2m}{\hbar^2}\right]^{3/2}\sqrt{E}\mathrm{d}E \tag{22}$$

The density-of-states energy curve for free electrons is parabolic in nature.

The Fermi energy is sometimes expressed as an equivalent temperature using the relation $E_F = kT_F$. T_F is called the Fermi temperature. For $E_F = 5$ eV, $T_F = 60{,}000$ K. It is worth noting that such a high Fermi temperature is a consequence of the Pauli exclusion principle.

3.3. *Fermi–Dirac distribution function*

In the previous section, we considered the ground-state properties of a free electron system incorporating the Pauli exclusion principle, which describes the properties at $T = 0$ K. At finite temperatures, thermal energy would excite some electrons with $k \leq k_F$ to states $k > k_F$ without violating the Pauli exclusion principle. We now consider the Fermi–Dirac distribution function, which determines the distribution of electronic states at finite temperatures. We also deduce the temperature dependence of several physical properties, such as electronic specific heat and electrical and thermal conductivities.

We are familiar with the velocity distribution law applicable to gas molecules, namely the Maxwell–Boltzmann statistics, wherein the molecules are distinguishable in principle. However, electrons are quantum particles with a spin of ½, obeying the Pauli exclusion principle, and are **indistinguishable**. These two attributes imposed by quantum mechanics lead to a new quantum statistics, namely the **Fermi–Dirac statistics** applicable to all quantum particles with nonintegral spin. The Fermi–Dirac distribution function, $f(E)$, which gives the probability that an orbital with energy E is occupied by an electron at temperature T, is given by

$$f(E) = \frac{1}{e^{(E-\mu)/k_BT} + 1} \tag{23}$$

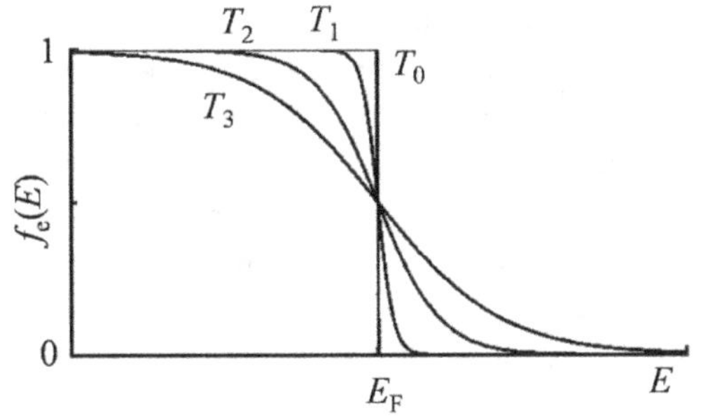

Figure 2. Fermi–Dirac distribution function as a function of temperature.

Here, μ is called **chemical potential.** It is chosen such that the total number of electrons (distributed in different energy states) equals N, the density of free electrons. Figure 2 gives a plot of the distribution function at $T = 0$ K and at higher temperatures.

At absolute zero, it is a step function, with $f(E) = 1$ for all energies $E < E_F$ and $f(E) = 0$ for all energies $E > E_F$. Furthermore, $\mu = E_F$ at $T = 0$. This means that all electronic states below E_F are occupied with a probability of 1, and all states above E_F are empty. At all temperatures, $f(E) = 1/2$ when $E = \mu$. To a first approximation, μ is close to Fermi energy.

It may be noted that for $E - \mu \gg k_B T$, the exponential term dominates in the denominator so that $\boldsymbol{f(E)} = \mathbf{e}^{\frac{\mu - E}{k_B T}}$, the familiar Maxwell–Boltzmann distribution. Thus, at very high temperatures, the Fermi–Dirac statistics approach the classical Maxwell–Boltzmann statistics.

The physical implications of quantum statistics are profound. First, electrons have a kinetic energy even at $T = 0$ K. The Fermi velocity V_F is high, of the order of 10^8 cm/s. It turns out that only those electron states near the Fermi energy play a dominant role in thermal and transport properties.

3.4. *Electrical conductivity*

In 1928, Sommerfeld revisited the Drude model of conductivity by replacing the Maxwell–Boltzmann classical statistics with the Fermi–Dirac statistics. The collision process required for the electron system to approach a steady state was treated in the same way by introducing a phenomenological relaxation time. The actual mechanism of interaction between the electrons and the lattice was not investigated, except for the fact that τ was assumed to depend only on the energy of the electron. We have seen earlier that the electron states with well-defined wave vectors form a lattice in $\boldsymbol{k}$-space, and electrons fill up to a wave vector k_F, the Fermi wave vector. The Fermi surface is a sphere in the wave vector space defined by k_X, k_Y, and k_Z. Figure 3(a) depicts the Fermi sphere in the equilibrium state. Here, the sphere is centered at the origin. Although electrons are moving at high velocities, there is no net current because, for every electron with wave vector $\boldsymbol{k}$, there is another electron with wave vector $-\boldsymbol{k}$. Thus, the total current vanishes due to pairwise cancelation of electron currents. Let us now apply an electric field $\boldsymbol{E}_X$ along the X-direction. Each of the wave vectors will change by $\delta k_X = \frac{eE_X}{\hbar} t$ after a time interval t. If there were no collisions, the Fermi sphere will continuously drift at a uniform rate. For a collision time $t = \tau$, $\delta k_X = \frac{eE_X}{\hbar}\tau$, and the entire Fermi sphere would be displaced by this quantity, and the electron system reaches a steady state (Figure 3(b)). The drift velocity is thus given by the relation

$$v_d = \frac{eE\tau}{m} \tag{24}$$

If n is the number of electrons per unit volume, then the current density J is given by the relation

$$J = nev_{\text{drift}} = \frac{ne^2\tau}{m}E \tag{25}$$

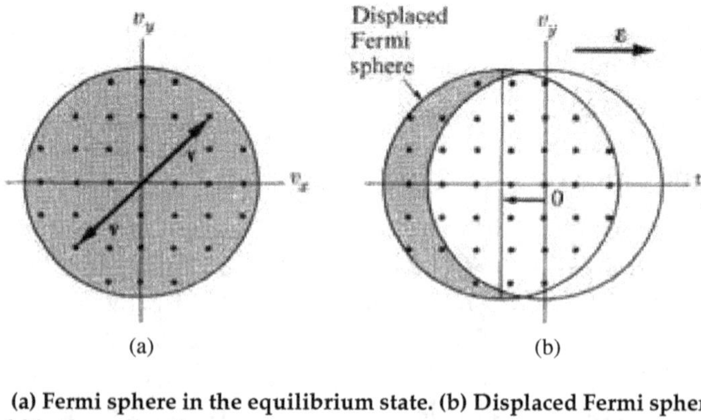

Figure 3. (a) Fermi sphere in the equilibrium state. (b) Displaced Fermi sphere due to an electric field in the x-direction.

The resulting expression for the electrical conductivity is

$$\sigma = \frac{ne^2\tau}{m} \tag{26}$$

This expression is the same as the one derived by Drude, employing classical statistics. However, the actual picture of electrical conduction is conceptually quite different from the one envisaged in the Drude model.

In the Drude model, current is carried by all electrons with a mean drift velocity v_d. Here, the shift of the Fermi sphere due to the applied field is pretty small, which makes the electrons in the overlapping area to completely cancel their individual currents. Only electrons in the shaded crescent area remain uncompensated and contribute to the current. The fraction of electrons that remain uncompensated is roughly v_d/v_F, where v_F is the Fermi velocity. Thus, the concentration of uncompensated electrons is $\sim n(v_d/v_F)$ and since each electron has a velocity close to v_F, the current density is given by

$$J = ne(v_d/v_F)v_F = nev_d \tag{27}$$

Thus, the current is carried by a few electrons located close to the Fermi surface and moving at high velocities of the order of v_F. As shown above, this is equivalent to a current being carried by all the electrons but moving with a mean drift velocity which is orders of magnitude smaller than the Fermi velocity. The other important change is that only the relaxation time relevant to electrons near the Fermi surface, i.e., τ_F, is part of the expression for conductivity. Thus,

$$\sigma = \frac{ne^2\tau_F}{m^*} = \frac{ne^2\Lambda_F}{m^* v_F} \tag{28}$$

where m^* is the effective mass of the electron. We also note that the mean free path $\Lambda_F = v_F\tau_F$. Since V_F is independent of temperature, the temperature dependence of conductivity is essentially hidden in the quantity τ_F. From an experimental measurement of electrical conductivity and from knowledge of Fermi energy, one can estimate the mean free path Λ_F. It turns out that at ambient temperatures, for many alkali and monovalent metals, the mean free path is several hundred angstroms. For silver metal at 0°C, $\Lambda_F \approx 570$ Å, a very large value as compared to interatomic distances. At low temperatures, in a pure metallic crystal, the mean free path would be in the range of 1 cm. This puzzle of electrons traveling distances in the range of 10^8 interatomic spacings remained unexplained in both the Drude and Sommerfeld models. This puzzle found a natural explanation in the band theory of solids developed by Bloch and others. We consider this aspect in some detail in a later section. We only note here that the electrical conductivity of a perfect crystal with the ions sitting in their pristine lattice positions is infinitely large. Any deviation from periodicity would be a source of electrical resistance. We can broadly classify these deviations into two categories:

(a) lattice vibrations (phonons), which correspond to vibration of the ions from their equilibrium positions;
(b) all static imperfections, such as impurities or crystal defects.

The probabilities of electrons being scattered by lattice vibrations (phonons) or by impurities are additive because they are independent processes. Noting that $\frac{1}{\tau}$ represents the probability per unit time that an electron is scattered, we have

$$\frac{1}{\tau} = \frac{1}{\tau_{\text{ph}}} + \frac{1}{\tau_i} \tag{29}$$

where the first term on the right is due to scattering from phonons and the second term is due to impurities. Thus, the resistivity can be written as

$$\rho = \rho_i + \rho_{\text{ph}}(T) = \frac{m^*}{ne^2}\frac{1}{\tau_i} + \frac{m^*}{ne^2}\frac{1}{\tau_{\text{ph}}} \tag{30}$$

The term ρ_i is due to impurity scattering and is independent of temperature. It is generally called **residual resistivity.** ρ_{ph} is due to scattering by phonons and depends on temperature. ρ_{ph} is also known as **ideal resistivity** because it is always present even in a pure sample. If we define ℓ_{ph} as the mean free path due to scattering of electrons due to lattice vibrations, we may write

$$\ell_{\text{ph}} = 1/(N_{\text{ion}}\sigma_{\text{ion}}) \tag{31}$$

where N_{ion} is the concentration of the positive ions in the lattice and σ_{ion} is the scattering cross-section of the vibrating ion. It is an effective area presented by the thermally vibrating ion to the electron. If x is the deviation of the ion from its equilibrium position, then it is straightforward to see that $\sigma_{ion} \approx \pi <x^2>$, where $<x^2>$ is the thermal average of x^2. $<x^2>$ can be easily computed by treating the vibrating ion as a quantum oscillator. Thus,

$$(1/2)m\omega^2<x^2> = E = \hbar\omega/[\exp(\hbar\omega/kT) - 1] \tag{32}$$

We first evaluate this expression at the high-temperature limit. For simplicity, we set $\hbar\omega \ll kT$. That is, we are essentially considering the high-temperature regime, $T > \theta_D = (\frac{\hbar\omega}{k})$, the Debye temperature. Then, Equation (33) simplifies to

$$<E> \approx \hbar\omega/(\hbar\omega/kT) = kT \tag{33}$$

This is simply the result of the equipartition theorem. Thus,

$$\rho_{ph}(T) \propto \left(\frac{\pi\hbar^2}{k\theta_D m}\right)\frac{T}{\theta_D} \tag{34}$$

The resistivity thus increases linearly with temperature, in general agreement with experiments.

In the low-temperature regime, the above equation is inadequate, and one has to take into account the full Debye spectrum of lattice vibrations. A detailed analysis shows that resistivity varies as T^5.

4. Band Theory of Solids

The Sommerfeld model is oversimplified because it completely neglects the periodic potential of the lattice while discussing electronic and thermal transport in metals. This model could not provide a clue to understanding why some chemical elements crystallize to form good metals, others to insulators, and some others to semiconductors with remarkable electronic transport behavior as a function of temperature. It is worth mentioning that among all the properties, electrical resistivity exhibits the widest variation of over 32 orders of magnitude between a good metal and an insulator. Another perplexing feature of note is that the magnitude of the mean free path for electrons in a pure metallic crystal at liquid helium temperatures can exceed 1 cm. Since the interatomic distance is of the order of a few Å, what this implies is that an electron encounters 10^7–10^8 positive ions on its way without being scattered. A realistic model of an electron interacting with the periodic potential of the lattice provides a natural explanation for these aspects. The wave aspect of the electron manifests itself remarkably in a periodic potential. First, the energy spectrum of the completely free electron gas in the Sommerfeld model acquires a new feature of **band gap** at well-defined points in $\boldsymbol{k}$-space. This feature provides a natural explanation for the distinction between metals, semiconductors, and insulators. The shape of the $\boldsymbol{E}$ versus $\boldsymbol{k}$ **dispersion curve** holds the key to the remarkable properties of electrons in a periodic lattice, wherein they respond to an external electric field as if the electrons are endowed with an effective mass, $\boldsymbol{m^*}$, which may be much larger or smaller than the free electron mass m and, more importantly, can even be **negative.**

The fact that an electron is a spin-½ particle obeying the Fermi–Dirac statistics formed the basis of the Sommerfeld model.

We now consider the other attribute of electrons, namely their wave nature, as they propagate in the periodic potential of the lattice. We show that the collective behavior of the interference of waves scattered from each ion in the lattice leads to **allowed** and **forbidden** regions in the energy spectrum.

4.1. *Bloch theorem*

The behavior of an electron in a crystalline solid is determined by solving the appropriate Schrodinger wave equation:

$$\left[-\frac{\hbar^2}{2m}\nabla^2 + V(r)\right]\psi(r) = E\psi(r) \tag{35}$$

where $V(r)$ is periodic and has the same translational symmetry as that of the lattice.

Thus, $V(\boldsymbol{r} + \boldsymbol{R}) = V(\boldsymbol{r})$, where $\boldsymbol{R}$ is a lattice vector. Bloch's theorem states that the wave function for a periodic potential $V(\boldsymbol{r})$ has the form

$$\psi_k(\boldsymbol{r}) = e^{ik.r}u_k(\boldsymbol{r}) \tag{36}$$

where the function $u_k(r)$ has the same translational symmetry as that of the lattice:

$$u_k(r + R) = u_k(r) \tag{37}$$

The vector k is related to the momentum of the electron in a way that is different from what we attribute to a free electron. A rigorous proof of the Bloch theorem can be found in standard texts on condensed matter physics. Here, we discuss the physical implications of the Bloch wave function. We note that in the limit of an infinitesimally small value of the periodic potential, the Bloch wave function approaches the free electron wave function, i.e., $\psi_k(r) = e^{ik.r}$.

The Bloch wave function incorporates the plane-wave character through this term, and the second term, $u_k(\mathbf{r})$, has the lattice periodicity. The effect of the function $u_k(\mathbf{r})$ is to modulate the traveling wave represented by the first term such that the amplitude oscillates periodically from one cell to the other.

4.2. *Concept of crystal momentum*

It is of great importance to learn how the physical quantities associated with the wave vector k in the Bloch wave differ from those derived from the free electron model. First, we note that $\hbar k$ is the eigenvalue of the momentum operator $\mathbf{i}\hbar\nabla$ in the free electron model. However, if it is operated on the Bloch wave function, we observe that $\hbar\mathbf{k}$ is no longer its eigenvalue. This is because the periodic ionic potential of the lattice exerts a force on the electron through the function $u_k(r)$. The Bloch wave for a position vector $r + L$, where L is a lattice vector, can be written as

$$\psi_k(r + L) = e^{ik\cdot L}\, \psi_k(r) \tag{38}$$

The reciprocal lattice vector G satisfies the relation $e^{(\pm iG\cdot L)} = 1$, where L is a lattice vector. We can thus replace the wave vector k of the Bloch wave with another wave vector satisfying the condition $k = k' \pm G$. Thus, we get the important result

$$\psi_{\mathrm{k}}(r + L) = e^{ik'\cdot L}\, \psi_{\mathrm{k}}(r) \tag{39}$$

The Bloch wave of wave vector k is equally describable in terms of the wave vector k' different from it by a reciprocal vector G.

This property is unique to the Bloch wave. If we multiply the wave vector relation by $\hbar$, we get

$$\hbar k = \hbar k' \pm \hbar G \tag{40}$$

This is the momentum conservation law for the Bloch electron, signifying that the Bloch electron **exchanges** its momentum with the lattice by an amount $\pm\hbar G$. An infinite array of identical lattice planes defines the reciprocal lattice vector G, and thus, $\hbar G$ is the momentum transferred to the lattice as a whole. **Thus, the momentum $\hbar k$ of the Bloch wave is not uniquely determined and involves momentum transfer to the crystal as a whole. We therefore call $\hbar k$ of the Bloch wave as the crystal momentum.**

We can now draw a parallel with Bragg's law of X-ray diffraction. The momentum conservation law is the same as the one for Bloch waves.

Just as X-rays go through a crystal unhindered when the Bragg condition is not met, the matter wave associated with the Bloch electron will propagate through the crystal without any attenuation for a general wave vector k not satisfying the momentum conservation law. **This remarkable feature of the Bloch wave explains why a perfect conductor at $T = 0$ K exhibits infinite conductivity.**

4.3. *Physical origin of band gap*

We now consider a nearly free electron model of a metal, such as sodium, wherein the deep periodic potential of the lattice is replaced by a shallow periodic potential called the **pseudopotential**. We note that the Bloch wave function has a largely smooth sinusoidal behavior in the region between the ion cores but exhibits a nodal behavior (with many wiggles) in the region of the ion cores. In many metals, the volume occupied by the ion cores is only a small part of the volume of the metal. The potential energy seen by the conduction electrons is thus weaker and smoothed over a large part of the volume. The wave function in these regions is more or less plane-wave-like, and their energy must depend on the wave vector approximately as $E_k = \frac{\hbar^2 k^2}{2m}$. We can treat the weak potential as a perturbation and study its effects on the energy spectrum of the electrons.

Let us consider a linear solid of lattice constant a and study the energy spectrum of an electron propagating in this periodic lattice. Figure 4(a) shows a plot of E versus k for a completely free electron,

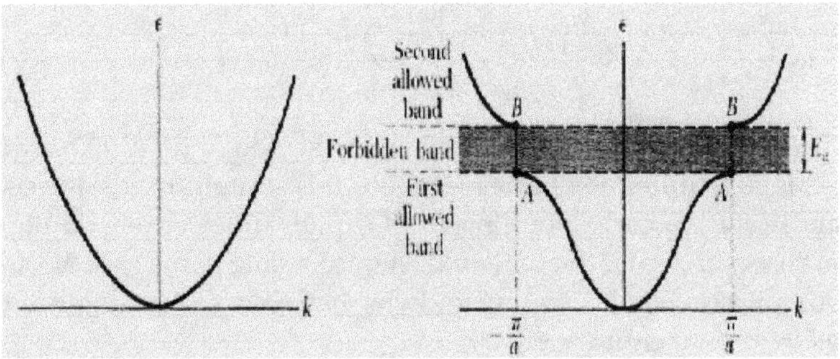

(a) (b)

Figure 4. Plot of E vs. k for (a) a free electron and (b) an electron in a monatomic linear lattice of lattice constant a. Energy gap E_g at $k = \pm\frac{\pi}{a}$.

wherein it exhibits quadratic behavior. Figure 4(b) gives a plot of E versus k for an electron in a monatomic linear lattice with lattice constant a. We observe discontinuities in the energy spectrum at well-defined values of $k = \pm\frac{\pi}{a}$. The Bragg condition of diffraction, i.e., $(k + G)^2 = k^2$, for this one-dimensional lattice turns out to be

$$k = \pm\frac{1}{2}G = \pm\frac{n\pi}{a} \tag{41}$$

We now consider the physical origin of the energy gap that occurs at the first Bragg reflection, wherein the wave vector satisfies the condition $k = \pm\frac{\pi}{a}$. Essentially, the wave reflected from one atom interferes constructively with the wave from the adjacent atom because the phase difference is $\pm 2\pi$.

Thus, at $k = \pm\frac{\pi}{a}$, an electron traveling along the $+x$ direction gets Bragg reflected along the $-x$ direction. This wave is again Bragg reflected, and the wave travels in the $+x$ direction. Each subsequent reflection reverses the direction of propagation of the electron. **This leads to a time-independent solution for a standing wave formed by a superposition of the forward and Bragg-reflected waves.** We now study the implications of this result on the energy versus wave vector plot for the electron.

First, we note that there are two ways of forming a standing wave out of the two propagating waves $e^{\frac{i\pi x}{a}}$ and $e^{\frac{-i\pi x}{a}}$:

$$\psi(+) = e^{\frac{i\pi x}{a}} + e^{\frac{-i\pi x}{a}} = 2\cos\left(\frac{\pi x}{a}\right) \tag{42a}$$

$$\psi(-) = e^{\frac{i\pi x}{a}} - e^{\frac{-i\pi x}{a}} = 2i\sin\left(\frac{\pi x}{a}\right) \tag{42b}$$

The (–) and (+) signs refer to whether they do or do not change sign when $-x$ is substituted for $+x$, respectively. It is straightforward to see that the standing wave corresponding to $\psi(+)$ piles up electrons on the positive ion cores (lowering the potential energy), while in the case of $\psi(-)$, the electrons pile up in the region midway between the neighboring ions (increasing the potential energy).

This difference in the potential energy occurring at $k = \pm\frac{\pi}{a}$, which is denoted by E_g, is the cause of the energy gap.

4.4. *Number of states in a band*

Consider a linear lattice of lattice constant $\boldsymbol{a}$. The allowed values of the electron wave vector $\boldsymbol{k}$ in the first Brillouin zone are given by $\boldsymbol{k} = 0$, $\boldsymbol{k} = \pm\frac{\pi}{L}$, $\boldsymbol{k} = \pm\frac{2\pi}{L},\ldots$, $\boldsymbol{k} = \pm\frac{N\pi}{L}$, where N is the total number of atoms in the lattice. Each primitive cell contributes exactly one independent value of $\boldsymbol{k}$ to each energy band. This result holds good even in three dimensions. ***N* atoms thus create *N* orbital states in a given band. Each of these orbital states can be occupied by two electrons of opposite spin, as dictated by the Pauli exclusion principle.**

4.5. *Distinction between metals, insulators, and semiconductors*

The existence of band gaps when the wave vector of the electron satisfies the Bragg condition of reflection leads to allowed bands separated by forbidden gaps in the $\boldsymbol{E}$ versus $\boldsymbol{k}$ dispersion curve for the Bloch electron.

Consider a solid containing N atoms. If each atom contributes one electron to the lattice, the band will be only half full because each orbital can be occupied by two electrons of opposite spin. We call this the conduction band of the metal, and the energy level of the highest occupied orbital corresponds to the Fermi energy. This is the general description of all monovalent metals, such as the alkali metals.

If, however, each of the N atoms contribute two electrons per atom, we have the situation wherein the band would be completely full. **A completely filled band cannot contribute to a current even when an external electric field is applied, as all the states in the band are full and the next available states lie in the forbidden band.** This is certainly true in a one-dimensional lattice, with each atom contributing two electrons. However, in three dimensions, along certain directions of the wave vector $\boldsymbol{k}$, the bands overlap, and electrons from the top of one band will spill over to the bottom states of the next band. Thus, divalent metals exist in nature. In some substances, the energy gap vanishes completely or the two bands overlap slightly, and these are known as "semimetals." Bismuth, antimony, and white tin all belong to this class of semimetals.

It is clear that a solid for which a certain number of bands are completely full and the bands above are completely empty would be an insulator. Typically, the forbidden energy gap would be very large for an insulator of the order of several electron volts, such as in diamond wherein it is around 7 eV.

Semiconductors have a band structure akin to that of an insulator, but they differ only in the magnitude of the band gap, which is typically of the order of 1 eV or less. Thermal excitation of electrons from a filled valence band to the empty conduction band at higher temperatures makes electrical conduction possible, and we would have what is called an **intrinsic semiconductor**. Silicon and germanium are typical semiconductors. It is worth mentioning that the distinction between a semiconductor and an insulator is only **quantitative in nature**. Extrinsic semiconductors, wherein the carrier concentration can be manipulated through "doping," form an important class of semiconductors.

4.6. *Velocity of the Bloch electron*

The study of the motion of a Bloch electron is of vital importance in understanding the effect of periodicity of the $\boldsymbol{E}$ versus $\boldsymbol{k}$ dispersion curve of a free electron. We have already seen that near a zone boundary, the shape of the dispersion curve changes markedly compared to the free electron dispersion curve. First, let us consider the case of a free electron. The velocity v is related to the wave vector $\boldsymbol{k}$ through the relation $v = \frac{\hbar k}{m}$. Clearly, the velocity is proportional to and parallel to $\boldsymbol{k}$. However, for the Bloch electron, the functional relationship is different, and since the wave aspect of propagation is important, the velocity is given by the group velocity of the wave packet. Thus,

$$v = \nabla_k \, \omega(\boldsymbol{k}) = \frac{1}{\hbar} \nabla_k E(\boldsymbol{k}) \tag{43}$$

The velocity of a Bloch electron in state $\boldsymbol{k}$ is directly proportional to the **gradient of the energy in *k*-space**. This relation means that the velocity is in the direction normal to the energy surface at the point of interest. In general, the energy contours are nonspherical; therefore, the velocity is not necessarily parallel to $\mathbf{k}$.

We now turn our attention to the seemingly peculiar behavior of a Bloch electron near the zone boundary. For a one-dimensional lattice, the expression for the velocity of a Bloch electron is given by

$$v = \frac{1}{\hbar}\frac{\partial \mathbf{E}}{\partial \mathbf{k}} \tag{44}$$

Thus, the velocity is proportional to the slope of the $\mathbf{E}$ versus $\mathbf{k}$ curve. As we move from the center of the zone, the velocity initially increases linearly (like in a free electron), reaches a maximum, and then **decreases to zero** at the zone boundary. We have already seen that the propagating electron wave becomes a standing wave at the zone boundary due to Bragg reflection, and the velocity is zero. Physically, as we approach the zone boundary, due to coherent scattering, a reflected wave develops, whose amplitude increases and equals that of the forward wave at the zone boundary. This process leads to the velocity of the Bloch electron decreasing to zero. The periodicity of the lattice thus leads to this strange behavior.

4.7. *Dynamical effective mass*

The Bloch electron exhibits strange behavior when subjected to an external electric field. The Bloch electron behaves as if it is endowed with an **effective mass m^*, which is very different from the normal electronic mass m**. This is also related to the nature of the $\mathbf{E}$ versus $\mathbf{k}$ dispersion curve. Let us calculate the acceleration of the electron a due to an applied electric field F in the one-dimensional case:

$$a = \frac{\mathbf{d}v}{\mathbf{d}t} \quad v = \frac{1}{\hbar}\frac{dE}{dk} \quad a = \frac{dv}{dk}\frac{dk}{dt} \quad \hbar\frac{\mathrm{d}k}{\mathrm{d}t} = F$$

This leads to

$$a = \frac{1}{\hbar^2}\frac{d^2E}{dk^2}F \tag{45}$$

This is equivalent to defining an effective mass m^*, given by the relation

$$m^* = \frac{\hbar^2}{\left(\dfrac{d^2E}{dk^2}\right)} \tag{46}$$

m^* is thus related inversely to the curvature of the dispersion curve. It must be emphasized that electrons physically weigh the same in these structures but respond to an impressed electric field as if endowed with an effective mass m^*.

The variation of the effective mass in a typical band structure can now be analyzed. Near the bottom of the band, the dispersion relation is akin to the free electron type, and m^* is close to the free electron mass m. However, as $\mathbf{k}$ increases, m^* also increases until one reaches the inflection point. Something dramatic happens beyond the inflection point. The curvature of the $\mathbf{E}$ versus $\mathbf{k}$ curve is now **negative**, and the effective mass becomes **negative. Negative effective mass implies that the acceleration is negative and opposite to the applied force *F*.**

This result implies that in this region of $\mathbf{k}$-space, the lattice exerts a retarding or breaking force on the electron of such magnitude that it overcomes the applied force F and produces a negative acceleration. Physically, an electron near the top of a band with a negative effective mass behaves as a positively charged particle. This is the genesis for the concept of a "hole." Instead of dealing with a negatively charged electron with a negative effective mass, we have a positively charged "hole" with a positive effective mass.

4.8. *Electrical conductivity of an intrinsic semiconductor*

It was noted earlier that a perfect semiconductor is essentially an insulator at $T = 0$. Thermal excitation of electrons from the top of the valence band to the electronic states at the bottom of the conduction band at higher temperature essentially determines the electrical conductivity of an undoped or intrinsic semiconductor. The electron concentration in the conduction band and the identical hole concentration in the valence band as functions of temperatures are determined by the Fermi–Dirac function. It may be noted that for $E - \mu \gg k_B T$, the exponential term dominates in the denominator of the Fermi–Dirac distribution function so that the term

1 can be neglected. The Fermi–Dirac statistics approache the classical Maxwell–Boltzmann statistics.

Figure 5 presents a schematic energy band diagram of an intrinsic semiconductor alongside the Fermi–Dirac distribution diagram. The significant difference when compared with the energy band diagram of a metal is that the Fermi energy or the chemical potential lies in the forbidden gap and is not a physically occupied electronic state. It is easy to visualize this by noting that at $T = 0$ K, the distribution function has a value of 1 for the valence band and 0 for the conduction band. Since the chemical potential μ is defined as that state for which the probability of occupation is ½, it is necessarily located in the forbidden gap. The entire energy distribution of the electrons in the conduction band falls in the Maxwell–Boltzmann tail of the distribution function.

It is conventional to choose the top of the valence band as the reference, or the zero of energy. The bottom of the conduction band is designated as the energy state E_g and corresponds to the energy gap of the material. Mathematically, the chemical potential is determined by the condition that at any temperature, **the electron concentration in the**

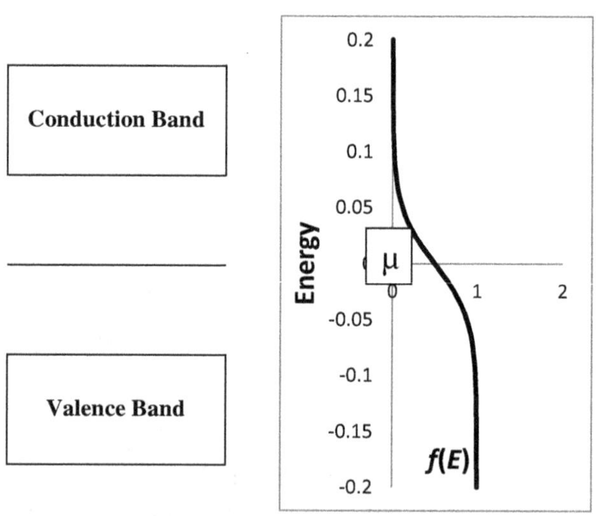

Figure 5. Schematic diagram of an intrinsic semiconductor along with the Fermi–Dirac distribution function. Note that the chemical potential μ or the Fermi energy lies in the forbidden band.

conduction band is identical to the hole concentration in the valence band:

$$f(E) = e^{\mu/kT} e^{-E/kT} \tag{47}$$

The electron concentration in the conduction band is given by the expression

$$n = \int_{E_g}^{\infty} f(E)Z(E)dE \tag{48}$$

Here, $Z(E)$ is the density of states relevant to the conduction electrons with an effective mass m_e and is given by the expression

$$Z(E) = \frac{1}{2\pi^2}\left(\frac{2m_e}{\hbar^2}\right)^{3/2}\left(E - E_g\right)^{1/2} \tag{49}$$

We have already noted that the relevant distribution function is given by the expression in Equation (47). The integral can be evaluated using these expressions, and one gets the result

$$n = 2\left(\frac{m_e kT}{2\pi\hbar^2}\right)^{3/2} e^{\mu/kT} e^{-E_g/kT} \tag{50}$$

Similarly, the hole concentration, p, can be evaluated and is given by

$$p = 2\left(\frac{m_h kT}{2\pi\hbar^2}\right)^{3/2} e^{-\mu/kT} \tag{51}$$

We now impose the condition $n = p$, as electrons in the conduction band originate from the thermal excitation of electrons in the valence band. The vacant states in the valence band are termed "holes," whose concentration is identical to the electron concentration in the conduction band. This conservation law leads to the following expression for the chemical potential. Thus, at a temperature T, the chemical potential μ is

$$\mu = \frac{1}{2}E_g + \frac{3}{4}\, kT \log\left(\frac{m_h}{m_e}\right) \tag{52}$$

Thus, at $T = 0$ K, the chemical potential is exactly midway between the valence and conduction bands. If the curvatures of these two bands are identical, with opposite signs leading to $m_e = m_h$, the chemical potential would not vary with temperature. Substituting this value of μ in the expression for n and p yields the following important result:

$$n = p = 2\left(\frac{kT}{2\pi\hbar^2}\right)^{3/2} (m_e m_h)^{3/4}\, e^{-E_g/2k_BT} \tag{53}$$

The significant feature of this expression is that the carrier concentration of an intrinsic semiconductor **increases exponentially with temperature**. The power-law dependence of $T^{3/2}$ is much weaker, and one can safely neglect this dependence. Thus, if one could measure n as a function of temperature, a plot of $\ln(n)$ against $1/T$ would yield a straight line with a slope of $-E_g/2k$. The energy gap E_g could thus be experimentally determined. This relation allows us to estimate the intrinsic carrier concentration in a typical semiconductor, such as silicon. With $E_g \cong 1.1$ eV, $m_e = m_h = m$ (free electron mass), and $T = 300$ K, the carrier concentration is $n \approx 10^{10}$ electrons/cm^3. It is worth noting that this is nearly 12 orders of magnitude lower than the carrier concentration in a good metal. The resistivity of intrinsic silicon near room temperature is ~0.26×10^6 Ωcm while that of a good metal is ~10^{-6} Ωcm and it correlates well with the electron concentration in these systems.

The expression for the electrical conductivity of a semiconductor is essentially the same as that of a metal, except that one has to take into account the contribution to the current from both electrons and holes. In general, the expression for the conductivity σ has the form

$$\sigma = \frac{ne^2\tau_e}{m_e} + \frac{pe^2\tau_h}{m_h} \tag{54}$$

In semiconductor physics, the conductivity is expressed in terms of another quantity called "**mobility,**" which is related to both the relaxation time τ and the effective mass of the current carrier involved. While discussing the Drude theory of electrical conductivity, we noted that the drift velocity v_d is given by the expression

$$v_d = -\frac{e\tau}{m}E \tag{55}$$

where E is the applied electric field, and the negative sign signifies that the electron moves in a direction opposite to the electric field. Mobility is defined as the drift velocity per unit electric field and is a measure of the ease with which electrons respond to an external electric field. The sign is disregarded in this definition. Mobility is expressed in cm²/Vs, and it can be experimentally measured. Thus, the mobilities of electrons and holes are given by the relations

$$\mu_e = \frac{e\tau_e}{m_e} \text{ and } \mu_h = \frac{e\tau_h}{m_h} \tag{56}$$

Typical values are $\mu_e \approx 1350$ cm²/Vs and $\mu_h \approx 475$ cm²/Vs for silicon and $\mu_e \approx 80{,}000$ cm²/Vs and $\mu_h \approx 750$ cm²/Vs for InSb, a compound semiconductor. The anomalously large electron mobility of InSb is due to $m_e \ll$ mass of the electron.

In this formulation, the expression for the conductivity turns out to be

$$\sigma = ne\mu_e + pe\mu_h \tag{57}$$

This relation is valid both for intrinsic semiconductors with $n = p$ and extrinsic semiconductors with $n \neq p$.

If we take a specific case of a doped semiconductor with predominantly one type of carrier, the measured Hall coefficient has a simple form of $R_H = 1/ne$ (discussed at the end of this chapter) so that the product $R_H\sigma = \mu_e$. Thus, an experimental measurement of the Hall coefficient and the electrical conductivity of a semiconductor as a function of temperature provides a wealth of information related to its energy band structure. It would suffice to mention here that a special technique called "Cyclotron resonance" is employed to get quantitative information about the effective mass of charge carriers in different directions in the lattice. These data, when combined with the experimentally determined mobility data, yield the corresponding values of the relaxation time.

4.9. *Electrical conductivity of an extrinsic semiconductor*

The traditional elemental semiconductors, germanium and silicon, belong to Group IV of the periodic table and crystallize into the diamond structure. This three-dimensional tetrahedral network features a Si atom covalently bonded to its four neighbors sharing two electrons in each bond. In the lattice, the ionic charge is completely balanced by the covalent electrons. This leads to an energy band structure with a completely filled valence band separated by a completely empty conduction band at $T = 0$ K. A significant advance in the semiconductor field took place when it was discovered that the carrier concentration and type of carriers can be controlled by doping with suitable impurities in required quantities. For silicon, if we substitute the Si atom in the lattice with a pentavalent impurity, such as phosphorus (P) or arsenic (As), four electrons participate in covalent bonding as in Si. The fifth electron is now loosely bound to the dopant impurity, which is now a positive ion. Typically, the binding energy of the extra electron is around 20–50 meV, and a small amount of thermal energy is sufficient to free this electron to the conduction band. The **donor** state thus lies just below the bottom of the conduction band within around 20–50 meV depending on the impurity. Typically, the doping concentration lies in the ppm–ppb range. Similarly, if a trivalent impurity, such as boron (B), aluminum (Al), or gallium (Ga), is substituted for the Si atom in the lattice, one of the covalent bonds would have an electron vacancy, which is filled up by an electron in the valence band. The valence band is now devoid of an electron whose dynamical properties are described in terms of holes. The energy level of this impurity, which is called an **acceptor**, lies just above the top of the valence band. Doped semiconductors containing predominantly itinerant or conduction electrons and a low concentration of holes are termed n-type semiconductors. Similarly, doped semiconductors containing acceptor impurities with majority carriers as holes are termed p-type semiconductors. We now briefly discuss the behavior of electrical conductivity of an n-type semiconductor as a function of temperature. At sufficiently low temperatures, the additional electron of the donor impurity would be bound to the ion and not take part in the conduction process. This is termed a "**frozen**" state (typically below 100 K), and the electrical conductivity would be very low as there are no free electrons in the conduction band. At temperatures above 100 K, all the bound electrons get delocalized, and the conduction electron density would attain a limiting value of N_d, the number of donor impurities per cm^3. The carrier concentration would be roughly independent of temperature (to within 10%) over a broad region above 100 K and extending to nearly 600 K. This region is termed "**extrinsic**," and the electrical conductivity exhibits weak temperature dependence, as in a metal, as determined by the power-law dependence of the mobility of electrons. At still higher temperatures, direct excitations of electrons from the valence band to the conduction band occur, leading to "**intrinsic**" behavior with the electrical conductivity **increasing** exponentially with temperature. It is clear that if one wants to extract the magnitude of the energy gap in a doped semiconductor, the fit of the conductivity data has to be done for the data points lying in the intrinsic region.

The position of the Fermi level is distinctively different in *n*- and *p*-type semiconductors. In an *n*-type semiconductor, there are more electrons in the conduction band compared to holes in the valence band. The probability of finding an electron near the conduction band edge far exceeds the probability of finding a hole near the top of the valence band. Thus, the Fermi level or chemical potential is located close to the conduction band edge in *n*-type semiconductors and close to the top of the valence band in *p*-type semiconductors. It is to be appreciated that the *n*- and *p*-type carrier concentrations cannot be large **simultaneously** due to the important relation that the product ***np* is a constant** for a given semiconductor at a given temperature. Combining the expressions for n and p, we have

$$np = 4\left(\frac{kT}{2\pi\hbar^2}\right)^{1/2} (m_e m_h)^{3/2}\, e^{-E_g/kT} \tag{58}$$

This expression is independent of chemical potential or Fermi energy and valid both for an **intrinsic semiconductor** and an **extrinsic semiconductor**. If n is very large, as in an *n*-type semiconductor, p has to decrease so that the product is held constant. The following example brings out this aspect clearly.

For silicon as an intrinsic semiconductor, $n = p$; therefore, $np = n_i^2$, where n_i is the intrinsic carrier concentration, which is ~10^{10} per cm^3 at room temperature. When it is doped with an *n*-type impurity with, say, $n = N_d \approx 10^{15}$ per cm^3, p can be estimated from the relation

$$p = \frac{n_i^2}{N_d} \approx 10^5 \text{ per cm}^3 \tag{59}$$

which is five orders of magnitude lower in value when compared with the corresponding value in an intrinsic semiconductor.

In summary, the technique of controlled doping in semiconductors paved the way for the development of numerous electronic and optical devices. We now consider some basic techniques involved in the measurement of electrical conductivity and the Hall coefficient at ambient and higher temperatures in both metals and semiconductors.

5. Resistivity Measurement Techniques for Bulk Samples, Thin Films, and Pellets and for Samples in Wire Form

In the realm of electronic properties, the measurement of the resistivity of a conducting sample in the form of bulk material, a thin or thick film, a homogeneous wire, or a pellet is of primary importance. The electrical resistivity, ρ, of a material is independent of its geometry and is characteristic of its electronic structure. It is worth noting that the resistivity is evaluated from a measurement of "resistance," which is dependent on the geometry of the sample.

In what follows, we investigate the profile of the potential distribution as one goes from three-dimensional systems (bulk sample) to essentially two-dimensional systems (thin films) and, ultimately, to systems close to one-dimensional systems, such as conducting wires.

5.1. *Bulk samples*

The electrical potential distribution in a conducting sample is essentially controlled by the linear dependence of the current density $\boldsymbol{J}$ on an applied electric field $\boldsymbol{E}$:

$$\boldsymbol{J} = \sigma \boldsymbol{E} = (1/\rho)\boldsymbol{E} \tag{60}$$

Consider the case of the injection of current I into an infinite bulk specimen of constant resistivity ρ through a single point-shaped probe somewhere in the interior. Charge conservation dictates that the total current, I, through a sphere of radius r centered on the probe must be constant, and thus, we have for the current density $J(r)$, the expression

$$J = \frac{\mathrm{I}}{4\pi r^2} = \frac{1}{\varrho}\left(-\frac{\mathrm{d}V}{\mathrm{d}r}\right) \tag{61}$$

where V is the potential. Integrating this expression leads to an expression for the spatial variation of the potential:

$$V(r) = I\rho/(4\pi r) \tag{62}$$

Since I is constant, the potential falls off as $1/r$. This formula for potential distribution can be used in conjunction with the **collinear four-probe arrangement** to arrive at a relation for the bulk resistivity.

5.1.1. Collinear four-probe resistivity technique

The schematic diagram of the collinear four-probe resistivity technique is given in Figure 6.

In this arrangement, the sample is contacted by four collinear probes of negligible size compared to any sample dimension. The outer two probes carry a current I, while the inner two probes measure a potential difference ΔV. From the measured potential difference and the known source current, the resistivity of the sample may be determined. We need only determine for various sample geometries how the current flows and the resultant potential difference. We now consider the specific case of a bulk sample. The four-probe attachment is pressed onto the surface of a semi-infinite bulk sample. The spacing between the various probes is indicated in the figure. It is clear that the potentials at V^+ and V^- due to the injection of current I^+ are given by

$$V^+ = \frac{I^+\rho}{2\pi a} \quad \text{and} \quad V^- = \frac{I^+\rho}{2\pi(a+b)} \tag{63}$$

Similarly, the potentials at the two voltage probes can be computed due to the current sink I^-. Since $I^+ = I^- = I$, it follows from the principle of

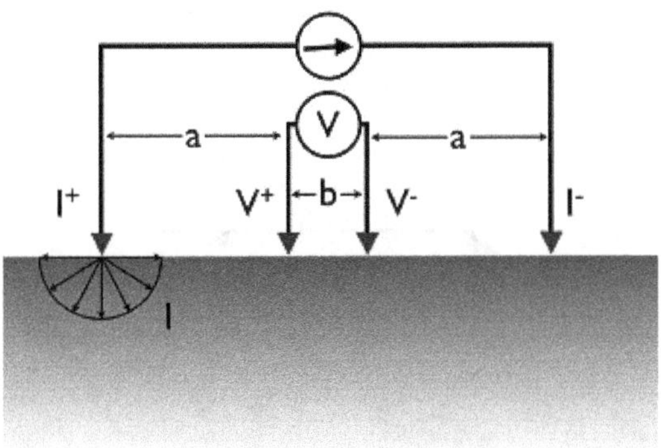

Figure 6. Collinear four-probe arrangement pressed on to the surface of a semi-infinite metallic block.

superposition that the total potential difference is simply doubling the potential difference due to a single current probe. Thus, we obtain the relation

$$\Delta V = V^{+} - V^{-} = \frac{I\rho}{\pi}\left(\frac{1}{a} - \frac{1}{a+b}\right) = \frac{I\rho b}{\pi a(a+b)} \tag{64}$$

The expression for the resistivity of a bulk specimen is thus given by

$$\rho = \frac{\pi a(a+b)}{b}\left(\frac{\Delta V}{I}\right) = \frac{\pi a(a+b)}{b} R \tag{65}$$

Here, $R = \frac{\Delta V}{I}$ is the experimentally measured resistance value. This expression becomes further simplified if we set $a = b$. This arrangement is termed the "**equidistant collinear four-probe**" method. Thus, we have the expression

$$\rho = 2\pi a R \tag{66}$$

The simplicity of this method makes it particularly attractive for checking the homogeneity of the bulk sample by probing its different regions.

5.2. *Thin-film geometry*

Specifically, let us consider an infinite two-dimensional sheet whose thickness is small compared to the probe spacing such that we may approximate the current density to be uniform in the direction perpendicular to the film plane. Figure 7 shows the current distribution in an idealized two-dimensional geometry.

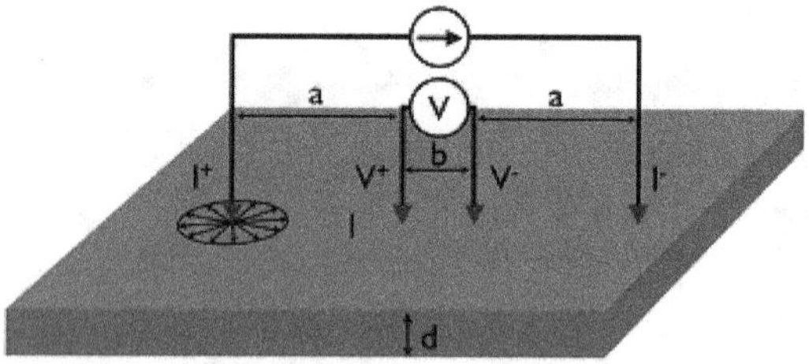

Figure 7. Linear four-point-probe measurement on a thin-film sample of thickness *d*.

We may note that the current is essentially confined within the film's thickness, d. In the plane of the film, the current spreads evenly, leading to circular equipotential surfaces. The crucial difference is that at a lateral distance r from a current probe, the total current I must pass through an area dictated by the circumference of a circle of radius r and the film thickness, and thus the current density is $J(r) = I/2\pi rd$. By definition, $J(r) = \sigma E$, where σ is the conductivity of the sample and E is the electric field. Furthermore, $E = -\partial V/\partial r$. Thus, the potential V at a distance r for the thin-film geometry is given by the expression

$$V(r) = \left(\frac{I\rho}{2\pi d}\right)\ln r \tag{67}$$

The potential distribution varying logarithmically with distance is a characteristic of a two-dimensional system. We now obtain an expression for the resistivity of a thin film or a thin sheet by employing the collinear four-probe arrangement described earlier. It may be noted that the outer and inner probes are separated by a distance a, while the inner probes are separated by b. It follows that the potential difference developed between the inner probes separated by b is given by the expression

$$\Delta V = \left(\frac{I\rho}{\pi d}\right)\ln\left(\frac{a+b}{a}\right) \tag{68}$$

The bulk resistivity of the film, ρ, is given by the expression

$$\rho = \left(\frac{\pi d \Delta V / I}{\ln\left((a+b)/a\right)}\right) = \left(\frac{\pi d R}{\ln\left((a+b)/a\right)}\right) \tag{69}$$

where $R = \Delta V/I$ is the experimentally measured resistance value.

For the special case of equally spaced collinear probes, wherein $a = b$, the above expression simplifies to

$$\rho = \left(\frac{\pi}{\ln 2}\right) dR \tag{70}$$

The bulk resistivity of a thin film can thus be evaluated through an experimental measurement of the resistance R and from the knowledge of the thickness d of the film. The main assumptions for the validity of the above relation are that the thickness of the film should be much smaller than the distance between the probes and that the other two dimensions, namely the length and breadth of the film, should be an order of magnitude larger than the distance between the probes.

5.3. *Wire or thin strip geometry*

This is perhaps the most commonly used geometry, wherein the sample is in the form of a wire or a thin strip of uniform area of cross-section. Here, the electrical conduction is essentially one-dimensional in nature, and the current I spreads out uniformly over the cross-sectional area A. The current density, $J = I/A$, is thus a constant. Thus, the electric field E is also a constant, and the potential V varies linearly with distance. Figure 8 presents the conventional four-probe arrangement for measuring the electrical resistivity of a sample. It is clear that the voltage V at a distance x from a current probe is given by

$$V = \rho\left(\frac{I}{A}\right)x \tag{71}$$

Thus, the potential difference ΔV, which is developed between the voltage probes separated by a distance b, and the resistivity are given by

$$\Delta V = \rho\left(\frac{I}{A}\right)b \quad \rho = \frac{\Delta V}{I}\frac{A}{b} \quad \rho = R\frac{A}{b} \tag{72}$$

It is to be noted that the potential difference ΔV is independent of the distance of the current probes a from the potential probes.

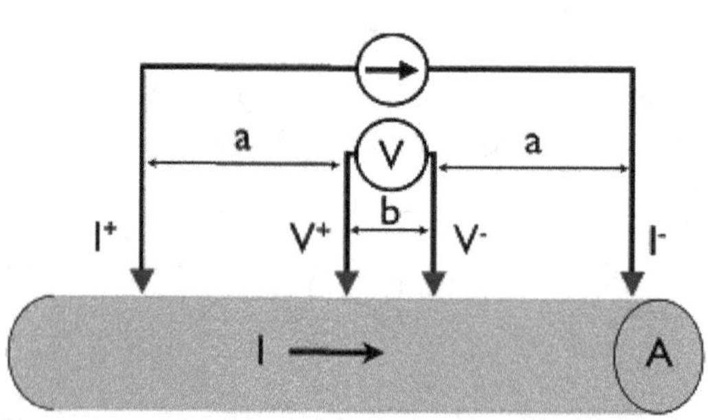

Figure 8. Conventional four-probe measurement technique.

5.3.1. Van der Pauw technique

The van der Pauw technique is one of the most extensively used measurement methods to characterize a thin wafer of a semiconducting material and extract all the important parameters, such as carrier concentration, electrical conductivity, and carrier mobility. The real power of this technique, invented by Van der Pauw (while working at the Philips Research Laboratories), lies in the fact that a thin, semiconducting sample of **any irregular shape** would suffice for measurement. The technique can equally be applied to the study of thin films and thin sheets of metals and alloys.

Van der Pauw solved the general problem of electrical potential distribution in a thin conducting sample of arbitrary shape but satisfying the following attributes. First, the sample has to be homogeneous and isotropic in its properties and should be continuous in the sense that there are no macroscopically observable holes or voids. In the van der Pauw method, electrical contacts are made on the perimeter of the sample to be measured. This means that the perimeter has to be sharply defined. The contacts cannot be on the edge of a region where the conductivity is merely dropping from a high value to a lower value, nor can they be placed substantially inside the edge of the sample. The basic measurement geometry is given in Figure 9.

The four contacts A, B, C, and D are located at the periphery of the sample, and their area of contact is negligibly smaller than the total area

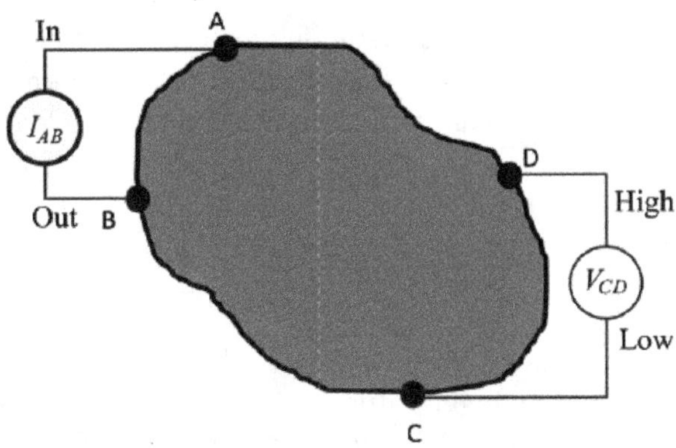

Figure 9. Schematic diagram of the basic experimental arrangement in the van der Pauw method.

of the sample. The sample thickness d is small compared to other dimensions of the sample under study. In the first measurement, current I_{AB} is passed through the contacts A and B, and voltage V_{DC} is measured across the other two contacts D and C. The ratio V_{DC}/I_{AB} is a characteristic resistance denoted as $R_{(AB,DC)}$. Thus,

$$R_{(AB,DC)} = \frac{V_{DC}}{I_{AB}} \tag{73}$$

The second measurement involves passing current I_{AD} through the contacts A and D and measuring the potential difference V_{BC}. This leads to another characteristic resistance $R_{(AD,BC)}$:

$$R_{(AD,BC)} = \frac{V_{BC}}{I_{AD}} \tag{74}$$

The van der Pauw method exploits the reciprocity theorem of electromagnetism, which states that the current in a detector divided by the voltage at the source remains constant when the source and detector are interchanged. This method of electrical images leads to the following relation of profound significance:

$$e^{-\frac{\pi d R_{(AB,DC)}}{\rho}} + e^{-\frac{\pi d R_{(AD,BC)}}{\rho}} = 1 \tag{75}$$

Here, ρ is the electrical resistivity of the sample, and d is the thickness of the sample. This equation has no analytical solution, and thus, one has to take recourse to numerical methods, such as the Newton–Raphson method, to extract the value of ρ. This expression can be recast into a more usable form in terms of what is known as the van der Pauw correction factor f:

$$\rho = \left(\frac{\pi}{\ln 2}\right) \frac{R_{(AB,DC)} + R_{(AD,BC)}}{2} df \tag{76}$$

The correction factor f depends only on the resistance ratio $Q = \frac{R_{(AB,DC)}}{R_{(AD,BC)}}$.

By convention, the lower value of the measured resistance is taken for the denominator so that Q is always greater than 1. f can be computed by numerical methods, and a plot of f against Q is given in Figure 10.

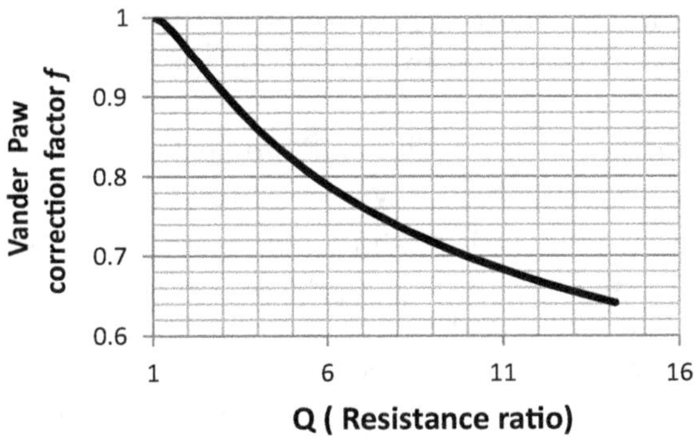

Figure 10. Van der Paw correction factor *f* against resistance ratio *Q*.

Table 1.

Shape	$R_{(AB,DC)}$ **(μΩ)**	$R_{(AD,BC)}$ **(μΩ)**	Q	f	ρ**(μΩcm)**
Square	3347	3340	1	1	83.5
Rectangle	16390	156	105	0.4	82
Irregular	5260	1956	2.689	0.924	83

The experimental method based on the van der Pauw technique thus reduces to the measurement of the two resistance values $R_{(AD,BC)}$ and $R_{(AB,DC)}$ and the reading of the correction factor f corresponding to the measured resistance ratio Q.

In a student laboratory, it is easier to handle thin metallic sheet-like stainless steel and make spot-welded contacts. The electronic instrumentation involves a DC current source and a DC microvoltmeter. Samples in the form of a square, rectangle, or any irregular shape can be cut, and measurements are carried out as detailed above. Typical experimental data are given in Table 1.

The material chosen was a stainless-steel sheet with a thickness of $d = 0.0055$ cm.

5.3.2. Resistivity measurement technique as a function of temperature

Resistivity measurement as a function of temperature is of prime importance in the study of the different scattering mechanisms in metallic systems, determining band gap in semiconductors, and serving as a powerful diagnostic tool in the study of temperature-driven phase transitions.

The method is based on the well-known four-probe technique for resistivity measurement, along with a chromel–alumel thermocouple probe spot-welded to the same sample for temperature measurement. In order to circumvent the effects of Seebeck voltages, which necessarily exist due to temperature gradients in the sample, an AC resistivity technique is employed. Figure 11 presents a block diagram of a high-resolution AC resistivity system, especially suitable for metallic samples.

The AC technique employs the **lock-in** or **phase-sensitive detection** technique to achieve high resolution in resistance measurement, an important requirement for probing **critical points** associated with phase transitions. The electronic system comprises a **quadrature oscillator** operating at 400 Hz. This circuit essentially produces two outputs: a sine wave and a cosine wave, which are always synchronized. The sine wave output voltage signal is transformed into an AC current signal in the AC constant

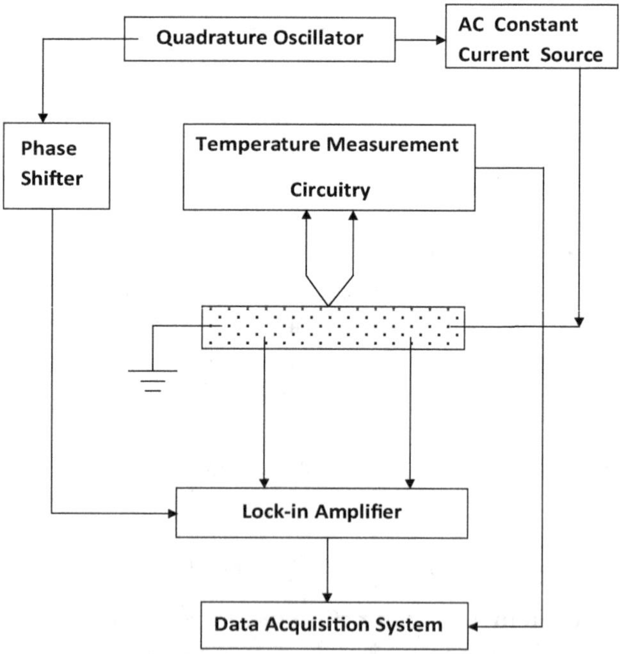

Figure 11. Block diagram of an AC resistivity setup.

current source circuit and drives the current through the sample under study. Typically, one can drive a constant current in the range of 5–50 mA through the sample. The cosine output, which is the quadrature component whose phase can be changed over the full 2π radians in the **phase shifter**, provides a reference signal for the **lock-in amplifier**. The low-level voltage developed across the voltage probes attached to the sample forms the input to the lock-in amplifier. It would suffice here for this signal to be "synchronously demodulated" after pre-amplification so that the resultant DC signal is directly proportional to the resistance of the sample. The lock-in amplifier could be either a commercial one or configured with a dedicated IC, such as AD640. The high signal-to-noise ratio of the system makes it possible to resolve the changes in the order of micro-ohms in the resistance of the sample. The temperature of the sample is measured through the chromel–alumel thermocouple in a way that is identical to that described in the section dealing with the DTA technique. The resistivity sample holder is inserted in the constant temperature zone of a compact vertical furnace. The temperature of the sample is controlled through a digital PID controller. It is possible to control the temperature of the sample to within ±0.5°C of the set point value over the entire temperature range from room temperature to 500°C. The experiment can also be completely automated with a programmed rate of heating, data acquisition, and real-time graphics of the resistance versus temperature plot. A typical measurement of the AC resistivity of nickel across the ferromagnetic–paramagnetic phase transition is discussed in the chapter on phase transitions.

6. Hall Effect in Semiconductors

The Hall effect is an important transport property of a conducting material subjected to an electric field and a transverse magnetic field. Basically, the Hall effect leads to the generation of a potential difference (Hall voltage) across a current-carrying conductor whose direction is transverse to both the electric current and the magnetic field. The basic experimental arrangement to observe the Hall effect is shown in Figure 12.

Note that if the current $\boldsymbol{I}$ is in the x-direction and the magnetic field $\boldsymbol{B}$ is in the z-direction, the Hall voltage V_H will develop across the y-direction. The sample has a width of w and a thickness of d. The physical mechanism responsible for the Hall effect is the Lorentz force acting on a moving electron subjected simultaneously to an electric and a magnetic field. It is

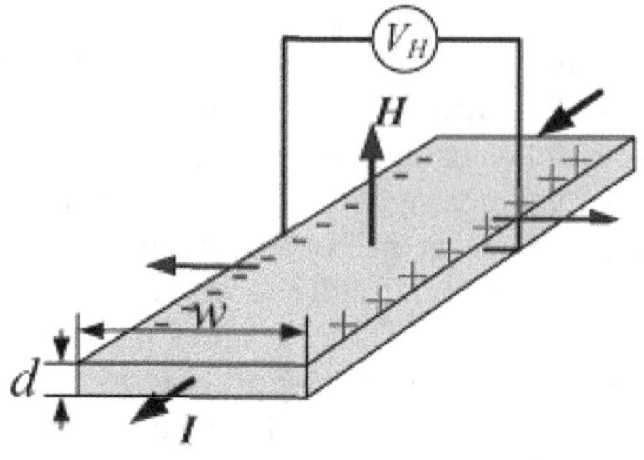

Figure 12. **Basic experimental arrangement in a Hall setup.**

straightforward to see that the Lorentz force $\boldsymbol{F}_L = -e(\boldsymbol{v}\boldsymbol{x}\boldsymbol{B})$ causes the electrons to bend in the y-direction, leading to an accumulation of negative charges on one side and the appearance of positive charge on the opposite side of the width, as shown in Figure 12. This separation of charges along the width leads to what is termed the Hall voltage V_H. The accumulation of charge proceeds until a steady state is reached when the Hall force completely balances the Lorentz force. In the diagram above, the conventional current is in the $+x$ direction, and in a typical metal or an n-type semiconductor, the electron current and hence the drift velocity v are along the negative x-direction. The polarity of the Hall voltage is as shown in the diagram. However, in a p-type semiconductor, wherein the majority carriers are holes, the current will be in the same direction as the conventional current and the sign of the Hall voltage will be reversed. This useful feature, wherein the sign of the Hall voltage is negative for electrons and positive for holes, makes it a powerful technique to determine the sign of the charge carrier and its concentration. The magnitude of the Hall voltage is determined by the condition that the Hall field is balanced by the Lorentz field in the steady state. Thus, one gets the relation

$$\frac{V_H}{w} = v_x B \tag{77}$$

The drift velocity v_x is related to the current density through the well-known relation $J_X = nev_x$, where n is the carrier concentration. In terms of the measured current $I = J_X\, wd$, the expression for the Hall voltage is given by

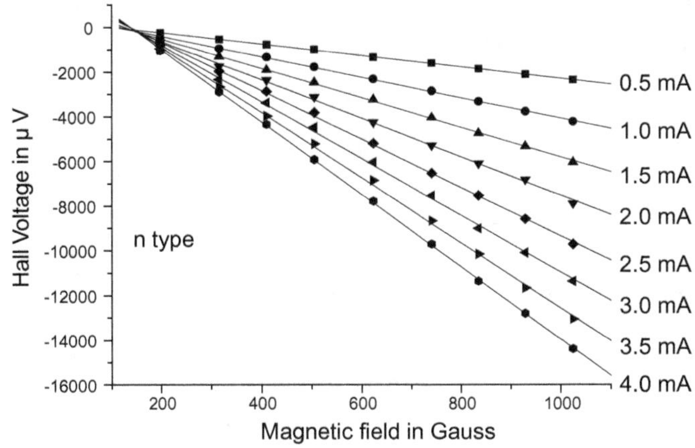

Figure 13. Variation of the Hall voltage as a function of the magnetic field for different values of current through an *n*-type sample.

$$V_H = v_x wB = (1/ne)\, IB/d \tag{78}$$

The instrumentation for measuring the Hall voltage is essentially the same as that employed in the setup for DC resistivity measurements. The current through the sample is driven through a constant current source, and the Hall voltage is measured with a precision DC microvoltmeter. Figure 13 presents typical data for an *n*-type semiconductor sample, showing the variation in the Hall voltage as a function of the magnetic field for different values of current through the sample. It may be noted that the Hall voltage is negative and characterized by a negative slope. The data correponding to a sample of thickness 0.5 mm lead to $n_e \approx 3 \times 10^{20}$ per m^3. Similar measurements can be carried out on a *p*-type sample.

References

1. Kittel, C. (1974). *Introduction to Solid State Physics* (4th Edition). Wiley Eastern Private Limited, New Delhi.
2. Ziman, J. M. (1972). *Principles of the Theory of Solids* (2nd Edition). Cambridge University Press, Cambridge.
3. Harrison, W. A. (1970). *Solid State Theory*. McGraw-Hill, New York.
4. Dekker, A. J. (1957). *Solid State Physics*. Prentice-Hall, Englewood Cliffs, N.J.

Chapter 14

SEEBECK COEFFICIENT IN METALS AND SEMICONDUCTORS

1. Introduction

The field of thermoelectricity is concerned with three fundamental effects, namely the Seebeck, Peltier, and Thomson effects. These effects are discussed in some detail. The thermodynamic analysis of these reversible phenomena by Kelvin is presented, which led to the discovery of the Thomson effect. A brief derivation of the Seebeck coefficient applicable to metals and semiconductors is presented based on their electronic band structure. The physical origin of the sign and magnitude of the Seebeck coefficient in *n*- and *p*-type semiconductors is discussed in some detail.

Two main experimental techniques for the measurement of the Seebeck coefficient as a function of temperature, namely the **integral method** of measuring the Seebeck coefficient applicable for wire-type metallic samples and the differential method particularly applicable to disks made of semiconductor material, are described in detail.

2. Seebeck Effect

In 1821, Thomas Johann Seebeck discovered that in an open circuit composed of two dissimilar metals, with the two junctions held at different temperatures, an electric voltage appears at the terminals, as shown in Figure 1. Furthermore, an electric current would flow continuously in a closed circuit made up of two dissimilar metals, A and B, if the junctions of the metals were maintained at two different temperatures.

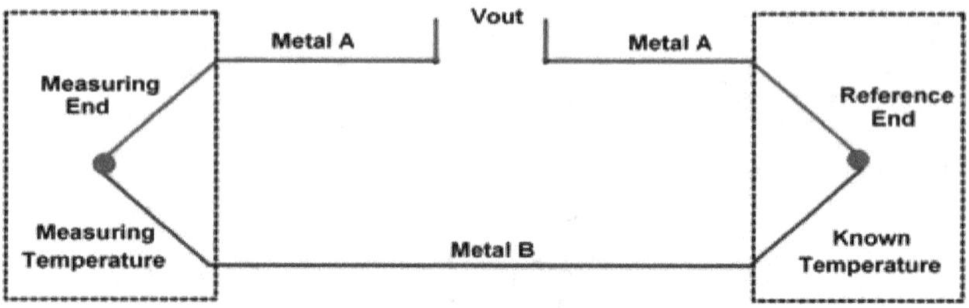

Figure 1. Conventional two-junction thermocouple arrangement of two dissimilar metals A and B.

If a voltage ΔV_{AB} is measured using a voltmeter for a temperature difference of ΔT between the two junctions, the Seebeck coefficient S_{AB} is defined by the relation

$$S_{AB} = \lim_{\Delta T \to 0} \frac{\Delta V_{AB}}{\Delta T} \tag{1}$$

It will be shown later that S_{AB}, the relative Seebeck coefficient of the thermocouple AB, is the difference in the individual contributions, S_A and S_B, of the two constituents and is given by the relation

$$S_{AB} = S_A - S_B \tag{2}$$

The Seebeck effect is reversible in the sense that the polarity of the measured voltage reverses its sign if the temperatures of the two junctions are reversed, keeping the magnitude the same.

It is appropriate to discuss the origin of the Seebeck effect from a phenomenological point of view, which prevailed before the advent of the modern theory of metals and semiconductors based on the Fermi–Dirac statistics. The view prevalent at that time was that a conductor is permeated by a gas of free electrons. Consider a single metal A subjected to a temperature gradient, as shown in Figure 2.

Electrons at the hot end possess higher kinetic energy than the electrons at the cooler end. This leads to the diffusion of electrons toward the cooler end and the accumulation of electrons (negative charges) at the cold end. This diffusion current disturbs the electrical neutrality, with the cold end at a **negative potential** and the hot end at a **positive potential**. This will set up an electric field in the body of the metal such that the current

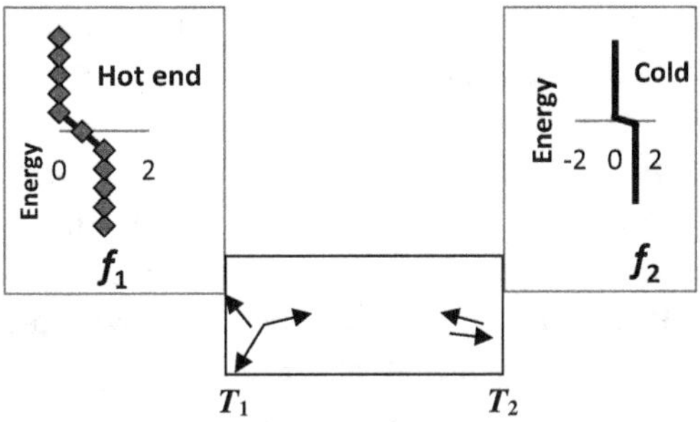

Figure 2. Schematic diagram of a conductor in a temperature gradient.

due to this voltage difference between the hot and cold ends completely opposes the diffusion current. In the steady state, a dynamic equilibrium is established between the current flow due to the temperature gradient and an opposing current due to the electric field. This is consistent with the boundary condition that, in an open circuit, the current is zero but a voltage difference or an electric field is created due to the temperature gradient. This electrical field will oppose the motion of the charge due to the temperature gradient. Thus, the **absolute Seebeck coefficient** of the constituent A can be defined by the relation

$$S_{\mathrm{A}} = \lim_{\Delta T \to 0} \frac{\Delta V_{\mathrm{A}}}{\Delta T} \tag{3}$$

In terms of the electric field $\boldsymbol{E}_{\mathrm{A}}$, which is the negative of the gradient of the potential (∇V_{A}), the above relation transforms to

$$\boldsymbol{E}_{\mathrm{A}} = -\boldsymbol{S}_{\mathrm{A}}\, \nabla \boldsymbol{T} \tag{4}$$

Thus, the absolute Seebeck coefficient of a single conductor is the electric field set up for a unit temperature gradient. Unlike electrical conductivity, which is always positive, the absolute Seebeck coefficient can be either negative or positive. The general convention chosen is that the polarity of the cold end of the conductor is the sign of the Seebeck coefficient. The simple electron gas model leads only to a **negative Seebeck coefficient**, with the electric field directed from the hot end to the cold end, coupled with a negative temperature gradient. Experimentally, many metals, such as copper, silver, and lead, have a **positive Seebeck**

coefficient, which can only be understood on the basis of the detailed energy band structure of metals based on the Fermi–Dirac statistics. We consider these aspects in a later section.

It is instructive to analyze the case of a symmetric circuit wherein both legs of the thermocouple are made from the same material A. Both legs will have the same voltage difference. If a unit charge is taken around the circuit, no work is done on it. The Seebeck emfs in both legs are identical, resulting in 0 V at the voltmeter. This necessitates that to observe a thermoelectric (TE) voltage, the two legs of the couple should be different, with different values of the absolute Seebeck coefficient.

The measurement of the absolute Seebeck coefficient of a conductor experimentally has been addressed in two ways. First, if one chooses a superconductor as one of the constituents of the thermocouple, say B, the measured thermo-emf contains only the contribution from the other leg, say A. This is because a superconductor cannot support an electric field, and its absolute Seebeck coefficient $S_B = 0$ over the temperature regime below its critical temperature. The measured emf $\boldsymbol{E}_{\mathbf{AB}}$ would directly yield S_A. However, the temperature regime covered by this is limited to around 120 K. The only practical method of determining the absolute Seebeck coefficient is through the Thomson effect, which requires only a single constituent and covers a very wide temperature regime.

The important relation between the Seebeck emf and the absolute Seebeck coefficients of the two constituents forms the basis for several laws of thermoelectricity, such as the law of successive temperatures and the law of intermediate metals.

According to the law of successive temperatures, the emf generated in a thermocouple AB operating between temperatures T_1 and T_3 is the algebraic sum of the emfs of the same couple operating between the temperatures T_1 and T_2 and the temperatures T_2 and T_3:

$$V_{\mathbf{AB}}(T_1, T_3) = V_{\mathbf{AB}}(T_1, T_2) + V_{\mathbf{AB}}(T_2, T_3) \tag{5}$$

The above law follows directly from the relation

$$\begin{aligned} V_{\mathbf{AB}}(T_1, T_3) &= \int_{T_1}^{T_3} (S_A - S_B)\,dT \\ &= \int_{T_1}^{T_2} (S_A - S_B)\,dT + \int_{T_2}^{T_3} (S_A - S_B)\,dT \end{aligned}$$

A direct fallout of this relation is the use of a single-junction thermocouple employed extensively for temperature measurement in the industrial environment. Here, the single-junction thermocouple is essentially operating between the ambient temperature and the measuring temperature. To the measured emf, one must add the emf value equivalent to the value that would arise from the thermocouple operating between 0°C and the ambient temperature. This is achieved through an electronic circuitry and generally referred to as **cold junction compensation**.

The other important law of thermoelectricity is the law of intermediate metals. According to this law, if a third metal is introduced into the thermocouple circuit, the measured emf will not be affected, provided the junctions with the third metal are kept at the same temperature. This law follows directly from the previous discussion, and noting that there is no temperature gradient in the third metal, hence it does not make any contribution to the emf of the thermocouple. This law is of great practical utility in connecting thermocouples to an electronic measuring system using lengthy cables.

Thermocouples, which work on the basis of the Seebeck effect, belong to an important class of temperature transducers for the measurement of temperature, covering a wide temperature range, from low temperatures up to 2000°C. The Seebeck emf as a function of temperature is tabulated in NBS thermocouple tables.

3. Peltier Effect

Jean Peltier, in the year 1834, made an important observation that when a current flows through a junction of two dissimilar metals, say A and B, heat is either **evolved** or **absorbed**, depending on the direction of the current. Normally, heat is always generated in a conductor carrying current as Joule heat. It is proportional to the square of the current and thus does not depend on the direction of the current flow. Figure 3 shows a typical arrangement of a Peltier setup, wherein heat is evolved when current flows from A to B and absorbed with the reversal of current from B to A (or vice versa). As in the Seebeck effect, this reversible generation of heat can only be observed at the junction of two dissimilar metals. We define the Peltier coefficient, π_{AB}, as the heat developed when a unit current passes per unit time:

π_{AB} = Heat evolved at the junction per unit time per unit current

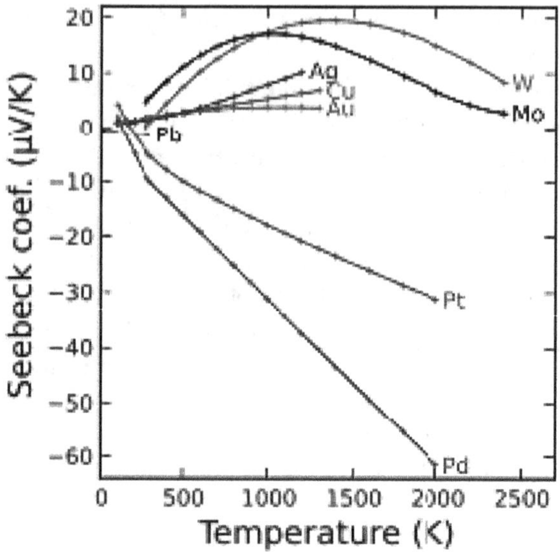

Figure 3. Temperature variation of absolute Seebeck coefficients of some metals (adapted from Wikipedia).

Thus, the Peltier coefficient is a measure of the change in heat content when one coulomb of charge crosses the junction. Peltier heat is independent of the shape or dimensions of the conductors composing the junction. Conceptually, the Peltier effect is easier to comprehend when one notes that an electron, apart from carrying electrical current, also carries heat current. The heat current is intimately related to the electronic specific heat, which is different in different metals. **Even though the electric current is the same across the junction, the heat current associated with the electric current is different. Thus, at the junction, the difference in the heat current between the two metals manifests as heat absorbed or evolved, depending on the direction of current flow.** It is a homogeneous effect manifesting at the junction. It is not a contact phenomenon and is in no way associated with the nature of the contact between the two conductors.

4. Thermodynamics Applied to Seebeck and Peltier Effects

Consider a closed circuit of two dissimilar metals A and B whose junctions are maintained at temperatures T_1 and T_2, respectively. TE current will flow in the closed loop due to the Seebeck effect. If $T_2 = T_1 + \mathrm{d}T$ and the open-circuit Seebeck voltage is $\mathrm{d}v_{AB}$, then the relative Seebeck coefficient for this thermocouple is $S_{AB} = \mathrm{d}v_{AB}/\mathrm{d}T$. The Peltier effect due to this current flow leads to heat absorption of π_1 per unit charge at temperature T_1 and heat evolution of π_2 per unit charge at temperature T_2. Kelvin reasoned that since both of these effects are perfectly reversible, one can apply the laws of thermodynamics to a Carnot engine operating between temperatures T_1 and T_2. Thus, the total entropy change is zero, leading to the relation

$$\pi_1/T_1 = \pi_2/T_2 \tag{6}$$

The first law of thermodynamics, which is the law of conservation of energy, stipulates that the heat energy change of $\pi_2 - \pi_1$ per unit charge should be equal to the work done for taking a unit charge round the circuit, i.e., $\mathrm{d}v_{AB}$. This leads to the result

$$\pi_2 - \pi_1 = \mathrm{d}V_{AB} \tag{7}$$

From relation (6),

$$\frac{\pi_2 - \pi_1}{\pi_1} = \frac{T_2 - T_1}{T_1} \tag{8}$$

Then, relation (7) reduces to

$$\frac{\mathrm{d}V_{AB}}{\mathrm{d}T} = \frac{\pi_1}{T_1} \tag{9}$$

Since $\frac{\pi_1}{T_1}$ is a constant, this result implies that the thermo-emf V_{AB} increases linearly with temperature. After this analysis, Kelvin came across an interesting experimental result pertaining to the behavior of thermo-emf as a function of temperature in an iron–copper thermocouple. Here, the behavior of the thermo-emf is highly nonlinear, leading to a neutral temperature and an inversion temperature. This experimental result led him to the remarkable conclusion that there must be another TE effect called the **Thomson effect**, which we discuss in the following.

5. Thomson Effect and Kelvin Relations

Thomson (Kelvin)'s remarkable discovery is perhaps the most fundamental TE effect in that the effect is observable in a single conductor subjected to a temperature gradient. When an electric current flows through a single conductor, the ends of which are maintained at different temperatures, heat is evolved or absorbed all along the length of the bar at a rate approximately proportional to the product of the current and the temperature gradient. The Thomson coefficient is the ratio of the power evolved per unit volume in the sample to the applied current and temperature gradient. Thomson recognized that an electric current in a conductor is always accompanied by a heat current related to the specific heat of the electron.

The Thomson coefficient γ is given by the relation

$$\gamma = \frac{\Delta Q}{I \Delta T} \tag{10}$$

Experimentally, the Thomson coefficient is small in magnitude and, being a thermal measurement, poses considerable challenges. Thomson himself verified his prediction experimentally and derived the famous Kelvin relations interconnecting the Seebeck, Peltier, and Thomson effects. The derivation considers the additional contribution of the reversible Thomson heat along with the Peltier heat. The three Kelvin relations are

$$\pi = TS \tag{11a}$$

$$\gamma = T \frac{\mathrm{d}S}{\mathrm{d}T} \tag{11b}$$

and

$$S = \int_0^T \frac{\gamma}{T} \mathrm{d}T \tag{11c}$$

It is worth emphasizing that among the three effects, the Seebeck coefficient is the easiest to measure as it involves the measurement of electrical voltage, unlike the other two coefficients, which involve sensitive thermal measurements.

6. Physical Basis for the Origin of the Seebeck Coefficient in Metals and Semiconductors

The early years of the study of the Seebeck effect were mainly confined to metals, and the low values of the Seebeck coefficient, i.e., of the order of 10–40 μV/°C forbade their usage as a TE power generator. The main use of thermocouples was confined to temperature measurement in both the laboratory and the industrial environment. The Peltier effect, which is an electrical means of cooling, is hardly discernible in a junction of two metals and was relegated to a subject matter of only academic interest. The great revival of the field of thermoelectricity took place with the discovery of semiconductors. Two prime reasons for this are (a) the order-of-magnitude higher values of the Seebeck coefficient, of the order of 200–300 μV/°C and (b) the fact that n-type semiconductors always have a **large negative Seebeck coefficient** and p-type semiconductors a **large positive Seebeck coefficient**. Thus, a thermocouple arrangement formed with an n-type and a p-type semiconductor has a large relative Seebeck coefficient of the order of 400–600 μV/°C. Although this is still a small voltage, fabrication techniques permit connecting many such basic units in such a way that they are electrically in series and thermally in parallel configuration. This basic arrangement forms the basis of a TE generator and a Peltier cooler.

Historically, Kelvin had reasoned that the Seebeck coefficient of a metal is directly related to the specific heat per electron. Thus, based on the equipartition theorem and classical statistics, one gets the result that the Seebeck coefficient of a metal is $S \approx -\frac{3}{2}\frac{k_B}{|e|} \approx -130\,\mu\text{V}/^\circ\text{C}$. Experimentally, this value is nearly two orders of magnitude higher than the observed value of 1–10 μV/°C. This anomaly was resolved later with the application of the Fermi–Dirac statistics. We note that due to the Pauli exclusion principle, only electrons near the Fermi energy within an energy width of ~k_BT are effective for both thermal and electronic transport so that the electronic specific heat turns out to be nearly two orders of magnitude lower than the classical value. Thus, one gets the right order of magnitude for the Seebeck coefficient $S \cong -3\frac{k_B}{|e|}\frac{k_BT}{E_F}$, where E_F is the Fermi energy. Typically, $\frac{k_BT}{E_F} \approx 0.01$, leading to values for S in the range of a few μV/°C. Furthermore, S would now vary with temperature and be different for different metals.

We now discuss the origin of the large magnitude of the Seebeck coefficient of a doped semiconductor as compared to a metallic system. The general energy band structure of metals, insulators, and intrinsic and extrinsic semiconductors has been adequately discussed in an earlier chapter on the electrical conductivity of metals and semiconductors. The most important aspects relevant to the current discussion on TE effects are (1) the role of the Fermi–Dirac distribution function in determining the energy distribution of electrons due to temperature effects and (2) the distinctive E versus k behavior exhibiting a positive curvature, leading to electron-like behavior in an n-type semiconductor and hole-like behavior (a negative curvature of the E versus k plot) in a p-type semiconductor.

The equilibrium Fermi–Dirac distribution function $f(E)$ is given by the relation

$$f(E) = \frac{1}{e^{\frac{E-\mu}{kT}} + 1} \tag{12}$$

where E is the energy of the state and μ is the chemical potential, which is a constant in the conducting system maintained in an isothermal environment of temperature T. It is to be noted that at any locale in the system, there are as many electrons moving in one direction as there are moving in the opposite direction with the same velocity distribution so that there is no current in the system. In other words, the distribution function is isotropic. Currents are essentially produced by a gradient in the potential, which, in the current case, is the chemical potential μ.

The constancy of the potential ensures zero current in the system. In the above expression, there are two external parameters, namely μ and T. We can keep the temperature constant and set up a gradient of the potential by simply connecting a battery with a voltage difference of V between the ends of a conductor. This electrostatic potential raises the entire energy distribution of the electrons at one end compared to the other, as depicted in Figure 2. The addition of an external electrostatic potential V to the chemical potential μ leads to the concept of what is termed **electrochemical potential**, i.e., $\mu + qV$, where q is the charge of the carrier. In this parlance, current flow in a closed circuit is associated with the gradient of the electrochemical potential. This current flow is well described by the relation

$$J = \sigma \mathbf{E} \tag{13}$$

where $\mathbf{E}$, the electric field, is the gradient of the electrochemical potential.

We now consider the general case of a conductor, one end of which is at temperature T_1 and chemical potential μ_1 and the other end is at temperature T_2 and chemical potential μ_2. The appropriate distribution functions at the two ends are given by the expressions

$$f_1(E) = 1 \Big/ e^{\left(\frac{E-\mu_1}{kT_1}\right)} + 1 \tag{14a}$$

and

$$f_2(E) = 1 \Big/ e^{\left(\frac{E-\mu_2}{kT_2}\right)} + 1 \tag{14b}$$

The expression for the current, I, in a conductor is generally derived based on the Boltzmann transport equation involving the relaxation time approximation. An equivalent description based on the Landauer–Datta–Lundstrom formalism, which is physically intuitive, is given by the general expression

$$J = \frac{1}{q} \int_{-\infty}^{+\infty} G(E)\left[f_1(E) - f_2(E)\right] dE \tag{15}$$

where J is the current density and $G(E)$ is the energy-dependent conductance of the system. The measured electrical conductance G is the average of $G(E)$ over the narrow energy window ~kT centered around the chemical potential or the Fermi energy. **The main point of interest for our discussion is that current is directly related to the difference in the Fermi functions at either end of the conductor.** It is clear from the above relation that current can be driven by either the difference in the electrochemical potential or the difference in the temperatures at either ends of the sample, or both. The interplay between these two currents with the appropriate boundary condition leads to a general expression for the Seebeck coefficient. The perturbation applied to the system, be it the temperature difference or the voltage difference, is small enough that the system is driven to a near-equilibrium state. The change in the equilibrium distribution function can thus be approximated by the Taylor series expansion. Thus, we have the relation

$$f_1(E) - f_2(E) = \left[\frac{\partial f}{\partial \mu}\right](\mu_1 - \mu_2) + \left[\frac{\partial f}{\partial T}\right](T_1 - T_2) \tag{16}$$

The derivatives are evaluated for the expression

$$f(x) = \frac{1}{e^x + 1} \qquad \text{with} \qquad x = \frac{E-\mu}{kT}$$

We note that $\frac{\partial f}{\partial E} = \frac{\partial f}{\partial x}\frac{\partial x}{\partial E} = \frac{\partial f}{\partial x}\frac{1}{kT}$

and
$$\frac{\partial f}{\partial \mu} = -\frac{\partial f}{\partial x}\frac{1}{kT} \quad \frac{\partial f}{\partial \mu} = -\frac{\partial f}{\partial E} \tag{17}$$

Furthermore,

$$\frac{\partial f}{\partial T} = \frac{\partial f}{\partial x}\frac{\partial x}{\partial T} = -\frac{\partial f}{\partial x}\frac{E-\mu}{kT^2}$$

$$\frac{\partial f}{\partial T} = -\frac{E-\mu}{T}\frac{\partial f}{\partial E} \tag{18}$$

Noting that $G_0 = \int_{-\infty}^{+\infty} G(E)[-\frac{\partial f}{\partial E}]\,\mathrm{d}E$ and $\mu_1 - \mu_2 = q(V_1 - V_2)$, the expression for J reduces to

$$J = G_0(V_1 - V_2) + G_S(T_1 - T_2) \tag{19}$$

The first term on the right represents the current flow due to a gradient in the electrochemical potential (Ohm's law), and the second term represents the current flow due to the temperature gradient. The main quantity of interest, G_S, turns out to be

$$G_S = \frac{1}{q}\int_{-\infty}^{+\infty}\left[\frac{E-\mu}{T}\right]G(E)\left[-\frac{\partial f}{\partial E}\right]\mathrm{d}E \tag{20}$$

Noting that $\left[-\frac{\partial f}{\partial E}\right]$ is a function with a strong peak at $E = \mu$, with an energy width of $\sim 3.5k_BT$, the above relation simplifies to

$$G_S \cong \frac{1}{q}\left[\frac{E-\mu}{T}\right]G(0) \tag{21}$$

Now, we apply the boundary condition that the current $J = 0$ for an isolated conductor. This leads to an important result for the Seebeck coefficient S:

$$S = \frac{V_1 - V_2}{T_1 - T_2} = \frac{\Delta V}{\Delta T} = -\frac{1}{q}\left[\frac{E-\mu}{T}\right] \tag{22}$$

The expressions for current J and voltage V ($= \mu/q$) refer to the electron current and electron voltage, respectively, whose sign is opposite to the conventional current and voltage. Thus, if G_S is positive, then the open-circuit voltage ΔV is negative for $T_1 > T_2$. Using this convention, the expression for the Seebeck coefficient turns out to be

$$S = -\frac{1}{|e|}\left[\frac{E-\mu}{T}\right] = -\frac{k}{|e|}\left[\frac{E-\mu}{kT}\right] \tag{23}$$

where $|e|$ is the magnitude of the charge on an electron. It is to be noted that E in this expression refers to the average energy of the electron. The above expression is valid both for the degenerate electron assembly, such as in a metal, and nondegenerate systems, such as semiconductors. The physical implications of the above important relation will now be explored. The magnitude of $k/e \cong 86\ \mu\text{V}/°\text{C}$. The magnitude and sign of the dimensionless quantity $\left[\frac{E-\mu}{kT}\right]$ are central to our understanding of the small magnitude of the Seebeck coefficient in metals compared to its orders-of-magnitude higher values in semiconductors.

7. Seebeck Coefficient in Metals

The electronic band structure of a metal has been discussed adequately in an earlier chapter. The band structure is shown in Figure 4(a). It would suffice here to note that only electrons near the Fermi energy or chemical potential within an energy window ~kT take part in electrical and thermal transport. Thus, in a metal $\text{E} - \mu \cong \pm\, kT$. It is important to note that the Fermi energy μ is a physically occupied state with a probability of occupation of ½ at any finite temperature. There are empty states available for both above and below μ for electrical conduction. It is clear from our

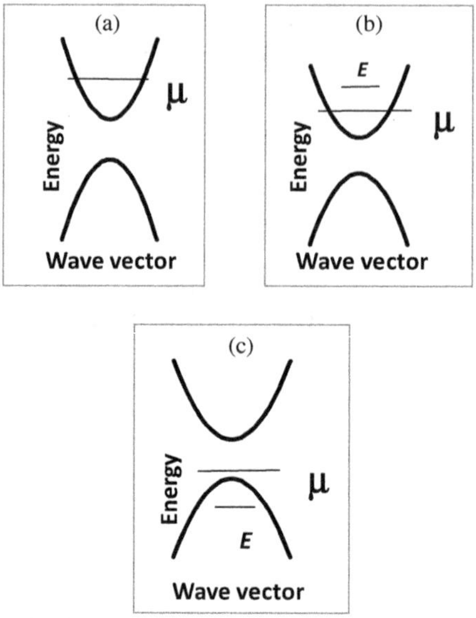

Figure 4. Energy band structure of (a) typical metal, (b) *n*-type semiconductor, and (c) *p*-type semiconductor.

previous discussion that the electron diffusion current is proportional to $f_1 - f_2$. Thus, the electron diffusion current is from the hot end to the cold end for states $E > \mu$ and reverses for states $E < \mu$. Normally, the current from states $E > \mu$ dominates, leading to a **negative Seebeck coefficient** for a metal. Further, the magnitude of $\left[\frac{E-\mu}{kT}\right] << 1$, considering the opposing contributions from electron energy states above and below μ. Thus, the typical value for the Seebeck coefficient of a metal lies in the range of 1–20 μV/°C. Many metals, such as copper and silver, however, have a **positive Seebeck coefficient**. It is worth mentioning that the Seebeck coefficient of a material contains two components. The first component arises due to the redistribution of electrons resulting from a temperature gradient. It is this aspect which has been covered in the previous discussion. The second contribution to the Seebeck coefficient arises from the energy dependence of the relaxation time. This involves detailed consideration of the scattering of charge carriers by thermal excitations, etc., and would not be considered here. It would suffice here to understand that the positive sign of the Seebeck coefficient in many metals, such as copper and silver, owes its origin to the energy dependence of the relaxation time.

But it needs to be emphasized that a positive sign for the Seebeck coefficient here does not mean that the charge carriers are holes. Figure 3 presents the data on temperature variation of the absolute Seebeck coefficient of some metals.

8. Seebeck Coefficient of an n-type Semiconductor

The energy band diagram of an *n*-type semiconductor is shown in Figure 4(b).

It is important to note that the chemical potential now is in the forbidden gap close to the bottom of the conduction band. It is not a physically occupied state but provides the reference level for measuring the energy of electrons in the relevant bands. The level corresponding to the average energy of electrons in the conduction band is significantly higher than μ, and typically, $E - \mu \cong +3$–4 *k*T. This leads to a large negative value for the Seebeck coefficient: ~ –300 μV/°C. Here, there is no ambiguity in the sign of the Seebeck coefficient. The majority carriers are electrons in the conduction band whose concentration is several orders of magnitude higher than the minority carriers in the valence band. It is thus clear that $E - \mu$ is always positive, and its magnitude is several times of kT, leading to a large negative value for the Seebeck coefficient. It is also worth noting that the dispersion curve, i.e., E versus k (energy vs. wave vector) plot has a positive curvature, which signifies that the charge carriers behave "electron-like."

9. Seebeck Coefficient of a p-type Semiconductor

The energy band diagram of a *p*-type semiconductor is shown in Figure 4(c). It is important to note that the chemical potential now is in the forbidden gap close to the bottom of the valence band. The majority carriers are holes in the valence band whose concentration is several orders of magnitude higher than that of the minority carriers, which are electrons in the conduction band. It is thus clear that $E - \mu$ is always negative, and its magnitude is several times that of kT, leading to a large positive value for the Seebeck coefficient: ~ +300 μV/°C. It is also worth noting that the dispersion curve, i.e., E versus k plot, relevant to the valence band has a negative curvature, signifying that the charge carriers behave "hole-like."

In summary, the large magnitude of the Seebeck coefficient of doped semiconductors owes its origin to the fact that the average energies of majority carriers are significantly different as compared to the chemical potential. The sign of the Seebeck coefficient is directly indicative of whether it is an *n*-type or a *p*-type semiconductor. Practical devices, such as TE power generators and Peltier coolers, are built around these doped semiconductors.

10. Integral Method of Measuring the Seebeck Coefficient of Metals and Alloys as a Function of Temperature

The integral method of measuring the absolute Seebeck coefficient of a metallic conductor is based on the relation

$$V_{AB} = \int_{T_1}^{T_2} [S_A - S_B] dT \tag{24}$$

The relative Seebeck coefficient $S_{AB} = S_A - S_B$ is the derivative of the experimentally measured thermo-emf V_{AB} versus temperature data. The method is well suited for metals or alloys in wire form. The reference junction can be either held at a well-defined temperature, such as at 0 °C (ice–water mixture) or at ambient temperature. If a single-junction thermocouple is employed, cold junction compensation must be incorporated. The temperature of the hot junction is varied over a wide temperature range, and the corresponding thermo-emf data are acquired. It is to be noted that the metallic conductors are subjected to large thermal gradients in this method. The entire set of emf versus temperature data is fitted to an appropriate order polynomial, and the derivative is evaluated for a temperature range of interest. One could also evaluate the derivative at sliding partial intervals, where the standard deviation is a minimum. The absolute Seebeck coefficient of metals, such as copper, platinum, and lead, has been evaluated experimentally through the Thomson effect over a wide temperature range. These data, coupled with the experimentally evaluated relative Seebeck coefficient, yield the absolute Seebeck coefficient of the metal or alloy of interest. In what follows, we describe a simple setup called the "Seebeck-emf analyzer" for determining the absolute Seebeck coefficient of a metal, such as iron, and an alloy, such as constantan.

10.1. *Seebeck-emf analyzer*

The Seebeck-emf analyzer is a simple system specially designed to study the temperature characteristics of standard and nonstandard thermocouples. Some typical experiments that can be carried out using this equipment are:

- calibration of any thermocouple against a standard thermocouple, namely, a chromel–alumel thermocouple;
- study of phenomena, such as "neutral temperature" and "inversion temperature," in nonstandard thermocouples, such as iron–copper and iron–silver;
- experimental verification of the law of intermediate metals;
- measurement of the "absolute Seebeck coefficient" of metallic samples in wire form as a function of temperature through an integral method.

The block diagram of the system to measure the Seebeck-emf of various thermocouples as a function of temperature is given in Figure 5. The system essentially consists of

- a high-temperature furnace capable of reaching temperatures up to 800°C;

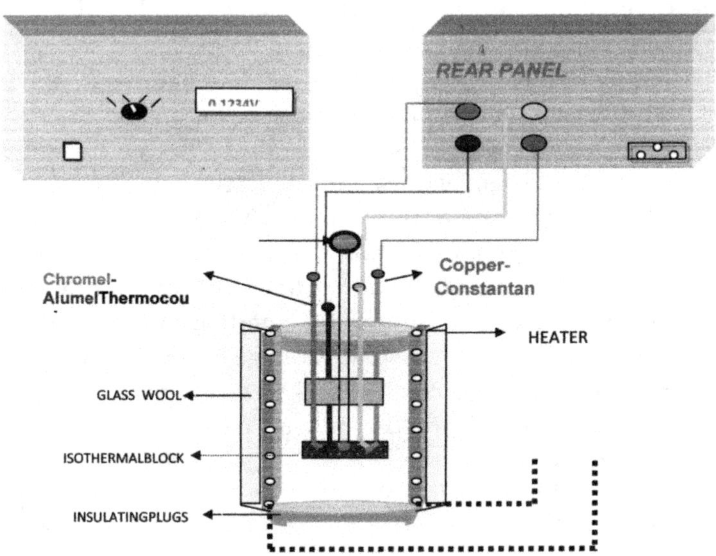

Figure 5. Block diagram of Seebeck-emf analyzer.

- a Seebeck-emf analyzer unit consisting of a precision DC microvolt amplifier for measuring low-level TE voltages;
- a two-pole, four-way band switch provided in the front panel of the instrument (TC selector switch), which is used to select various combinations of thermo-emfs generated in a pair of dissimilar metals/alloys, such as chromel–alumel, copper–constantan, chromel–copper, and copper–alumel;
- an AC power source to power the furnace;
- a specially designed thermos-well arrangement housing a reference or a standard thermocouple (chromel–alumel) and a sample thermocouple, such as copper–constantan.

10.2. *Measurement of absolute Seebeck coefficient of constantan as a function of temperature*

Chromel–alumel and copper–constantan thermocouples are kept in close proximity in a cell, and the thermowell assembly is then inserted into the center of the furnace. The heating coil has a nominal resistance of 6 Ω. Typically, 40–50 W is sufficient to attain temperatures around 500°C at the center of the furnace. The thermo-emf developed across the chromel–alumel thermocouple (which is a standard thermocouple) is fed to Channel 1 and amplified using an instrumentation amplifier (of gain = 100). The amplified signal is digitized and read on a digital panel meter directly in millivolt. Similarly, the thermo-emf developed across the copper–constantan thermocouple, which is fed to Channel 2, can be read directly by shifting the TC selector switch to position labeled 2.

Normally, thermocouple voltages are measured with the reference junction held at 0°C. These values are tabulated in the NBS tables and used to read out the temperature. In the current experimental arrangement, the concept of a single-junction thermocouple is used, where it is understood that the reference junction is at room temperature. Hence, to use the NBS tables, a voltage value corresponding to the room temperature must be added to the measured voltage from the single-junction thermocouple. This can be carried out by measuring the room temperature on a thermometer, and the thermo-emf corresponding to this temperature can be read off from the tables.

Typically, for chromel–alumel thermocouples, a room temperature of 25°C corresponds to 1.0 mV. For example, a 7 mV read from a single-junction chromel–alumel thermocouple corresponds to 8 mV (with the reference junction at 0°C), and the temperature as read from the NBS tables is 196.58°C.

Similarly, for the experimental copper–constantan thermocouple, an emf of the order of 1 mV corresponding to the room temperature value of 25°C must be added to the measured value.

10.3. *Results*

The Seebeck-emf of copper–constantan is measured against the temperature as read through the NBS tables for the chromel–alumel thermocouple. The typical plots of these data are given in Figure 6.

The thermo-emf, V, is related to the absolute Seebeck coefficient of copper, S_{Cu}, and that of constantan, $S_{Constantan}$, through the integral relation

$$V = \int_0^T [S_{Cu} - S_{Constantan}] dT \tag{25}$$

Here, $[S_{Cu} - S_{Constantan}]$ is referred to as the "relative Seebeck coefficient" of the copper–constantan thermocouple. Thus, the slope of the thermo-emf versus temperature curve gives the relative Seebeck

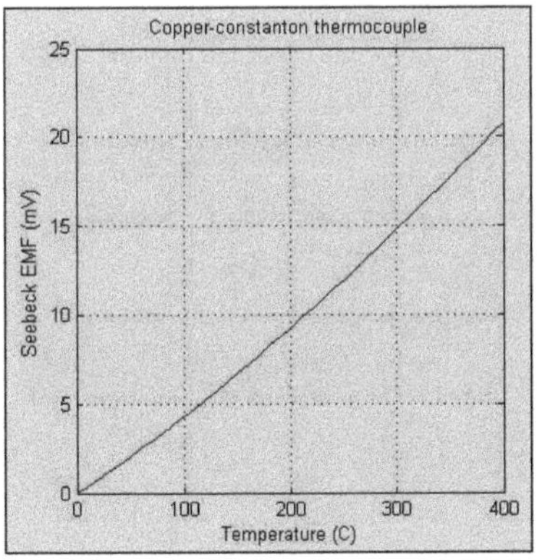

(a)

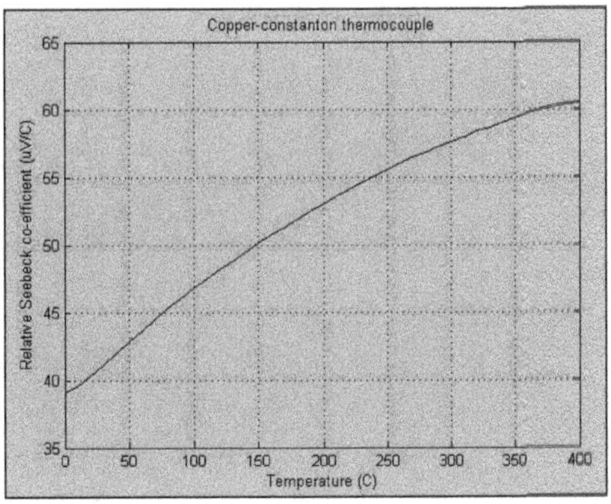

(b)

Figure 6. (a) Seebeck emf of copper–constantan vs. temperature. (b) Relative Seebeck coefficient vs. temperature.

coefficient of the copper–constantan thermocouple. This quantity can thus be experimentally evaluated as a function of temperature, covering the range of 25°C–400°C, as shown in Figure 6(b).

S_{Cu}, through the Thomson effect studies, has been experimentally determined and can be approximated through a linear relation:

$$S_{Cu}(\mu V/°C) = 1.7 + [4.989 \times 10^{-3}]T \tag{26}$$

where T is in °C. The experimental determination of the relative Seebeck coefficient of the copper–constantan thermocouple, coupled with the data on the absolute Seebeck coefficient of copper, thus provides a method of evaluating the absolute Seebeck coefficient of constantan.

Figure 7 presents the variation with temperature of the Seebeck coefficient of constantan along with that of copper.

The main results that emerge out of these experimental studies are as follows:

1. The absolute Seebeck coefficient of constantan is negative, and its magnitude is large, of the order of −40 μV/°C near ambient temperature and increases to nearly −60 μV/°C near 400°C.

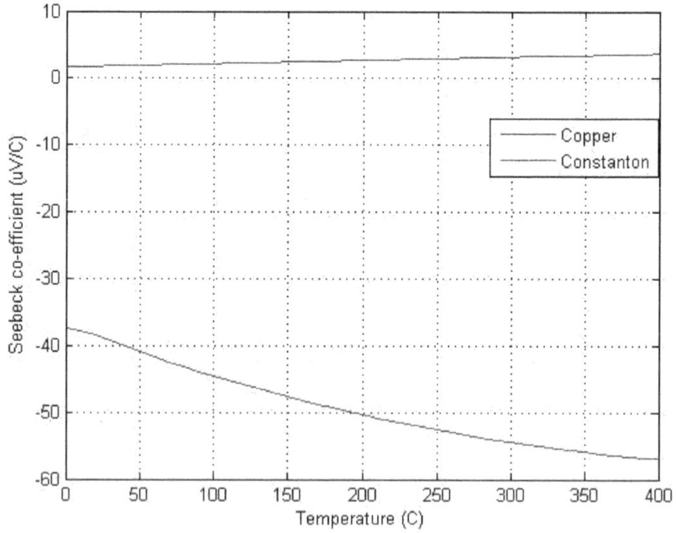

Figure 7. Absolute Seebeck coefficient of constantan and copper.

2. The relative Seebeck coefficient, which is the sensitivity of the copper–constantan thermocouple, exhibits a marked increase in its magnitude over the temperature range 0–400°C.
3. The Seebeck emf of the copper–constantan thermocouple exhibits a strong **nonlinear** behavior with respect to temperature. A polynomial of sixth order is required to fit the experimental data.

Among the alloys, constantan has the highest negative Seebeck coefficient and thus forms the negative leg of standard thermocouples, such as iron–constantan and chromel–constantan.

This "integral method" of determining the absolute Seebeck coefficient is simple and can be used for any metallic system which can be processed in wire form.

Primarily, constantan is composed of nickel and copper. It contains 60% nickel and 40% copper. The large negative value of the Seebeck coefficient in this material is related to the electronic band structure of this alloy, where both *s*-type and *d*-type states coexist near the Fermi energy. The gradual filling up of the holes in the d-band due to the alloying of nickel with copper leads to a situation wherein the Fermi energy of the alloy system probes the strongly decreasing density of states of the d-band. This leads to a large negative value of the Seebeck coefficient of constantan.

10.4. *Neutral and inversion temperature in iron–copper thermocouple*

In general, the thermo-emf of a thermocouple will monotonically increase with temperature, as seen in Figure 7, for the copper–constantan thermocouple. But for iron–copper and iron–silver thermocouples, the behavior is significantly different. The thermo-emf goes through a maximum value at a certain temperature. At this temperature, the relative Seebeck coefficient becomes zero. This temperature is called the neutral temperature.

The experimental procedure is the same as described earlier, except that in the thermowell, the copper–constantan thermocouple is replaced by an iron–copper thermocouple.

Figure 8 shows how the thermo-emf of an iron–copper thermocouple varies with temperature.

The thermo-emf $V_{\text{Fe–Cu}}$ versus temperature is fitted to a second-degree polynomial:

$$V_{\text{Fe - Cu}} = A + B_1 T + B_2 T^2 \tag{27}$$

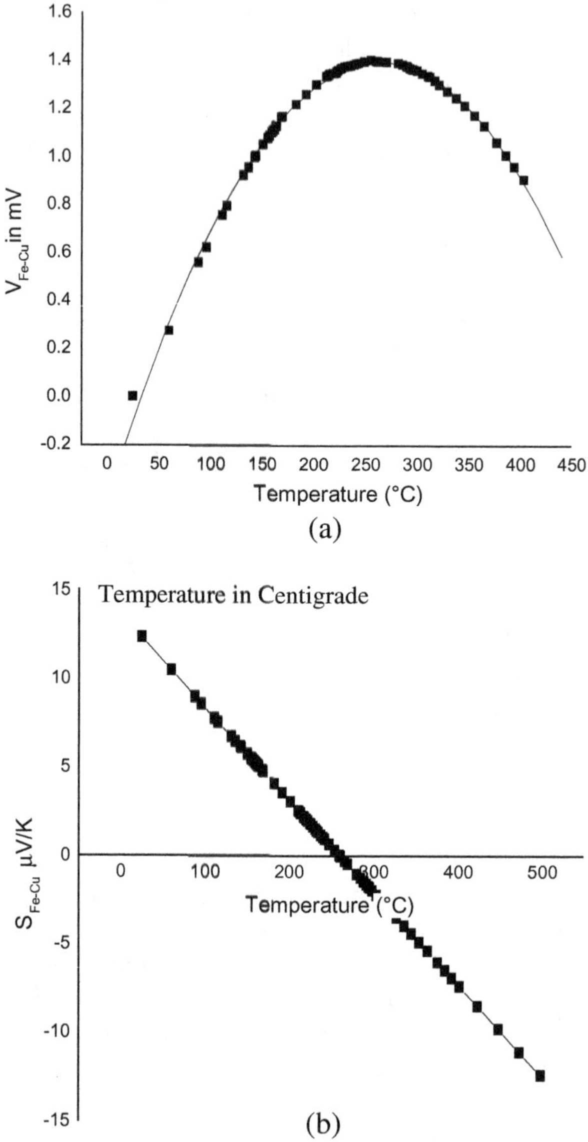

Figure 8. (a) Thermo-emf vs. temperature and (b) relative Seebeck coefficient vs. temperature of iron–copper thermocouple.

The fit is shown by the continuous curve in Figure 8. The relative Seebeck coefficient S_{Fe-Cu} is obtained by differentiating V_{Fe-Cu} with respect to T giving

$$S_{Fe-Cu} = B_1 + 2B_2T \tag{28}$$

Figure 8(b) presents the variation of S_{Fe-Cu} with respect to temperature.

The Seebeck coefficient S_{Fe-Cu} varies from positive to negative values. It becomes zero at around 260°C. This is the neutral temperature of the iron–copper thermocouple. Using Equation (26), one can calculate S_{Cu} at different temperatures and then calculate S_{Fe} as a function of temperature. A plot of the absolute Seebeck coefficients, S_{Fe} and S_{Cu}, as a function of temperature is shown in Figure 9. We see that the two curves intersect at the neutral temperature.

This experiment is of historic significance. From thermodynamic arguments, Thomson showed that the relative Seebeck coefficient for any pair of metals should be a constant and independent of temperature if the only TE effects are the Seebeck and Peltier effects. This will imply that for any thermocouple, the thermo-emf will vary linearly with temperature. This experiment showed clearly that this is not so. So, Thomson postulated a third TE effect, namely the Thomson effect.

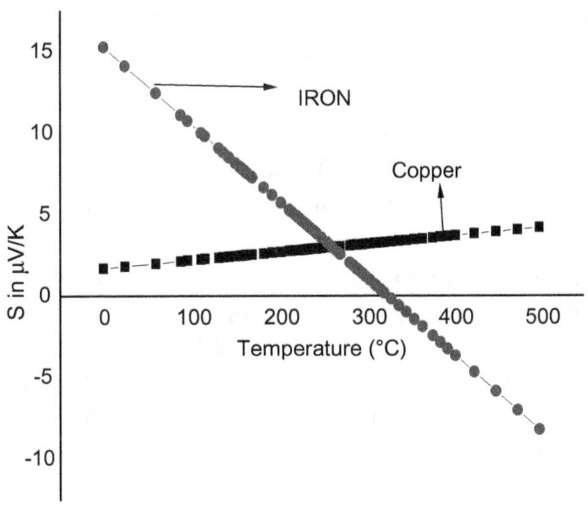

Figure 9. Seebeck coefficients S_{Fe} and S_{Cu} as functions of temperature.

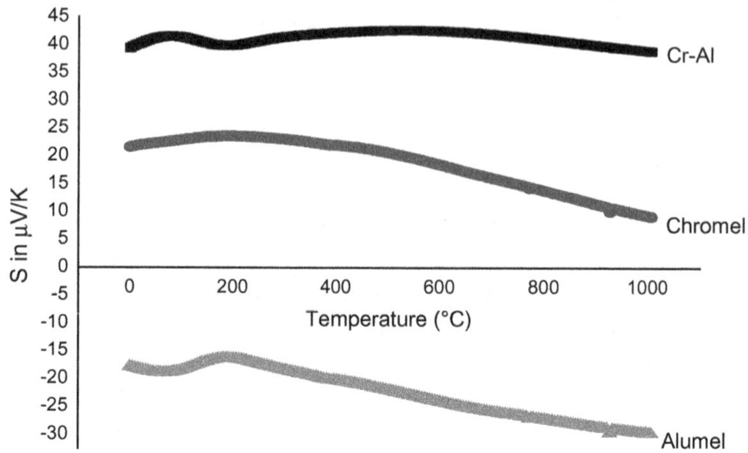

Figure 10. Absolute Seebeck coefficients of chromel and alumel and the relative Seebeck coefficient $S_{\text{Cr-Al}}$ as a function of temperature.

Figure 10 shows the temperature dependence of the Seebeck coefficient for chromel and alumel, as well as their relative Seebeck coefficient $S_{\text{Cr-Al}}$. We see that the Seebeck coefficient of chromel is positive, while the Seebeck coefficient of alumel is negative. The relative Seebeck coefficient, which is the difference between the Seebeck coefficients for the individual metals is therefore large (about 40 μV/K) and nearly independent of temperature over a wide range. It is worth noting that this feature is the main reason for the widespread use of chromel–alumel as a standard thermocouple.

11. Differential Method for Measuring the Absolute Seebeck Coefficient of n-type and p-type Bismuth Telluride

Semiconductors offer a unique TE advantage in the sense that their Seebeck coefficients are **orders of magnitude higher** than those of metals. Unlike in metals, the sign of the absolute Seebeck coefficient in a semiconductor is determined by the sign of the majority charge carriers. An *n*-type semiconductor is characterized by a **negative** Seebeck coefficient, while a *p*-type semiconductor exhibits a **positive** Seebeck coefficient. Apart from the Hall coefficient studies, the sign of the Seebeck coefficient provides a convenient tool to decipher whether the majority carriers are electron-like or hole-like.

11.1. *Preparation of n-type and p-type Bi_2Te_3 by mechanical alloying technique*

The ability of direct thermal to electrical energy conversion of a TE material has received enormous attention in the past two decades, considering its potential application in eco-friendly, solid-state energy harvesting. In the TE family, bismuth telluride (Bi_2Te_3) and its alloys are the most widely used TE materials around room temperature. Unidirectional crystal growth methods, such as the Bridgman method and the traveling heater method (THM), are widely used for their preparation. The unidirectional grown materials are inclined to cleavage fracture due to their weak van der Waals bonds between hexagonally stacked five-atomic layers, Te–Bi–Te–Bi–Te.

Powder metallurgical methods, which produce a randomly oriented polycrystalline microstructure with a grain-strengthening effect to avoid incidental cleavage failure, are commonly used to improve their mechanical properties. The conventional powder metallurgical process is devalued for the preparation of TE materials by its high energy consumption and impurity contamination. Mechanical alloying by high-energy ball milling is a powerful powder processing method. It has been applied to prepare powders of various alloys and compounds in a very short time. Mechanical alloying (MA) has also proved to be a versatile alternative to other processing routes in preparing nanostructured materials with a broad range of chemical compositions. Although there are numerous techniques to produce nanostructured materials, this method has become more popular due to its simplicity, capacity for large-scale production, and cost-effectiveness. The MA process is greatly affected by a number of factors that play very important roles in the preparation of homogeneous materials. The properties of the milled powders (particle size distribution, degree of disorder or amorphization, and final stoichiometry) depend on the milling conditions. A very large number of grain boundaries are generated when mechanical alloying is employed to prepare nanosized powder. Low dimensions and nanostructure significantly enhance the TE performance of a material.

Bi_2Te_3 with a stoichiometric composition can be prepared by high-energy ball milling using a planetary ball mill. The elemental granules of bismuth (Bi) and tellurium (Te) with 4N purity were crushed down to pieces smaller than 10 mm. The starting materials were weighed according to their composition and then subjected to mechanical alloying in the planetary ball mill under an inert atmosphere. The bigger particles were milled to a fine powder by high-energy ball milling at a speed of 450 rpm using a hardened tungsten carbide vessel and grinding balls. The prepared powder of Bi_2Te_3 was analyzed through X-ray diffraction (XRD) and surface electron microscopy for phase formation and microstructural study. Later, it was hot-pressed at 700 K and 50 MPa using a cylindrical graphite die.

11.2. *Differential method of measuring Seebeck coefficient*

The rather simple integral method of measuring the Seebeck coefficient in metallic wires cannot be employed in the case of semiconductors owing to the difficulty of drawing them in wire form. Semiconductors, for the current experiments, are prepared through the MA route and are available in pellet form.

The experimental arrangement consists of attaching the chromel–alumel thermocouple to both the top and bottom surfaces of the pellet. The pellet is sandwiched between a cooling platform below and a heating platform above, as shown in Figure 11. The cooling platform is constructed from a Peltier cooler powered by a 0–15 V, 0–5 A power supply. The heating stage is constructed out of a nichrome heater powered by a suitable power supply. The main objective of this experimental arrangement is to create a temperature difference varying between 1°C and 10°C between the top and bottom surfaces of the pellet. The mean temperature of the sample can be maintained to be close to the ambient

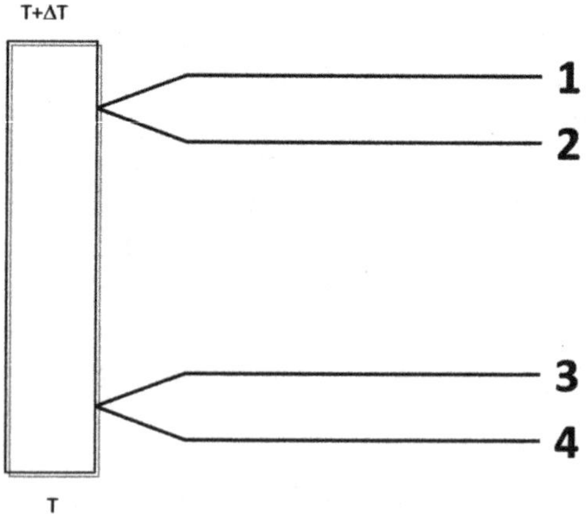

Figure 11. Schematic diagram of the differential technique.

temperature. The schematic diagram of the sample with thermocouple probes at either end is shown in Figure 11.

The Seebeck coefficient $S(T)$ of the semiconducting sample at temperature T is given by the relation

$$S_{\text{Chromel}}(T) - S(T) = \frac{\Delta V_{13}}{\Delta T} \tag{29}$$

Where S_{chromel} is the Seebeck coefficient of the chromel probe and ΔV_{13} is the voltage developed between the differential thermocouple formed out of the chromel–sample–chromel probes subjected to a temperature difference of ΔT. In a similar way, the differential voltage ΔV_{24} developed across the alumel–sample–alumel thermocouple for the same temperature difference is given by the expression

$$S_{\text{Alumel}}(T) - S(T) = \frac{\Delta V_{24}}{\Delta T} \tag{30}$$

where $S_{\text{Alumel}}(T)$ is the Seebeck coefficient of the alumel probe. It is easy to see that the temperature difference ΔT is given by the expression

$$\Delta T = \frac{[\Delta V_{13} - \Delta V_{24}]}{[S_{\text{Chromel}} - S_{\text{Alumel}}]} \tag{31}$$

The temperature difference can thus be more accurately measured, unlike the difference of two large quantities, T_1 and T_2. $[S_{\text{Chromel}} - S_{\text{Alumel}}]$ is the relative Seebeck coefficient of the standard chromel–alumel thermocouple and well tabulated as a function of temperature in a thermocouple handbook.

Thus, the Seebeck coefficient of the sample under study $S(T)$ is given by the relation

$$S(T) = S_{\text{Chromel}}(T) - \left\{ \frac{\Delta V_{13}}{(\Delta V_{13} - \Delta V_{24})} \right\} [S_{\text{Chromel}} - S_{\text{Alumel}}] \tag{32}$$

The four low-level voltages ΔV_{12}, ΔV_{34}, ΔV_{13}, and ΔV_{24} are measured using a precision DC microvoltmeter. The voltages ΔV_{12} and ΔV_{34} are used to arrive at the mean temperature T of the sample. Experiments can be carried out for different values of ΔT, keeping the mean temperature T constant. From the slope of the plot of ΔV_{13} against $(\Delta V_{13} - \Delta V_{24})$, the absolute Seebeck coefficient of the semiconducting sample can be evaluated.

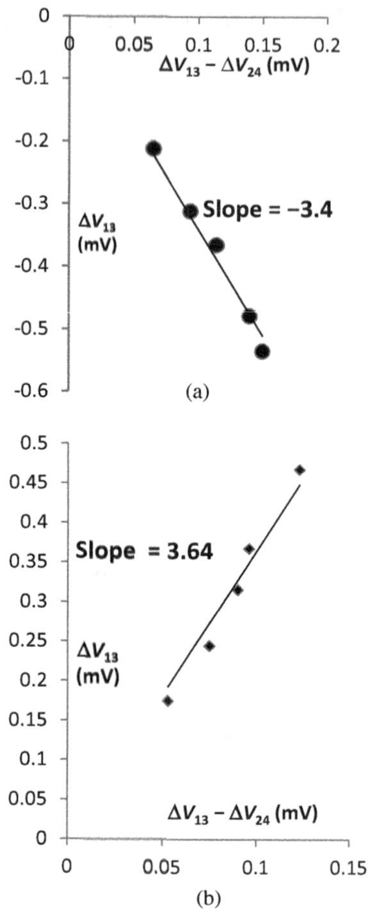

Figure 12. **Typical data for (a) *p*-type Bi_2Te_3, S = 158 μV/°C and (b) *n*-type Bi_2Te_3, S = –124 μV/°C.**

The value for $S_{\text{Chromel}} \approx +22\ \mu\text{V}/°\text{C}$ and $[S_{\text{chromel}} - S_{\text{Alumel}}] \approx 40\ \mu\text{V}/°\text{C}$ are to be plugged in for the calculation. It may be noted that a negative slope corresponds to a *p*-type semiconductor and vice versa. Figure 12(a) presents typical data of a semiconductor sample with measurements carried out near room temperature, and plugging in the value of the slope of –3.4 in the above expression yields the value of S as 158 μV/°C appropriate for *p*-type Bi_2Te_3. Similarly, the data presented in Figure 12(b) has a positive slope of 3.64, yielding S = –124 μV/°C appropriate for *n*-type Bi_2Te_3.

Chapter 15

DIELECTRIC PROPERTIES

1. Introduction

Dielectric materials find wide applications in electronics, aerospace, industrial materials, food and agriculture, industrial appliances and machinery, forestry, mining, and pharmaceuticals. An overview of dielectric measurement techniques has been presented by Tereschenko *et al.* (Ref. 1) and Nelson (Ref. 2).

Before dealing with dielectric properties, we give an introduction to the basic concepts and definitions in the electromagnetic theory of materials in the following paragraphs.

In the presence of a charge density distribution, $\rho(\boldsymbol{r})$, in a vacuum, an electric field $\boldsymbol{E}(\boldsymbol{r})$ is created. This electric field is related to charge density through the first Maxwell's equation:

$$\text{div}\ \boldsymbol{E} = \rho/\varepsilon_0 \tag{1}$$

The electric field $\boldsymbol{E}$ is in V/m, the charge density ρ is in C/m^3, and ε_0, called the permittivity of free space, has a value of 8.854×10^{-12} F/m.

Associated with the electric field $\boldsymbol{E}$ is the electric displacement vector $\boldsymbol{D}$, which in vacuum is given by

$$\boldsymbol{D} = \varepsilon_0 E \tag{2}$$

The electric displacement vector is expressed in C/m^2

The electric field $\boldsymbol{E}$ exerts a force on a test charge q, given by

$$\boldsymbol{F} = q\boldsymbol{E} \tag{3}$$

Consider a charge distribution in which charges q_j are located at positions $\boldsymbol{r}_j$. For such a charge distribution, we define the position of the effective center of the electrical charge by

$$\boldsymbol{R} = \Sigma q_j \boldsymbol{r}_j / \Sigma q_j \tag{4}$$

The total charge Q in such a charge distribution is

$$Q = q^+ - q^- \tag{5}$$

Here, q^+ and q^- are the total amounts of positive and negative charges in the distribution. If q^+ and q^- are equal, the total charge Q is zero. Such a charge distribution is said to be neutral.

Using Equation (4), one can calculate the effective center separately for all the positive charges and all the negative charges, comprising the total charge distribution. Let $\boldsymbol{R}^+$ be the center of the positive charges, and let $\boldsymbol{R}^-$ be the center of the negative charges. Then, the charge distribution is said to have an electric dipole moment $\boldsymbol{p}$ given by

$$\boldsymbol{p} = q^+\boldsymbol{R}^+ - q^-\boldsymbol{R}^- \tag{6}$$

In a neutral charge distribution, $q^+ = q^- = q$, and $\boldsymbol{p}$ is given by

$$\boldsymbol{p} = q(\boldsymbol{R}^+ - \boldsymbol{R}^-) = q\boldsymbol{d} \tag{7}$$

In Equation (7), $\boldsymbol{d}$ is the vector distance between the centers of the positive and negative charge distributions. The dipole moment is expressed in Cm. For example, the molecule HF is composed of a positively charged hydrogen atom and a fluorine atom with an equal but negative charge. It will have a dipole moment, as given by Equation (7).

When such a dipole is placed in a uniform electric field $\boldsymbol{E}$, it experiences a torque $\boldsymbol{\tau}$, given by

$$\boldsymbol{\tau} = \boldsymbol{p} \times \boldsymbol{E} \tag{8}$$

This torque tries to align $\boldsymbol{p}$ in the direction of $\boldsymbol{E}$. The energy of a dipole $\boldsymbol{p}$ in an electric field $\boldsymbol{E}$ is

$$U = -\boldsymbol{p}.\boldsymbol{E} \tag{9}$$

The energy U is at its minimum when the dipole $\boldsymbol{p}$ is aligned parallel to the electric field $\boldsymbol{E}$. This energy is at its maximum when the dipole is aligned antiparallel to $\boldsymbol{E}$.

Molecules constituting any dielectric are basically neutral as far as total charge is concerned. Some molecules can be dipoles and, hence, are called polar molecules. On the contrary, some may not have any charge separation, in which case they are called nonpolar molecules. In the case of molecules, when an external electric field is applied, in addition to realigning the molecular dipoles, the action of the force $\boldsymbol{F}$ due to the electric field on the positive and negative charges will cause an increase in the separation of the centers of the positive and negative charge distributions and induce an additional dipole moment $\boldsymbol{p}_{\text{induced}}$. In the case of polar molecules, this additional dipole moment will be proportional to the electric field $\boldsymbol{E}$ when the electric field is small. If the molecules are nonpolar, the applied electric field will cause only the induced dipole moment $\boldsymbol{p}_{\text{induced}}$.

In a material medium, the total dipole moment per unit volume of the material is called the electric polarization. This polarization is denoted by $\boldsymbol{P}$. The electric displacement $\boldsymbol{D}$ in such a medium, in the presence of an applied electric field $\boldsymbol{E}$, is given by

$$\boldsymbol{D} = \varepsilon_0 \boldsymbol{E} + \boldsymbol{P} \tag{10}$$

If the applied electric field is not large (linear region), the polarization $\boldsymbol{P}$ is seen to vary linearly with $\boldsymbol{E}$ and is expressed as

$$\boldsymbol{P} = \varepsilon_0 \chi \boldsymbol{E} \tag{11}$$

where χ is the called the dielectric susceptibility of the material. The vector $\boldsymbol{D}$ can then be written as

$$\boldsymbol{D} = \varepsilon_0(1 + \chi)\boldsymbol{E} = \varepsilon_0 \varepsilon_r \boldsymbol{E} \tag{12}$$

Here, ε_r is a dimensionless number, called the dielectric constant of the medium.

For an isotropic medium, such as a gas, liquid, or an amorphous solid, ε_r is a scalar quantity. In these materials, the displacement vector $\boldsymbol{D}$ is always parallel to the applied electric field vector $\boldsymbol{E}$, whatever the direction of the electric field. However, in a single crystal, which is anisotropic, the direction of $\boldsymbol{D}$ will, in general, be different from the direction of $\boldsymbol{E}$. The dielectric constant ε_r is a second-rank symmetric tensor. This implies that each component, D_i, of the displacement vector depends linearly on all three components, E_j, of the electric field vector with different proportionality constants, $\varepsilon_{I,j}$, as follows:

$$D_1 = \varepsilon_0[\varepsilon_{r11}E_1 + \varepsilon_{r12}E_2 + \varepsilon_{r13}E_3] \tag{13a}$$

$$D_2 = \varepsilon_0[\varepsilon_{r21}E_1 + \varepsilon_{r22}E_2 + \varepsilon_{r23}E_3] \tag{13b}$$

$$D_3 = \varepsilon_0[\varepsilon_{r31}E_1 + \varepsilon_{r32}E_2 + \varepsilon_{r33}E_3] \tag{13c}$$

with the added restrictions of

$$\varepsilon_{r12} = \varepsilon_{r21},\ \varepsilon_{r23} = \varepsilon_{r32},\ \varepsilon_{r31} = \varepsilon_{r13} \tag{13d}$$

Thus, a sample of a triclinic crystal is characterized by the six independent parameters ε_{r11}, ε_{r22}, ε_{r33}, ε_{r12}, ε_{r23}, and ε_{r31}, which represent the dielectric behavior of the crystal. Depending on the various symmetry elements in the crystal, the number of independent components of the dielectric constant tensor gets reduced, as shown in Table 1.

In a monoclinic system, there is either a two-fold axis of symmetry along the y-axis, or a mirror plane of symmetry, the normal to which is in the y-direction. The subscripts 1, 2, and 3 for ε in Table 1 refer to the x-, y-, and z-axes of the orthogonal coordinate system. Crystals belonging to the

Table 1. Number of independent components of ε_r in different crystal systems.

Crystal system	No. of independent constants	Independent constants
Triclinic	6	$\varepsilon_{11}, \varepsilon_{22}, \varepsilon_{33}, \varepsilon_{12}, \varepsilon_{23}, \varepsilon_{31}$
Monoclinic	4	$\varepsilon_{11}, \varepsilon_{22}, \varepsilon_{33}, \varepsilon_{13}, \varepsilon_{12} = \varepsilon_{23} = 0$
Orthorhombic	3	$\varepsilon_{11}, \varepsilon_{22}, \varepsilon_{33}, \varepsilon_{12} = \varepsilon_{23} = \varepsilon_{12} = 0$
Tetragonal	2	$\varepsilon_{11} = \varepsilon_{22}, \varepsilon_{33}, \varepsilon_{12} = \varepsilon_{23} = \varepsilon_{12} = 0$
Trigonal	2	$\varepsilon_{11} = \varepsilon_{22}, \varepsilon_{33}, \varepsilon_{12} = \varepsilon_{23} = \varepsilon_{12} = 0$
Hexagonal	2	$\varepsilon_{11} = \varepsilon_{22}, \varepsilon_{33}, \varepsilon_{12} = \varepsilon_{23} = \varepsilon_{12} = 0$
Cubic	1	$\varepsilon_{11} = \varepsilon_{22} = \varepsilon_{33}, \varepsilon_{12} = \varepsilon_{23} = \varepsilon_{12} = 0$
Isotropic	1	

monoclinic system are characterized by four independent values of ε_r. In the orthorhombic system, there are either two-fold axes of symmetry or normal to mirror planes of symmetry along all three axes 1, 2, and 3. In this system, the number of independent components of ε_r is three. In tetragonal systems, there is a four-fold axis of symmetry along *z* (labeled as 3). In trigonal and hexagonal systems, there is a three-fold or a six-fold axis of symmetry along the *z*-axis. In these three systems, the number of independent constants is two. In the cubic system, there are three four-fold axes of symmetry along the *x*-, *y*-, and *z*-axes. This results in a single independent component of the tensor ε_r. Thus, crystals belonging to the cubic system behave as an isotropic material as far as the dielectric properties are concerned.

It is also useful to represent the tensor ε_r as a 3×3 symmetric matrix, with six elements: ε_{11}, ε_{22}, ε_{33}, ε_{12}, ε_{23}, and ε_{31}. In this representation, ε_r is a diagonal matrix for crystals belonging to orthorhombic, tetragonal, trigonal, hexagonal, and cubic systems, which greatly simplifies the theoretical analysis.

2. Contributions to Electric Polarization P

If we consider a gas or a liquid material comprising polar molecules, then the molecules can easily align along the applied electric field since they have permanent dipole moments. In long-chain polymers, there can be groups with a permanent dipole moment, which can orient themselves upon application of an electric field. The orienting influence of the electrical field will be countered by thermal energy, which tries to disorient the molecules. The fraction, $f(\theta)$, of molecules, for which the dipole moment $\boldsymbol{p}$ makes an angle θ with the electric field $\boldsymbol{E}$, will be proportional to the Boltzmann factor:

$$f(\theta) \ \alpha \ \exp(-U/kT) = \exp\left(pE\cos(\theta)/kT\right) \tag{14}$$

where U is the potential energy of the dipole in an electric field. In a gas, the density is low. So, the additional contribution to the electric field coming from the dipole moments of the molecules can be neglected. Then, the effective polarization $\boldsymbol{P}_{\mathbf{Orient}}$ in an applied electric field $\boldsymbol{E}$ at a temperature T is given by

$$P_{\text{Orient}} = Np^2\mathrm{E}/3kT \tag{15}$$

Here, N is the number of molecules per unit volume. P_{Orient} is the orientational contribution to the electric polarization arising from the alignment of the dipoles due to the applied DC electric field.

In addition, one can have a vibrational contribution P_{vib}. Neutral molecules have an equal amount of positive and negative charges q. We may consider the centers of the two charges to be connected by a spring-like force with a spring constant of k. When an electric field is applied, there will be an equal and opposite force acting on the charges to further separate the positive and negative charges. This is countered by the spring. If the resultant displacement of the charges is d, then

$$F = qE = kd \tag{16}$$

The induced dipole moment is

$$p_{\text{vib}} = qd = (q^2/k)\, E \tag{17}$$

In a medium of low density, such as a gas, with N molecules per unit volume, this contributes a dipole moment of P_{vib} to the induced polarization, given by

$$P_{\text{vib}} = Np_{\text{vib}} = N(q^2/k)\, E \tag{18}$$

The stronger the spring, the larger the value of k, and the higher the vibrational frequency v of the molecule. The electronic contributions coming from the electron distributions around the nucleus in the atoms in a molecule can be modeled similarly with a much higher value of k.

Thus, the total contribution to the polarization of the gas is given by

$$P = P_{\text{Orient}} + P_{\text{vib}} + P_{\text{elec}} \tag{19}$$

The orientational contribution shows a strong temperature dependence, varying as $1/T$, while the vibrational and electronic contributions are only weakly dependent on temperature. For gaseous molecules, the DC dielectric constant will be given by

$$\varepsilon_r = D/(\varepsilon_0 E) = (E + P)/E = 1 + N\alpha \tag{20}$$

where $$\alpha = (\alpha_{\mathrm{Orient}} + \alpha_{\mathrm{vib}} + \alpha_{\mathrm{electronic}}) \tag{21}$$

α is called the polarizability of the molecule. The contribution arising from the alignment of molecules, $\alpha_{\mathrm{Orient}} = p^2/3kT$, is strongly temperature dependent.

In dense media, such as liquids and solids, one will have to add the field due to electric dipole moments of the surrounding molecules to the applied electric field $\boldsymbol{E}$ to get the local electric field $\boldsymbol{E}_{\mathrm{local}}$, acting on the molecule under consideration. For an **isotropic** dense medium, the local electric field $\boldsymbol{E}_{\mathrm{local}}$ at a molecule is given by

$$\boldsymbol{E}_{\mathrm{local}} = \boldsymbol{E} + \boldsymbol{P}/3\varepsilon_0 \tag{22}$$

and $$\boldsymbol{P} = N\alpha\boldsymbol{E}_{\mathrm{local}} \tag{23}$$

Combining Equations (22) and (23), we obtain

$$(\varepsilon_{\mathrm{r}} - 1)/(\varepsilon_{\mathrm{r}} + 2) = N\alpha/3 \tag{24}$$

This is called the Clausius–Mossotti relation.

In a solid composite sample, which is prepared using grains of two materials with different dielectric constants, the application of an electric field will cause surface charges to appear at the interface of the grains with different dielectric constants. This surface charge will also contribute to the polarization $\boldsymbol{P}$ of the mixture. This is called Maxwell–Wagner surface polarization.

We have so far dealt with the DC electric field and the DC dielectric constant. If the applied electric field is sinusoidal with an angular frequency ω relaxation plays an important role in determining the value of the dielectric constant at a given frequency. This will be discussed in the following paragraph.

3. Frequency Dependence of Dielectric Constant

The dielectric constant of a material varies with frequency. The frequency dependence is different for the orientational contribution compared to that for the vibrational and electronic contributions.

In the case of materials subjected to DC electric fields, it is seen that when the applied DC electric field is suddenly reduced to zero, the orientational contribution to polarization P decays with time to zero according to the equation

$$\mathrm{d}P_{\text{Orient}}/\mathrm{d}t + P_{\text{Orient}}/\tau = 0 \tag{25}$$

τ is called the relaxation time for the orientational contribution to polarization. If the applied electric field E varies as $\exp(\mathrm{i}\omega t)$, then Equation (25) is to be modified by adding the orientational contribution of polarization due to the instantaneous electric field:

$$\mathrm{d}P_{\text{Orient}}/\mathrm{d}t + P_{\text{Orient}}/\tau = \chi_{\text{Orient}}(0)\, E\exp(\mathrm{i}\omega t)/\tau \tag{26}$$

Here, $\chi_{\text{Orient}}(0)$ is the orientational contribution to the dielectric susceptibility in a DC electric field and is equal to $N\alpha_{\text{Orient}}$ for a gas. The solution to the above equation is

$$P_{\text{Orient(t)}}(t) = [\chi_{\text{Orient}}(0)E/(1 + \mathrm{i}\omega\tau)]\exp(\mathrm{i}\omega t) \tag{27}$$

This is called the Debye relaxation formula. The solution indicates that the orientational polarization lags behind the electric field by an angle φ given by

$$\tan\varphi = \omega\tau \tag{28}$$

and

$$\chi_{\text{Orient}}(\omega) = \chi_{\text{Orient}}(0)/(1 + \omega^2\tau^2)^{1/2} \tag{29}$$

The phase lag of the polarization with respect to E leads to a loss of energy from the electric field. This loss of energy turns into heat that warms the material. This is called dielectric loss.

Similarly, for the vibrational contribution to the polarization, we may write the following equation:

$$\mathrm{d}^2P_{\text{vib}}/\mathrm{d}t^2 + \gamma\mathrm{d}P_{\text{vib}}/\mathrm{d}t + \omega_0{}^2P_{\text{vib}} = \omega_0{}^2\chi_{\text{vib}}(0)E\exp(\mathrm{i}\omega t) \tag{30}$$

Equation (30) is similar to the equation for the displacement of a simple harmonic oscillator with a natural frequency ω_0 and a damping constant γ oscillating under the action of an external force due to the applied electric field. The solution to this equation is

$$P_{vib}(t) = \{\omega_0^2\chi_{vib}(0)E/\ [(\omega_0^2 - \omega^2) + i\omega\gamma)]\}\ \exp(i\omega t) \quad (31)$$

The vibrational contribution to the polarization also lags behind the electrical field by the angle ξ given by

$$\tan\xi = \omega\gamma/(\omega_0^2 - \omega^2) \quad (32)$$

The vibrational contribution to dielectric susceptibility varies as

$$\chi_{vib}(\omega) = \omega_0^2\,\chi_{vib}(0)\ /[(\omega_0^2 - \omega^2)^2 + \omega^2\gamma^2]^{1/2} \quad (33)$$

The dielectric constant $\varepsilon_r(\omega)$ is

$$\varepsilon_r(\omega) = 1 + \chi_{orient}(\omega) + \chi_{vib}(\omega) \quad (34)$$

The dielectric constant $\varepsilon_r(\omega)$ is a complex number. It may be written as

$$\varepsilon_r(\omega) = \varepsilon'_r(\omega) - i\varepsilon''_r(\omega) \quad (35)$$

For any dielectric material, we measure the real part of the dielectric constant $\varepsilon'_r(\omega)$ and tan δ defined by

$$\tan\delta = \varepsilon_r''(\omega)/\varepsilon_r'(\omega) \quad (36)$$

tan δ is the ratio of the energy lost (as heat) to the energy stored per cycle in a unit volume of the material due to the applied electric field.

The variations in the real part ε_r' and imaginary part ε_r'' of the dielectric constant as a function of the frequency f (= $\omega/2\pi$) are shown in Figure 1.

When small ions, such as Li^+, are present in a solid, ionic oscillations can also occur, causing dielectric loss in the solid. This energy loss will decrease and become smaller in the radio frequency range. Molecular rotations occur in the microwave (MW) frequency range. Molecular vibrations are in the infrared range, and electronic transitions are in the visible

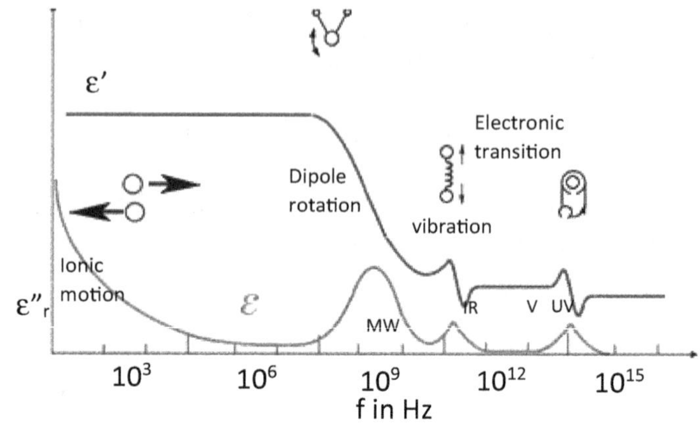

Figure 1. Variation of ε'_r and ε''_r with frequency.

and ultraviolet ranges. Thus, one will have to make measurements over a large range of frequencies to get complete information about the dielectric properties of a material.

Naturally, the techniques one uses will differ for the different ranges of frequencies. The techniques for the microwave, IR, and UV–visible ranges of frequencies are based on the propagation of electromagnetic waves in a material medium. This will be briefly discussed in the following section.

4. Propagation of Electromagnetic Waves in an Unbounded Medium

An oscillating electric field produces an alternating magnetic field in a material medium. In an infinite material medium, the oscillating electric and magnetic fields propagate as waves with a phase velocity of

$$v = c / \sqrt{\varepsilon_r \mu_r} \tag{37}$$

In Equation (37), ε_r is the dielectric constant and μ_r is the magnetic permeability of the medium relative to vacuum. If the material is non-magnetic, μ_r is 1.

The refractive index $\boldsymbol{n}$ of any medium is defined as the ratio of the phase velocity c of the electromagnetic wave in vacuum to the phase velocity v of the same electromagnetic wave in the medium. So, for a non-magnetic material, $\boldsymbol{n}$ is given by

$$\boldsymbol{n} = c/v = \sqrt{\varepsilon_r} \tag{38}$$

Since ε_r is a complex number, $\boldsymbol{n}$ is also a complex number given by

$$\boldsymbol{n} = n - \mathrm{i}\kappa \tag{39}$$

The electric field of a plane electromagnetic wave of frequency ν, traveling along the z-axis in an infinite material medium, may be written as

$$\begin{aligned} E(z, t) &= E_0 \exp[\,\mathrm{i}2\pi\nu(t - \boldsymbol{n}z/c)] \\ &= E_0 \exp[-(2\pi\kappa z/\lambda] \exp[\mathrm{i}2\pi\nu(t - nz/c)] \end{aligned} \tag{40}$$

The amplitude of the electric field decreases as the wave travels along the z-axis in the medium, and the phase velocity of the wave in the medium is c/n. λ is the wavelength of the electromagnetic wave in vacuum.

The intensity of the electromagnetic wave is proportional to $|E(z, t)|^2$. So, it will decay with z as

$$I(z) = I(0)\exp(-\beta z) \tag{41}$$

where β, called the absorption coefficient, is given by

$$\beta = 4\pi\kappa/\lambda \tag{42}$$

Thus, by measuring n and κ, one can obtain ε_r' and ε_r''.

If an electromagnetic wave is incident normally on a plane boundary between a vacuum and the material medium, the wave will get partially reflected and partially transmitted. This is shown in Figure 2.

The amplitude of the reflected wave will be less than the amplitude of the incident wave, and there will be a phase change on reflection. The transmitted wave will decay with distance and also suffer a phase change with respect to the incident wave at the boundary. By measuring the amplitude and phase of the reflected and transmitted waves in a sample using a vector network analyzer (VNA), one can measure the real and

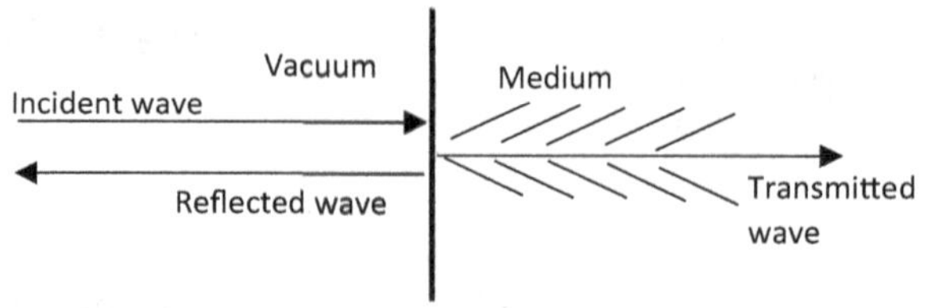

Figure 2. Reflection and transmission of an EM wave incident normally on a face separating medium from vacuum.

imaginary parts of the dielectric constant. This will be discussed in a later section.

5. Measurement of the Dielectric Constant ε_r of Materials in the Audio and Radio Frequency Ranges

Audio and radio frequencies refer to a range from 0 to about 100 MHz. In this range, one may use an LCR meter or an impedance analyzer. We fabricate a capacitor between two electrodes with the material as the dielectric medium between the electrodes. Two simple configurations of the electrodes are used. In one, the two electrodes are parallel metal plates separated by a distance d. This is called a parallel-plate capacitor. If d is very small compared to the length and width of the plates, the capacitance C is given by

$$C = \varepsilon_0 \varepsilon_r A/d \tag{43}$$

The other configuration, called the cylindrical capacitor, consists of two coaxial electrodes of radii r_1 and r_2 and of length ℓ. The region between the two cylinders is filled with the dielectric material. If length ℓ is large compared to r_1 and r_2, then

$$C = 2\pi\varepsilon_0 \varepsilon_r \ell/\mathrm{Ln}(r_2/r_1) \tag{44}$$

Ln(x) is the logarithm of x to the base e. The two configurations are shown schematically in Figures 3(a) and 3(b), respectively.

In Figure 3(a), the black bands represent metallic plates and lead wires. In Figure 3(b), the thick black lines represent the lead wires. The thin lines in Figure 3(b) represent two coaxial cylindrical metal tubes.

Figure 3. **(a) Parallel-plate capacitor and (b) cylindrical capacitor (extracted from Ref. 3).**

The inner cylinder can be a solid rod. The impedance of such a capacitor at a frequency ω is

$$Z = 1/\mathrm{i}\omega C = -\mathrm{i}/\omega C \tag{45}$$

Since $\varepsilon_r(\omega)$ is complex, Z is also a complex number and may be written as

$$Z = R + \mathrm{i}X \tag{46}$$

R corresponds to the resistive loss of the capacitor and is related to $\varepsilon''_r(\omega)$. X corresponds to the reactance of the capacitor and is related to $\varepsilon'_r(\omega)$. Thus, a measurement of Z gives the values of the real and imaginary parts of the dielectric constant. Such a description is called the lumped circuit parameter description and is valid if the thickness of the sample is less than the wavelength of the electromagnetic wave propagating through the sample. The wavelength is given by $c/(\nu\sqrt{\varepsilon'_r})$. Such a lumped parameter model is adequate up to a frequency of 100 MHz with the usual sample cell dimensions.

Levitskaya and Sternberg (Radio Science 35, 371–383, 2000) have described the construction of a parallel-plate capacitor for disk-shaped samples with diameters of 5 cm or less and variable thickness. Figure 4 shows the details of the sample holder. The figure is self-explanatory. The electrodes are silver-coated brass plates with a thickness of 2 mm. The gap d between the electrodes is adjusted using a Teflon screw. The sample material is loaded into the gap between the electrode plates.

Figure 5 shows a cylindrical capacitor constructed by the same authors.

The dimensions of the cylindrical capacitor are as follows: diameter of the outer tube = 0.71 cm, diameter of the inner tube = 0.31 cm, and

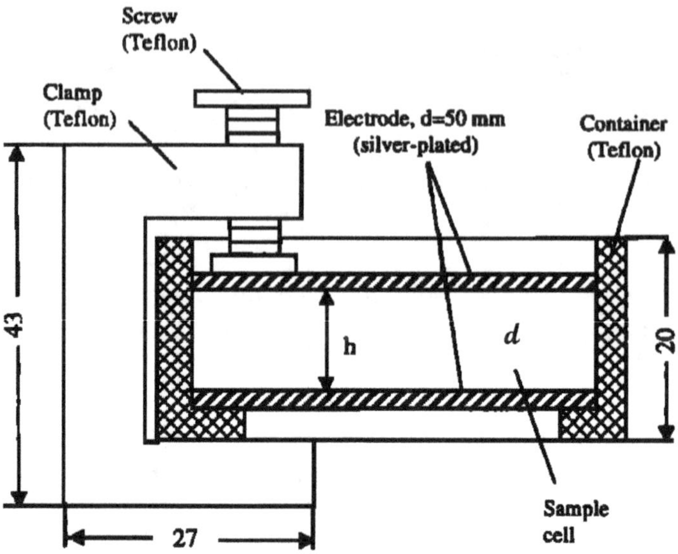

Figure 4. Sample holder of the parallel-plate capacitor configuration.

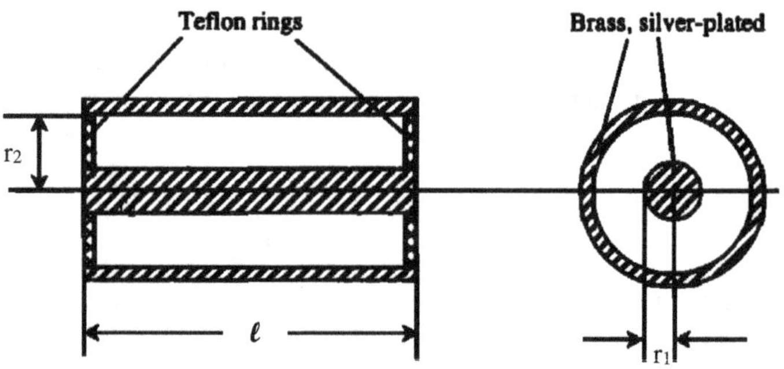

Figure 5. Coaxial cylindrical capacitor (reprinted with permission from Levitskaya and Sternberg, Radio Science, 35, 371–383, 2000).

$\ell = 3$ cm. With such dimensions, one can carry out measurements up to 1 GHz.

The LCR meter or the impedance analyzer can be used to measure the complex impedance of the sample. This is done by applying a small, fixed voltage to the sample and measuring the amplitude and phase of the current through the sample. In an LCR meter, the measurement is done at a few fixed frequencies. In the impedance analyzer, the frequency can be swept continuously, and the instrument displays graphically the variation in the real and imaginary parts of the impedance as a function of frequency.

The basic operation of a four-terminal LCR meter is shown in Figure 6.

SG is a signal generator that produces a sinusoidal fixed AC voltage. There are four terminals. The terminal L_1 applies the AC signal from the signal generator to the sample S under test in the capacitor holder. The two terminals L_2 and L_3 measure the voltage drop across the sample. The terminal L_4 measures the current through the capacitor. The signals from these leads are further processed electronically to obtain the amplitude of the current *I* and the phase difference between the applied voltage and the current. One can display the real and imaginary parts of the impedance or the real and imaginary parts of the dielectric constant on a computer using appropriate software.

In the experiments by Levitskaya and Sternberg, no guarded electrodes were used. Hence, they had to apply corrections for errors arising from the nonuniform electric field at the edges of the capacitance cell. These corrections are described in their paper.

Figure 7 shows the results they obtained for ε'_r and ε''_r for Teflon and air with both the disk (parallel plates) and coaxial tube setups. A good agreement was found between the results obtained with the two different sample holders.

In Figure 8, we show the results for pure n-octanol liquid.

Two results reported in the literature are also shown in the figure. The graphs for n-octanol show typical relaxation behavior. In such cases, one plots ε''_r against ε'_r at different frequencies, as shown in Figure 9. Such a plot is called the Cole–Cole plot.

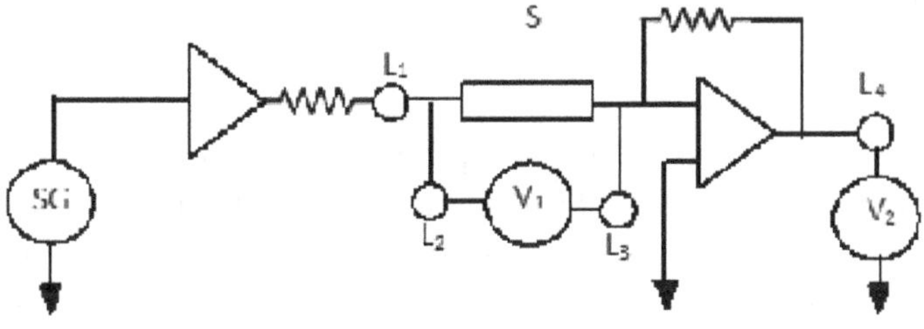

Figure 6. Principle of operation of a four-terminal LCR meter/impedance analyzer. The terminals are L_1, L_2, L_3, and L_4.

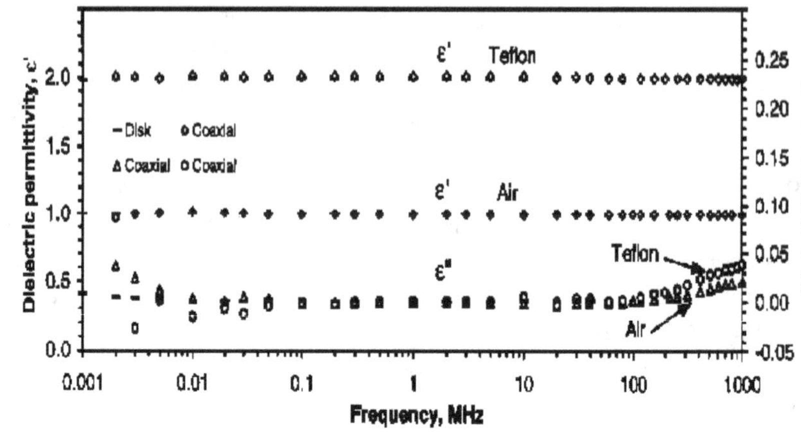

Figure 7. Real and imaginary parts of the dielectric constant for Teflon and air using both the disk and coaxial tube sample holders (reprinted with permission from Ref. 1).

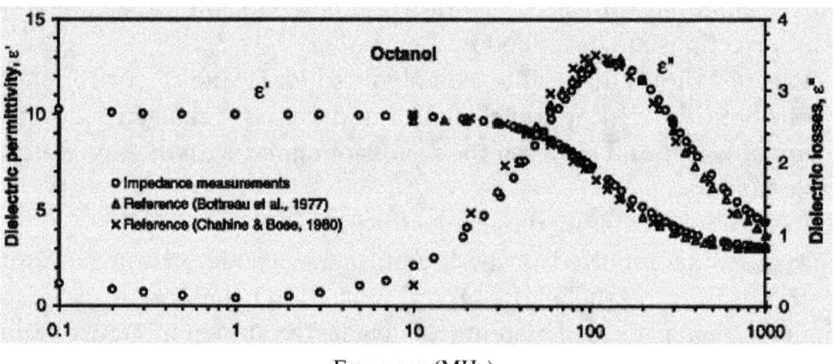

Figure 8. Real and imaginary parts of the dielectric constant of pure n-octanol (reprinted with permission from Ref. 1).

If there is a single relaxation process with relaxation time τ, the Cole–Cole plot will be a semicircle with its center on the x-axis (or the ε'_r axis). For n-octanol, one obtains such a semicircle, except for a small deviation at the highest frequencies. The relaxation time τ from the fit to the data came out as 1.26 ns. In a material, there can be more than one process of relaxation with very different relaxation times. In such

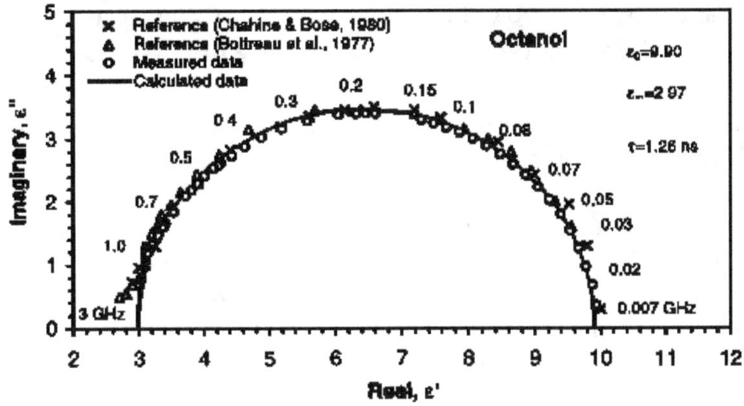

Figure 9. Cole–Cole plot for n-octanol (reprinted with permission from Ref. 1).

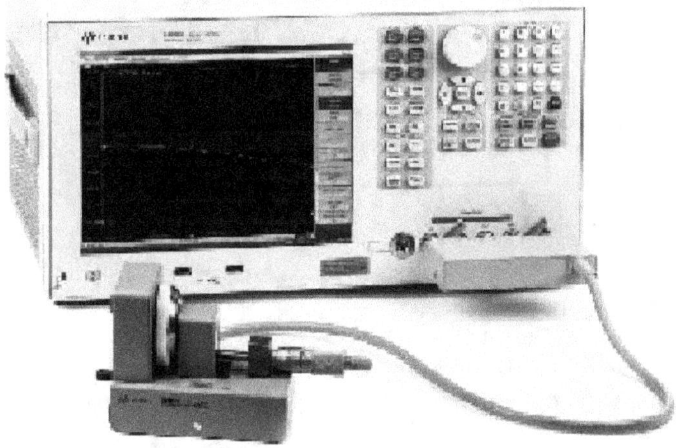

Figure 10. Commercial impedance analyzer Keysight E4990A with a sample holder to measure dielectric properties of solids.

cases, the Cole–Cole plot is an arc of a circle with its center below the x-axis.

Figure 10 shows a commercial impedance analyzer, Keysight E4990A, along with the sample holder for measuring the dielectric properties of solids in the frequency range up to 30 MHz (from Ref. 2).

6. Measuring Dielectric Parameters at High Frequencies using the S-Parameters and a Vector Network Analyzer

In Section 3, we discussed the reflection and transmission of electromagnetic waves at the interface of two dielectric media. The VNA measures the amplitude and phase of the reflected and transmitted signals in relation to the amplitude and phase of the incident signal. A block diagram illustrating the operation of VNA is shown in Figure 11.

The signal source produces an AC voltage signal of fixed amplitude. The switch SW can be used to direct the signal to the splitter on either the left or right. The splitter splits the signal and sends one part to the receiver and the other part to the directional coupler 1 or directional coupler 2. The signal comes out of the corresponding port on the VNA and is carried by a precision cable to either port P_1 or port P_2 of the holder in which the

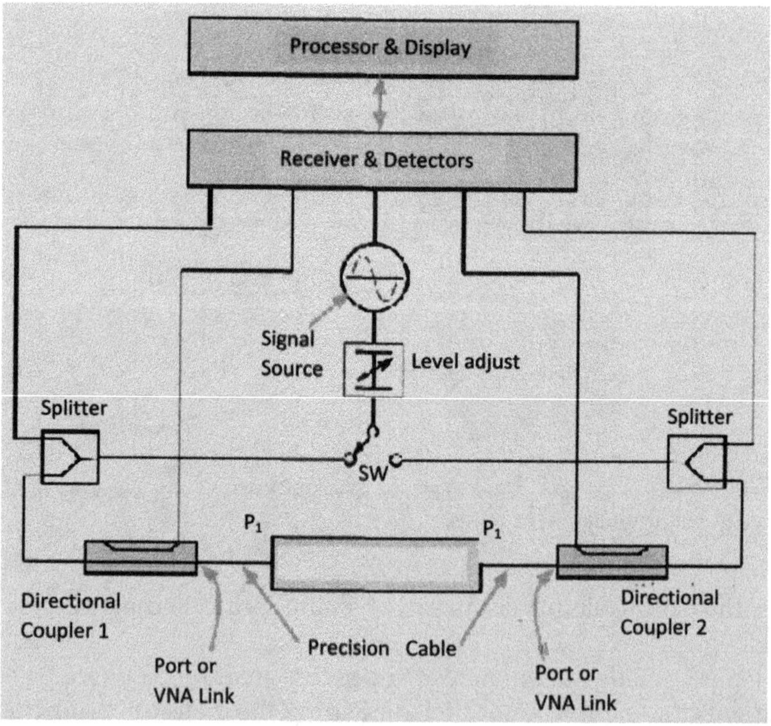

Figure 11. Block diagram of a vector network analyzer.

sample under test is mounted. Suppose the signal is incident on port P_1. A part of it will be reflected, and this part is picked up by the directional coupler 1 and sent to the receiver. This signal is called S_{11}. The rest of the signal is transmitted through the sample and exits at port P_2. This signal is called S_{12}. This is picked up by the directional coupler 2 and sent to the receiver.

If the switch is put in the other position, the signal is directed to port P_2 of the sample holder. The reflected signal at this port is S_{22}. It is picked up by the directional coupler 2 and sent to the receiver. The signal transmitted through the sample1 now exits at port P_1, and this signal is called S_{21}. It is picked up by directional coupler 1 and sent to the receiver. The detectors in the receiver measure the amplitudes and phases of S_{11} and S_{12} (or S_{22} and S_{21}) relative to the signal from the source received from the splitter. The data analysis is quite complex and is carried out on a computer to obtain the real and imaginary parts of the dielectric constant.

Different techniques of measurement are possible. These are summarized in Table 2 (from Ref. 3).

A brief description of the different methods is given in the following (Ref. 3).

6.1. *Transmission line technique*

This is illustrated in Figure 12 (from Ref. 3). In this technique, the sample tightly fits into the sample holder without leaving any air gaps. The VNA is connected with precision cables at both ends of the sample holder. The VNA is first calibrated to get the phase of the incident signal at the two

Table 2. Techniques of measurement using VNA from Reference (3).

Measurement techniques	Materials	S parameters	Dielectric properties
Transmission/Reflection Line	Coaxial Line, Waveguide	S_{11},S_{21}	ε_r, μ_r
Open-ended coaxial probe	Liquids, biological specimens semi-solids	S_{11}	ε_r
Free space	High temperature material, large flat solid, gas, hot liquids	S_{11}, S_{21}	ε_r, μ_r
Resonator Method(Cavity)	Rod-shaped solid materials, waveguides, liquids	Frequencies, Q factors	ε_r, μ_r

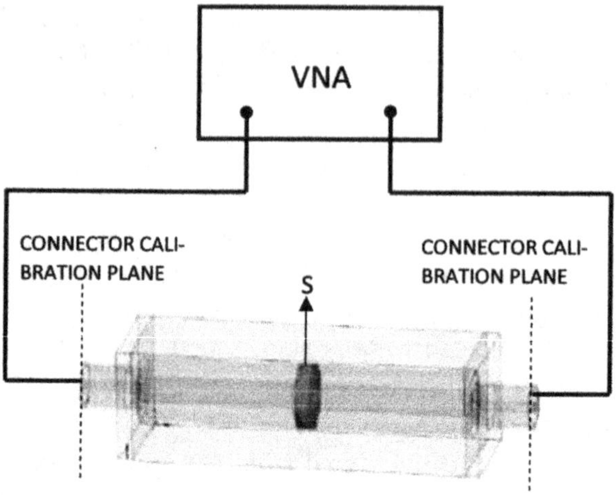

Figure 12. Schematic diagram of the transmission line method.

calibration connector planes. To determine the phase shift of the reflected and transmitted signals, we need the phase of the incident wave at the sample surface. This is obtained by adding the phase shift corresponding to the transit time of the wave from the connector plane to the sample surface to the phase at the connector plane. The S-parameters are then measured and processed by the VNA.

This method has the advantage that coaxial lines and wave guides for use as transmission lines are commercially available, and the method can be used for gases, liquids, and solids. Any air gap introduces an error in the results, and the accuracy of the method is low.

6.2. *Open-ended coaxial probe technique*

This is a nondestructive method. Figure 13 (from Ref. 3) illustrates this technique schematically. In this method, an open-ended coaxial probe is pressed against the solid sample under measurement or is immersed into the liquid sample. This method is especially suitable for soft materials, such as biological tissues. Only the reflected signal is measured, and this gives the value of the dielectric constant. Here again, we may calibrate the VNA to obtain the amplitude and phase of the signal at the connector plane and use a simulated model for the probe to get the amplitude and phase at the aperture plane. We may also directly obtain this information

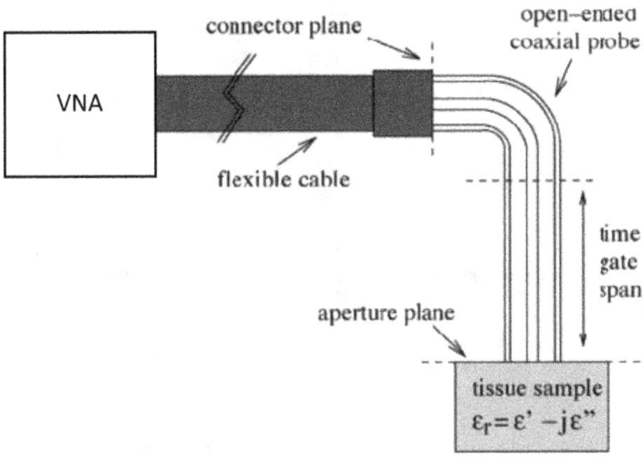

Figure 13. Open-ended coaxial technique (reprinted with permission from Ref. 3).

at the aperture plane on the surface of the sample by using standard liquids, such as water and methanol, whose dielectric constants are known for the phase measurement. The main advantages of this method are as follows: (1) The sample preparation is easy, and (2) once the standard liquid has been used to get the information of amplitude and phase at the aperture plane, we can measure any number of samples one after the other. One can also make measurements at different temperatures by heating the sample.

6.3. *Free space method*

Here, we have two horn antennae, A, separated by a distance *d* and connected to the VNA. The sample S is placed on a sample holder SH at the midpoint between the two antennae. The sample must be large and flat. The VNA needs to be calibrated first without the sample holder and the sample. After calibration, the *S*-parameters of the empty sample holder are measured. Then, the sample is placed in the holder, and the *S*-parameters are measured. The data are then processed to yield the dielectric parameters for the sample. This is a broadband technique. It allows the measurement of the parameters at different temperatures of the sample and in different environments for the solid. The technique is illustrated schematically in Figure 14. This method has the advantages that: (1) it can be used at high frequencies; (2) it is nondestructive; and (3) both the

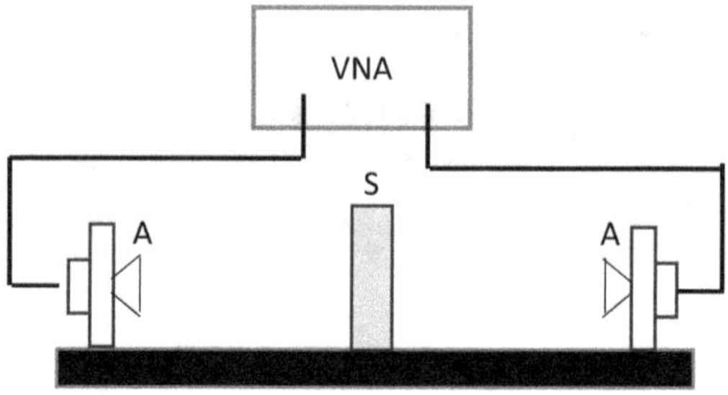

Figure 14. Free space technique.

dielectric constant and magnetic permeability can be determined. The technique requires large and flat samples. Multiple reflections between the antenna and sample surface cannot be avoided, and diffraction effects arise at the edges of the sample.

6.4. *Resonator technique*

A microwave cavity resonator resonates at a frequency determined by the dimensions of the cavity. If a sample is placed inside the cavity, the resonance frequency is slightly altered, depending on the real part of the dielectric constant of the sample. The width of the resonance line will increase due to the losses in the sample. The quality Q of the resonator is defined as $v/\Delta v$, where v is the frequency at which the intensity is at its maximum and Δv is the width of the resonance curve at points where the intensity is three decibels below the maximum. So, if the sample material is lossy, the Q-factor will be lower. In the resonator method, the shift in the resonance frequency and the change in quality factor of a resonator are measured when the sample is placed inside the cavity.

Figure 15(a) shows a schematic diagram of the split-cylinder, 10 GHz cavity resonator Model 85072A of Keysight. The figure is self-explanatory. Figure 15(b) shows a photograph of the resonator.

This method yields highly accurate values of the dielectric constant, but the disadvantage is that the measurement can be done only at one frequency.

Baker-Jarvis *et al.* (Ref. 5) have written a tutorial article on high-frequency dielectric measurements, which may be consulted for further details.

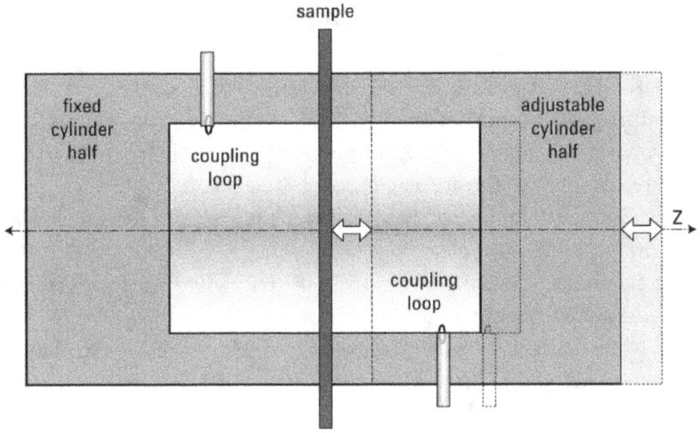

(a)

(b)

Figure 15. (a) Schematic diagram and (b) photograph of a 10 GHz, split-cylinder cavity resonator (extracted from Ref. 4).

7. Conclusion

There is extensive literature on the measurement of the dielectric properties of various materials. It is not possible to give all the references within the limited length of this article. An attempt has been made to present the essential principles involved in such measurements over a wide range of frequencies, from a few Hz to about 100 GHz. The interested reader must refer to the extensive literature available.

References

1. Tereschenko, O. V., Buesink, F. J. K., and Leferink, F. B. J. (2011). An overview of the techniques for measuring the dielectric properties of materials. *XXX URSI General Assembly and Scientific Symposium,* https://www.academia.edu>An_overview_of_the-tech.
2. Nelson, S. O. (1999). Dielectric properties measurement techniques and applications. *Trans. ASAE* **42**, 523.
3. Rohde & Schwarz. Measurement of Dielectric Material Properties, https://www.rohde-schwarz.com › file.
4. Basics of Measuring Dielectric Properties of Materials — Application Note, https://www.keysight.com › assets › application-notes.
5. Baker-Jarvis, J., Janezic, M. D., and DeGroot, D. C. High-Frequency Dielectric Measurements: A Tutorial | NIST, https://www.nist.gov › publications › high-frequency-di.

Chapter 16

MAGNETIC PROPERTIES

1. Introduction

Basic magnetic measurement methods are reviewed by Urbaniak (Ref. 1).

The magnetic field $\boldsymbol{H}$ is expressed in ampere-turn/m. When a magnetic field is applied to any medium, a magnetic induction field $\boldsymbol{B}$ (also called flux density) develops in the material. This induction field is related to $\boldsymbol{H}$ by

$$\boldsymbol{B} = \mu_0 \mu_r \boldsymbol{H} \tag{1}$$

Here, μ_0 has a value of $4\pi \times 10^{-7}$ N/A^2, and μ_r is a dimensionless number called the magnetic permeability of the medium. μ_r is unity for a vacuum medium. For a material medium, μ_r is different from unity. The magnetic induction field $\boldsymbol{B}$ is measured in Tesla (T).

When a material medium is placed in a magnetic field $\boldsymbol{H}$, a magnetization $\boldsymbol{M}$ is generated in the material. Magnetization is the magnetic moment per unit volume of the material. $\boldsymbol{M}$ is proportional to $\boldsymbol{H}$. So,

$$\boldsymbol{M} = \chi \boldsymbol{H} \tag{2}$$

χ is a dimensionless number and is called the magnetic susceptibility of the material. The relation between $\boldsymbol{B}$, $\boldsymbol{H}$, and $\boldsymbol{M}$ is expressed by

$$\boldsymbol{B} = \mu_0 (\boldsymbol{H} + \boldsymbol{M}) \tag{3}$$

So,

$$\mu_r = (1 + \chi) \tag{4}$$

Materials can be diamagnetic or paramagnetic. In diamagnetic materials, χ is negative, and $\boldsymbol{M}$ is in a direction opposite to $\boldsymbol{H}$. Some common examples of diamagnetic materials are copper, silver, gold, mercury, bismuth, and water. The diamagnetic susceptibility is usually very small. For copper, it has a value of -1×10^{-5}. However, for a superconductor under an applied magnetic field below its critical magnetic field, χ_{dia} has a value of −1. **So, from Equation (3), *B* is zero in a superconductor as long as the applied magnetic field *H* is below a critical value.**

In paramagnetic materials, χ is a positive number and $\boldsymbol{M}$ is in the direction of the applied field $\boldsymbol{H}$. The atoms of a paramagnetic material have nonvanishing magnetic moments, $\boldsymbol{m}$, which try to align themselves in the direction of the field. Thermal energy tries to create disorder in the alignment. So, the paramagnetic susceptibility increases as the temperature T decreases, thus following the Curie law:

$$\chi = C/T \tag{5}$$

where C is a constant of the material.

In solid-state materials, there is an exchange interaction between two electrons on neighboring sites. This exchange interaction tries to align the magnetic moments of the electrons on neighboring sites either parallel to each other or antiparallel to each other. As the temperature is lowered below a temperature T_C, the exchange interaction overcomes the disordering tendency of thermal energy. Then, the magnetic moments of neighboring atoms tend to become aligned parallel or antiparallel, as shown in Figures 1(a) and (b) for a one-dimensional chain of atoms.

In Figures 1(a) and (b), the lattice contains one type of atom. In Figure 1(a), the magnetic moments of all atoms are aligned in parallel. There is a net magnetization per unit cell even in the absence of an applied magnetic field. Such a material is called ferromagnetic. In Figure 1(b), the magnetic moments of adjacent atoms are aligned in opposite directions. Such a material is called antiferromagnetic. There is no net spontaneous magnetization in the unit cell. Note that the length of the unit cell in the antiferromagnetic chain is twice the distance between adjacent atoms.

In both the ferro- and antiferromagnetic cases, above the Curie temperature T_C, the magnetic moments are arranged in random orientations,

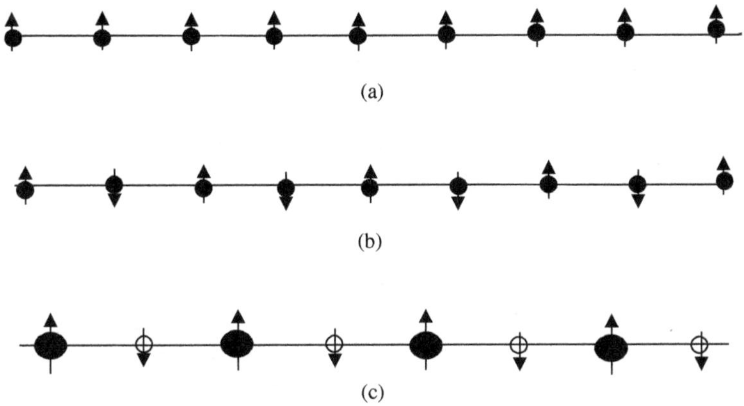

Figure 1. (a) Ferromagnetic, (b) antiferromagnetic, and (c) ferrimagnetic orderings in a one-dimensional lattice.

and the material is paramagnetic. At temperatures above T_C, the magnetic susceptibility varies as

$$\chi = C/(T - \theta) \tag{6}$$

This relation is valid until the temperature T of the material comes within a few degrees of the value of T_C. θ is positive for a ferromagnetic material and negative for an antiferromagnetic material. The magnitude of θ is close to that of T_C. The behavior of magnetic susceptibility as a function of temperature for a paramagnetic, ferromagnetic, and antiferromagnetic material is shown in Figure 2.

In Figure 1(c), we have a lattice containing two different atoms with different magnetic moments. The exchange interaction is antiferromagnetic between the nearest neighbors and ferromagnetic between the second-nearest neighbors. The antiferromagnetic interaction is the stronger of the two. So, neighboring atoms arrange themselves with their magnetic moments pointing in opposite directions. However, because the magnetic moment of one type of atom is greater than the magnetic moment of the other type, there is a net spontaneous magnetization of the unit cell. Such a material is called ferrimagnetic.

In a three-dimensional ferro- or ferrimagnetic material, the magnetization in different regions may be aligned in different directions. These

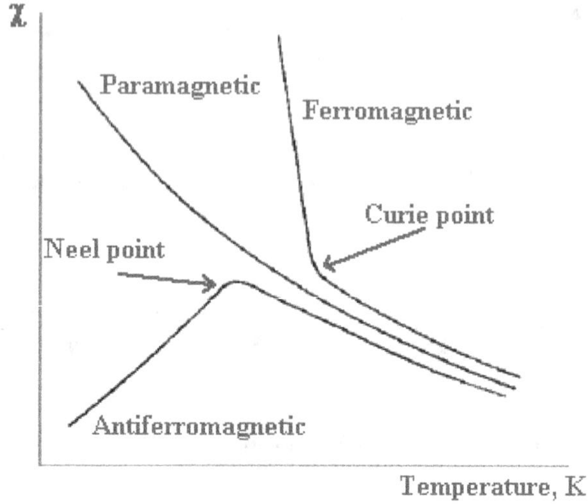

Figure 2. Variation of susceptibility with temperature in a paramagnetic, ferromagnetic, and antiferromagnetic material.

regions are called domains. The bulk material does not show spontaneous magnetization, though each domain has a spontaneous magnetization. When the direction of the spontaneous magnetization changes from one domain to the neighboring domain, there is a narrow region separating the two domains in which the change in direction of magnetization takes place slowly. This region is called a domain wall. Also, the exchange interaction is anisotropic. There are easy and hard directions of magnetization. If a bulk material is placed in a magnetic field, at low fields the domain walls move so that domains with a favorable orientation of magnetization relative to the magnetic field grow, whereas the unfavorably oriented domains shrink in size. This happens until the entire material becomes a single domain, and the magnetization in all parts of the material is oriented parallel to the magnetic field. If the field is decreased, the domain walls move back, but not by the same amount. So, the bulk magnetization shows hysteretic behavior. If one plots B versus H in a ferro- or ferrimagnetic material, one gets a curve as shown in Figure 3.

When the magnetic field H is initially applied to a material which shows no bulk magnetic moment, the induction field B follows the dashed curve and reaches a saturation value, B_s, as H is increased. If H is now decreased, B follows the upper part of the curve. When $H = 0$, B is not zero. This value, denoted by B_r, is called retentivity. When B

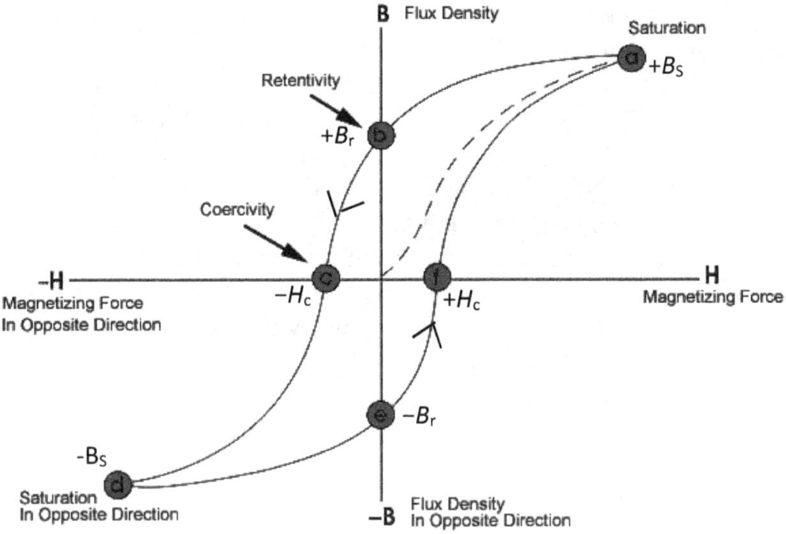

Figure 3. ***B–H*** **curve in a ferro- or ferrimagnetic material.**

is decreased further and reaches a large negative value, B reaches a saturation value, which is denoted by $-B_S$. Now, if H is increased, the value of B follows the lower part of the curve. The remanent induction B_r is destroyed either by heating the material or by mechanical treatment of the material.

The slope of the initial dashed part of the B–H curve is called the initial permeability of the material. The values of retentivity B_r and coercivity H_C are characteristic of the material. Materials with a low value of H_C are called soft magnetic materials. Those for which H_C is high are called hard magnetic materials. When B_R and H_C are large, the hysteresis curve will enclose a large area. This area represents a dissipation of magnetic energy into heat energy per unit volume of the material per cycle due to the motion of the domain walls.

In soft magnetic materials, the domain walls move easily with the application of a magnetic field. These materials have a high initial permeability and a low coercive field. In hard magnetic materials, the domain walls are strongly pinned. They do not move easily in a magnetic field. The coercivity is high in such a material, and the hysteresis loop is wide.

Hard magnetic materials are used to make permanent magnets. On the other hand, in building transformers for AC applications, one must use a soft ferromagnetic material, which has a narrow hysteresis loop, as the core. This will reduce the hysteresis loss. However, if a solid metallic core is used, eddy currents will be set up, and there will be a resistive loss due to these eddy currents. To avoid this loss, the soft iron core is made of thin laminae of iron bonded together by an insulating varnish. The high resistance of the thin laminae reduces the eddy current loss.

Ferrites are ferrimagnetic materials. As they are oxides, their resistivities are high. Nickel–zinc ferrites with a certain composition are soft magnetic materials. The hysteresis loss is low, and because of their high resistivity, eddy current losses are very low even if the material is used as a solid core. They are used as cores in RF coils.

In a magnetic material, one needs to measure the following properties:

(1) magnetic permeability,
(2) Curie or Neel temperature,
(3) initial permeability in a ferro- or ferrimagnetic material,
(4) coercivity and retentivity in a ferro- or ferrimagnetic material.

In the following sections, we describe methods to measure these properties.

2. Methods Based on the Force Exerted on a Sample by the Magnetic Field

2.1. *Gouy balance*

The Gouy balance is used to measure the magnetic susceptibility of a sample of the material. The principle of the method is illustrated in Figure 4.

In this method, the sample is in a glass tube and hangs from one arm of a balance between the pole pieces of an electromagnet. One end of the sample is between the pole pieces near the center and the other end of the sample is outside the pole pieces in a region where the magnetic field is zero when current flows through the coils of the magnet.

The weight of the sample and tube is balanced without switching on the magnetic field. When the current through the coils of the electromagnet is switched on, a magnetic field $H(z)$ is produced, which has a value

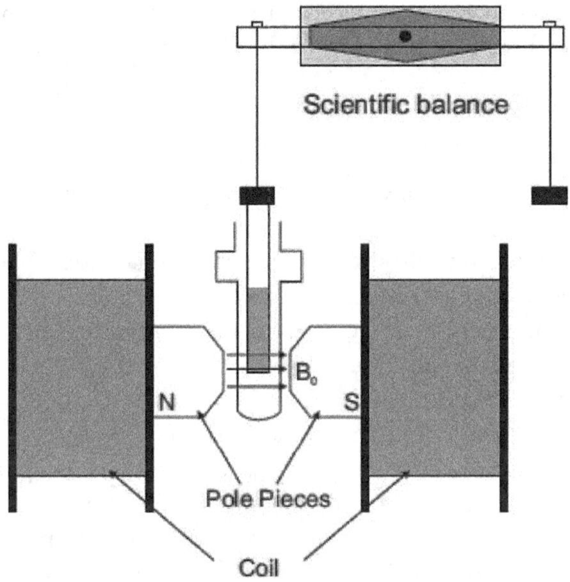

Figure 4. Schematic diagram of the Gouy balance method.

of H_0 at the center of the pole pieces ($z = 0$) and zero at the top of the sample ($z = h$). The field creates a magnetic moment of M per unit volume of the sample. The energy per unit volume due to this magnetization is

$$U = -1/2\ \chi H^2 \tag{7}$$

Since H varies with z, U varies with z. So, there is a force per unit volume at z given by

$$F(z) = -\mathrm{d}U/\mathrm{d}z = +(1/2)\chi\ \mathrm{d}(H^2(z))/\mathrm{d}z \tag{8}$$

Let A be the area of cross-section of the sample. Then. the total force acting on the sample is

$$F = (1/2)\chi A \int_0^h F(z)\ \mathrm{d}z = -(1/2)\chi A H_0^2 \tag{9}$$

This force is in the downward direction. To balance this force, we have to add an additional mass m to the right pan of the balance. Then,

$$F = mg = (1/2)\chi A H_0^2 \tag{10}$$

Knowing A and H_0, one can calculate the magnetic susceptibility of the sample.

Figure 5 shows a photograph of a commercial Gouy balance manufactured by Holmarc Opto Mechatronics Pvt. Ltd.

The electromagnet and the sample are in a closed enclosure to prevent the effect of air drafts. The electronic balance is on top of this enclosure. A constant current source at the left energizes the coils of the electromagnet to produce a magnetic field. The magnetic field can be varied by changing the current. A digital Gaussmeter using a Hall probe measures the magnetic field at the center of the pole pieces. One can measure the susceptibility per unit volume, χ, to an accuracy of 1×10^{-7}. Since the sample is long, it is necessary to ensure that the sample, in the form of a powder, is packed uniformly in the tube.

2.2. *Faraday balance*

It is a modification of the Gouy balance and is described in Ref. 2. In the Faraday balance, the pole pieces of the magnet are shaped so that the

Figure 5. Gouy balance manufactured by Holmarc Opto Mechatronics Pvt. Ltd., India.

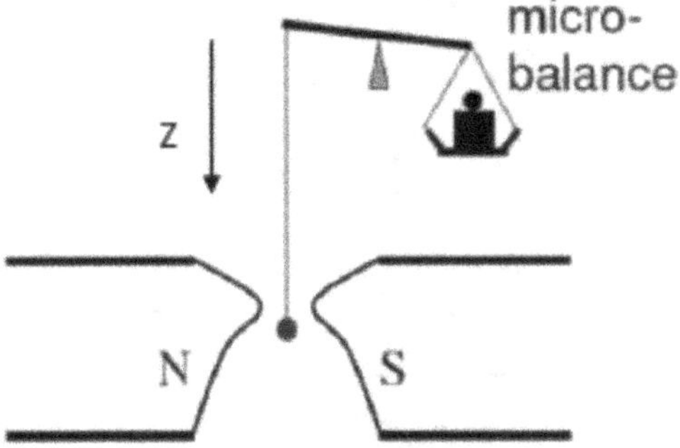

Figure 6. Schematic diagram of the Faraday balance (reprinted with permission from Ref. 2).

value dH^2/dz is a constant over a region in which the sample is placed. This is shown in Figure 6.

The magnetic force on the sample is measured, as was done with Gouy's method. Because the entire sample is in the uniform region of the force, the packing of the material is not important. The sample is smaller in volume than what one uses in a Gouy balance. The force depends on the total mass, M, of the material. Here, we measure the susceptibility of unit mass, χ_M, of the material of mass M using the formula

$$mg = (1/2)\chi_M M(dH^2/dz) \tag{11}$$

The susceptibility per unit volume χ can be calculated from χ_M using the formula

$$\chi = \chi_M \rho$$

where ρ is the density of the sample.

2.3. *Torsion balance*

Here, a pair of permanent magnets are fixed to a frame, which hangs by a wire W from a fixed support.

The permanent magnets (PM) are colored red for the north and black for the south poles. When the sample S (blue) is placed between one pair of magnets, the magnets experience a force perpendicular to the plane of the paper due to the inhomogeneous magnetic field. This causes a torque

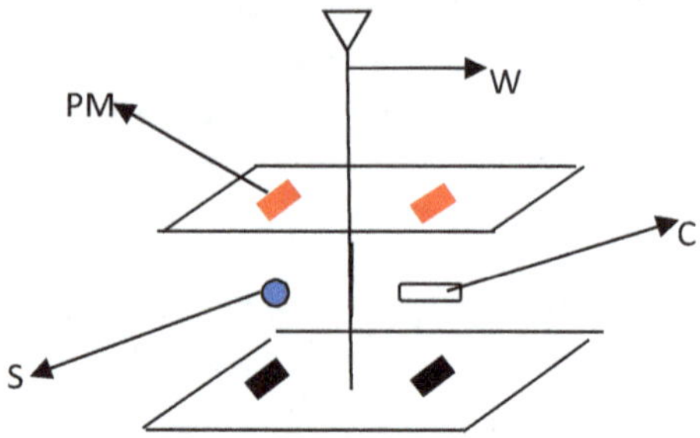

Figure 7. Schematic diagram of the Evan's torsion balance.

which twists the suspension W. This twist is detected by a photosensitive device. It is compensated by sending a current through a coil C (indicated by a rectangle) placed between the second pair of magnets. This current is a measure of the force on the magnets. The instrument is calibrated using standard samples (Figure 7).

3. AC Susceptibility

The above-mentioned balances measure the magnetic susceptibility in DC magnetic fields. In AC magnetic fields, the susceptibility χ is a complex number, indicating that the magnetization produced by the magnetic field will lag in phase from the applied magnetic field. χ will have a real part χ' and an imaginary part χ''. If the applied magnetic field is $H\sin(\omega t)$, the magnetization M will vary as

$$M = H[\chi' \sin(\omega t) - \chi'' \cos(\omega t)] \tag{12}$$

Both χ' and χ'' are functions of frequency. As the frequency increases, χ' decreases because the atomic magnetic moments will find it more and more difficult to follow the rapid variation in H. χ'' is related to the absorption of energy from the magnetic field and the conversion of this energy into heat due to the relaxation of the magnetic moments. If the magnetization changes by dM in a field H, the work done by the field on the magnetization is $-H\mathrm{d}M$. So, the work done in a time dt is

$$\mathrm{d}W = -[H(\mathrm{d}M/\mathrm{d}t)]\,\mathrm{d}t$$

Substituting for dM/dt from (11) and $H(t) = H\sin(\omega t)$ and integrating over one cycle (i.e., t from 0 to $2\pi/\omega$), we get

$$W = (1/2)\, 2\pi\chi'' H^2 \tag{13}$$

This work goes into heating up the specimen.

To measure AC susceptibility, one uses the arrangement shown in Figure 8. There is a primary coil wound on a former of length a few centimeter and diameter about 1 cm. Inside the primary coil and tightly fitting into it is a second former on which two identical secondary coils are wound in opposition to one another. The primary coil is shown in blue in Figure 8. The secondary coil 1 (SC1) is shown in red, and the secondary coil 2 (SC2) is shown in green. The leads of the primary coil are connected to a signal generator in series with a resistance R. The voltage across R

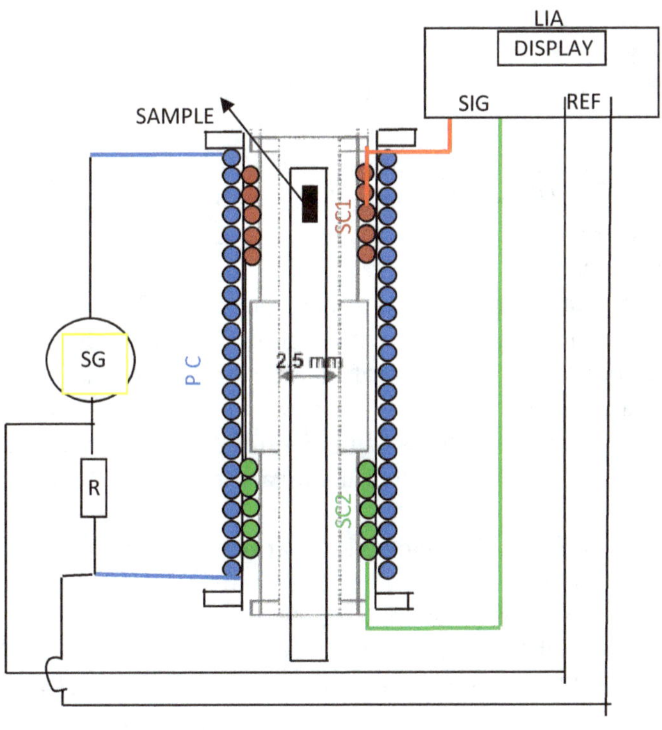

Figure 8. Experimental arrangement for measuring AC susceptibility.

will be in phase with the primary current and is used as the reference REF in the lock-in amplifier. The output of the two secondary coils in opposition comes out at the red and green lines. This is the signal SIG connected to the lock-in amplifier. First, without the sample, the small output signal that is in phase and 90° out of phase with the reference is nullified within the lock-in amplifier. If the sample shown in black is now placed in the SC1, the output of SC1 becomes more than the output of SC2. The component of the unbalanced output in phase with the reference and the component 90° out of phase with the reference are both shown on the display as DC voltages V′ and V″. V′ and V″ are proportional to the frequency ν and the RMS value I_p of the primary current.

If the volume of the sample is V_S, then the susceptibility components χ' and χ'' can be obtained from V′ and V″ using the equations

$$\chi' = \alpha\,(V'/[(\nu\,I_p)V_S]) \tag{14a}$$

and

$$\chi' = \alpha\,V''/[(\nu I_p)V_S]) \tag{14b}$$

Here, α is a calibration constant. If the RMS voltage, V_R, across R is measured, $I_p = V_R/R$. For finding α, the calibration constant, one may use the oxides of the rare-earth metals Nd, Gd, Dy, and Erbium (Ref. 3) as standard materials.

One can attach a heater and a thermometer to the sample holder to do high-temperature measurements. One may connect the sample thermally to a cold finger of a cryorefrigerator and make measurements at low temperatures.

There are many references to homebuilt AC susceptibility setups in the literature. One may refer to Chakravarti *et al.* (Ref. 4) for one such homebuilt setup.

Balanda (Ref. 5) has reviewed the application of AC susceptibility studies for phase transitions and magnetic relaxations in conventional, molecular, and low-dimensional magnetic materials.

Figure 9 shows the temperature variation of χ' and χ'' from AC magnetic susceptibility measurements in $Fe^{II}(4H_2O)_2Nb^{IV}(CN)_84H_2O$ (from Ref. 5). The red curve is DC magnetic susceptibility. $Fe^{II}(4H_2O)_2$ $Nb^{IV}(CN)_84H_2O$ undergoes a ferromagnetic-to-paramagnetic transition at 43.1 K. This phase transition is signaled by a sharp drop in χ' from the AC

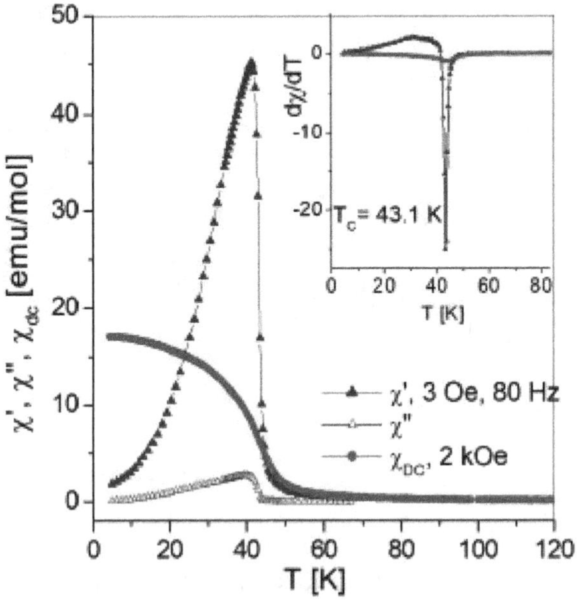

Figure 9. **Temperature variation of χ' and χ'' from AC magnetic susceptibility measurements in $Fe^{II}(4H_2O)_2Nb^{IV}(CN)_8\ 4H_2O$. The red curve is DC magnetic susceptibility (reprinted with permission from Ref. 5).**

susceptibility data at 43.1 K with increasing temperature. The magnetic field is given in Oersted in this diagram. One Oersted (Oe) is 79.58 ampere-turn/m. The magnetic susceptibility is given in emu/mole. To get the susceptibility in SI units, multiply emu/mole by $4\pi \times 10^{-6}$ and divide by the molar volume in m^3. Note that the AC measurements are done with a magnetic field amplitude of 3 Oe. On the other hand, the DC magnetic susceptibility measurement was done with a magnetic field of 2 kOe. This shows the sensitivity of the AC susceptibility technique.

4. Vibrating Sample Magnetometer

A brief account of the vibration sample magnetometer (VSM) is given in Ref. 1. Figure 10 illustrates its principle. The sample, kept in a uniform DC magnetic field produced by an electromagnet, develops a magnetic moment. There are two pick-up coils (yellow), one above and one below the axis of the magnet, connected in opposition. The sample can be

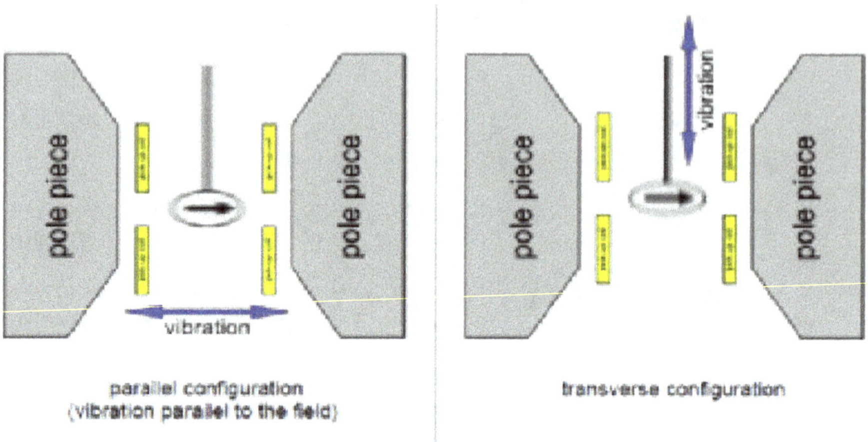

Figure 10. Schematic diagram of the VSM.

vibrated in the z direction (horizontally) or x direction (vertically), at a frequency ν, which is a few tens of Hertz. When the sample is not magnetized, there may be a small output from the two coils connected in opposition. This can be nullified. When the DC field is switched on and the magnet develops a moment m, the vibration of the sample produces an out-of-balance signal, which can be amplified and detected using a lock-in amplifier.

When the sample is oscillated sinusoidally with an amplitude a, it has a velocity v varying with time as

$$v(\mathrm{t}) = 2\pi\nu a\cos(2\pi\nu t) \tag{15}$$

The signal will be proportional to the product of the magnetic moment of the sample, a spatial distribution function $G(a/r_0)$, where r_0 is the mean radius of the coil, and the velocity of the sample. Figure 11 shows the function G plotted against a/r_0 for various values of inter-coil distance, d_0/r_0 for thin coils. For vibration along z, the plot is shown on the right-hand side of the figure, and for vibration along x, the plot is shown on the left-hand side. If the amplitude is small relative to the inter-coil separation, G is 1. As the amplitude increases, G comes down. Note that if d_0/r_0 is 0.8660, the factor G is slightly less than 1, but this factor is flat over a larger range of the amplitude a/r_0.

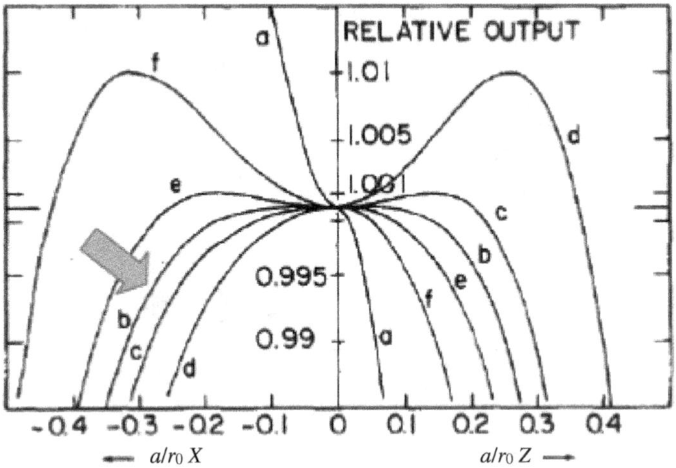

Figure 11. Spatial distribution function $G(a/r_0)$ for various values of d_0/r_0 (a: 0.5, b:0.8660, c: 0.8841, d: 0.9244, e:0.8444, and f:0.7992.

In some VSMs, more pairs of coils are used in different configurations. G has to be computed for each coil configuration.

In a finite sample, there is a demagnetizing field which opposes the applied DC field. The actual magnetic field in the sample is the difference between the applied field and the demagnetizing field due to the magnetic moment of the sample. The demagnetization factor can be calculated exactly from the dimensions of the sample if the sample is ellipsoidal in shape. So, the sample is usually ellipsoidal or spherical. The sample size should be small compared to the inter-coil separation. In such a case, the sample can be considered a point dipole.

There is a vibration head which is connected to a glass rod, on which the sample is mounted at the other end. In homebuilt magnetometers, one may use a loudspeaker as the vibration head. The vibration head is connected to a signal generator. The vibration frequency and the vibration amplitude (usually about 1–3 mm) can be set. The signal from the coil is fed to a lock-in amplifier, to which a reference signal is sent from the signal generator. The output DC signal is shown on the display. This is schematically shown in Figure 12.

One can measure the magnetic moment as a function of temperature by a suitable design of the sample holder. For temperatures ranging from 300 to 1000 K, a heater made of a resistive material is lithographically

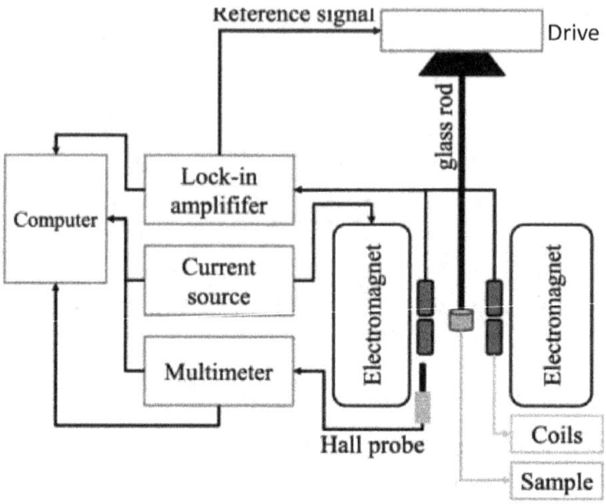

Figure 12. Schematic diagram of the complete VSM setup.

patterned on the sample holder. By applying a current to it, the sample can be heated. For measurements below room temperature, suitable refrigeration techniques are used.

The VSM usually comes with a standard nickel sphere with a known magnetic moment. This is used for calibrating the VSM.

In the literature, many homebuilt VSM setups are described. One such work is cited in Ref. 6. Figure 13 shows a commercial VSM setup.

The VSM measures the magnetic moment of the sample. It can have a sensitivity as low as 5×10^{-7} emu (1 emu = 10^{-3} A/m^2). Dividing the magnetic moment by the volume of the sample gives the magnetization.

There is a provision for ramping the DC magnetic field up or down at a steady rate. In ferro- or ferrimagnetic samples, one can measure magnetization as a function of the magnetic field to get the hysteresis curve.

5. SQUID Magnetometer

The superconducting quantum interference detector (SQUID) is a device which is based on the Josephson effect. In the SQUID, we have two weak links WL (Josephson junctions) connecting superconductors A and B

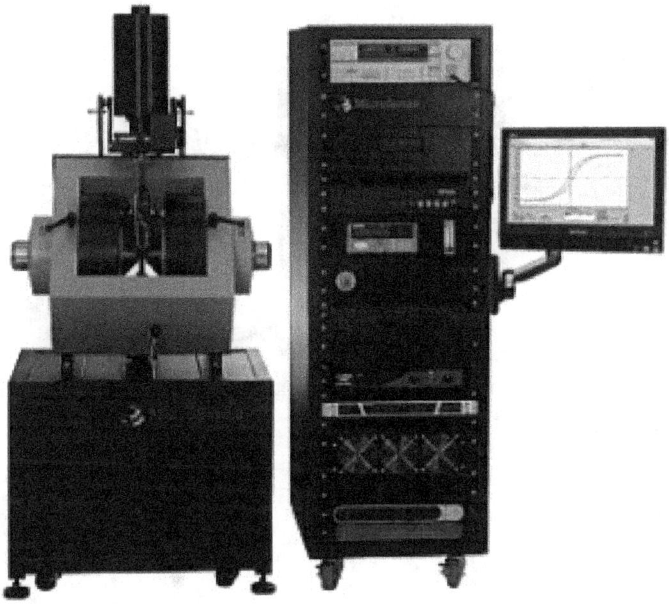

Figure 13. Vibration sample magnetometer setup manufactured by Microsense LLC.

(Figure 14). A magnetic flux Φ is enclosed by the SQUID loop. Then, the Josephson current I through the loop will vary with the magnetic flux as

$$I = I_0 \cos(2\pi\Phi/\Phi_0) \tag{16}$$

Here, $$\Phi_0 = h/2e = 2.067 \times 10^{-17}\ \text{Tm}^2$$

Φ_0 is called the flux quantum.

When the flux changes, the current changes. This is converted into a voltage in the SQUID.

The principle of operation of the SQUID magnetometer is shown in Figure 15.

A superconducting magnet at liquid helium temperature produces a DC magnetic field pointing upward. The sample is placed within a second-order gradiometer. This gradiometer consists of four loops of a superconducting wire, as shown in the bottom left of the figure. The loops

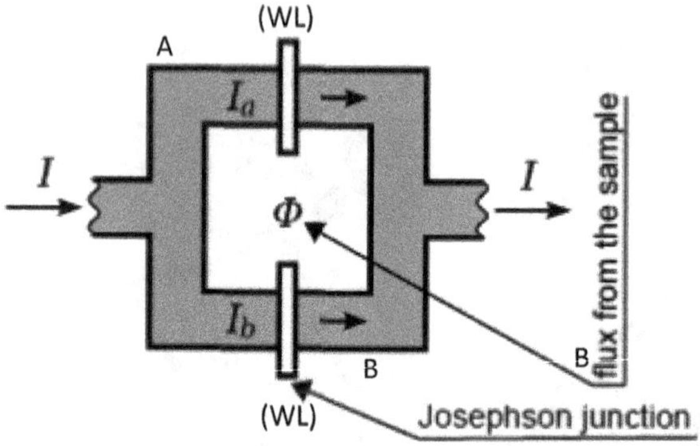

Figure 14. Principle of the Josephson junction device (reprinted with permission from Ref. 1).

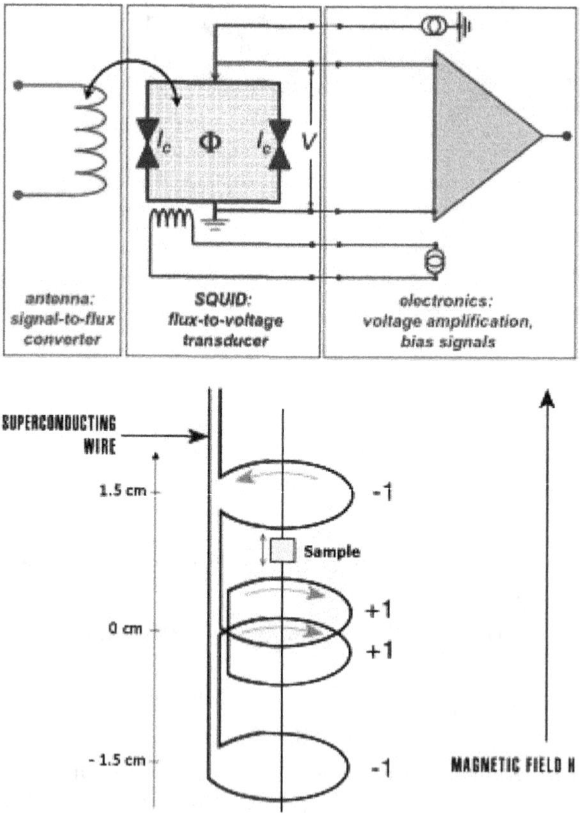

Figure 15. Principle of operation of the SQUID magnetometer.

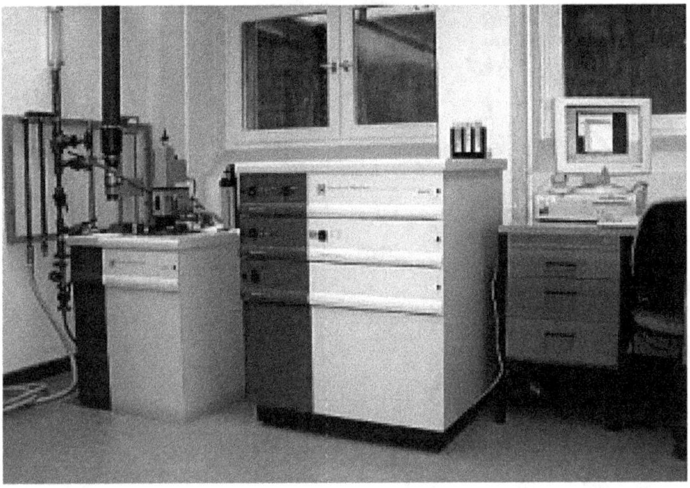

Figure 16. Quantum Design MPMS XL7 SQUID magnetometer.

are in series. The first loop from the top is wound anticlockwise and the second clockwise. The third loop is wound clockwise, and the fourth is wound anticlockwise. This is called a second-order gradiometer. With this gradiometer, the maximum signal will be obtained if the sample is positioned exactly at the midpoint between the second and third loops. The signal decreases as the sample moves away from this position. The vibrating sample produces AC currents in the second-order gradiometer. This is picked up by an antenna kept away from the sample. This antenna will produce a periodic change in the magnetic flux associated with the DC squid. The antenna and DC SQUID are in liquid helium but at a distance from the sample. The change in current through the SQUID is converted into an AC voltage, which is detected by a lock-in amplifier to produce an output DC signal. A commercial SQUID magnetometer can detect a magnetic moment as small as 10^{-11} Am^2.

The commercial SQUID magnetometer of Quantum Design is shown in Figure 16.

6. B–H Curve Tracer

The retentivity, coercive field, and energy loss due to hysteresis per cycle are measured using the curve tracer, the circuit for which is shown in Figure 17 (from Ref. 7).

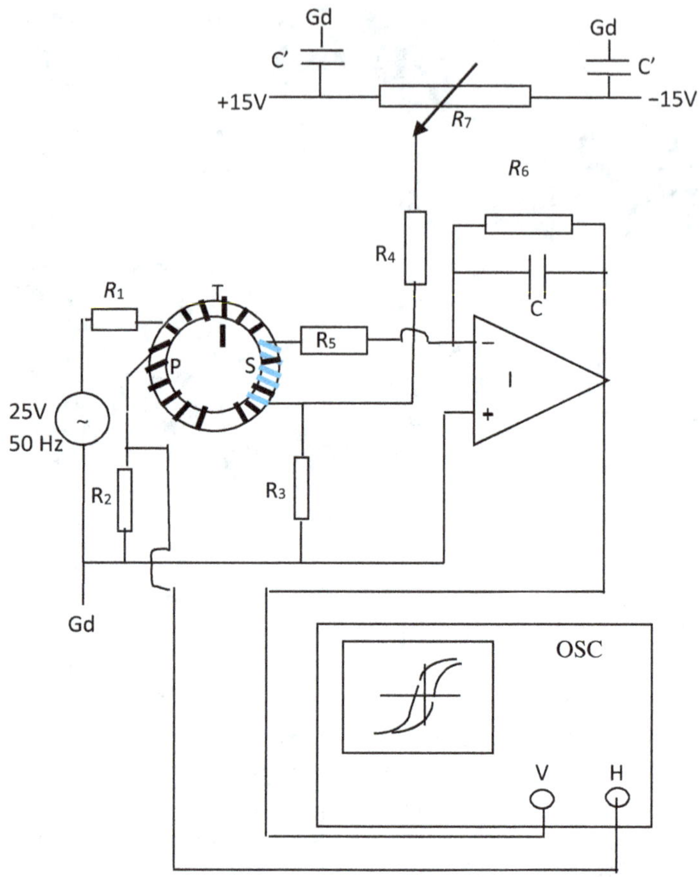

Figure 17. Circuit for a *B–H* loop tracer (reprinted with permission from Ref. 7).

The ferrite material is in the form of a toroidal core T with an outer diameter of approximately 20 mm and an inner diameter of approximately 10 mm. On this core, we wind around the toroid an enamel-coated copper coil P (SWG 23) of about 25 turns. This is the primary coil. On a part of the primary coil, a secondary coil S of SWG 36 of about 25 turns is wound. The primary turns are shown in black and the secondary in blue.

The primary coil is energized by a 50 Hz, 25 V supply, which is connected to a rheostat R_1 of 20 Ω with a current-carrying capacity of 2 A. The other end of the coil P is connected through a resistance R_2 of 1 Ω 2 W to the grounded end of the AC power supply. The IC chip I is powered by a

DC +15V–0–15V power supply. A ten-turn potentiometer R_7 of 100 kΩ is connected between the +15 V and −15 V terminals. The variable point of the potentiometer is connected to the ground through a resistance R_4 of 1 MΩ and a resistance R_3 of 10 Ω in series. One end of the secondary S is connected to the common point between R_4 and R_3. The other end of the secondary is connected through a resistance R_5 of 1 kΩ to the inverting input terminal of the op-amp I. The noninverting input of the op-amp is connected to the ground. The op-amp is used as an integrator. Between the inverting input and output terminals of the op-amp, a polymer capacitor C of 1μF is connected in parallel with a resistor R_6 of 10 M Ω. C′ denotes 100 nF ceramic capacitors to bypass any AC ripples in the DC power supply.

The oscilloscope (OSC) has a vertical (V) and a horizontal (H) input terminal, as well as a ground terminal. The output of the integrator is connected to the vertical terminal. The voltage across R_2 is connected to the horizontal terminal. The ground terminal is connected to the common ground of the AC and DC power supplies.

The AC supply is switched on, and the current is adjusted by moving the sliding contact of the rheostat to about 0.3 A. The *B–H* loop is seen on the oscilloscope. By adjusting the pot R_7, one can get a stable symmetric *B–H* loop.

The voltage excursion V_H along the horizontal axis from the left to the right end of the loop is proportional to twice the amplitude of the AC current I_P through the primary:

$$I_P = V_H/2R_2 \tag{17}$$

The maximum magnetic field H_{max} is

$$H_{max} = N_P I_P/\ell_e \tag{18}$$

N_P is the total number of turns in the primary coil, and ℓ_e is the equivalent length of the primary winding, which may be taken as

$$\ell_e = \pi(d_o + d_i)/2 \tag{19}$$

Here, d_o and d_i are the outer and inner diameters of the torus, respectively.

The output voltage V_S of the secondary is a function of time. It is given by

$$V_s = -N_s \, d\Phi/dt \tag{20}$$

according to Faraday's law of induction, where Φ is the magnetic flux linked with one turn of the secondary and N_S is the total number of turns in the secondary. V_S generates a current i which is

$$i = V_s/R_5 \tag{21}$$

This charges the capacitor C. If $V_0(t)$ is the voltage across C, at time t,

$$C dV_o/dt = -(N_S/R_5) \, d\Phi/dt \tag{22}$$

So, the voltage across the capacitor V_o is

$$V_o = (N_S/CR_5) \, \Phi \tag{23}$$

This voltage V_0 is the vertical voltage measured on the oscilloscope. From Φ, one can calculate B using the equation

$$B = \Phi/A \tag{24}$$

Here, A is the area of cross-section of the torus.

Thus, from the curve on the screen, one can deduce B_S, B_R, H_C, and the area of the curve. If B is in T and H is in ampere-turn/m, the area of the loop gives the energy lost per unit volume of the material of the torus in one cycle.

References

1. Urbaniak, M. Basic Magnetic Measurement Methods, http//www.ifmpan.poznan.pl> Wyklady2014 >u.
2. Adeyemi, N. Faraday Balance Magnetometer, https://faculty.sites.iastate.edu › files › inline-files.

3. Fukuma, K. and Torii, M. (2011). Absolute calibration of low- and high-field magnetic susceptibilities using rare earth oxides. *Geochem. Geophys. Geosystems* **12**, July, https://doi.org/10.1029/2011GC003694.
4. Chakravarti, A., Ranganathan, R., and Raychaudhuri, A. K. (1991). An automated AC magnetic susceptibility apparatus. *Pramana-J. Phys.* **36**, 231.
5. Balanda, M. (2013). AC susceptibility studies of phase transitions and magnetic relaxation: Conventional, molecular and low dimensional magnets. *Acta Phys. Pol.* **124**, 964.
6. Niazi, A., Poddar, P., and Rastogi, A. K. (2000). A precision low-cost vibration sample magnetometer. *Curr. Sci.* **79**, 99.
7. Plotting Magnetization Curves, https://info.ee.surrey.ac.uk>advice>coils>BHCkt.

Part IV

Spectroscopic Techniques

Chapter 17

NMR AND EPR SPECTROSCOPY

1. Introduction

In an atom, an electron has an orbital angular momentum of $\ell(h/2\pi)$ and a spin angular momentum of $s(h/2\pi)$, where s is ½. This gives rise to a total angular momentum $j(h/2\pi)$, where j can have two values, $(\ell + ½)$ or $(\ell - ½)$. If there are many electrons in the atom, first sum all the orbital angular momentum quantum numbers ℓ to yield a total value of $\boldsymbol{L}$ and all the spin quantum numbers $\boldsymbol{s}$ to yield a total value of $\boldsymbol{S}$. Then, the total angular momentum $\boldsymbol{J}$ is obtained as a vector sum of $\boldsymbol{L}$ and $\boldsymbol{S}$.

The electron is charged. So, the angular momentum gives rise to a magnetic moment for the atom. This magnetic moment $\boldsymbol{\mu}_e$ is given by

$$\boldsymbol{\mu}_e = -g_J \boldsymbol{J} \mu_B \tag{1}$$

Here, μ_B is the Bohr magneton, given by

$$\mu_B = eh/4\pi m_e = 1.53 \times 10^{-24}\ \text{J/T} \tag{2}$$

e is the magnitude of the electronic charge, h is the Planck's constant, and m_e is the mass of the electron. g_J is a dimensionless number called the Lande splitting factor and has a value between 1 and 2. Note that the mass of the electron occurs in the denominator of μ_B. The negative sign in Equation (1) arises from the negative charge of the electron.

A nucleus contains protons and neutrons. The nucleus of an atom will also have a total spin angular momentum, given by $\boldsymbol{I}(h/2\pi)$. The total nuclear spin, $\boldsymbol{I}$, will depend on the numbers of neutrons and protons in the nucleus:

(i) If the number of protons is **even** and the number of neutrons is **odd**, or vice versa, the spin I will be a half-integral (1/2, 3/2, 5/2,...).
(ii) If the number of protons **and the** number of neutrons are both odd, then the nuclear spin I will be a nonzero integer.
(iii) If the number of neutrons **and** the number of protons are both **even**, $I = 0$ because a particle with a spin of ½ will be paired with an identical particle with a spin of −½.

Thus, ^{1}H will have a nuclear spin of ½, ^{2}H (deuterium) will have a nuclear spin of 1, and ^{16}O (eight protons and eight neutrons) will have a nuclear spin of zero.

In nuclear magnetic resonance (NMR), one often uses the ^{1}H and ^{13}C nuclei.

Associated with the nuclear spin, there will be a nuclear magnetic moment $\boldsymbol{\mu}_N$ given by

$$\boldsymbol{\mu}_N = \gamma \boldsymbol{I} \tag{3a}$$

$$\gamma = geh/4\pi m_p \tag{3b}$$

Here, m_p is the mass of the proton. The value of g depends on the nucleus.

The value of g for a proton is 5.586 and that of a neutron is −2.002. Though the neutron is uncharged, it carries a magnetic moment. The magnitude of γ is three orders of magnitude smaller than the value of μ_B, the Bohr magneton, because of the heavier mass of the proton, which appears in the denominator of the expression for γ.

2. Principle of Magnetic Resonance

When a nucleus (or atom) with a magnetic moment $\boldsymbol{\mu}_N$ (or $\boldsymbol{\mu}_e$) is placed in a magnetic field $\boldsymbol{B}$ applied along the z-axis, it has an additional energy of $-\boldsymbol{\mu}_N \cdot \boldsymbol{B}$ (or $-\boldsymbol{\mu}_e \cdot$B). In addition, the magnetic moment experiences a torque

$$\boldsymbol{\mu}_N \times \boldsymbol{B} \text{ (or } \boldsymbol{\mu}_e \times \boldsymbol{B})$$

Because $\boldsymbol{\mu}_N$ (or $\boldsymbol{\mu}_e$) is proportional to the angular momentum $\boldsymbol{I}$ (or $\boldsymbol{J}$), this torque causes the angular momentum vector (and hence the magnetic moment vector) to process about the direction of $\boldsymbol{B}$ with the Larmor precession frequency ω. This is shown in Figure 1 for the nuclear magnetic moment.

The Larmor precession frequency is given by the magnetic field times the constant of proportionality between the magnetic moment and the angular momentum. So, in the case of the nuclear magnetic moment, the precession frequency is $\omega = \gamma B$, and in the case of the electron magnetic moment, $\omega = g_J \mu_B B$.

In the case of the nucleus, the projection m_I of the angular momentum quantum number $\boldsymbol{I}$ along the magnetic field will take $(2I + 1)$ values: $I, (I - 1), (I - 2),\ldots,-(I - 1), -I$. In the case of the electron, the projection m_J of the angular momentum quantum number along the magnetic field will take $(2J + 1)$ values: $J, (J - 1),\ldots -(J - 1), -J]$. In the absence of a magnetic field, the $2I + 1$ levels of a nucleus (or the $(2J + 1)$ levels of an electron) are degenerate. In the presence of a magnetic field, the additional contribution to the energy of a nucleus will depend on m_I, or that of an electron on m_J, as follows:

$$E_{Mi} = -\gamma B m_I \text{ (for the nucleus)} \tag{4a}$$

or

$$E_{Mj} = g_J \mu_B B m_J \text{ (for the atom)} \tag{4b}$$

Note the difference in sign in Equations 4(a) and (b). The nuclear charge is positive, whereas the electronic charge is negative.

So, the energy levels for $I = ½$ (or $J = ½$) will split, as shown in Figure 2(a) (and (b)).

The energy splitting in the nuclear case will be three orders of magnitude smaller compared to the energy splitting in the case of the atom

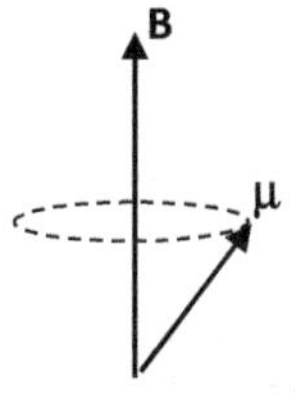

Figure 1. Precession of the magnetic moment about a magnetic field.

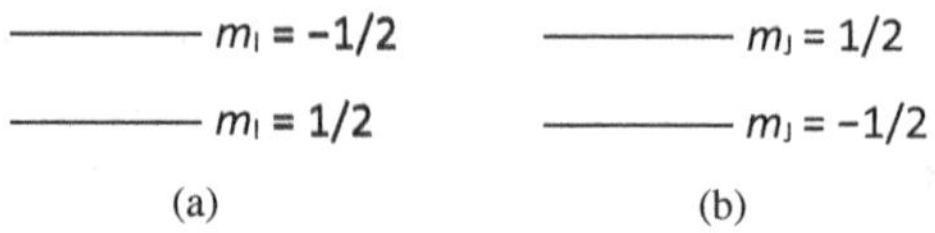

Figure 2. Splitting of the energy levels in a magnetic field: (a) for a nucleus of angular momentum quantum number $I = ½$; (b) for an atom of angular momentum quantum number J = ½.

because the magnetic moment of the nucleus will be three orders of magnitude smaller than the magnetic moment of the atom. If the applied field is of the order of 1 T, the energy difference, divided by h, in the nuclear case will be in the radio frequency range (MHz), while the energy difference, divided by h, in the case of the atom, will be in the microwave range (GHz).

Following the Boltzmann distribution, the population of the energy levels will be proportional to $\exp(-E_m/kT)$, where E_m is the energy of the mth level.

If we apply radiation with a frequency of v given by

$$hv = E_{m+\Delta m} - E_m = \Delta E \tag{5}$$

a transition will take place from the lower energy state (E_m) of the nucleus (or atom) to the upper energy state $E_{m+\Delta m}$. ΔE is the difference in energy between the two states. $\Delta m = -1$ for the nucleus and +1 for the electron. This will result in the absorption of energy from the source providing the radiation. The frequency of absorption will depend on the applied magnetic field. If we keep the magnetic field fixed and sweep the frequency, then absorption will occur at a specific frequency. One can also keep the frequency constant and sweep the magnetic field to obtain resonance at a specific magnetic field. This is the principle of magnetic resonance.

3. Nuclear Magnetic Resonance

A schematic diagram of a nuclear magnetic spectrometer is shown in Figure 3.

The sample is in a tube, which is placed in a magnetic field generated by an electromagnet or a superconducting magnet. On the sample tube, an RF coil is wound, which is connected to an RF generator, the frequency

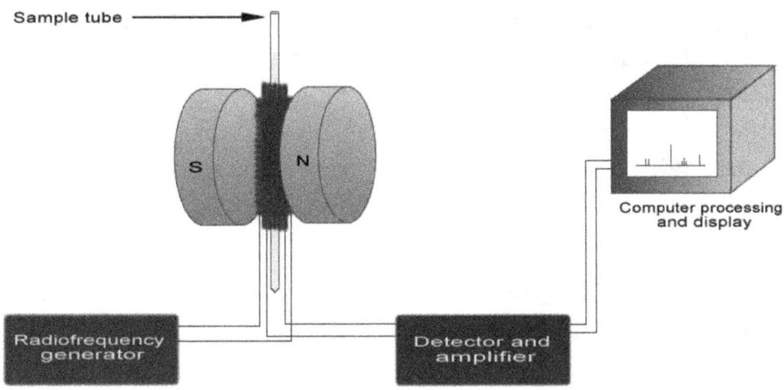

Figure 3. Schematic diagram of a nuclear magnetic resonance setup.

Table 1. NMR frequency v per T for some in nuclei.

Nucleus	Abundance %	Spin	v MHz/T
^{1}H	99.98	½	42.60
^{2}H	0.02	1	6.53
^{13}C	1.1	½	10.72
^{14}N	99.64	1	3.07
^{15}N	0.365	½	4.33
^{19}F	100	½	40.08
^{31}P	100	1	17.25
^{35}Cl	75.4	$^{3}/_{2}$	4.17
^{37}Cl	25.6	$^{3}/_{2}$	3.46

of which can be ramped over a certain range about a mean frequency, called the spectrometer frequency. A pick-up coil senses the absorption of RF energy by the sample, and this signal is amplified and shown on the screen of a computer as a function of the frequency of the RF generator. The frequency increases from left to right. The resonance frequency of a proton in a magnetic field of 4.7 T is 200 MHz and that of a ^{13}C nucleus in the same field is 50 MHz. Table 1 gives the resonance frequency v in MHz/T for some nuclei.

One may perform an NMR experiment using an electromagnet. An electromagnet cannot generate a field higher than 1.5 T. Use of a superconducting magnet allows a higher field, 5–7 T, to be generated. This raises the frequency of the spectrometer. If a superconducting coil is used, it will be cooled either by a liquid helium bath or by a cryocooler.

If the sample is cooled, the relative proportion of the nuclei in the ground state will increase, making the absorption signal larger.

3.1. *Features of NMR*

NMR is extensively used by chemists in the study of organic molecules. The features of NMR which make it so useful are discussed in the following.

3.1.1. Chemical shift

The resonance frequency will vary for the same nucleus in different compounds. This is called the chemical shift. The shift in frequency for resonance in the compound from the value for resonance in a standard is expressed in parts per million, as defined in the following:

$$\text{Shift in ppm} = [(\nu_{\text{Sample}} - \nu_{\text{Standard}})/\nu_{\text{spectrometer}}] \times 10^{6} \tag{6}$$

Expressed this way, it is independent of the frequency of the spectrometer.

For proton resonance, the standard used is tetramethyl silane (TMS; Si $(CH_3)_4$). Usually, the sample is dissolved in a solvent, which may have its own shift. A commonly used solvent is deuterated chloroform. The deuteration is never 100%, and the residual protons will have a shift which is 7.24 ppm relative to TMS.

The chemical shift in a compound arises from the electron cloud in the atom, which produces a magnetic field at the nucleus which opposes the applied field. So, for a given applied field, magnetic resonance will occur at a lower frequency. This shielding effect is maximum in TMS. In other compounds, the shielding is reduced due to factors described as follows. **Because of the reduction in shielding, the actual magnetic field seen by the nucleus in the compound will increase relative to the field seen by the nucleus in the standard. At a given frequency of resonance, one will have to lower the applied magnetic field to see resonance in the compound. At a given applied magnetic field, the resonance frequency will increase because of the chemical shift compared to that in the standard.** This is shown in Figure 4 for proton resonance.

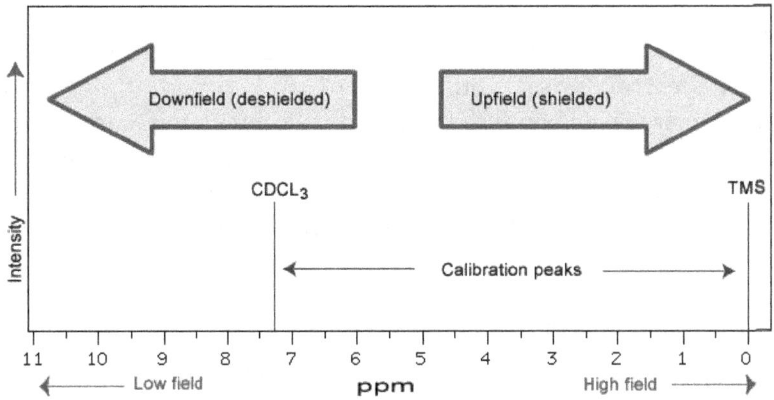

Figure 4. The calibration peaks in TMS and $CDCl_3$ and their chemical shifts.

Table 2. De-shielding effect on proton resonance due to electronegativity.

Compound CH_3X	CH_3F	CH_3OH	CH_3Cl	CH_3Br	CH_3I	CH_4	TMS
Electronegativity of X	4	3.5	3.1	2.8	2.5	2.1	1.8
Chemical shift *d* (ppm)	4.26	3.4	3.05	2.68	2.16	0.23	0

This shift will depend on the following factors:

1. The chemical shift depends on the electronegativity of the atom neighboring the proton for which the NMR is observed. The larger the electronegativity, the greater the charge cloud drawn away from the proton, and the larger the de-shielding effect.
2. The larger the number of electronegative atoms or groups in the molecule, the greater is the de-shielding effect. For example, the chemical shift of the proton resonance in CH_2Cl_2 is 5.30 ppm compared to 3.05 ppm in CH_3Cl (Table 2).
3. In a chain of hydrogen atoms as in $-CH_2-CH_2-CH_2Br$, the chemical shift of the proton in the CH_2 group nearest to Br is 3.30 ppm. In the next nearest CH_2 group, the chemical shift is 1.69 ppm, and in the third CH_2 group, it is 1.25 ppm. Thus, the presence of bromine causes a larger chemical shift in the proton nearest to it, and this shift decreases as the distance of the proton from bromine increases.

4. It is difficult to predict the chemical shift when hydrogen bonds are present. The hydrogen bonds have a range of lengths, and there are other effects such as solvation.
5. When there is a π bond, as in benzene, the circulating currents in the bond produce a field at the proton, which adds to the applied magnetic field. The chemical shift shows an increase. This shows that the circulating currents in the π bond increase the de-shielding at the nucleus.

3.1.2. Integrated intensity

Consider the compound, methyl terbutyl ether ($CH_3OC(CH_3)_3$). Here, we have the protons in the single CH_3 group attached to oxygen, and the protons in the three CH_3 groups are attached to C. The three protons in the first CH_3 group are equivalent and will have one chemical shift. The nine protons in the other three CH_3 groups will be equivalent and will have a different chemical shift. The proton resonance spectrum will show two peaks with different chemical shifts, as shown in Figure 5.

We expect that the protons in the CH_3 group attached to the O atom will have a larger chemical shift than the protons in the CH_3 group attached to C because of the larger electronegativity of the O atom and the shorter distance of the proton from O. If we integrate the intensity under

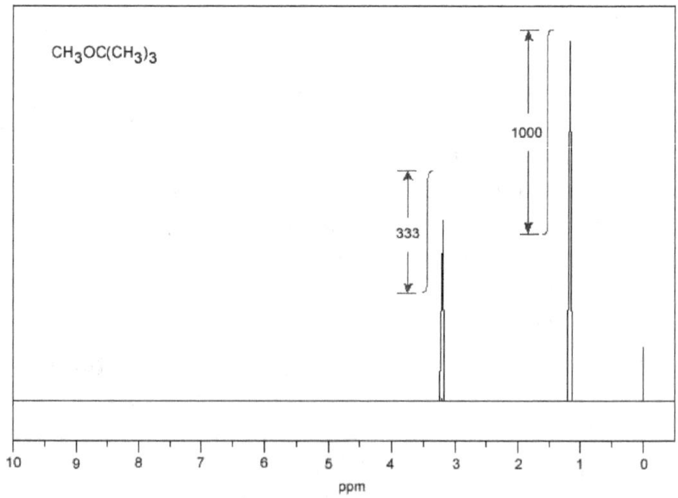

Figure 5. Proton resonance spectrum of $CH_3OC(CH_3)_3$.

the two resonance peaks and compare the ratio, it is 333:1000 (or 1:3). Thus, the integrated intensity is proportional to the number of equivalent protons in a group and can be used to distinguish the signals from protons in different groups.

3.1.3. Multiplicity of peaks

Consider the compound $BrCH_2CHBr_2$. There are two groups of equivalent protons coming from the CH_2 and CH groups. We expect the chemical shift of the proton in the CH group to be more than that of the proton in CH_2 because of the proximity of two Br atoms to the former proton. Also, the ratio of integrated intensities from the proton in $CHBr_2$ to that from the proton in CH_2Br should be 1:2. Both features are seen in the proton resonance of the compound shown in Figure 6.

In addition to the above features, we see that each peak is a multiplet. There is a triplet peak and a doublet peak. In the triplet peak, the relative intensities of the components are in the ratio of 1:2:1. There is a doublet peak in which the two components are equal in intensity.

The multiplicity arises as follows. The single proton in $CHBr_2$ can have its spin up or down. It will produce a small magnetic field at the protons in the $BrCH_2$ group, which will aid the applied magnetic field or

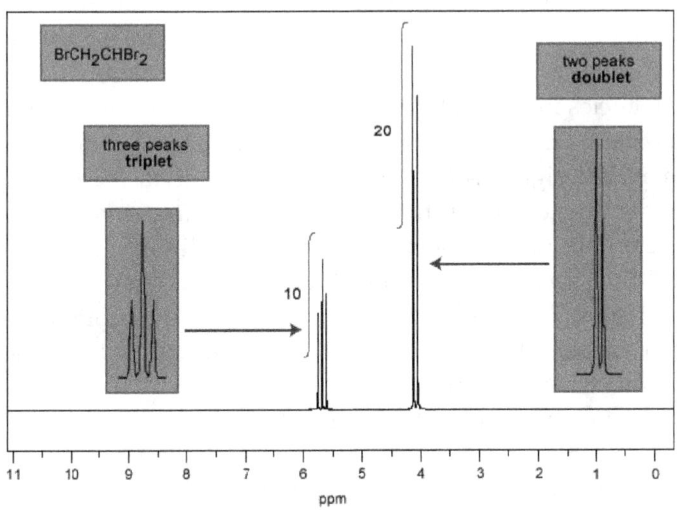

Figure 6. Proton resonance in $BrCH_2CHBr_2$.

oppose it, producing two different chemical shifts in the resonance of these protons. Since both spin orientations of the proton in $CHBr_2$ are equally probable, the two different peaks in the resonance of the proton in $BrCH_2$ will be equal in intensity. On the other hand, consider the situation of the proton in $CHBr_2$. The magnetic field at this proton will be affected by the orientation of the spins of the two protons in $BrCH_2$. There will be three possible spin orientations, as follows: (i) Both protons have spin up (total spin of +1); (2) one of the protons has spin up and the other spin down (total spin of zero); and (3) both protons have spin down (total spin of −1). So, we should get a triplet spectrum with three equally spaced peaks. The probability that the spin is 1, 0, or −1 is in the ratio of 1:2:1. So, the central peak in the triplet must be twice as intense as the other two peaks. This is found to be true.

These three features, namely the chemical shift, integrated intensity ratio, and multiplicity, of the NMR spectrum make it very useful in organic chemistry.

Continuous-wave NMR, which was described above, has a low sensitivity and is time-consuming. Pulsed Fourier transform (FT) NMR is more sensitive and less time-consuming.

3.2. *Fourier transform NMR*

If we use a square pulse, we know that it contains a range of Fourier components around the mid-frequency. The range $\Delta\nu$ of the Fourier components will depend on the pulse width. The wider the pulse width, the narrower the range $\Delta\nu$, and vice versa. Each of the closely spaced resonance frequencies in the molecule, due to the presence of the nonequivalent nuclei and its multiplet structure, will produce its own signal. In the time domain, the signal will be a superposition of all the free induction decay signals due to the multiplets. A FT of the time-domain signal will give different Fourier components with their amplitudes. This is called FT NMR. This is especially useful and time-saving in analyzing biological molecules. Figure 7 shows how the time-domain signal containing different close frequencies looks and how the FT gives all the frequencies present in the time-domain signal.

The advantage of pulsed FT NMR is that all the free induction decay signals at different frequencies are recorded as a time signal at the same time. This saves time, especially with biological molecules containing

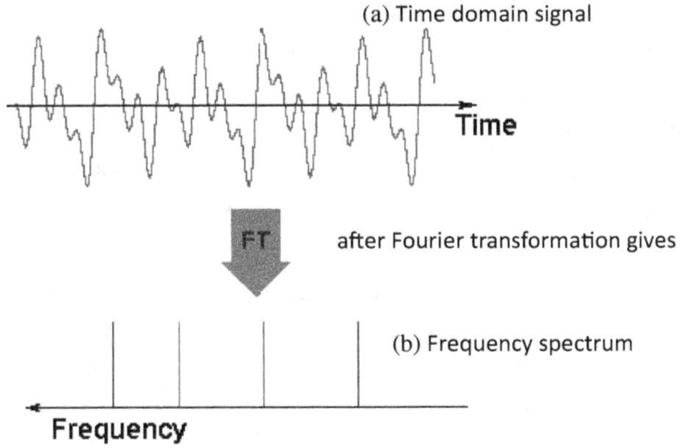

Figure 7. Time-domain signal after Fourier transformation gives all the frequencies in the signal (Ref. 2).

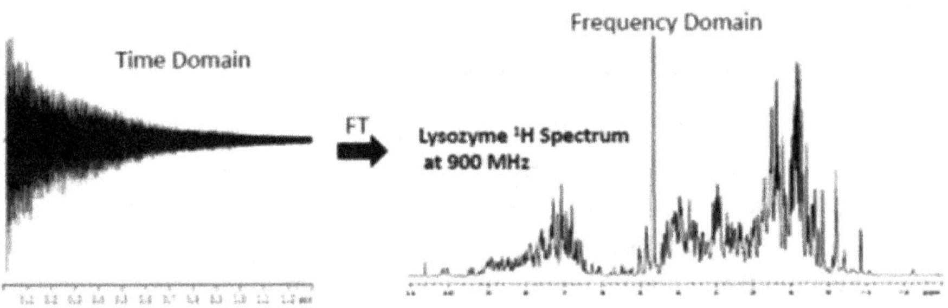

Figure 8. Time-domain proton NMR signal in lysozyme (left) and its FT (right).

many inequivalent protons at different sites. As an example, we show in Figure 8 the time-domain proton NMR signal in lysozyme and its FT.

3.3. *Relaxation phenomenon in NMR*

When the nucleus is excited to a higher energy state, it will have to come back to the ground state over a certain period of time. This is called relaxation.

In the case of optical absorption, the atom comes spontaneously down from the excited state to the ground state, emitting radiation. The probability for such a transition is proportional to the cube of the frequency of absorption (or emission). The optical absorption frequencies are of the order of 10^{13} or 10^{14} Hz, whereas the NMR absorption frequencies are in the range of 10^7–10^8 Hz. So, the transition probability for spontaneous emission is negligibly low at NMR frequencies.

In NMR, the excited nuclei are in the presence of other moving nuclei in the medium. For example, if we study the NMR of a compound in solution, the molecules of the solvent will be moving around. They produce fluctuating magnetic fields at the nucleus under study. If the frequency of one of the Fourier components of the fluctuation agrees with the resonance frequency, there will be a transfer of energy between the nucleus and the fluctuating component, and the nucleus will relax from the excited state by induced emission. In a solid medium, this process is very slow, and the time of relaxation is thus long. On the other hand, in a liquid medium, this happens quickly, thus the time of relaxation is short. The time of relaxation will depend on the size of the molecule, the environment of the nucleus, and temperature.

3.3.1. Principle of pulsed NMR to measure relaxation time

In a DC magnetic field along the z-axis, the nuclear moments are aligned along the z-axis to give a magnetic moment component M_{z0} of the sample. In addition, the magnetic moment $\boldsymbol{\mu}$ of the nucleus will experience a torque of $\boldsymbol{\mu}_N \times \boldsymbol{B}$. Because the magnetic moment is proportional to the angular momentum, it causes the magnetic moment to precess around the direction of the field, with the Larmor precession frequency of $\omega_0 = \gamma B$ in the nuclear case or $\omega_0 = g_J \mu_B B$ in the case of the electron.

Instead of the fixed coordinate system of xyz, we choose a rotating coordinate system of $x'y'z$, by setting xyz to rotate about z with the Larmor precession frequency ω_0. In this coordinate system, the components of $\boldsymbol{\mu}$ will not change with time.

Let us now apply an oscillating magnetic field, which is along x' at time $t = 0$. This oscillating field can be applied by having a coil with its axis perpendicular to z and applying an RF voltage to it. Let the frequency of the RF voltage be ω' and the amplitude of the oscillating magnetic field be B_1. We may consider this field oscillating along x' to be made up of two magnetic fields of amplitude $B_1/2$, one rotating clockwise in the xy-plane and the other anticlockwise at the frequency ω'. If ω' is equal to the Larmor precession frequency ω_0, then one of the rotating components will be stationary in the rotating frame $x'y'z$, while the other will be rotating at an angular velocity of $2\omega_0$ in that frame. The stationary component $B_1/2$ along x' and will cause the z component of the magnetic moment to precess around the x' direction in the rotating frame. The frequency of precession will be $\omega'' = \gamma B_1/2$. In time τ, the magnetic moment component along z would have rotated through an angle $\omega''\tau$. If $\omega''\tau = \pi/2$, the z component would have turned through 90° to come into the y'-direction. If we make the RF pulse of current to last for this period of time, it is called a $\pi/2$ pulse. At the end of the pulse, the component of the magnetization of the material in the z-direction would have turned through 90° to lie in the y'-direction. If the pulse is switched off now, this component of the magnetization will slowly relax back to the z-direction, with a relaxation time of T_1. At the same time, the y' component keeps precessing around the steady field B with the Larmor precession frequency ω_0.

The magnetization along the z-axis will slowly revert back from 0 to its full value of M_{z0} in an exponential fashion characterized by the relaxation time T_1:

$$M_z(t) = M_{z0}\,[1 - \exp(-t/T_1)] \tag{7}$$

The component $M_\perp$, in the xy-plane will therefore decrease with time t as $\exp(-t/T_1)$. T_1 is called the spin–lattice relaxation time. If we measure the magnetization as a function of time, we can determine T_1.

Suppose we measure the magnetization along the x-direction as a function of time. We can do this by placing a detector coil with its axis along x. We recall that the magnetization $M_\perp$ is rotating in the xy-plane with a frequency of ω_0. If at time $t = 0$, the direction of M_{z0} was along y, then at time t,

$$M_x(t) = M_\perp(t)\sin(\omega_0 t) \tag{8}$$

We have already seen that $M_\perp(t)$ comes down exponentially due to the spin–lattice relaxation. The output of the probe coil will be periodic with a frequency of ω_0, with its amplitude decaying exponentially with time. This is called free induction decay.

There is a second reason for $M_\perp(t)$ to decrease with t. The magnetic field seen by the nucleus when it precesses about the z-axis after the RF pulse is switched off fluctuates due to the spins of neighboring protons. So, different protons in the sample will precess around the applied magnetic field at slightly different rates, leading to the magnetic moments pointing in different directions in the *xy*-plane. This relaxation mechanism is called spin–spin relaxation, and it causes $M_\perp(t)$ to further decrease exponentially with time T_2. Then, there could be inhomogeneities in the DC magnetic field B in the specimen. This could also lead to an exponential decay of $M_\perp(t)$ with time T_B. So, we may write

$$M_\perp(t) = M_\perp(0)\exp(-t/T^*) \tag{9}$$

where

$$1/T^* = 1/T_1 + 1/T_2 + 1/T_B \tag{10}$$

The signal picked up by the coil will then oscillate at frequency ω_0 but with its amplitude decaying with time as represented by Equation (9).

The relaxation times T_1 and T_2 vary with the size of the molecule, as shown in Figure 9. The relaxation time is a measure of the size of the molecule. In small molecules, the relaxation times T_1 and T_2 are comparable. As the size of the molecule increases, T_1 becomes much larger than T_2.

A schematic diagram of the pulsed NMR arrangement is shown in Figure 10. In pulsed NMR, we measure the relaxation times T_1 and T_2 at the Larmor precession frequency ω_0. The pulse programmer (1) decides

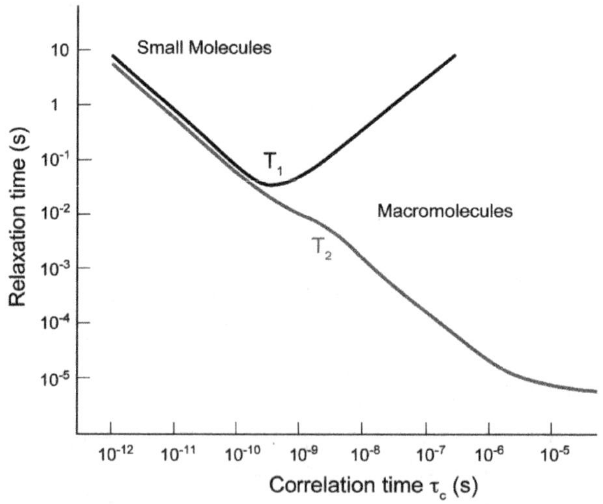

Figure 9. Variation of T_1 and T_2 depending on the size of the molecule (extracted from Ref. 1).

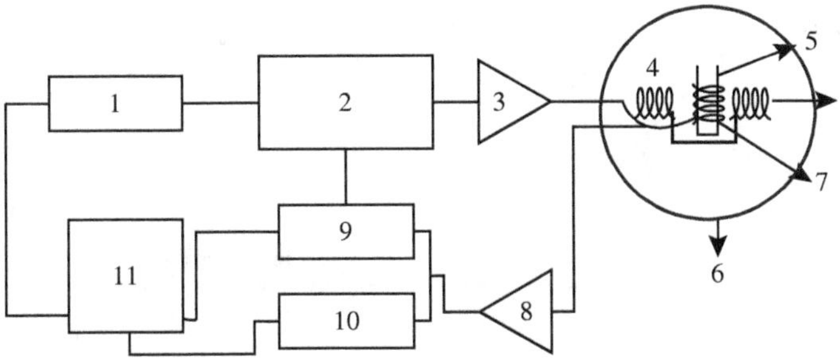

Figure 10. Schematic diagram of the arrangement for pulse NMR (redrawn from Ref. 4).

the width and repetition rate of the pulse. The RF oscillator (2) produces RF oscillations at different frequencies. The output of the oscillator will be a pulse of RF oscillations at the frequency of the RF oscillator. This pulse is amplified (3) and fed to a pair of horizontal Helmholtz coils (4) to produce an RF magnetic field along the axis of the coils. The sample is kept in a tube (5), which is located between the pole pieces of a magnet (6).

The electromagnet produces a DC magnetic field perpendicular to the plane of the paper. The probe coil (7) produces an output proportional to the magnetization perpendicular to the direction of the DC field and perpendicular to the axis of the RF coils. This output is amplified in the receiver (8). The output of the RF oscillator and the receiver go to a mixer (9) to produce beats. The output of the receiver is amplified and detected by the detector (10). This gives a DC output proportional to the probe signal. The outputs of the pulse programmer, mixer, and detector are fed to an oscilloscope (11).

If the pulse frequency is different from ω_0, we will see beats in the output of the mixer, i.e., periodic variations in the output at the beat frequency, $(\omega' - \omega_0)$, where ω' is the frequency of the RF pulse. The frequency of the RF pulse is adjusted so that the beats disappear. This is shown in Figure 11.

To measure T_1, we use two pulses separated by a time interval. The first pulse is a π pulse, which makes the z component of magnetization change from $+M_{z0}$ to $-M_{z0}$. After a time t, the magnetization along the z direction will have a value of

$$M_z(t) = M_{z0} - 2M_{z0}\exp(-t/T_1) \tag{11}$$

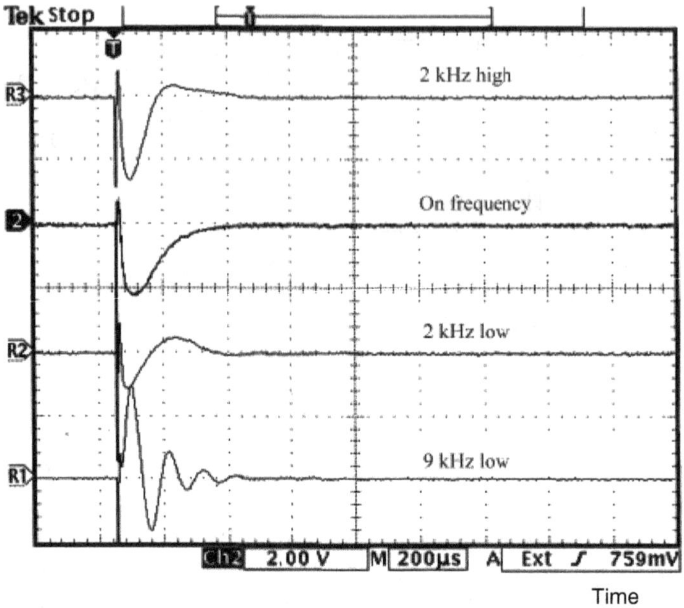

Figure 11. Beats from the mixer signal when $\omega' \neq \omega_0$. At resonance, there are no oscillations in the mixer output (extracted from Ref. 4).

When we apply a $\pi/2$ pulse after time t, $M_z(t)$ along the z-axis will be rotated by 90°, producing an oscillatory signal in the probe coil. The rectified output of the probe signal will vary as given by Equation (11). There will be a time delay t_0 between the π and $\pi/2$ pulses, for which the rectified output will go to zero. That time t_0 will be given by

$$\exp(-t_0/T_1) = ½ \tag{12}$$

From a measurement of t_0, T_1 can be calculated using Equation (12).

To measure T_2, we use a $\pi/2$ pulse first to turn the z component of the magnetization to the xy-plane. Then, we apply a π pulse after a time τ. This produces a spin echo at time 2τ after the initial $\pi/2$ pulse. The mechanism of the production of spin echo is illustrated in Figure 12. Due to fluctuations in the magnetic field arising from the spins of neighboring nuclei, the perpendicular components of the magnetic moments of different nuclei will precess at different speeds around the z-axis. If we look in the $x'y'z$ frame, the faster precessing spins will be rotating, say, counterclockwise, and the slower precessing spins will be rotating clockwise.

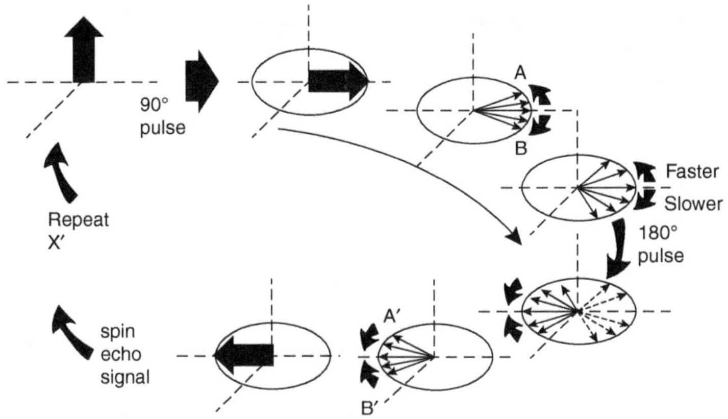

Figure 12. Mechanism of generation of spin echo (redrawn from Ref. 4).

The spread in the perpendicular components of spin will increase with time. After a time τ, we apply a π pulse. Let us say that in the rotating frame, B_1 is along the x'-axis. The π pulse will cause the z and y' components of the magnetic moments to change sign. So, the magnetic moment component OA, which was rotating clockwise, will now come to position OA′ and will continue to rotate clockwise. The magnetic moment component OB rotating anticlockwise will come to position OB′ and will continue to rotate anticlockwise. The spread in the spins will decrease, and the spins will come into alignment after a time τ from the π pulse. This is shown in Figure 12.

If we have a probe to measure the induction in the xy-plane, it will show a peak at $t = 0$ and another peak at time $t = 2\tau$, where τ is the time interval between the $\pi/2$ and π pulses. This is called a spin echo and is shown in Figure 13.

To measure T_2, we repeat the π pulse, as shown in Figure 14.

We see that the amplitude of the spin echo pulse decreases with time. One measures the time T for the decay of the amplitude to half its value. Then,

$$\exp(-T/T^*) = \tfrac{1}{2} \tag{13}$$

If T_1 and T_B are long compared to T_2, then $T^* \approx T_2$. From this, T_2 can be found.

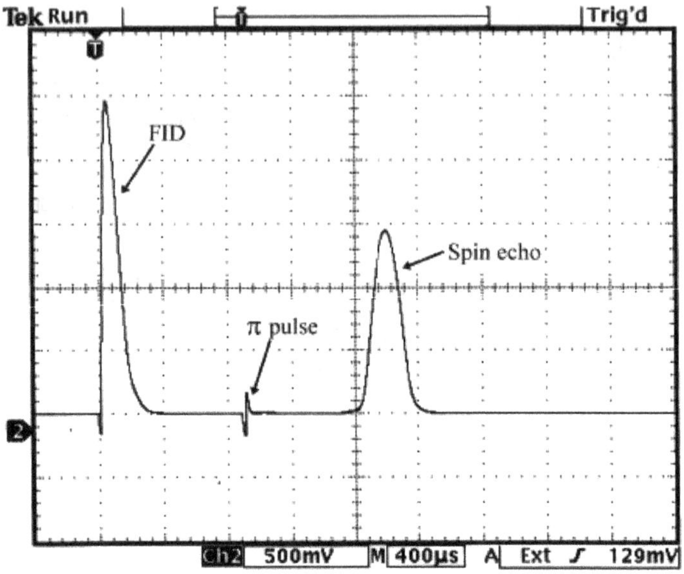

Figure 13. Signal at $t = 0$ and the spin echo at $t = 2\tau$ (extracted from Ref. 4).

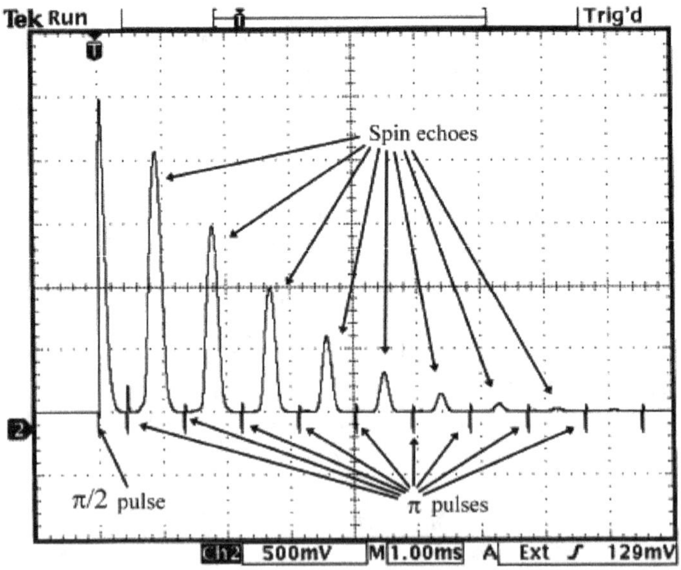

Figure 14. Sequence of spin echoes when π pulse is applied repeatedly (extracted from Ref. 4).

3.4. *Applications of NMR spectroscopy*

NMR spectroscopy finds wide applications in the following areas:

(1) identifying small molecules;
(2) elucidating the structure of macromolecules;
(3) probing molecular motions on a wide variety of time scales; and
(4) magnetic resonance imaging in healthcare.

4. Electron Paramagnetic Resonance

For an electron for which $L = 0$, $J = S$, and only the spin angular momentum contributes to the magnetic moment, resonance in a magnetic field will occur at the frequency

$$\nu = g_e \mu_B B/h \tag{14}$$

Here, g_e, the gyromagnetic ratio for free spin, is 2.00232. So, in a magnetic field of 0.1 T, the resonance frequency is 2.8×10^9 Hz. This is in the microwave range. All discussions regarding continuous- and pulsed-wave techniques and relaxation phenomena pertaining to NMR are also valid for EPR.

The range of microwave bands is given in Table 3.

The mid-frequency in the band is given in column 2 of the table, and the magnetic field, in Tesla, for free electron resonance at this frequency is given in column 3. The commonly used bands are X and Q.

Table 3. Microwave bands.

Band	Mid-frequency in GHz	*B* in Tesla for resonance
L	1.1	0.0392
S	3	0.107
X	9.5	0.3389
K	23	0.82
Q	35	1.2485
W	90	3.2152
J	270	9.6458

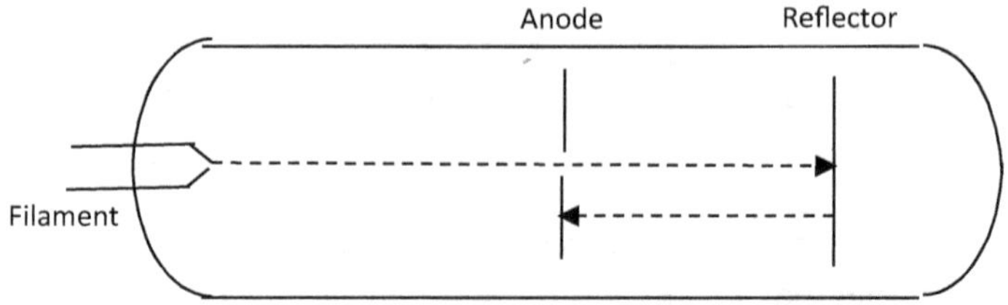

Figure 15. Schematic diagram of a klystron.

Microwaves are transmitted through wave guides whose dimensions are comparable to the wavelength of the microwave. Thus, in the X band, the length of the wave guide will be about 3 cm, and in the Q band, it will be about 1 cm.

Microwaves are generated by a klystron. In the klystron, electrons from a thermionic source are accelerated and passed through a hole in the anode. They are then reflected back to the anode by the reflector. This is shown in Figure 15.

The frequency of the klystron depends on the time taken by the electron beam to go from the anode to the reflector and return back to the anode. The frequency can be changed a little by deforming the klystron to alter the distance between the anode and the reflector. This tuning can be done only over a narrow range of frequencies.

In EPR, one therefore keeps the frequency constant and changes the magnetic field to observe resonance.

4.1. *EPR spectrometer*

A schematic diagram of an EPR spectrometer is shown in Figure 16. Microwaves at a frequency v are generated in the klystron. The power of the beam is controlled by the attenuator. The beam coming out of the attenuator passes into the circulator. The beam moves anticlockwise around the circulator. Then, the beam enters a cavity. The dimensions of the cavity causes it to resonate at the microwave frequency, resulting in a large amplitude of the oscillating magnetic field component at the sample placed inside the cavity. This oscillating component of the magnetic field is perpendicular to the DC magnetic field produced by an electromagnet. EPR can be performed with an iron-cored electromagnet, which can produce a magnetic field of up to 1.5 T. Of course, one could also use a superconducting magnet and adjust the current to produce a field of this

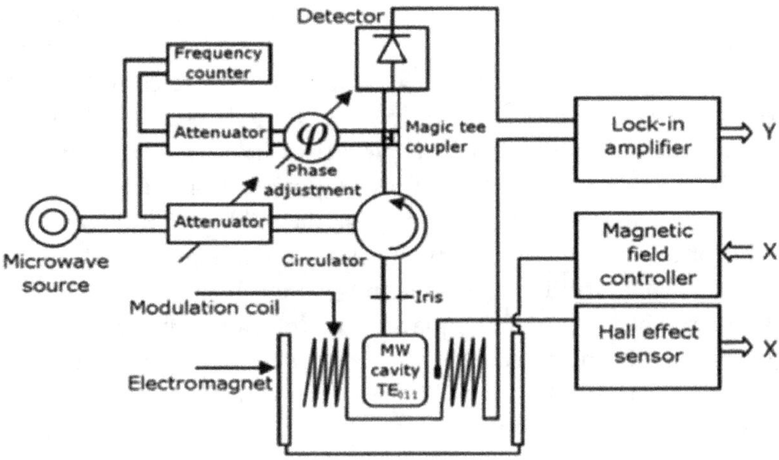

Figure 16. Schematic diagram of an EPR spectrometer.

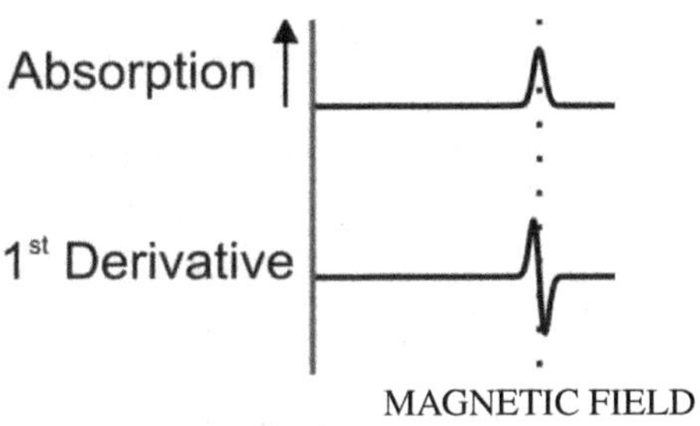

Figure 17. ESR absorption spectrum and its first derivative.

magnitude. One must be able to sweep the magnetic field to observe resonance at the fixed frequency of the microwaves. The field is modulated by placing Helmholtz oil between the pole pieces of the magnet and passing an AC current at a frequency of 100 kHz. The reflected microwave beam then goes through the circulator to the detector. Here, a diode will produce a current proportional to the intensity of the microwave field. The detector system will include a signal preamplifier and a phase-sensitive detector tuned to the modulating frequency of the magnetic field. The modulated output of the phase-sensitive detector appears as a first derivative of the absorption signal. The absorption and first derivative signals are shown in Figure 17.

The field at which the derivative goes to zero will correspond to maximum absorption.

The design of a pulsed EPR spectrometer is more complex. In general, one uses a pulse programmer to generate pulse sequences of microwave power. The low-power microwave pulses are sent to a traveling-wave tube to produces $\pi/2$ pulses of kilowatt power in a duration of a few nanoseconds. The free induction decay signals obtained are weak, i.e., of the order of a few nano- to millivolts. The signal is amplified by a low-noise GaAs FET amplifier or a traveling-wave tube and fed to a quadrature detector. The output of the quadrature detector provides the real and imaginary parts of the signal. The detection is followed by further amplification and filtering. The signal is stored in a fast digital oscilloscope for digitizing and further processing.

In order to obtain information which does not depend on the frequency of the spectrometer, one calculates the g value from the absorption frequency:

$$g = hv/\mu_{\beta}B \tag{15}$$

This g value varies from 1.99 to 2.01 for organic radicals and from 1.4 to 3.0 for transition metal compounds. g will show anisotropy if the magnetic ion is in a frozen liquid or solid medium.

There are two effects affecting the EPR spectrum:

(1) Hyperfine interaction: A nucleus has a spin angular momentum I and a magnetic moment. So, a neighboring nucleus will produce a magnetic field at the electron, and this will add to or subtract from the applied magnetic field. This interaction is called the hyperfine interaction. The energy of this interaction is given by $AI.S$. Here, A is the hyperfine constant, which depends on the nucleus. $\boldsymbol{I}$ and $\boldsymbol{S}$ are vectors representing the nuclear and electron angular momentums, respectively, with the components M_I and M_S directed along the applied magnetic field. Each electronic Zeeman level in the magnetic field, characterized by a given value of m_S, will split into $(2I + 1)$ closely spaced hyperfine levels characterized by different values of m_I. Upward electronic transitions among the Zeeman levels will be characterized by $\Delta m_S = 1$ and $\Delta m_I = 0$.

Thus, the ESR absorption line will be split into $(2I + 1)$ closely spaced components. This is shown in Figure 16 for the case of $I = ½$. On the left are shown the Zeeman levels for $m_S = \pm½$ without hyperfine splitting.

There is a single ESR line shown as a derivative below. When hyperfine interaction with a nucleus of spin $I = ½$ is taken into account, each of the two levels, $m_S = ±½$, splits further into two hyperfine levels: $m_I = ±½$. Transitions are allowed only between two hyperfine states of the same m_I value. This gives rise to two closely spaced ESR lines, as shown in Figure 18.

The hyperfine interaction will provide information on the number of neighboring nuclei and their distances from the unpaired electron. The number of lines in the spectrum is $(2NI + 1)$, where N is the number of equivalent nuclei and I is the nuclear spin. If only one nucleus interacts, all lines are of equal intensity. If there is more than one equivalent nucleus, the relative intensities of the lines vary in a specific fashion. If the nuclear spin is ½, and there are two equivalent nuclei, we get three lines with a ratio of their relative intensities given by 1:2:1 (corresponding to two nuclear spins pointing up, one up and one down, or two down).

As an example, we consider a benzene radical anion shown in Figure 19.

The extra electron is delocalized and sees the hyperfine interaction due to the six protons equally. So, there should be $(2 \times 6 \times ½ + 1) = 7$ lines,

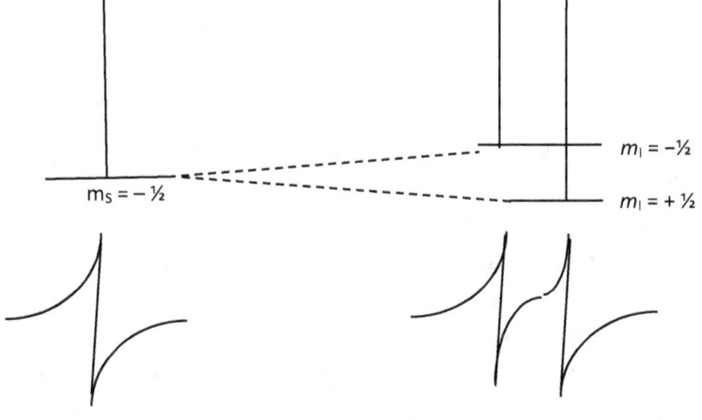

Figure 18. Nuclear hyperfine splitting of the ESR line.

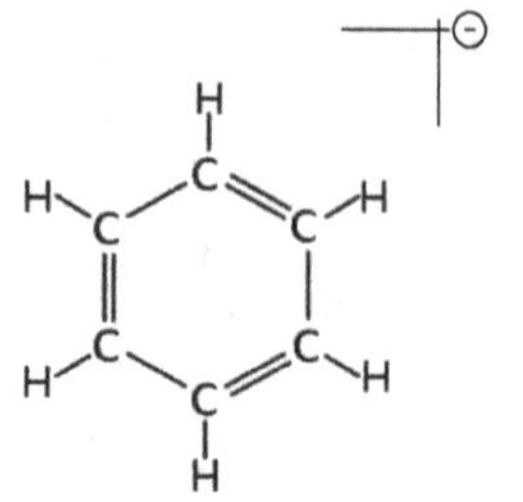

Figure 19. Benzene anion radical.

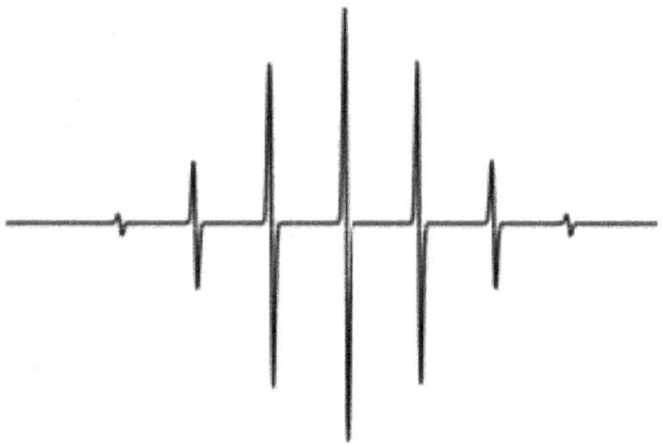

Figure 20. Hyperfine lines in the EPR spectrum of benzene radical anion (reprinted with permission from Ref. 6).

and the relative intensities must be in the ratio of 1:6:15:20:15:6:1. The hyperfine lines in the EPR spectrum of the benzene anion radical is shown in Figure 20.

In the investigation of free radicals, one gets the following information from the EPR spectrum:

(a) The hyperfine coupling pattern gives the number and type of nuclei with which the electron spin interacts.
(b) The spacing of the lines gives the hyperfine coupling constant, and the center of gravity of the spectrum gives the g value.

(2) Spin–orbit interaction: As mentioned in the introduction, the orbital angular momentum of the electron $\ell(h/2\pi)$ will also produce a magnetic field. This will add or subtract from the applied magnetic field, changing the actual magnetic field experienced by the spin magnetic moment of the electron and altering the resonance frequency. This will cause a value of g different from the free electron value.

4.2. *Anisotropy in g value due to the anisotropy in surrounding ligands*

We consider a molecule in which the paramagnetic metal ion is coordinated with two equivalent ligands along the z-direction and four equivalent ligands of a different type in the xy-plane. This is shown in Figure 21.

Here, the resonance frequency will depend on the orientation of the magnetic field relative to the z-axis. When the field is parallel to z, the g value is $g_{\parallel}$, and when the field is perpendicular to z, the g value is $g_{\perp}$, which is different from $g_{\parallel}$. Such a molecule is said to be axially symmetric. There will be two cases, as follows: (i) $g_{\parallel} > g_{\perp}$ and (ii) $g_{\parallel} < g_{\perp}$. We could also have a situation in which the environment is such that the three principal values of g are different. Such a situation is called orthorhombic.

If the paramagnetic species is in a liquid, the molecule is tumbling around. If the liquid is frozen, different molecules are oriented in different directions. So, the EPR spectrum will be a superposition of the spectra for all orientations of the magnetic field relative to the axes of the molecule. This will also be the situation when we take the EPR spectrum of a powder of the crystalline solid in which the paramagnetic ion is present. Such a powder spectrum will appear as shown in Figure 22 for different anisotropic types of g values.

Figure 23 shows the anisotropy in g value of Cu^{2+} ion in the powder EPR spectrum of $Cu(ClO_4)_2$.

We see a large anisotropy in the g values, $g_{\parallel}$ being more than $g_{\perp}$. This means that, at a fixed microwave frequency, resonance occurs at a smaller value of the magnetic field when it is parallel to the unique axis than when the field is perpendicular to the unique axis. Cu^{2+} has a nuclear spin

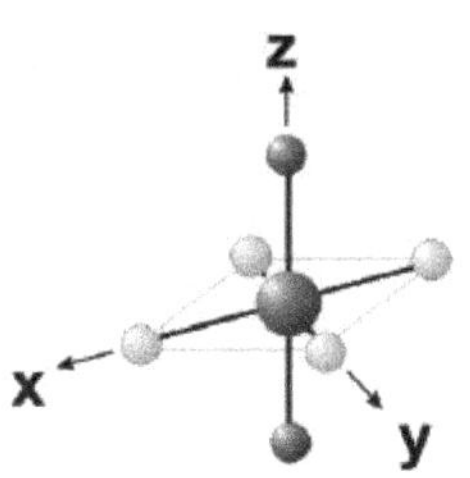

Figure 21. Central paramagnetic metal atom with different coordination along *z* and perpendicular to *z*.

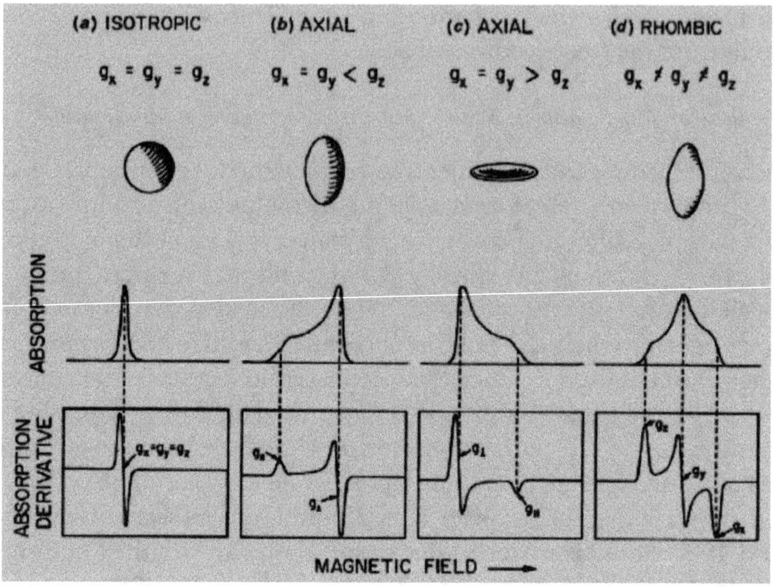

Figure 22. Powder spectrum for different anisotropic situations in g values (reprinted with permission from Ref. 8).

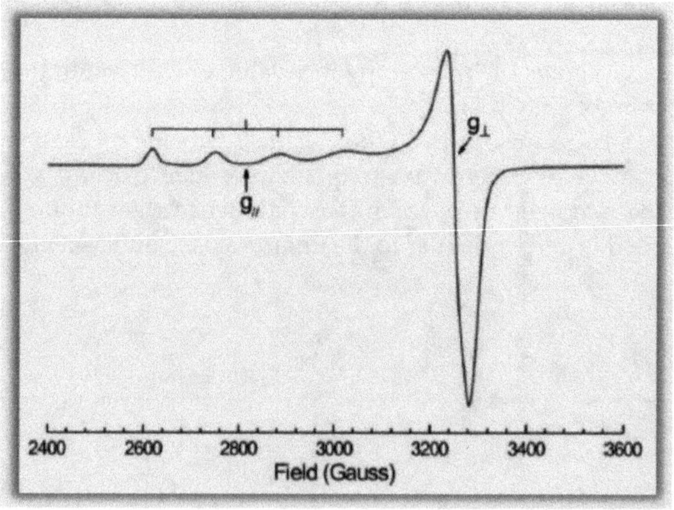

Figure 23. Anisotropy in g value of Cu^{2+} in $Cu(ClO_4)_2$ (reprinted with permission from Ref. 8).

of 3/2, and hence, there are four nuclear hyperfine levels associated with each electronic energy level. This will give four nuclear hyperfine lines, which are clearly seen when the magnetic field is parallel to the unique axis. The hyperfine coupling constant is also anisotropic and is small when the field is perpendicular to the unique axis. So, hyperfine lines are not seen when the magnetic field is perpendicular to the unique axis.

4.3. *Improvement in resolution at higher frequencies*

When one uses higher microwave frequencies, the resolution becomes better. This is shown in Figure 24.

4.4. *Electron–nuclear double resonance*

It may often happen that the nuclear hyperfine splitting is not clearly resolved and is submerged in the EPR signal. Then, one can use the electron–nuclear double resonance (ENDOR) technique. Figure 25 illustrates the principle of the ENDOR technique.

The electronic Zeeman effect splits the energy level into $m_S = \pm½$ to give two energy levels separated by $g\mu_B B$. The nuclear Zeeman effect splits the nuclear spin levels $m_I = \pm 1/2$ into two levels separated by γB.

The hyperfine interaction splits the nuclear levels further. For EPR absorption, the selection rule is $\Delta m_s = +1$ and $\Delta m_I = 0$. For NMR, the selection rule is $\Delta m_I = -1$ and $\Delta m_s = 0$. The dashed red arrow indicates the EPR

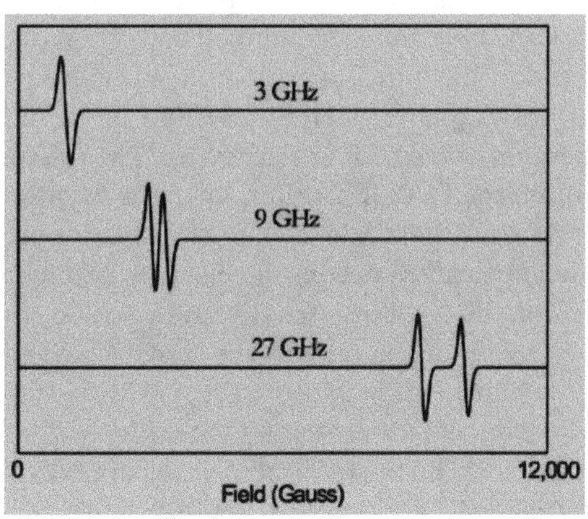

Figure 24. EPR spectrum at three different frequencies (reprinted with permission from Ref. 8).

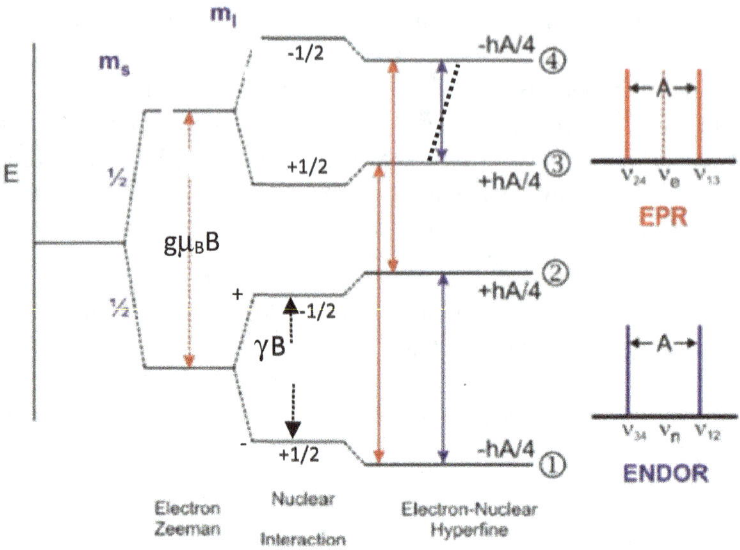

Figure 25. Splitting of electronic and nuclear energy levels of a system with $S = ½$ and $I = ½$ in a magnetic field.

absorption line in the absence of hyperfine interaction, and the two red arrows indicate the hyperfine split lines in the ESR. The dashed blue arrow indicates the NMR line in the absence of hyperfine interaction, and the two blue arrows indicate the split NMR absorption lines arising from hyperfine interaction. One must remember that the red lines are in the microwave region and the blue lines are in the radio frequency region.

The ENDOR instrument has an RF coil and probe in the microwave cavity of a continuous-wave EPR spectrometer. The microwave power is increased to saturate the EPR absorption line. The frequency of the RF is swept. When it matches the nuclear magnetic absorption frequency, the population in the nuclear hyperfine levels will be changed. This will cause the microwave absorption signal to move away from saturation, resulting in the recovery of the microwave signal. The absorption is narrower in ENDOR than in EPR spectroscopy, which enables one to see weak hyperfine lines in metalloproteins. One can also identify the nucleus by its Larmor frequency. The major disadvantage of ENDOR is its reduced sensitivity compared to EPR.

A more detailed discussion of EPR spectroscopy is beyond the scope of this book.

4.5. *Applications of EPR*

EPR finds the following applications:

(1) estimation of metal ions, such as Cu^{2+}, Cr^{2+}, Mn^{2+}, Gd^{3+}, and Fe^{3+}, in various environments;
(2) study of free radicals in healthy and diseased tissues;
(3) study of biological systems and photosynthesis;
(4) reaction kinetics;
(5) active sites of metalloproteins;
(6) organic conducting polymers;
(7) active sites on the surface of catalysts;
(8) to analyze hole and electron centers in semiconducting materials.

References

1. Maciejiewski, M. (2016). Basics of NMR Spectroscopy, markm.uchc.edu.
2. 14.2: Fourier Transform NMR-Chemistry Libre Texts, https://chemistrylibretexts.org>...>14:NMR Spectroscopy.
3. NMR Spectroscopy Theory, https://teaching.shu.ac.uk>hwb>chemistry>tutorials>molspec>nmr1.
4. UCSB Physics. Pulsed Nuclear Magnetic Resonance, https//web.physics.ucsb.edu>nmr>PULSED_NMR.
5. Dikanov, S. A. and Crafts, A. R. Chapter 3. Electron paramagnetic resonance spectroscopy, https://www.life.illinois.edu/crofts/pdf_files/Dikanov_Crofts_EPR review_Chapter-3-2006.
6. CNS Sites. Principles of Electron Spin Resonance, https://sites.cns.utexas.edu/sites/default/files/epr_facility/files/epr_intro.ppt.
7. Electron spin resonance tutorial, www.phys.ubbcluj.ro>~vchis>lab>esr1.
8. Duin, E. Electron Paramagnetic Resonance Theory, http://webhome.auburn.edu/~duinedu/epr/1_theory.pdf.

Chapter 18

IR, VISIBLE, AND UV SPECTROSCOPIES

1. Introduction

Figure 1 shows the range of the electromagnetic spectrum.

We see that the range of frequencies covered in the spectrum is spread over 16 orders of magnitude, increasing from radio-frequency waves to gamma rays. The corresponding wavelengths vary from 1 km to 10^{-13} m.

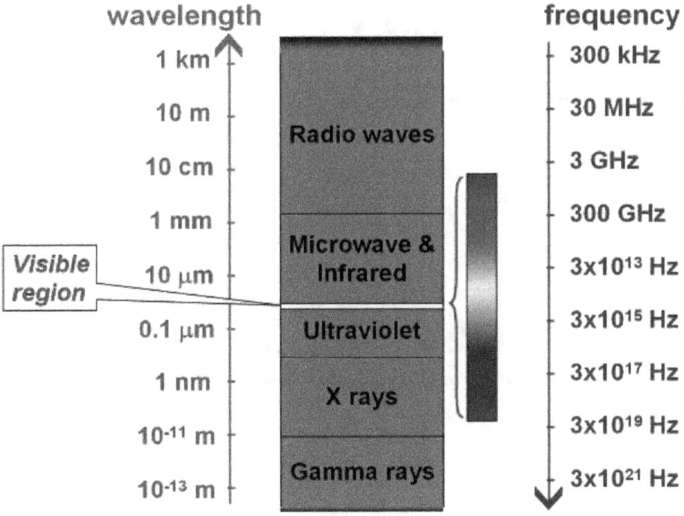

Figure 1. Range of the electromagnetic spectrum.

2. Infrared Spectroscopy

Infrared (IR) spectroscopy covers the regions of wave numbers $\bar{\nu} = 1/\lambda$ (where λ is the wavelength) as shown in Figure 2.

Molecular rotations and vibrations as well as lattice vibrations are in the far-IR and mid-IR regions. These are the regions of interest to chemists and physicists.

A system of N particles in three dimensions has $3N$ degrees of freedom. Of these, three dimensions correspond to translations in which the entire system moves together in a certain direction. There are three independent directions of motion, namely, the X, Y, and Z directions. Translation in these three directions accounts for the three degrees of freedom. The entire system can rotate about an axis passing through its center of mass. This axis can be parallel to the X-, Y-, or Z-axis. The rotations of the system around these three axes passing through the center of mass account for another three degrees of freedom. The remaining $3N - 6$ degrees of freedom correspond to oscillations of the particles against one another without a motion of the center of mass or rotation of the molecule as a whole.

For example, consider the molecule AB_2, in which the two AB bonds make an angle different from 180° with each other. Figure 3(a) shows the molecule at rest. A is heavier than B, as shown by the sizes of the circles. The center of mass is then on the bisector of the angle BAB and close to A. This is shown by a cross. The axes X, Y, and Z pass through the center of mass.

The three translational modes correspond to the motion of the molecule as a whole along the X-, Y-, or Z-axis, which are mutually orthogonal and thus independent. The three rotational modes correspond to rotation about an axis through the center of mass, parallel to the X-, Y-, or

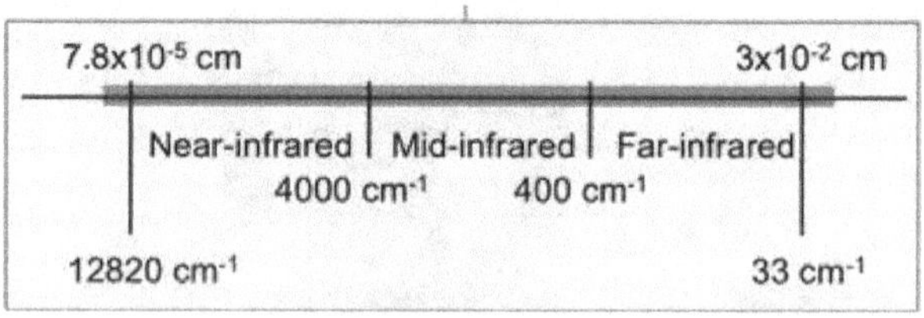

Figure 2. Range of infrared wavenumbers indicated in cm^{-1} The wavelength is indicated in cm at the top.

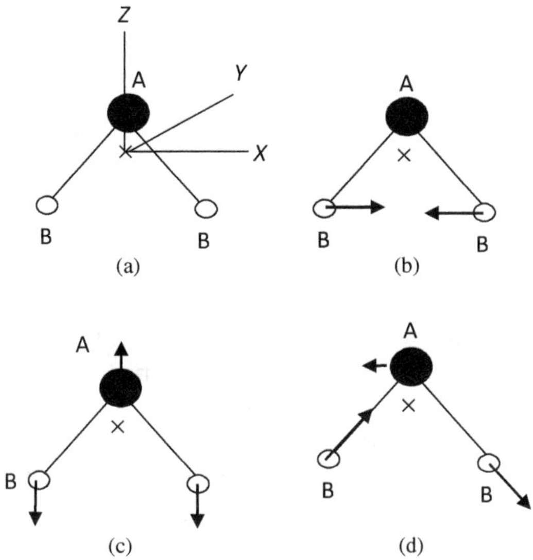

Figure 3. Vibrational modes of molecule AB_2.

Z-axis. That leaves three vibrational modes, which are shown in Figures 3(b)–(d).

In Figure 3(b), atoms B move against each other along the X-axis with equal displacements at all times. Atom A does not move. During this vibration, the center of mass is at rest. In Figure 3(c), atoms B have the same displacements along the Z-axis, whereas atom A has an opposite displacement along Z so that the center of mass does not move. In Figure 3(d), one atom B moves along AB in a direction to reduce the length of AB. The other atom B moves along its AB bond to increase its length by the same amount. The result of these two displacements is to move the center of mass along X. This is compensated by the motion of A along X in the opposite direction.

If the N atoms are arranged in a line (say along the X-axis), then there are three translations, two rotations along the Y- and Z-axis, and $(3N - 5)$ vibrational modes. This is shown for the diatomic molecule A_2 in Figure 4.

Figure 4(a) shows the diatomic molecule A_2 at rest. The line joining the two atoms is taken as the X-axis. The center of mass is at the midpoint between the two atoms and indicated by a cross. The axes X, Y, and Z passing through the center of mass are shown in Figure 4(a). The three translational degrees of freedom correspond to independent motions of

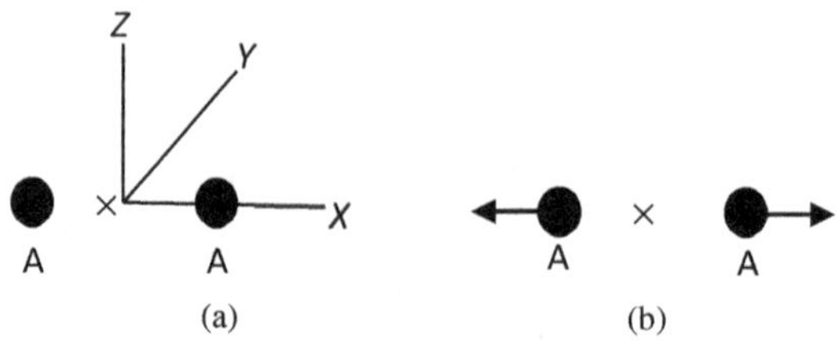

Figure 4. Diatomic molecule.

the molecule as a whole along the X-, Y-, or Z-axis. There are two rotational motions around the Y- and Z-axes that pass through the center of mass. The atoms are considered points along the line joining them. Rotation along the X-axis does not change anything. There should be one vibrational degree of freedom. This is shown in Figure 4(b). Here, the two atoms are displaced by equal and opposite amounts along the line joining them.

A molecule has a moment of inertia I about the axis of rotation. The moments of inertia for rotation around the X-, Y-, and Z-axes can be, in general, different. If the angular frequency of rotation is ω, the angular momentum is $I\omega$. This is quantized in units of $(h/2\pi)$, i.e.,

$$I\omega = n(h/2\pi) \tag{1}$$

The kinetic energy

$$E_n = (I\omega)^2/2I = n^2\,(h^2/(8\pi^2 I) \tag{2}$$

One can have a transition from a state n to a state $(n + 1)$, giving rise to the absorption of photon. Since the moment of inertia of a molecule is large, the rotational absorption frequencies are in the far-IR region.

A vibrational mode will have a frequency of ν. It will correspond to a linear harmonic oscillator with energy levels of $(m + ½)h\nu$, where m is an integer. When radiation of frequency ν falls on it, the mode will change from the ground state to the first excited state, absorbing a photon of energy $h\nu$. The vibrational frequency will be much larger than the rotational absorption frequency. Each vibrational energy level will have a number of closely spaced rotational energy levels. A vibrational transition will have a number of associated rotational transitions, giving rise to an absorption band.

Of the vibrational degrees of freedom, only those modes of vibration in which an electric dipole moment changes with the vibration will appear in IR absorption. If we consider the modes in Figures 3(b)–(d), only those shown in Figures 3(c) and (d) will cause a change in the dipole moment. The mode shown in Figure 3(b) is symmetric to a reflection in a mirror plane parallel to the *YZ*-plane and passing through the center of mass. Such a mode will not cause a change in the electric dipole moment along the *X*-axis. So, it will not be seen in IR absorption. This is also true of the mode of molecule A_2 shown in Figure 4(b). **Thus, we see that not all molecular vibrational modes will be seen in IR absorption.**

3. Absorption Spectrometer

Any absorption measurement, whether made in the IR or visible region, will need: (1) a source of continuous radiation covering a given spectral range, (2) a dispersing element to separate the wavelengths in the radiation, (3) a sample holder in which the sample is kept, and (4) a detector. These are arranged as shown in Figure 5.

In Figure 5, S is a source of radiation. Two types of sources are commonly used for IR radiation: It can be a globar source, which consists of silicon carbide rod heated to about 1000°C, or a Nernst glower, made of a mixture of refractory oxides heated to a temperature higher than that to which a globar is heated. These are sources which give radiation over a wide, continuous range of wavelengths in the IR region.

M is a mirror to reflect the rays from the source on to a grating G. For IR radiation, the grating G is a reflection grating with lines ruled on a polished mirror surface. Depending on the orientation of the grating, one wavelength of the radiation from the source will fall on the sample holder SH.

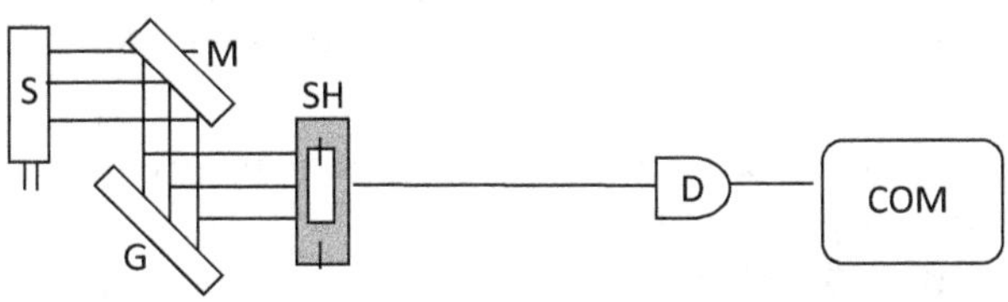

Figure 5. Schematic diagram of arrangement used to measure absorption.

The sample is held in the sample holder SH. For IR absorption studies, the sample holder consists of two polished potassium bromide, cesium chloride, or sodium chloride plates, between which the sample, either in the form of a liquid or a solid powder, is kept. **Note that glass will not transmit IR radiation in the region of interest.** These alkali halide plates are susceptible to attack by humidity and so should be kept in dry conditions.

The radiation passing through the sample holder is detected by a detector. For IR radiation, two types of detectors, thermal and photonic, are used.

In a thermal detector, there is a change in temperature caused by the absorption of radiation by the detector. This change produces either an emf in a thermopile or a change in resistance in a bolometer. The material of the bolometer is germanium or lead selenide.

One can also have photonic detectors, which are narrow bandgap semiconductors. The absorbed photons create electronic excitations, leading to photoconductivity. These detectors have a faster response than thermal detectors.

The output of the detector is fed to a computer. Here, the detector output is plotted against the wavelength of the radiation.

The sources of IR radiation are not stable over time. Water vapor in the atmosphere also produces absorption in the range of wavelengths studied; this absorption will change depending on the humidity. These will cause the intensity of the IR beam to change with time, and this will cause a change in the IR spectrum. To overcome these effects, a double-beam IR absorption instrument is used. Figure 6 shows a schematic diagram of the double-beam instrument.

Here, the light from the grating is split into two beams by the partially reflecting mirror M_2. One of the split beams passes through the sample holder SH, which contains the sample under study. The other beam goes through an identical blank holder BH. The beam passing through SH is brought to a detector, D_2. Similarly, the beam passing through BH is brought to a detector, D_1. The difference in the signals from the two detectors is fed to the computer.

The instability of the source and the change in environmental conditions affect the two beams in the same way. Since a difference between the outputs of the two detectors is measured, these changes cancel out.

However, IR absorption measurement using a grating is time-consuming, as the grating setting has to be changed to get the absorption

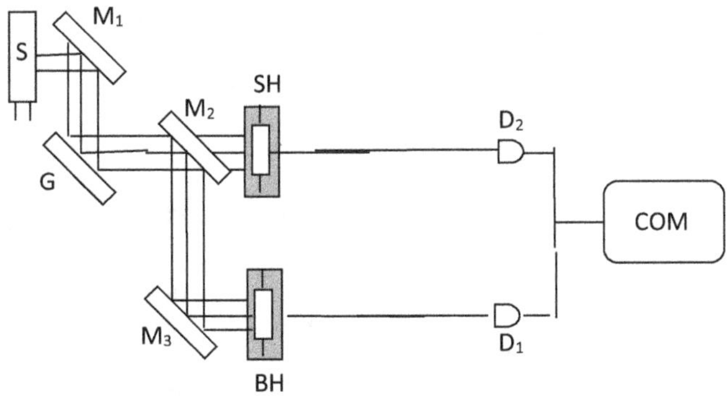

Figure 6. Double- beam absorption instrument.

at different wavelengths. To overcome this problem, a Fourier transform (FT) IR instrument is used.

4. Fourier Transform IR Instrument

In the FT IR instrument, we use a Michelson interferometer instead of a grating and move one of the mirrors periodically back and forth. The schematic diagram of an FT IR instrument is shown in Figure 7.

Light from the source is split into two beams. One beam falls on a fixed mirror of a Michelson interferometer, while the other falls on a movable mirror. The light after reflection from these mirrors will undergo a path difference depending on the position of the movable mirror and will give an interference pattern depending on the wavelength of the IR radiation. By moving the mirror back and forth, this interference pattern at the sample changes periodically.

The interferogram recorded appears as shown in Figure 8. Each point of the interferogram contains information from all wavelengths. The interferogram is Fourier transformed using software to obtain the IR absorption spectrum over the complete wavelength range.

The advantages of FT IR are as follows:

(a) Speed: The entire interferogram is recorded in a matter of minutes.
(b) Sensitivity: Photodetectors are used, which respond fast and are more sensitive than thermal detectors.
(c) The only moving part is the mirror of the interferometer, so maintenance is easy.

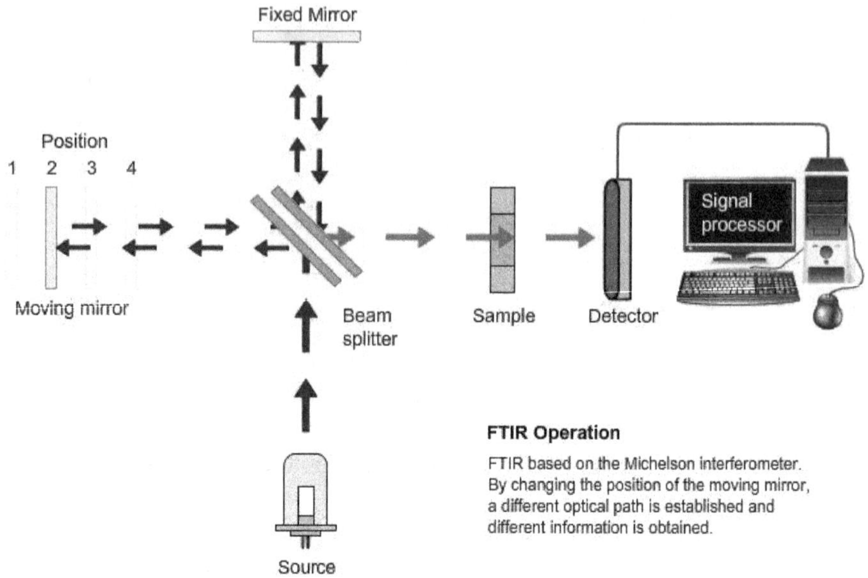

Figure 7. Schematic diagram of an FT IR instrument.

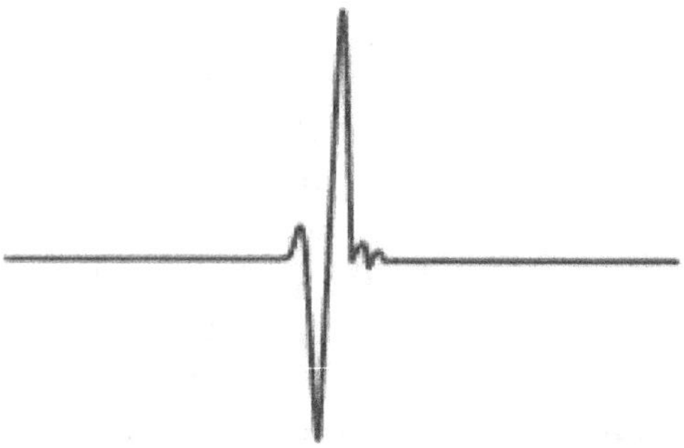

Figure 8. Interferogram recorded in an FT IR spectrometer.

A complete IR transmission spectrum of a sample appears as shown in Figure 9.

The dips in transmission correspond to absorption peaks. One may identify the peaks with the stretching or bending frequencies of different types of bonds in organic samples, as listed in Table 1.

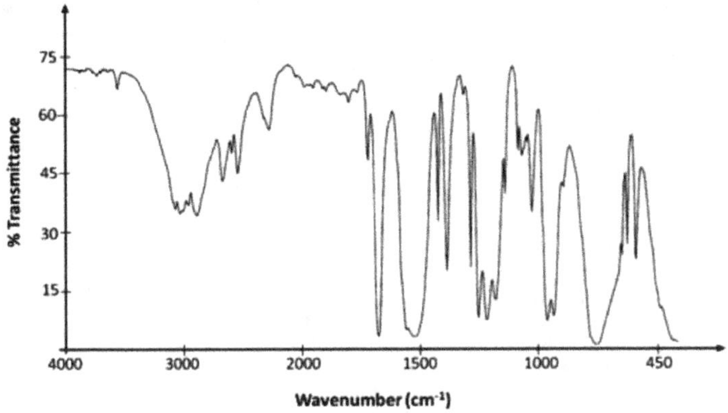

Figure 9. IR transmission spectrum in a sample.

The IR spectra are of great utility in identifying the types of atoms in the compound and the types of bonds. IR absorption gives partial information about the structure of the molecule.

In a semiconductor, the IR absorption spectrum gives the energy bandgap. IR absorption is useful in studying optical modes of vibration in ionic crystals.

5. Visible and Ultraviolet (UV) Spectroscopy

The visible region of the electromagnetic spectrum covers the range of wavelengths from 700 to 400 nm, the near-ultraviolet region from 400 to 200 nm, and the far-ultraviolet region from 200 to 10 nm. Many different types of spectroscopies can be conducted in the visible–UV region. These different types of spectroscopies are

(a) absorption spectroscopy,
(b) fluorescence spectroscopy, and
(c) Raman spectroscopy.

We briefly discuss each of these types of spectroscopies in the following sections.

Table 1. Wavenumbers of stretching and bending vibrations in organic compounds.

Bond	Compound	Wavenumber (cm^{-1})
—C—H	alkanes	2800–3100
—C—D	alkanes	~2200
=C—H	alkenes and arenes	3000–3100
≡C—H	alkynes	3200–3350
	esters 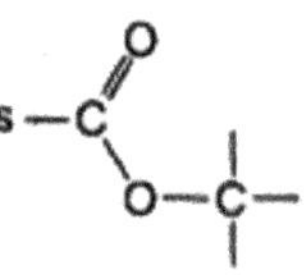	1035–1300
>C=O	aldehydes 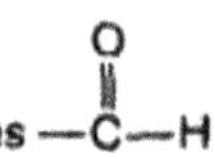	1690–1740
>C=O	ketones 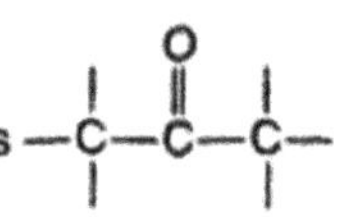	1650–1730
>C=O	acids 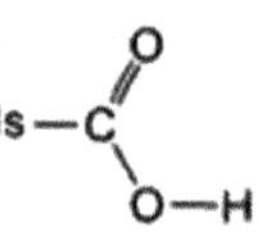 esters 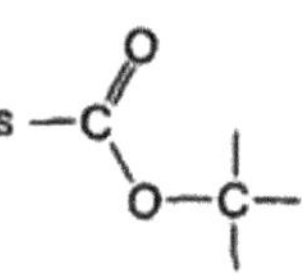	1710–1780
—C≡C—	alkynes	2050–2260
—C≡N	nitriles	2200–2400
—C—O—	alcohols 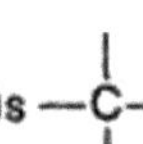, ethers 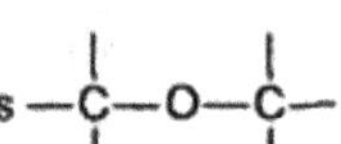	980–1250
	carboxylic acids 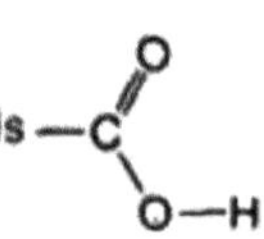	1350–1440 1210–1320
—C—C—	alkanes	750–1200*
>C=C<	alkenes	1600–1680

5.1. *Absorption spectroscopy*

In absorption spectroscopy, the electron in an atom, molecule, or solid is raised from a lower energy state to a higher energy state, absorbing an incident photon of an appropriate frequency.

Visible absorption spectroscopy deals with the transitions of the outer valence electrons in an atom or in the molecular orbitals in a molecule.

For example, the sodium atom has the electronic structure $1s^2 2s^2 2p^6 3s^1$. Its inner shells, $n = 1$ and $n = 2$, have the full complement of electrons. In the outermost $n = 3$ state, it has one electron in the S shell. This outer electron in the energy level $3S_{1/2}$ can be raised to one of the two higher energy levels, $3P_{1/2}$ or $3P_{3/2}$. These two levels are slightly different in energy, as shown in Figure 10.

If a photon of energy *hv*, equal to the energy difference between $3S_{1/2}$ and one of the 3P levels, is incident on the atom, then the photon will be absorbed, and the electron is raised to the corresponding 3P level. This gives rise to absorption at two closely spaced wavelengths: 589.6 and 589.0 nm. There are selection rules for the transition. For example, a transition between the S and P levels is allowed, while a transition between the S and D levels is forbidden.

In a molecule, each of the electronic levels is associated with closely spaced vibrational levels, as shown in Figure 11 for the molecule I_2.

The minimum electronic energy for the ground and excited states occur at different internuclear distances. So, the vibrational frequency in the ground state is different from that in the excited state. The vibrational levels have energies of $(v + ½)hv$, where v is the vibrational frequency associated with the given electronic state. The electron can be excited from one vibrational state in the ground electronic energy level to a different vibrational state in the excited electronic level, giving rise to an absorption

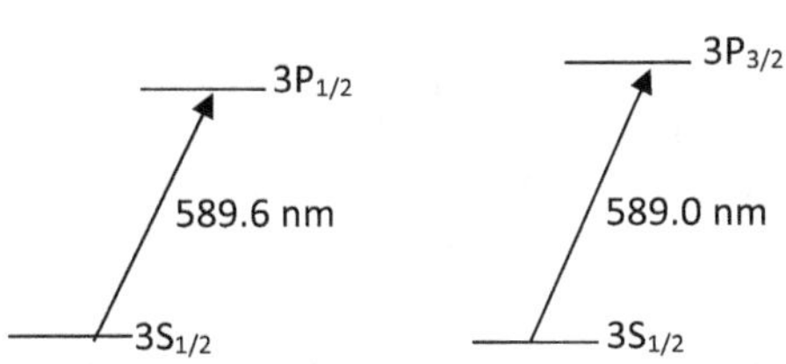

Figure 10. **D_1 and D_2 absorption lines in sodium.**

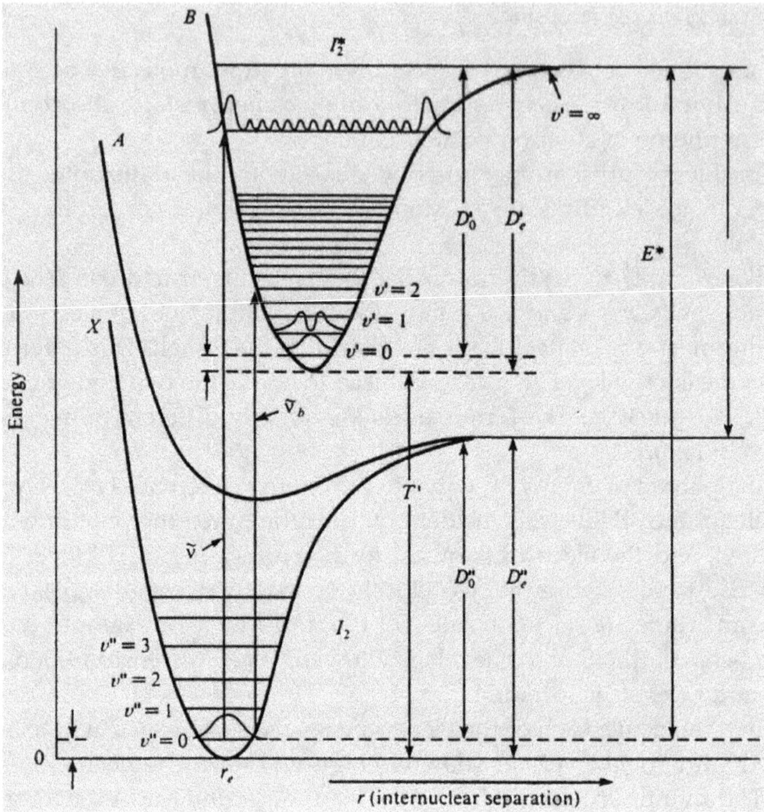

Figure 11. Energy of the ground and excited electronic levels of I_2 molecule as a function of internuclear separation. Vibrational energy levels associated with each electronic level are shown as horizontal lines.

band. However, electronic transitions in a molecule are determined by the Franck–Condon principle. This principle states that:

(a) during a transition, the nuclear coordinates do not change; and
(b) the transition probability depends on the overlap of the vibrational wave functions in the ground and excited states, and transition probability is maximum when the overlap is at its maximum.

In the ground electronic state, the ground vibrational state v = 0 has the maximum number of molecules. Its wave function has the maximum overlap with the wave function corresponding to the vibrational state $v' = 20$ in the excited state. So, the absorption will be at a maximum corresponding to the energy difference between these two states.

UV radiation has higher energy photons and will cause the transition of the inner electrons in the atoms to the outer, unoccupied shells. This is shown in Figure 12.

Absorption studies require a source of continuous radiation. For the visible range, this is a tungsten filament lamp or a halogen lamp. For the near-UV region, it is a deuterium lamp. The most common detector used is a photomultiplier. The sample chambers are made of fused quartz, which transmits UV–visible radiation.

The arrangement for absorption studies is the same as that shown in Figures 5 and 6.

Absorption studies in the visible region give information about the nature of chemical bonds. Absorption studies in the far-UV region give information about the elements present in the material.

5.2. *Fluorescence and phosphorescence*

Figure 13 shows (1) a ground electronic state, S_0, with associated vibrational energy levels and (2) two excited electronic states, S_1 and S_2, with associated vibrational states of a molecule. As mentioned earlier, the vibrational frequencies in the different electronic states will be different. So, the spacing of the vibrational energy levels is different in the electronic states S_0, S_1, and S_2. The dipole selection rule permits a transition from S_0 to S_1 but not from S_0 to S_2.

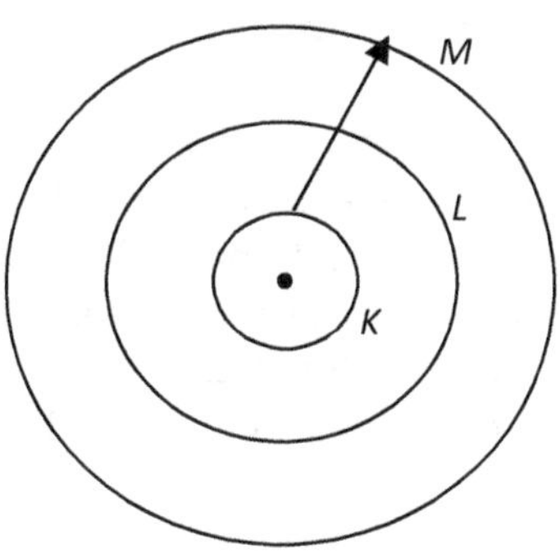

Figure 12. UV absorption resulting from a transition of an electron from the K shell to the M shell.

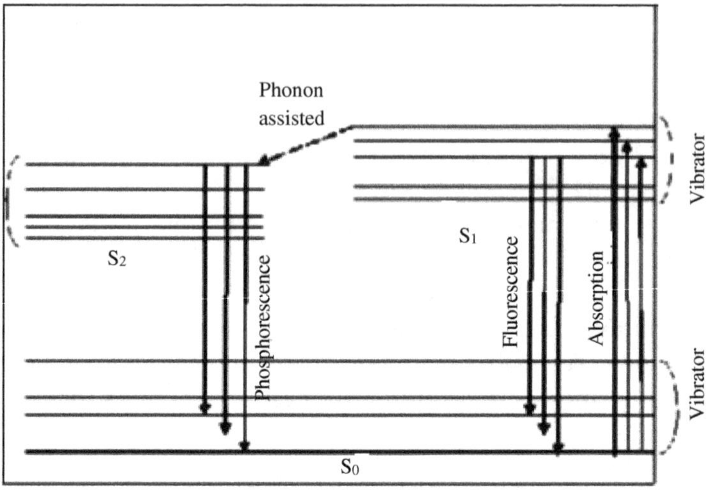

Figure 13. The ground electronic state S_0 of a molecule with associated vibrational levels and two excited electronic states S_1 and S_2.

The molecule goes to the higher electronic energy level, S_1 (a singlet state), from the ground electronic energy level, S_0 (again a singlet state), on absorption of a photon of appropriate frequency. Due to collisions with other molecules, it may lose a part of its energy to come to a lower vibrational level in the higher excited electronic state S_1. After a few hundred picoseconds, the molecule will make a transition from this higher electronic state S_1 to the ground electronic state S_0, emitting a photon. This is shown in Figure 13 on the right. This phenomenon is called fluorescence.

Thus, the absorption band will extend from the wavelength λ_0 to shorter wavelengths, while the emission will extend from λ_0 to longer wavelengths.

Figure 14(a) shows a schematic diagram of the arrangement for a fluorescence spectrometer. There are two monochromators, one to choose the wavelength of the incident light and the other to analyze the wavelength distribution of the fluorescence emission. Figure 14(b) shows the absorption and fluorescence spectra of anthracene.

After losing a part of its energy to collisions, the molecule may come to a vibrational state in the excited electronic state S_2, which is a triplet state. The dipole selection rule will not permit a transition from S_2 to the ground electronic state S_0. The molecule stays in the excited state S_2 for a few hundred microseconds and ultimately comes to the

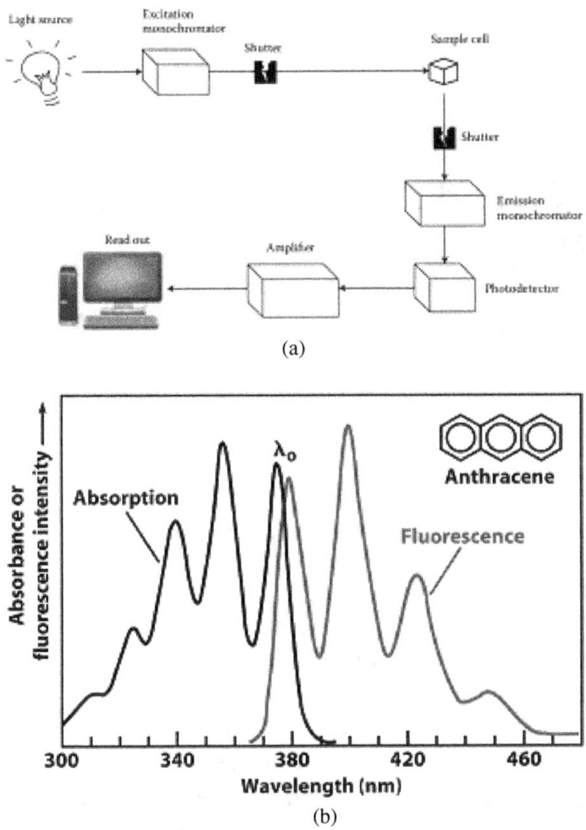

Figure 14. (a) Schematic diagram of a fluorescence spectrometer. (b) Absorption and fluorescence spectra of anthracene.

ground electronic state, emitting a photon. Thus, phosphorescence is an emission which takes place after a delay of a few hundred microseconds after absorption.

Molecular fluorescence spectroscopy finds applications in biochemical and medical research and in organic chemistry.

In fluorescence spectroscopy, one can use a halogen lamp to excite the fluorescence.

5.3. *Raman spectroscopy*

In Raman spectroscopy, we use incident **monochromatic** light obtained from a laser source. The incident light has an oscillating electric field

$E_0\sin(2\pi v_0 t)$. The molecule has polarizability, denoted by α. The electric field in the incident light causes an induced oscillating dipole moment:

$$P = \alpha E_0 \sin(2\pi v_0 t) \tag{3}$$

The molecule has several vibrational modes with frequencies v_j. So, we may write

$$\alpha = \alpha_0 + \Sigma_j(\partial\alpha/\partial q_j)q_j \sin(2\pi v_j)t \tag{4}$$

Here, q_j is the amplitude of the jth normal mode and v_j is its frequency. Substituting (2) in (1), we get

$$\begin{aligned} P &= \alpha_0 E_0 \sin(2\pi v_0 t) + E_0 \Sigma_j(\partial\alpha/\partial q_j)q_j \sin(2\pi v_0 t)\sin(2\pi v_j t) \\ &= \alpha_0 E_0 \sin(2\pi v_0 t) + (½E_0)\Sigma_j(\partial\alpha/\partial q_j)q_j[\cos(2\pi(v_0 - v_j)t) \\ &\quad - \cos(2\pi(v_0 + v_j)t))] \end{aligned} \tag{5}$$

The oscillating dipole moment is responsible for the scattering of light. So, the scattered light contains the unmodified incident frequency v_0 and the modified frequencies. The component of the scattered light at a frequency $(v_0 - v_j)$, which is lower than the incident frequency, is called the Stokes component, and the component at a frequency $(v_0 + v_j)$, higher than the frequency of the incident light, is called the anti-Stokes component.

In quantum mechanical language, the normal mode of frequency v_j has a set of energy levels $(\mathrm{v} + ½)hv_j$, where v is the vibrational quantum number. In Stokes scattering, the incident photon of energy $\mathrm{h}v_0$ excites the normal mode from its ground state v = 0 to the excited state v = 1 and gets scattered as a photon with energy $\mathrm{h}(v_0 - v_j)$. This is shown in Figure 13(a). In anti-Stokes scattering, a molecule in the vibration state v = 1 comes down to the ground state v = 0 in the presence of an incident photon. The scattered photon then has an energy $h(v_0 + v_j)$. Since the ratio of the number of molecules in the state v = 1 to that in the state v = 0 is $\exp(-hv_j/kT)$, the intensity of anti-Stokes to the Stokes Raman lines is also in this ratio. So, the two lines will be equally spaced about the incident frequency, but the Stokes line will be more intense than the anti-Stokes line.

If $\partial\alpha/\partial q_j = 0$, i.e., the vibrational mode j does not change the polarizability of the molecule, the normal mode will not cause Raman scattering.

So, all vibrational modes of the molecule do not appear in the Raman effect. In IR absorption, only modes which will cause a change in the dipole moment of the molecule will be seen. In the Raman effect, only the modes which change the polarizability of the molecule will appear. Some modes cause a change either in the dipole moment or in polarizability. They will appear either in IR absorption or in the Raman effect. Some modes will cause a change in both the dipole moment and polarizability. They will appear in both IR absorption and the Raman effect. Thus, to get complete information about the vibrational modes of the molecule, one should study both IR absorption and Raman scattering.

The advantage of the Raman effect is that we can get information about the frequency of the normal mode of vibration using visible or near UV incident light. It is much easier to build a visible-near UV spectrometer than an IR spectrometer.

In Raman spectroscopy, one uses a stable monochromatic laser source. The scattered light is picked up by an optical fiber, which will exclude, as much as possible, the scattered incident light. The scattered light is analyzed by a high-resolution spectrometer. A schematic diagram of a Raman spectrometer is shown in Figure 15.

As an example, the Raman scattering for carbon di-sulfide is shown in Figure 16 at two different temperatures.

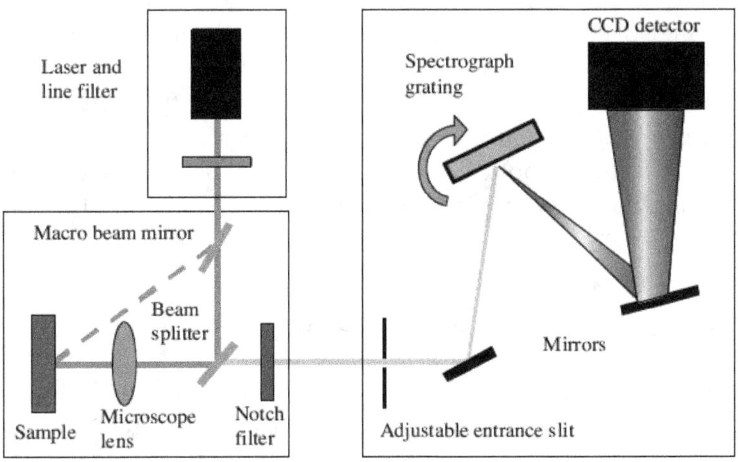

Figure 15. A schematic diagram of a Raman spectrometer.

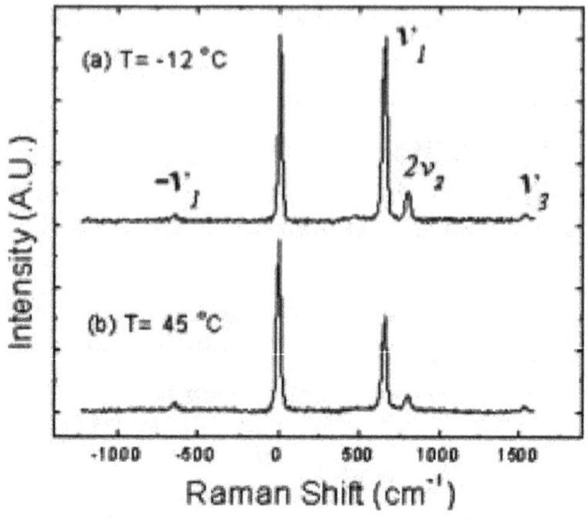

Figure 16. Raman spectrum of CS_2 at two different temperatures: A positive Raman shift indicates a Stokes line and a negative Raman shift indicates an anti-Stokes line.

We note that the ratio of intensities of the anti-Stokes line to the Stokes line at v_1 increases as temperature increases. Also note the Stokes line at the frequency $2v_2$. This corresponds to a transition from the ground state v = 0 to the second excited state v = 2 of the normal mode 2. This is called second-order Raman scattering. The Stokes line with a shift of v_2 is not seen. Thus, some modes which are forbidden to appear in the first-order Raman scattering by the selection rule governing the change in polarizability with the amplitude of vibration of the mode may appear in the second-order scattering.

The Raman spectrum has applications in identifying the vibrations of molecules and the nature of a chemical bond. Crystals in different phases have different optical mode frequencies. So, if a material is in a mixed phase, the microscopic Raman spectrum can identify the different phases.

6. Conclusion

In this chapter, the salient features of IR, UV, and visible absorption, fluorescence, and Raman spectroscopies were described. These techniques are useful in getting information about the vibrational frequencies of molecules and crystalline solids and the energy bandgap of semiconductors.

They give information about chemical bonds and the structure of molecules and can be used for the identification of different crystalline phases.

For further study, the following references may be consulted.

References

1. Introduction to Infrared Spectroscopy, www.chromacademy.com> Introduction_To_Infrared_Spectroscopy.
2. Colthup, N. B., Daly, L. H., and Wiberley, S. E. (1975). *Introduction to Infrared and Raman Spectroscopy*. Academic Press.
3. UV-Visible Spectroscopy: chem.www.uzh.ch> dam> UV/VIS-HS17.
4. Raman Spectroscopy, Isa.physics.instructor.umich.edu> Raman Spectroscopy.

Chapter 19

MOSSBAUER SPECTROSCOPY

1. Introduction

Mossbauer spectroscopy is the *recoilless* emission and resonant absorption of a photon. We have to understand what recoilless means.

Let us consider a free atom of mass m with a ground state of energy E and an excited state of energy $E + \Delta E$. The atom comes down from the excited state to the ground state and emits a photon with an energy of $h\nu$, where h is the Planck's constant and ν is the frequency. Before emission, let us assume that the atom is at rest. The photon of energy $h\nu$ has a momentum of $h\nu/c$. The law of conservation of momentum requires the atom to recoil with a momentum of $p = h\nu/c$ in a direction opposite to the direction of emission of the photon. The recoil kinetic energy, E_R, of the atom after emission is

$$E_R = p^2/2m = (h\nu)^2/2mc^2 \tag{1}$$

Let us take a sodium atom emitting a photon of wavelength 5890 Å. Such a photon has a frequency of 5.09×10^{14} Hz. The photon's energy, $h\nu$, is 2.109 eV. The rest mass energy mc^2 of sodium is 21.6×10^9 eV. So, E_R is 1.03×10^{-10} eV. The conservation of energy requires

$$\Delta E = h\nu + E_R \tag{2}$$

So, the energy of the emitted photon is less than ΔE by the recoil energy.

Let us consider a photon of energy $h\nu$ absorbed by the same atom to raise it from the ground state E to the excited state $E + \Delta E$. The incident photon must provide for both the energy difference ΔE and the recoil energy E_R. So, for absorption,

$$h\nu = \Delta E + E_R \tag{3}$$

The photon emitted by one atom cannot be resonantly absorbed by another atom of the same kind because the emitted photon will fall short in energy by $2E_R$ for absorption. This means that, if the energy levels are precisely defined, resonant absorption of a photon will not be possible.

However, the excited energy level has a lifetime of τ before the atom in the excited state comes down to the ground state by spontaneous emission. According to the uncertainty principle, the energy of the excited state is not precise and has a width of $h/(2\pi\tau)$. The lifetime τ for the excited state in sodium is 10^{-8} s, and the uncertainty in the energy of the excited state is 6.6×10^{-8} eV. This uncertainty in the excited energy state is more than $2E_R$. **So, the photon emitted by one sodium atom will be resonantly absorbed by the other sodium atom, as the half-width of the excited state in the sodium atom is much more than twice the recoil energy of the atom.**

Let us consider the case of the nucleus of Fe^{57}, the iron atom with a mass number of 57. Any nucleus has a spin I. The ground state of Fe^{57} nucleus corresponds to $I = ½$. The excited states of the nucleus correspond to $I = 3/2, 5/2, \ldots$. Consider the excited state $I = 3/2$. The energy difference, ΔE, between the ground state, $I = ½$, and the first excited state, $I = 3/2$, is 14.4 keV, and so, the emitted γ-ray photon will have an energy approximately equal to 14.4 keV. The mass of Fe^{57} atom is 9.47×10^{-23} g and its rest mass energy is 53.6 GeV. The recoil energy E_R of the iron atom when it emits a γ-ray photon is 1.95×10^{-3} eV. **Note that the recoil energy is seven orders of magnitude higher than the recoil energy of the sodium atom calculated above.** This is because the energy of the γ-ray photon is three orders of magnitude higher than the energy of the photon emitted by the sodium atom. The lifetime of the excited state in the Fe^{57} nucleus is 97 ns. The uncertainty in the energy of the excited state is 6.75×10^{-9} eV. In this case, the uncertainty in energy is much less than twice the recoil energy of the atom. **So, resonant absorption of the γ-ray photon is not possible.**

If the absorbing Fe^{57} atom is in a vapor state, the atoms will be moving around at a temperature T with a Maxwellian distribution of velocities. If the atom is moving with a velocity v in a direction that makes an angle θ with respect to the direction of propagation of the incoming photon, the frequency of the incoming photon will appear to have a value $\nu' = \nu[1 - v\cos\theta/c]$ in the rest frame of the atom. This is the Doppler shift. For atoms moving with $\theta > 90°$, the frequency ν' will be greater than ν. So, a small fraction of Fe^{57} atoms in the vapor will have the right velocities for the Doppler shift to compensate for the difference in energy $2E_R$ required for the resonant absorption of the incident photons. This fraction of atoms will resonantly absorb the γ-ray photons coming from the Fe^{57} source. The Doppler width of the absorption line will increase as the temperature increases. So, a rise in temperature will cause a larger number of γ-ray photons to be resonantly absorbed.

Suppose we put the Fe^{57} atom in a solid. This atom will be connected to other atoms through interactions, which can be modeled in terms of springs. These atoms will have different vibrational modes in a solid. Let us take the simple Einstein model in which all the modes have the same vibrational frequency, υ. This frequency, υ, will be of the order of 10^{12} Hz in solids, and a quantum of energy, $h\upsilon$, will be of the order of 10^{-2} eV. When the atom comes down from the excited state of energy $E + \Delta E$ to E, it may be accompanied by a change in the vibrational quantum number v of the lattice vibrations of frequency υ. The change in vibrational quantum number Δv could be 0 or $\pm n$, where n is an integer. If the change in vibrational quantum number is 0, then the emitted photon $h\nu$ has the energy ΔE, and the photon is called a recoilless photon. If $\Delta \mathrm{v} \neq 0$, then the emitted photon has a frequency given by $\Delta E = h\nu' - \Delta \mathrm{v} h\upsilon$. The photon frequency is now lower or higher than $\Delta E/h$, depending on the sign of Δv. One can calculate the probability of photons of different frequencies emitted by the nucleus of an ^{57}Fe atom embedded in a solid. This probability plotted against $(\nu' - \nu)$ is shown in Figure 1, where ν is the frequency of the recoilless photon. In the recoilless case, it is as if the entire solid in which the Fe^{57} nucleus is embedded absorbs the recoil. Since the recoil energy is inversely proportional to the mass of the solid, it becomes negligible. In the simple Einstein model for the solid, which we have assumed, the recoilless fraction $f(0)$ is given by

$$f(0) \propto \exp(-k^2 < X^2 >) \tag{4}$$

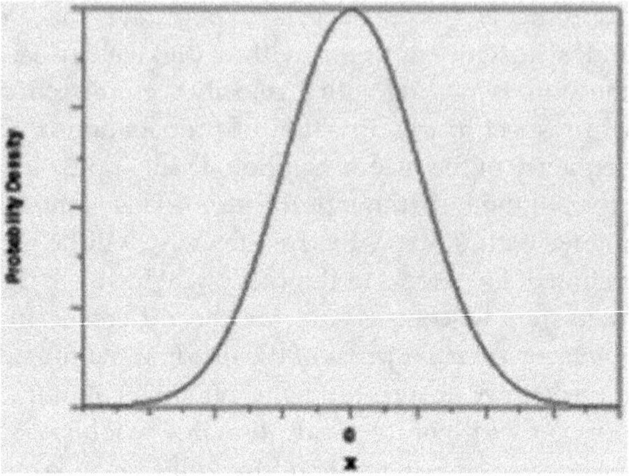

Figure 1. Probability of emission of a photon at a frequency ν'. ν is the frequency of recoilless photon and $x = (\nu' - \nu)$.

Here, k is the wave vector of the photon ($k = 2\pi\nu/c$) and $<X^2>$ is the mean square displacement of the atom in the solid.

The mean square displacement will depend on the temperature of the solid. If the temperature is low, the mean square displacement will be small. So, when the Fe^{57} atom is embedded in a solid, the fraction of recoilless photons emitted or absorbed will increase when the solid is cooled. If we have two identical solids containing Fe^{57} atoms, then a large fraction of the emitted photons from the Fe^{57} atoms in the emitter will be resonantly absorbed by the Fe^{57} atoms in the absorber. This fraction will increase if the emitter, or the absorber, or both are cooled. This, in essence, is the basis of the Mossbauer effect.

If the emitter iron atoms are in a different environment from the absorber iron atoms, there will be small shifts in ΔE of the two sets of atoms. If ν_E and ν_A are the recoilless frequencies of the emitter and absorber iron atoms, respectively, there will be a slight difference in the frequencies, as shown in Figure 2.

In Figure 2(a), we have an emitter embedded in a solid, shown by a red vertical slab. The emitter can be moved back and forth with different velocities. The absorber is stationary and is shown by the blue vertical slab. The recoilless emission frequency is ν_E, shown by a dashed red line in Figure 2(b). The recoilless absorption frequency, ν_A, is shown by a blue line in Figure 2(b). The two frequencies are different because the two

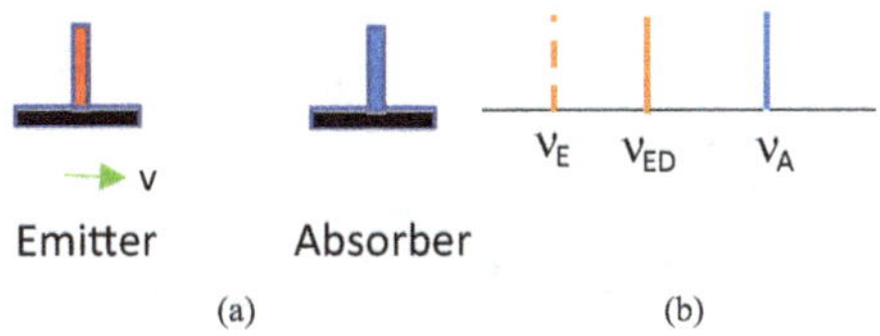

Figure 2. Illustrating how Doppler shift can be made to compensate for $\nu_A - \nu_E$.

solids in which the ^{57}Fe atoms are embedded are different. In Figure 2(b), ν_A is shown to be greater than ν_E. When the emitter is moved toward the absorber with a velocity of v, the emitter frequency suffers a Doppler shift. In the rest frame of the absorber, the emitter photons appear to have a frequency $\nu_{ED} = \nu_E(1 + v/c)$. This Doppler-shifted frequency is shown by the red vertical line in Figure 2(b). If v is adjusted so that $\nu_{ED} = \nu_A$, then the absorber will resonantly absorb the emitter photons. Thus, by measuring the velocity at which the absorption is at its maximum, we will get the value of $(\nu_A - \nu_E)$. If $\nu_A < \nu_E$, the emitter will have to be moved away from the absorber. This is the principle of Mossbauer spectroscopy. If v is 1 mm/s, the shift in the energy of the γ-ray photon from Fe^{57} will be 4.8×10^{-8} eV, which is more than ten times the half-width of the line. One can therefore measure such small shifts in energy using Mossbauer spectroscopy.

2. Emitter Source in Mossbauer Spectroscopy

The γ-ray source for Mossbauer spectroscopy must satisfy the following conditions:

(a) It should be a nucleus formed in the excited state of the source.
(b) It should have a large lifetime so that the half-width of the excited state is of the order of nanoelectron volts.
(c) The γ-ray energy should be sufficiently low to gain a good signal-to-noise ratio.

$_{26}Fe^{57}$ (mass number of 57 and atomic number of 26) is a commonly used source. $_{26}Fe^{57}$ nucleus in the excited state $I = 5/2$ is obtained when a $_{27}Co^{57}$ nucleus decays by capturing a K electron. The Fe^{57} nucleus comes down from the excited state $I = 5/2$ to $I = 3/2$ by emitting a γ photon of

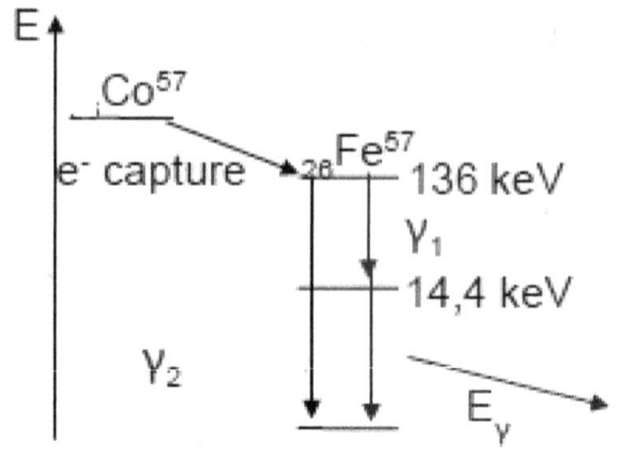

Figure 3. Mossbauer transition in a source of $_{27}Co^{57}$.

136 keV. It makes a second transition from $I = 3/2$ to the $I = ½$ ground state, emitting the required photon of 14.4 keV for Mossbauer spectroscopy. Only about 11% of the decay from $I = 3/2$ to $I = ½$ occurs with the γ-ray emission. This is shown in Figure 3.

The half-life of Co^{57} is 271 days. Co^{57} is diffused into a matrix material, such as rhenium. Rhenium has a face-centered cubic lattice, and cobalt atoms are added as dilute substitutional impurities in rhenium. **Every cobalt atom has the same surroundings in the pure rhenium metal. So, the γ rays emitted by all Fe^{57} have precisely the same energy.** Though this source (cobalt diffused in rhenium) is radioactive and requires protection for the handlers of the source, the samples (absorbers) used in the measurement are not radioactive.

The γ-ray photons from the isotopes Sn^{119}, Sb^{121}, and I^{129} are also used in Mossbauer spectroscopy.

3. Experimental Setup for Mossbauer Spectroscopy

Figure 4 shows a schematic diagram of the arrangement for Mossbauer spectroscopy.

The velocity drive is an electromagnetic drive which moves the source with constant acceleration so that the velocity varies linearly with time. A feedback mechanism maintains the constant acceleration. The velocity varies from −v to +v mm/s back and forth at a frequency of the order of 20 Hz. A velocity range of up to 10 mm/s covers the shifts to be observed commonly in Mossbauer spectroscopy. The data are collected

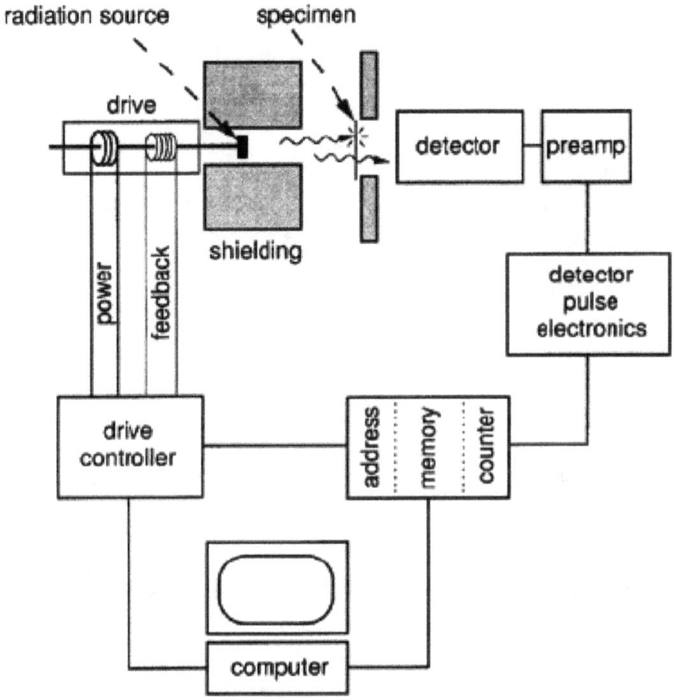

Figure 4. A schematic diagram of the arrangement for Mossbauer spectroscopy (reprinted with permission from Ref. 2).

repetitively in a multichannel detector. Each channel collects data over a 50 μs interval, corresponding to a definite velocity.

Conventional X-ray detectors can be used with the relatively low-energy γ-ray photons from the Fe^{57} nuclei. Thin scintillation detectors, gas-filled proportional counters, and solid-state detectors can be used.

The absorption spectrum is a convolution of the absorption signal with the half-width Lorentzian spectrum of the source. Figure 5 shows the absorption as a function of the velocity of the source.

One can achieve a resolution of the order of 0.3 mm/s in the velocity, i.e., a resolution of 1.5×10^{-8} eV in energy.

4. Factors Responsible for Shift in Energy Levels

The energy levels of the same nucleus in different environments will be shifted slightly. This shift is responsible for absorption at different

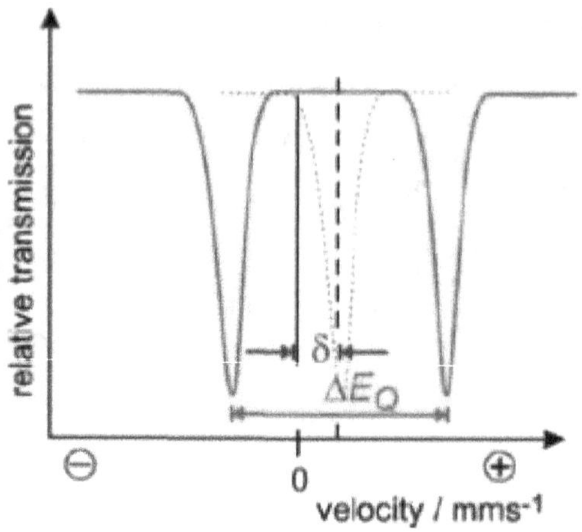

Figure 5. Mossbauer spectrum of Fe^{57} showing quadrupole splitting (reprinted with permission from Ref. 3).

velocities in the Mossbauer spectrum. The various contributions to such a shift are briefly described in the following.

4.1. *Isomer shift*

The different excited states of a nucleus are called isomers. Isomers have slightly different radii. The s electrons in the atom have a nonzero electron density at the nucleus. The interaction energy between the positively charged nucleus and the negative charge density of the s electrons will be different for isomers with different radii.

In Fe^{57}, the radius in the state $I = 3/2$ is more than the radius in the state $I = ½$. The isomer shift is negative. On the other hand, in Sn^{119}, the isomer in the excited state has a smaller radius than the isomer in the ground state. In Sn^{119}, the isomer shift is positive.

In Fe, the 3d electrons screen the 4s electrons from the nucleus. If the number of 3d electrons is reduced, this screening effect is reduced, increasing the s electron density at the nucleus. This makes the isomer shift more negative. So, when the oxidation state of Fe changes, the s electron density at the nucleus changes. The isomer shift in the energy of the γ-ray photon arising from the transition from $I = 3/2$ to $I = ½$ will change depending on the oxidation state of Fe. This is shown in Figure 6. In K_2FeO_4, Fe is in the oxidation state 6+, in Na_4FeO_4, it is in the oxidation state 4+, and in $KFeO_2$, it is in the oxidation state 3+. The isomer shifts

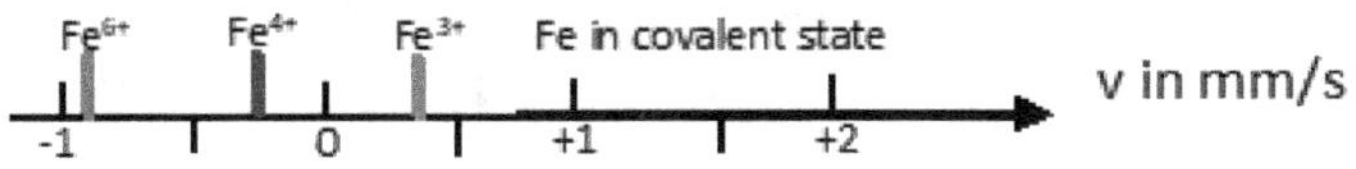

Figure 6. Isomer shift in Fe^{57} for different oxidation states.

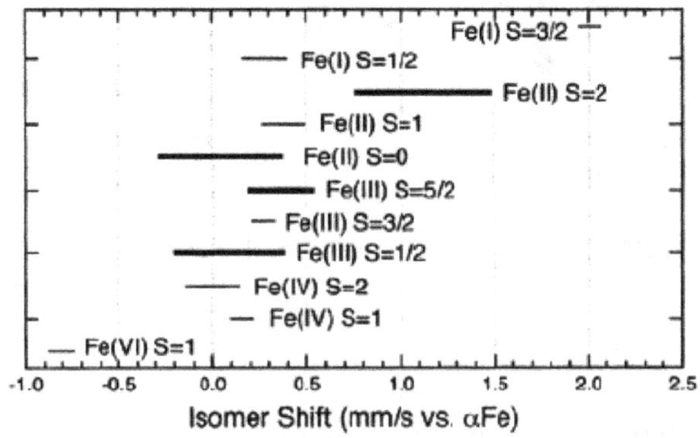

Figure 7. Range of isomer shifts for iron atoms in different oxidation states compared to BCC Fe (*α*Fe) (reprinted with permission from Ref. 2).

measured in velocity in mm/s occur at different values. Thus, the isomer shift depends on the oxidation state of the iron atom. The higher the oxidation state, the larger the magnitude of the negative isomer shift. Fe in its covalent state has a positive isomer shift.

The range of isomer shifts for the different oxidation states of Fe is shown in Figure 7.

If the same iron atom is in a different oxidation state due to different sites occupied in the material, the isomer shift may be used to find the fraction of atoms in different oxidation states. The isomer shift has proven most useful for studies of ionic or covalently bonded materials, such as oxides and minerals.

4.2. *Nuclear quadrupole splitting*

When the nuclear spin is more than ½, the nuclear charge distribution is asymmetric. There is no nuclear electric dipole moment. So, there is no interaction of the dipole moment with the electric field produced by the electron cloud. The nucleus will have a quadrupole moment. The charge

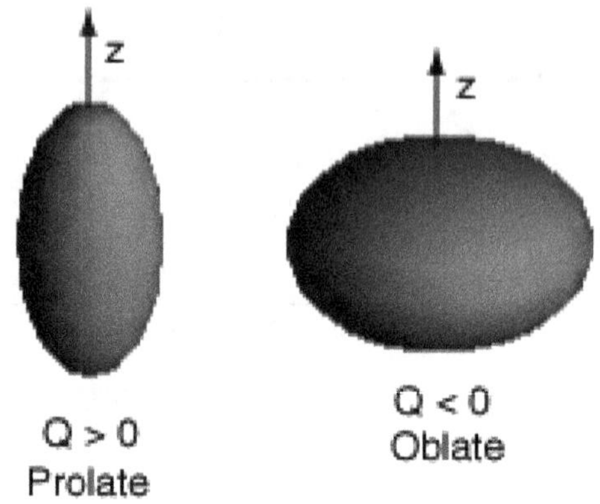

Figure 8. Charge distribution in a nucleus with $I > 1/2$.

distribution for a nucleus with a spin of $I > 1/2$ is either a prolate or an oblate spheroid, as shown in Figure 8.

Such a charge distribution is defined by a parameter, Q, which depends on the projection m_I of the spin quantum number I. For a prolate spheroid, the parameter Q is positive, while for an oblate spheroid, Q is negative. The electronic charge distribution in the atom, as well as in its nearest neighbors, will produce a nonuniform electric field $\boldsymbol{E}$ on the nucleus. The nuclear quadrupole moment will interact with the gradient of the electric field. This field gradient is again a second-rank symmetric tensor. One can choose a set of principal axes so that only the components $\partial E_x/\partial x$, $\partial E_y/\partial y$, and $\partial E_z/\partial z$ are nonzero and the cross components, such as $\partial E_x/\partial y$, are zero. Since div($\boldsymbol{E}$) = 0, the electrical field gradient is specified by only two parameters, $\partial E_z/\partial z$ and $\eta = (\partial E_x/\partial x - \partial E_y/\partial y)$. For the Fe^{57} nucleus, the excited state $I = 3/2$ is split into two states, with $m_I = \pm 3/2$ as one state and $m_I = \pm 1/2$ as the other. The energy level $I = 3/2$ is displaced by

$$\Delta E = \pm(1/4)\,(\partial E_z/\partial z)eQ[1 + (\eta^2/3)]^{1/2} \tag{5}$$

The + sign is for $m_I = \pm 3/2$, and the – sign is for $m_I = \pm 1/2$.

Figure 9 shows the quadrupole split levels in Fe^{57}. In the ground state, $I = ½$, the nuclear charge distribution is spherical. There is no nuclear quadrupole interaction. But there is a small isomer shift. The two states, $m_I = \pm 1/2$, have the same energy. In the excited state $I = 3/2$, the

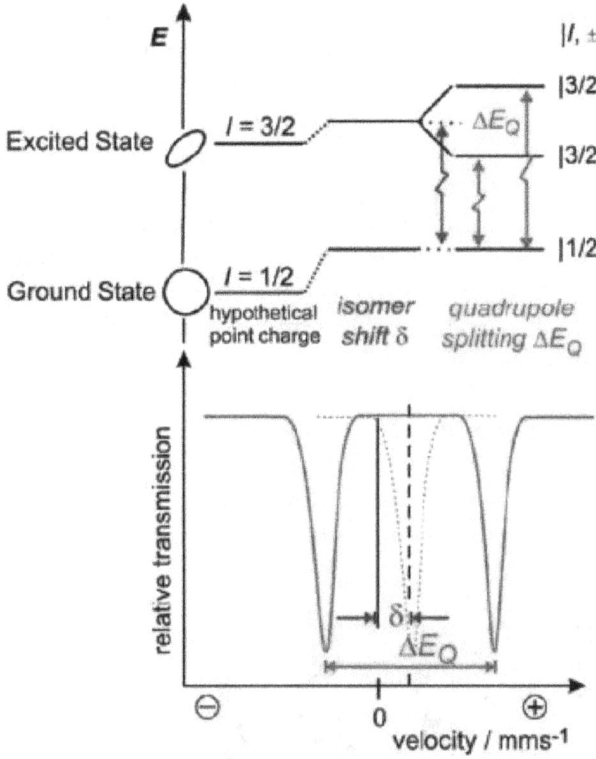

Figure 9. Illustrating nuclear quadrupole splitting in Fe^{57} (reprinted with permission from Ref. 3).

charge distribution is asymmetric. All the four states $m_I = \pm 3/2, \pm 1/2$ have the same isomer shift, which is different from the shift in the ground state. In addition, the nuclear quadrupole interaction splits the states $m_I = \pm 3/2$ from the states $m_I = \pm 1/2$. The resulting Mossbauer spectrum is not symmetric about the velocity $v = 0$ because of the isomer shift, as shown in the lower portion of the figure.

The electrical field gradient will arise both from the unsymmetrical electronic charge distribution in the atom and from the charge distribution of its neighbors. The Fe^{2+} ion has six 3d electrons. In the free ion, the six 3d electrons are distributed in the five degenerate 3d levels with a total spin of zero. Suppose it is in a complex in which it is bound to six identical ions at the corners of an octahedron. The crystalline electric field in which the 3d electrons move will now have an octahedral symmetry, which has three four-fold axes of symmetry directed from the center to the corners of the octahedron. In such a symmetry, the five degenerate d states of the free ion will be split into a set of two levels, one of which is doubly degenerate and the other triply degenerate. The doubly degenerate levels are labeled e_g and the triply degenerate levels t_{3g}. The e_g levels have a higher energy than the t_{3g} levels. The energy-level separation is called Δ. In octahedral symmetry, the electrical field gradient is zero. There will be no quadrupole splitting. The six electrons of Fe^{2+} will be distributed in the three degenerate t_{3g} levels. In each level, the electrons will have opposite spins. If Δ is small, this arrangement of spins will be stable. The total spin of Fe^{2+} will be zero. This is called a low-spin (LS) state. In this state, the Mossbauer spectrum will exhibit only the isomer shift, which will be small. This is the case in $K_2Fe(CN)_6$, as shown in Figure 10.

Suppose we replace one of the CN groups with NO. Then, the molecule has only one four-fold axis of symmetry along the line joining Fe and NO. The structure of this molecule is shown in Figure 11(a). This

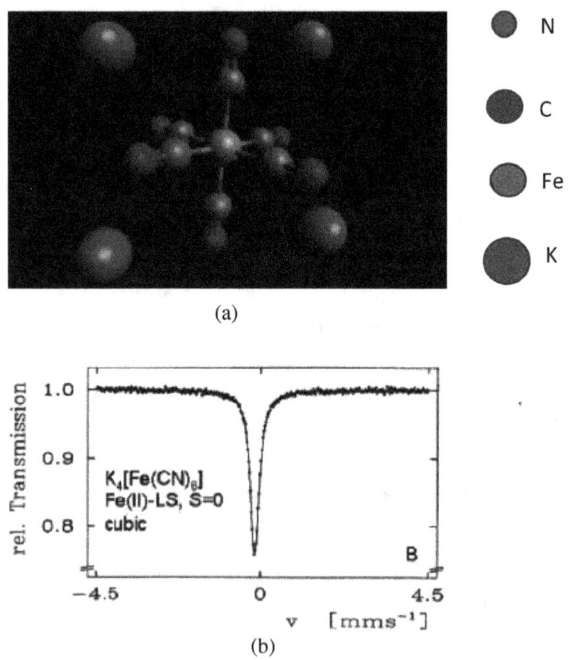

Figure 10. **(a) Structure of $K_4Fe\ (CN)_6$. (b) Mossbauer spectrum of Fe (reprinted with permission from Ref. 3).**

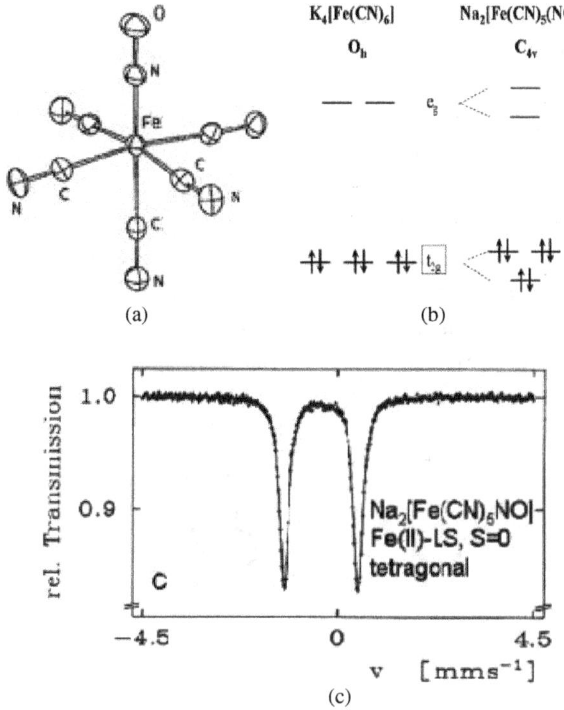

Figure 11. (a) Structure of Fe $(CN)_5(NO)^{2-}$. (b) Energy-level diagram. (c) Mossbauer spectrum (reprinted with permission from Ref. 3).

structure has tetragonal symmetry. In this tetragonal structure, the two e_g levels of the same energy in the octahedral coordination will split into two different energy levels. The three degenerate t_{2g} levels in the octahedral coordination will split into two levels, one singly degenerate and the other doubly degenerate. This is shown in Figure 11(b). The Fe^{2+} is still in the LS state because the six d electrons occupy the three t_{2g} levels in pairs. This arrangement of the valence 3d electrons does not produce an unsymmetric electronic charge distribution. The electrical field gradient arises from the charge distribution around the inequivalent neighbors of Fe. This field gradient is small. The Mossbauer spectrum of $Na_2Fe(CN)_5NO$ will be a doublet with a small shift between the two lines due to the nuclear quadrupole interaction. This is shown in Figure 11(c). There are two lines which are not symmetric about $v = 0$ due to the isomer shift.

Now consider $FeSO_4$ $7H_2O$. Here again, the six H_2O groups can be arranged at the corners of an octahedron with the Fe^{2+} ion at the center.

In this arrangement, the separation between the e_g and t_{2g} levels due to the electric field of the six H_2O groups is large. This arrangement is unstable. It undergoes a Jahn–Teller distortion, which is a contraction of the octahedral arrangement along one of its axes, say z. In this Jahn–Teller distorted state, the symmetry is tetragonal.

The degenerate energy levels e_g split into two separate energy levels, and the degenerate energy levels t_{2g} split into a singly degenerate and a doubly degenerate level. Now, the six electrons are distributed among the levels in the Jahn–Teller distorted structure, as shown in Figure 12(a). The total spin is 2, and it is a high-spin (HS) state. The d electron charge distribution is now asymmetric and produces a large electric field gradient.

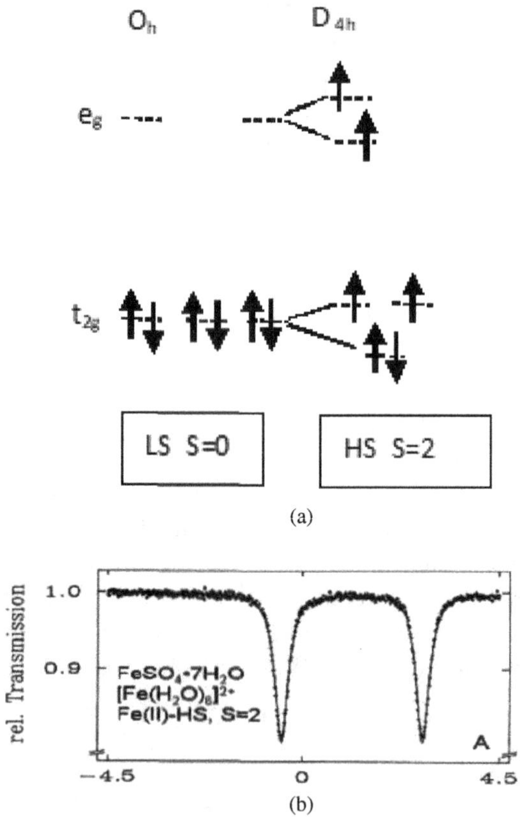

Figure 12. (a) Energy levels in the Jahn–Teller distorted state. (b) Mossbauer spectrum of $FeSO_4.7H_2O$ (reprinted with permission from Ref. 3).

The resulting Mossbauer spectrum is shown in Figure 12(b), and the Mossbauer splitting is very much more than in Figure 11(c).

It was mentioned that when ions of different oxidation states exist in a material, one may distinguish between them using the isomer shift. It may happen that the isomer shift for the same ion in different oxidation states may be very close. Fe^{3+} has five electrons. In an octahedral or tetrahedral environment, each electron may occupy each of the five states ($2e_g + 3t_{2g}$) with spin up. This will be a HS state with $S = 5/2$. When all five states are fully occupied, the electrical field gradient due to the d electron charge distribution will be zero. However, the presence of the neighboring ions will create a small electrical field gradient. On the other hand, the HS state of Fe^{2+} will have a large field gradient. So, one can distinguish the peaks in the Mossbauer spectrum due to the Fe^{3+}, which is in the HS state, and Fe^{2+}, which in the LS state, if they occur together.

As the temperature rises, the electron populations in the different crystal field levels will change. This will cause the quadrupole coupling to change as the temperature increases.

4.3. *Application to mixed valence states*

As an example of the application of Mossbauer spectroscopy to mixed valent states, we consider the case of $Li_{0.6}FePO_4$ investigated by *Dodd et al.* (Ref. 4). Normally, when Li is introduced or removed from $LiFePO_4$, we get a mixed phase system. One phase is the lithiated triphylite phase ($LiFePO_4$), and the other phase is the de-lithiated heterosite phase (FePO4). In both phases, Fe is surrounded by oxygen in the octahedral coordination. In the triphyllite phase, Fe is in the Fe^{2+} HS state, while in the heterosite phase, iron is in the Fe^{3+} state. However, when the lithium concentration is 0.6, the disordered phase, prepared above 200°C, is stable when cooled to room temperature. The structure of $Li_{0.6}FePO_4$ is shown in Figure 13. The blue circles are lithium ions forming chains. The planes of the octahedron of oxygen surrounding the iron atom are shown in brown. The planes of the phosphate tetrahedra are shown in grey.

In this material, electron transport occurs by small polaron hopping. The electron is trapped by the Fe^{3+} atom, making it Fe^{2+}. It also causes a local contraction of the octahedral Fe–O bonds. Then, Fe^{3+} becomes Fe^{2+} in the HS state. After some time, the electron hops over to a neighboring Fe^{3+} site, carrying with it the deformation of the octahedron and converting Fe^{3+} to the Fe^{2+} HS state. The vacated Fe^{2+} HS site returns to the Fe^{3+} state.

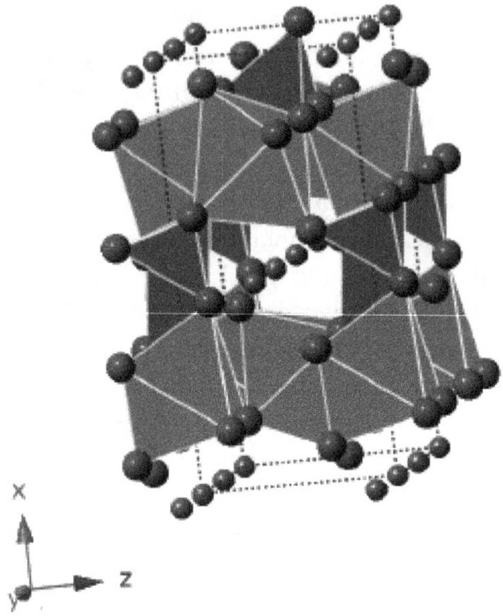

Figure 13. Structure of $Li_{0.6}FePO_4$.

The mobility of this polaron is small, leading to a very small electrical conductivity. The hopping rate follows an Arrhenius law with an activation energy. The hopping rate increases as the temperature increases. Therefore, the time τ between two sequential hopping events decreases as the temperature increases.

The Mossbauer spectra of the two-phase sample of Li_xFePO_4 (from Ref. 4) are shown at different temperatures in the bottom panel of Figure 14. In the top panel, the Mossbauer spectra taken at different temperatures of the stable, disordered $Li_{0.6}FePO_4$ are shown. In the two-phase sample spectrum in the bottom panel, the electrical quadrupole peaks of Fe^{2+} in the lithiated triphylite phase and Fe^{3+} in the dilithiated heterosite phase are marked. The low velocity peak of Fe^{3+} falls on the low velocity peak of Fe^{2+}. In the two-phase sample, the electrical quadrupole splitting of Fe^{2+} at 25°C is about 3 mm/s while that for Fe^{3+} is half as large. The centroid of the two quadrupole peaks gives the isomer shift of the Fe^{2+} and Fe^{3+} ions.

In Figure 15, the temperature variation of the quadrupole splitting and the isomer shift for the two valence states of iron in the two-phase sample are shown by dotted lines. Because the temperature of the sample

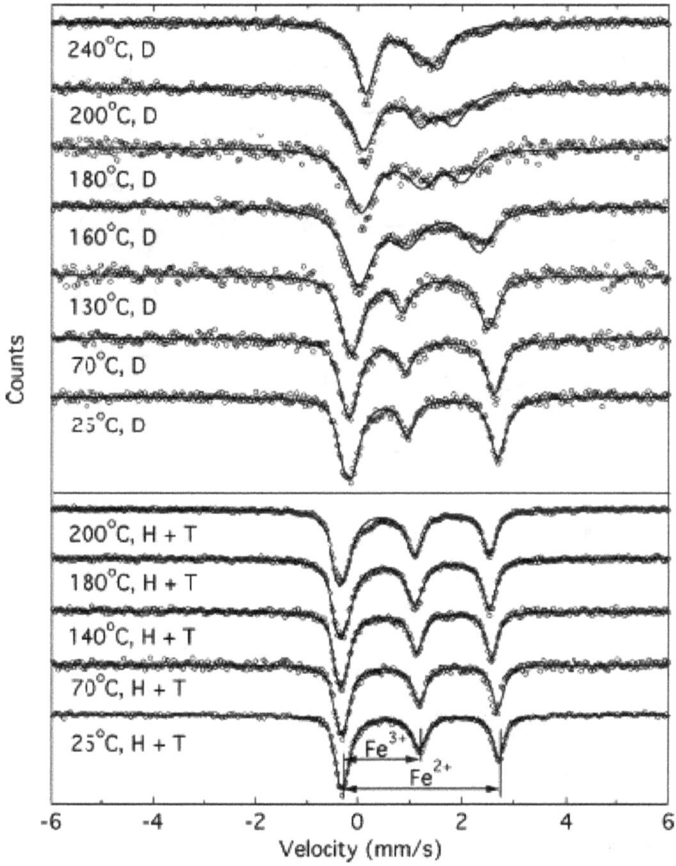

Figure 14. Mossbauer spectra in the stable, disordered phase of $Li_{0.6}FePO_4$ (upper panel) and the two-phase sample (lower panel) at different temperatures (reprinted with permission from Ref. 4).

is higher than the temperature of the source, there will be a small shift in the absorption lines. This is called the thermal redshift, which is plotted in Figure 15. If we take this redshift into account, the isomer shifts for the two states of iron are almost independent of temperature in the two-phase samples. The quadrupole splitting for both states of iron comes down as the temperature increases. This is because thermal energy raises the electrons from the lower-energy t_{2g} states to the higher-energy e_g states, causing a reduction in the electrical field gradient.

From Figure 15, we see that, up to 130°C, the isomer shift and the quadrupole splitting in the disordered stable phase of $Li_{0.6}FePO_4$ show a

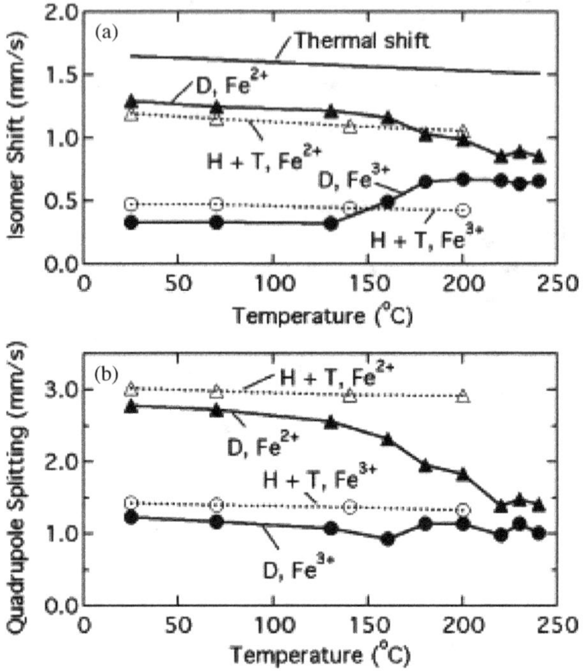

Figure 15. Temperature dependence of the isomer shift and quadrupole splitting in the two-phase samples (dashed lines) and the stable disordered phase of $Li_{0.6}FePO_4$ (reprinted with permission from Ref. 4).

behavior similar to that in the two-phase samples. Beyond this temperature, there is a sudden change in their behavior in the spectra for the disordered sample. The isomer shifts of Fe^{2+} and Fe^{3+} tend to approach each other in value. The quadrupole splitting also shows a similar behavior. The hopping rate of the electron increases as the temperature increases. There is a window of time for Mossbauer absorption ($=h/\Delta E_{QS}$). It is of the order of a few tens of nanoseconds. If the hopping time τ is much more than the Mossbauer window, one will see the difference in the two valence states of Fe distinctly. When the hopping time is much less than the Mossbauer time window, Fe^{2+} will relax to Fe^{3+} within the Mossbauer window. One cannot clearly differentiate between the two states. One will then see only the average effect of the two valence states. Thus, Mossbauer spectroscopy is of great utility in the study of valence fluctuations and mixed valent states.

4.4. *Splitting due to hyperfine interaction*

There is a hyperfine interaction between the nuclear spin and the electron spin, given by the Hamiltonian

$$H = -A\,\boldsymbol{I}.\boldsymbol{S} \tag{6}$$

A is the hyperfine constant, and $\boldsymbol{S}$ is the electron spin of the atom. We may interpret it as the energy of the magnetic moment $\gamma \boldsymbol{I}$ of the nucleus in the magnetic field of the electron with a magnetic moment of $g\mu_B\boldsymbol{S}$. The internal magnetic field of the electron will be related to A by

$$B = A/\gamma g\mu_B \tag{7}$$

Here, g is the Lande splitting factor for the electron and μ_B is the Bohr magneton.

The state m_I will have energy in the magnetic field given by $-\gamma m_I B$. The γ value will be different for the isomeric states. In Fe^{57}, γ is positive for the ground state and negative for the excited state $I = 3/2$. The ground state $I = ½$ will split into two states with $m_I = ½$ and $-½$. The excited state $I = 3/2$ will split into four equally spaced states, with $m_I = +3/2, +1/2, -1/2$, and $-3/2$. Since the γ value is different for the ground and excited states, the spacing between the magnetic ground states will be different from the spacing between the magnetic excited states. For the absorption of a photon, the selection rule is $\Delta m_I = 0$ or ± 1. So, there will be six possible transitions, giving rise to six Mossbauer absorption lines. This is shown in Figures 16(a) and (b).

One can determine the effective internal magnetic field at the nucleus from this splitting. In iron metal, this internal field is 33 T at 300 K. If we apply an external magnetic field, the internal field decreases. The external field produces a lattice magnetization in the specimen. The decrease implies that the internal field of 33 T is opposite in sign to the lattice magnetization.

This high value of the internal magnetic field of the nucleus originates from a special type of interaction called the Fermi contact interaction. The effective magnetic field at the nucleus due to this interaction can be written as

$$B_{eff} = -(2\mu_0/3)\, g_e\mu_B S\,|\Psi(0)|^2 \tag{8}$$

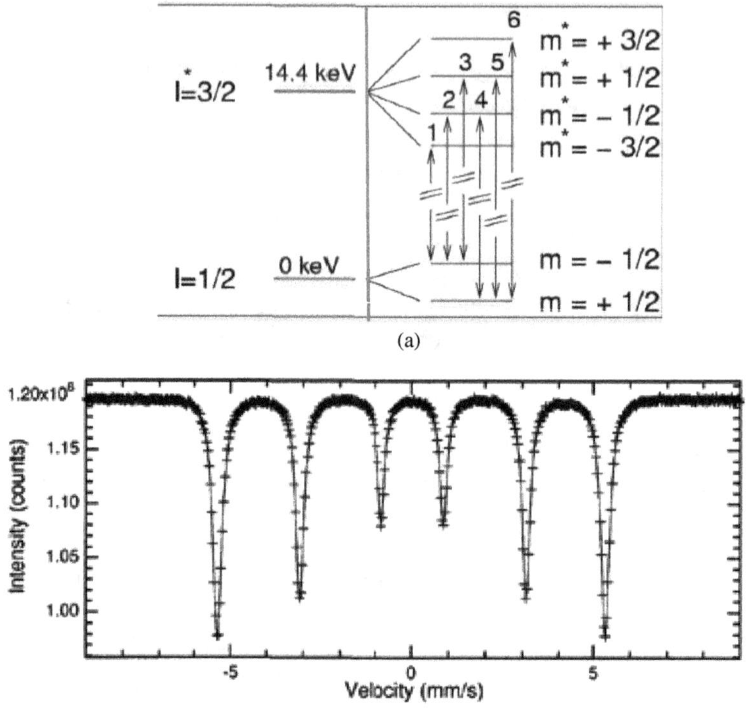

Figure 16. (a) Splitting of the levels due to internal magnetic field *B* in BCC Fe metal. (b) Resulting Mossbauer spectrum in BCC Fe at 300 K (reprinted with permission from Ref. 3).

Here, $\mu_0 = 4\pi \times 10^{-7}$ N/A^2, g_e is approximately 2, and $S\,|\Psi(0)|^2$ is the electron spin density at the nucleus. The internal field will exist only when S ≠ 0. In Fe, it is the 2s electrons which make the largest contribution to the spin density at the nucleus. But the 2s electrons are paired. So, one should not have any internal magnetic field. However, the 2s electrons have an exchange interaction with the 3d electrons. The magnetic moment of the iron atom arises from unpaired 3d electrons. The exchange interaction between a d up-spin electron and an s up-spin electron reduces the repulsive interaction between the two electrons. A reduction in the repulsive interaction causes the wave function of the 2s electron to expand outward, reducing the spin density at the nucleus. Similarly, a down-spin d electron reduces the spin density of the down-spin 2s electron nucleus. Since there are more up-spin d electrons compared to down-spin d electrons in the ferromagnetic iron atom, the 2s up-spin electron density is reduced more than the 2s down-spin electron density. This is the reason for a nonzero Fermi-contact interaction at the nucleus of the iron atom.

Mossbauer spectroscopy has been used to study a number of iron alloys and the variation of the hyperfine field due to alloying elements found. As an example, let us take Fe–Ni alloys. Below a concentration of 30% nickel, the alloys have the BCC α-Fe structure. Above a concentration of 40% nickel, the alloys have the FCC γ-Fe structure. In between, we have both BCC and FCC structures. The Mossbauer spectrum of three alloys with nickel compositions of 30%, 35%, and 40% taken at 80 K by Lehlooh and Mahmood (Ref. 5) is shown in Figure 17. In each case, we get

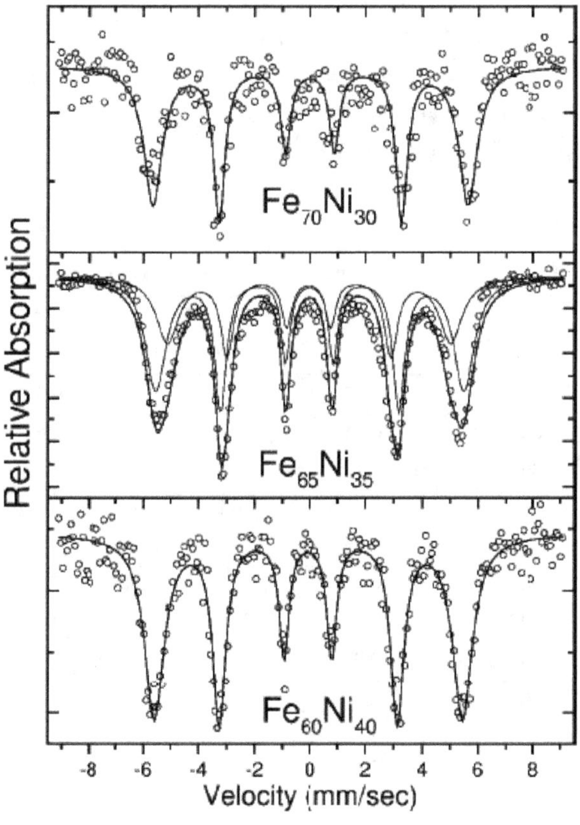

Figure 17. Mossbauer spectra of Fe–Ni alloys with the Ni concentrations of 30%, 35% and 40% at 80 K (reprinted with permission from Ref. 5).

a six-component Mossbauer spectrum. In the case of Ni 30 and Ni 40 concentrations, the spectrum was fitted with a single six-component hyperfine spectrum to get the hyperfine fields. In the case of a 35% Ni concentration, X-ray diffraction indicates the presence of both BCC and FCC phases. The Mossbauer spectrum was therefore fitted with two six-component spectra. The results are given in Table 1.

In the two extreme cases where there was only one phase, the hyperfine field increases slightly as the temperature is lowered. In the case of Ni 35%, the spectrum was fitted with two sextet components. One of the components showed a hyperfine field similar in value to those observed in the two extreme compositions. But the hyperfine field shown by the second component decreases drastically from 27.2 T at 80 K to 16.3 T at 300 K. The authors attribute this to fast relaxations in this sample at 300 K. In the sample with 30% composition, they saw a single component superposed on the sextet in the 300 K spectrum. This single component is due to a paramagnetic phase with $S = 0$ and a negative isomer shift of –0.08 mm/s. This component vanishes at 80 K.

Table 2 gives the hyperfine fields in some oxides and oxyhydroxides of iron.

Table 1. Magnetic hyperfine fields in some Fe–Ni alloys (Ref. 5).

Ni%	Temperature	Isomer shift (mm/s)	B HF (T)	Width	%	
30	80 K	0.16	35.4	0.8	100	
	300 K	0.04	34	0.8	86	
35	80 K	0.12	31.8	0.98	39	Comp I
	300 K	–0.03	16.2	1.32	39	
	80 K	0.15	34.6	0.89	61	Comp 2
	300 K	–0.01	27.3	1.13	61	
40	80 K	0.11	34.6	0.85	100	
	300 K	–0.01	30	0.94	100	

Table 2. Hyperfine magnetic field, quadrupole splitting, and isomer shift in some oxides and oxyhydroxides of iron (Ref. 2).

Compound (Fe Site)	HMF (T)	Q.S.	I.S. (vs. Fe)	Temp. (K)
α-FeOOH	50.0	–0.25		77
α-FeOOH	38.2	–0.25	+0.61	300
β-FeOOH	48.5	0.64	+0.38	80
β-FeOOH	0	0.62	+0.39	300
γ-FeOOH	0	0.60	+0.38	295
δ-FeOOH (big xtls.)	42.0		+0.35	295
FeO		0.8	+0.93	295
Fe_3O_4(Fe(III),A)	49.3		+0.26	298
Fe_3O_4(Fe(II,III),B)	46.0		+0.67	298
α-Fe_2O_3	51.8	+0.42	+0.39	296
γ-Fe_2O_3(A)	50.2		+0.18	300
γ-Fe_2O_3(B)	50.3		+0.40	300

[a]Abbreviation: HMF, hyperfine magnetic field: I.S., isomer shift; Q.S., quadrupole splitting; T, tesla.

5. Conclusion

Mossbauer spectroscopy finds applications in the following:

(1) solid-state reactions,
(2) structural properties and bonding,
(3) mixed valency,
(4) magnetic properties.

It is an important tool for studies in materials science. For further information, the following references may be consulted.

References

1. Gütlich, P. Mossbauer Spectroscopy — Principles and Applications, www.blogs.uni-mainz.de>files>Mossbauer_lectures.

2. Schultz, B. Introduction to Mossbauer Spectrometry, www.aps.ani.gov> pdf > MossbauerSpectrometry_Fulz.
3. Lennert, A., Bronnbauer, C., Kränzlein, E., and Krebs, K. Mössbauer Spectroscopy Advanced Inorganic Chemistry Seminar Presentation in WS 2011/2012, https://hugepdf.com> mssbauer-spectroscopy_pdf.
4. Dodd, J. L., Halevy, I., and Fultz, B. (2007). Valence fluctuations of Fe^{57} in disordered $Li_{0.6}FePO_4$. *J. Phys. Chem. Lett.* **111**, 1563.
5. Lehloo, A.-F. D. and Mahmood, S. H. (2002). Mossbauer spectroscopy study of Fe-Ni alloys. *Hyperfine Interact.* **139–140**, 387.

Part V

Phase Transition

Chapter 20

PHASE TRANSITIONS

1. Introduction

Phases are states of matter in thermodynamic equilibrium and are characterized by uniform macroscopic properties, such as pressure and temperature. The most common example familiar from everyday experience is the states of aggregation of H_2O, namely ice (solid), water (liquid), and water vapor (gas). The melting of ice to water and the boiling of water to steam are typical examples of a phase transition involving solid–liquid and liquid–gas transformations. There are numerous examples of solid–solid phase transitions involving a change in crystal symmetry. In the realm of magnetism, there is a rich variety of phase transitions. If we take a piece of nickel which is magnetized, one is familiar with the observation that it attracts other nickel or iron pins. However, if we heat nickel to around 360°C, it will lose all its magnetic properties and the attached pins will fall away. This is the well-known **ferromagnetic–paramagnetic** phase transition occurring at the Curie temperature T_C. In the class of materials called ferroelectrics, we have the analogous **ferroelectric–paraelectric** phase transition, occurring at a characteristic Curie temperature. In both of these cases, the magnetic moments or the electric dipoles are disordered above T_C and order sets in when we cool the system below T_C. Another example of such an order–disorder phase transformation occurs in binary alloy systems consisting of, say, A and B types of atoms. If they are distributed at random over the lattice sites, we have a disordered alloy at high temperature, and as we cool the alloy, order develops at a characteristic ordering temperature, wherein the positions of the two types of atoms alternate in the lattice. In all these phase transitions, the ordering is in real space with well-defined spatial coordinates. Another type of phase transition occurs in quantum systems, where the ordering is in momentum space, or *k*-space. The **superfluid** phase transition in liquid helium near 2.17 K is associated with the transformation from a normal liquid to a new phase, exhibiting remarkable flow properties such as zero viscosity. In the superfluid phase, the atoms are in a coherent zero-momentum state, which leads to frictionless flow. The **superconducting** phase transition in some metals is again dramatic in that the electrical resistivity completely vanishes below a characteristic temperature. Here, electrons of opposite spin near the Fermi surface condense to form what are called "Cooper pairs" in a zero-momentum state. This ordering in the momentum space, or *k*-space, leads to the frictionless flow of electron pairs, resulting in an electric current without the appearance of a potential difference. A comprehensive treatment of the theory of phase transitions can be found in Refs. 1 and 2.

This chapter provides a basic introduction to the vast field of phase transitions centered around the concept of a "chemical potential." Thermodynamics provides a comprehensive framework to understand phase equilibrium and the criterion for a phase transition. Differential thermal analysis and differential scanning calorimeter, which are used extensively in the study of phase transitions, are described in some detail.

2. Gibbs Free Energy and Chemical Potential

In thermodynamics, we construct several state functions out of basic quantities, such as internal energy ***U***, pressure ***P***, temperature ***T***, and entropy ***S***. Among them, the Gibbs free energy ***G*** is central to our discussion on phase transitions due to its most important feature that it is dependent on the experimental variables, such as pressure and temperature, which can be easily controlled in a laboratory. We define the Gibbs free energy ***G*** as

$$G = U - TS + PV \tag{1}$$

In the differential form, the above expression leads to the relation

$$\mathrm{d}G = \mathrm{d}U - T\mathrm{d}S - S\mathrm{d}T + P\mathrm{d}V + V\mathrm{d}P \tag{2}$$

The first law of thermodynamics leads to the expression

$$dU = TdS - PdV \quad (3)$$

Thus,

$$dG = -SdT + VdP \quad (4)$$

The natural variables of G are thus temperature T and pressure P. The relevant partial derivatives of G correspond to physical quantities, such as entropy and volume, through the relations

$$S = -\left(\frac{\partial G}{\partial T}\right)_P \quad \text{and} \quad V = \left(\frac{\partial G}{\partial P}\right)_T \quad (5)$$

In thermodynamics, we define a "**phase**" as a region of space throughout which the properties of a material are uniform. This is used synonymously with the state of matter. As a typical example, we are familiar with the three equilibrium phases of a single-component system, e.g., for H_2O, they are ice, water, and water vapor. The condition of equilibrium of a phase can be derived using the cardinal principle that the total entropy change according to the second law of thermodynamics is given by the relation

$$dS_{\text{total}} = dS_{\text{system}} + dS_{\text{Surroundings}} \geq 0 \quad (6)$$

Thus, for a system at constant pressure and temperature, the stable equilibrium state is given by the condition that the Gibbs free energy function G_0 is at its minimum. Hence, one can derive the necessary stability condition against temperature variations as $\left(\frac{\partial G}{\partial T}\right)_v = 0$ and the second derivative $\left(\frac{\partial^2 G}{\partial T^2}\right) \geq 0$.

Similar stability conditions can be derived for variations with pressure.

3. Chemical Potential

We now consider another important concept called "chemical potential," which determines the stability of phases, the tendency to chemically react to form new species, or to provide a driving force for species to diffuse from one locale to another. It is closely related to Gibbs free energy generalized to cases wherein the number of species/particles can change. The generalized Gibbs free energy in differential form has the form

$$dG = VdP - SdT + \mu dN \tag{7}$$

where N is the total number of particles in the system. μ is called the chemical potential and is related to the Gibbs free energy through the relation

$$\mu = \left(\frac{\partial G}{\partial N}\right)_{T,P} \tag{8}$$

It is important to note that G is an "extensive" property as it scales with the size of the system under study. However, μ is an "intensive" property like pressure and temperature. To establish a link between μ and G, let us consider a simple system whose internal energy U can be considered to be consisting of three contributions, namely, thermal energy U_T, mechanical energy U_M, and chemical energy U_C. Thus, we have the relation

$$U = U_T + U_M + U_C \tag{9}$$

It is straightforward to show that $U_T = TS$, and the temperature T ($=U_T/S$) plays the role of a "thermal potential," which is the thermal energy content of one unit of thermal matter with one unit of entropy S. Similarly, $U_M = -PV$. Here, pressure, P, plays the role of a "mechanical potential," which is the mechanical energy stored in one unit of volume ($P = -U_M/V$). The chemical energy term U_C is directly proportional to the number of moles N of the substance under study, and one can write it as $U_C = \mu N$, where μ is defined as the chemical potential of the species under consideration. Thus, one gets the relation

$$U = TS - PV + \mu N \tag{10}$$

Since the Gibbs free energy, $G = U - TS + PV$, we get the important relation for the chemical potential μ, i.e., $\mu = \frac{G}{N}$. Therefore, the physical significance of the chemical potential is that it corresponds to the Gibbs free energy per mole of the substance and has the unit of J/mole. In analogy with electrical potential, we note that chemical potential should be distinguished from chemical energy.

4. Phase Equilibrium in a Single-Component System

It is instructive to discuss the stability and phase transitions between different phases in a typical single-component or -species system comprising H_2O molecules. For simplicity, let us specify that we are dealing with the case where the pressure is held constant at, say, the ambient pressure of 1 bar. Then, at temperatures lower than 0°C, one has **ice** as the stable phase, while **water** is the stable phase in the temperature interval of 0–100°C, and **water vapor** is the stable phase beyond this temperature range. We will now show that it is the **chemical potential** and its variation with temperature that provide an adequate description of the stability of the phases and the phase transition between the phases. Since phase equilibrium is dictated by the condition that the chemical potential is at minimum, we have the following conditions for the various phases:

If $\mu_S\,(T, P) < \mu_L\,(T, P)$, then ice is the stable phase.
If $\mu_L\,(T, P) < \mu_S\,(T, P)$, then water is the stable phase.
If $\mu_L\,(T, P) = \mu_S\,(T, P)$, then water and ice coexist and are in equilibrium.

The fundamental expression for μ is the same as the one for Gibbs free energy, thus

$$d\mu = -SdT + VdP \tag{11}$$

Since pressure is held constant, μ depends only on temperature, and we have

$$\left(\frac{d\mu}{dT}\right)_P = -S \tag{12}$$

Here, S is the entropy of the system, and from the third law of thermodynamics, we know that it is always a positive definite quantity. Thus, the slope of the μ versus T plot would always be negative. We also note that entropy is a measure of the "disorder" in the system, and hence, S_{gas} >

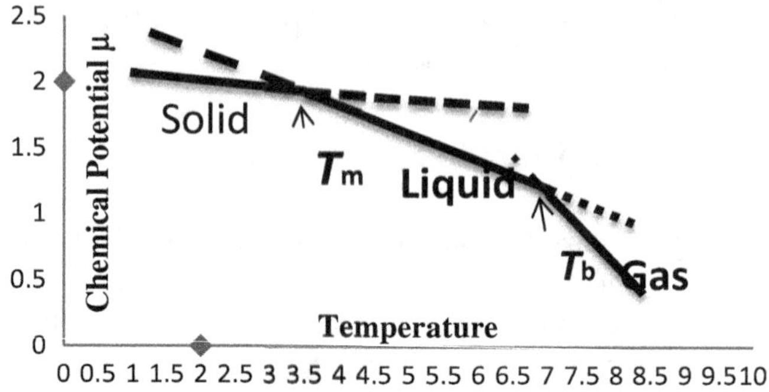

Figure 1. Schematic diagram of the variation in the chemical potential with temperature in the solid, liquid, and gaseous states of matter.

$S_{\text{liquid}} > S_{\text{solid}}$. Thus, the slope of the μ versus T plot has the largest value for the gas phase, an intermediate value for the liquid phase, and the least value for the solid phase. The variation in the chemical potential with temperature for the solid, liquid, and gaseous phases of matter is shown schematically in Figure 1. The intersection of the solid and liquid curves at a temperature T_{m} (called the melting temperature) signifies that their chemical potentials are the same, and hence, the two phases coexist and are in equilibrium. Similarly, the intersection of the liquid and gaseous phase curves signifies the phenomenon of "boiling point" at a characteristic temperature of T_{b}, wherein the water and vapor phases coexist and are in equilibrium.

5. Phase Diagram

The delineation of different phases in the pressure–temperature plane leads to the phase diagram of the system under study. Figure 2 depicts a typical phase diagram of a single-component system, wherein the different lines signify the coexistence of the relevant phases. For example, the "melting line" signifies that the solid and liquid phases coexist all along this line, and the chemical potential for the two phases would be identical. Thus, we have

$$\mu_{\text{S}}(T, P) = \mu_{\text{L}}(T, P) \tag{13}$$

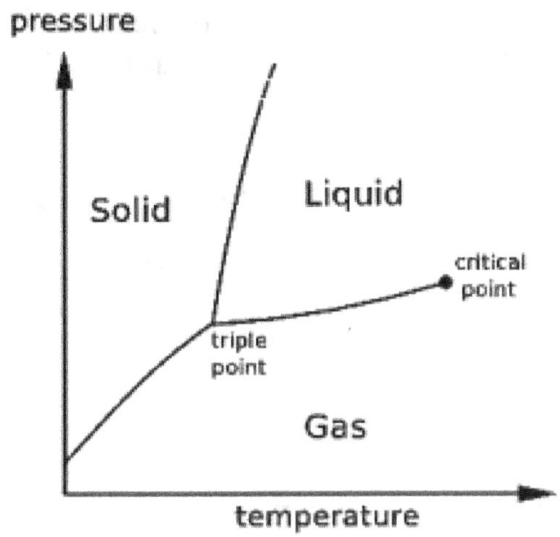

Figure 2. Typical phase diagram of a single-component (species) system depicting the sublimation, melting, and boiling phenomena. Note that there are two special points in the diagram, namely "triple point" and "critical point."

This central equation with two variables (T, P) signifies that the coexistence of the two phases can be described by either $T(P)$ or $P(T)$ along the melting line.

At the triple point, all three phases coexist with identical chemical potential. The triple point at (T_t, P_t) is a unique point in the phase diagram, and for H_2O, $T_t = 273.16$ K and $P_t = 0.006$ bar.

The **critical point** in the phase diagram is the termination of the liquid–gas coexistence curve. The physical significance of the critical point is that beyond this point, the liquid and gaseous phases lose their identities and merge into a single phase called the fluid phase.

6. Clausius–Clapeyron Relation and the Shape of the Coexistence Curves

The Clausius–Clapeyron relation is a powerful equation relating the slope of the coexistence curves in the P–T diagram to thermodynamic state functions, such as entropy and molar volume. Let A and B be the two phases, and their chemical potentials would be identical at a given point (T, P) on the coexistence curve. Thus,

$$\mu_A(T, P) = \mu_B(T, P) \tag{14}$$

Moving along the coexistence curve to a neighboring point ($T + dT$, $P + dP$) and noting that the same relation holds good for the new coordinates, we get

$$d\mu_A = d\mu_B \tag{15}$$

Using the differential form of the chemical potential,

$$d\mu = -SdT + VdP \tag{16}$$

one gets the relation

$$-S_A\, dT + V_A dP = -S_B dT + V_B dP \tag{17}$$

leading to the Clausius–Clapeyron equation:

$$\left(\frac{dP}{dT}\right) = \left[\frac{S_A - S_B}{V_A - V_B}\right] = \left[\frac{\Delta S}{\Delta V}\right] \tag{18}$$

where ΔS and ΔV are the entropy and volume difference between phases A and B at the point of coexistence, respectively. This equation has profound implications for the slopes of the coexistence curves in the phase diagram, which we now analyze.

The solid (S) – liquid (L) phase boundary has a very steep **positive** slope due to the fact that the molar volumes for the solid and liquid are nearly the same, with $V_L \geq V_S$ and $S_L > S_S$. However, for the important case of ice–water phase boundary, the slope is negative because $V_L < V_S$ (i.e., water is denser than ice).

The liquid (L) – gas (G) phase boundary has the **smallest positive** slope due to the fact that $S_G > S_L$ and the molar volume of the gas is much larger than the liquid $V_G >> V_L$.

The solid (S) – gas (G) phase boundary, called the sublimation curve, also has a **positive** slope, whose magnitude is intermediate between that of the other two phase boundaries and arises due to the large magnitudes of both the entropy change and the volume change. Here, $S_G >> S_S$ and $V_G >> V_S$.

6.1. *Latent heat*

The Clausius–Clapeyron equation can be recast into a more useful form by introducing the "**latent heat**," L. The expression for the chemical potential in terms of the "**enthalpy**" H has the form $\mu = H - TS$, and in equilibrium, $\mu_A = \mu_B$, leading to the equality

$$H_A - TS_A = H_B - TS_B \tag{19}$$

Hence, the discontinuous change in enthalpy is given by

$$\Delta H = H_A - H_B = T[S_A - S_B] \tag{20}$$

The quantity ΔH is called the **latent heat**. It is clear that the latent heat is the change in enthalpy per molecule. If we supply heat to water at, say, 25°C, its temperature will increase by an amount depending on its specific heat, and it will continue to increase until we reach the boiling point. At the boiling point, the heat supplied will not increase the temperature but will bring about a change of phase from liquid to vapor. The latent heat is thus the heat energy given to a system to change its phase. If the change in enthalpy is positive, the system under study absorbs heat from the surroundings, leading to an endothermic phase change. On the other hand, if heat is given out to the surroundings, then one has an exothermic phase change. Physically, one would also expect that the magnitude of the latent heat for the melting of ice should be much smaller than the latent heat for the vaporization of water for the simple reason that the entropy change in melting is much smaller as compared to the entropy change in vaporization. A typical value of L_{melting} is 6 kJ/mole, while for vaporization, $L_{\text{vapor}} = 44$ kJ/mole.

7. Symmetry Aspects of Phase Diagram

It is important to note that the liquid–gas phase boundary always terminates at a "**critical point**," whereas this feature is not present in the solid–liquid phase boundary, which runs all the way to infinity or is interrupted by a solid–solid phase boundary. Landau enunciated the profound symmetry principle that, during a phase change, there is either no change in symmetry or a change in symmetry. We now apply this principle to the liquid–gas phase boundary. The symmetry of both liquid and gas is the same, i.e., they are isotropic, and the difference between them is purely **quantitative** in the sense that their densities are different, but qualitatively both are disordered phases. The density difference between the two phases decreases continuously as one approaches the critical point and is precisely zero at the critical point. It is clear that one can choose a path in the P–T plane in such a way that one can go from the gaseous phase to the liquid phase without crossing the coexistence curve. The solid–liquid phase boundary cannot terminate at a critical point because the solid has a symmetry that is quite distinct from that of the liquid phase, and there is no way of going continuously from one symmetry to the other.

8. Classification of Phase Transitions

Ehrenfest proposed a classification scheme for phase transitions based on the degree of non-analyticity involved in the chemical potential or Gibbs free energy. In this scheme, phase transitions were labeled according to the lowest derivative of the free energy that is **discontinuous** at the transition. Thus, for a melting transition, such as ice–water, the chemical potentials for the two phases are identical at the melting temperature. The **first derivative** of the chemical potential, which corresponds to physical quantities, such as **entropy** and **volume**, are **discontinuous** at the melting temperature (Figure 3(a)). Such a type of phase transition is called "**first-order transition**," signifying the discontinuous change in

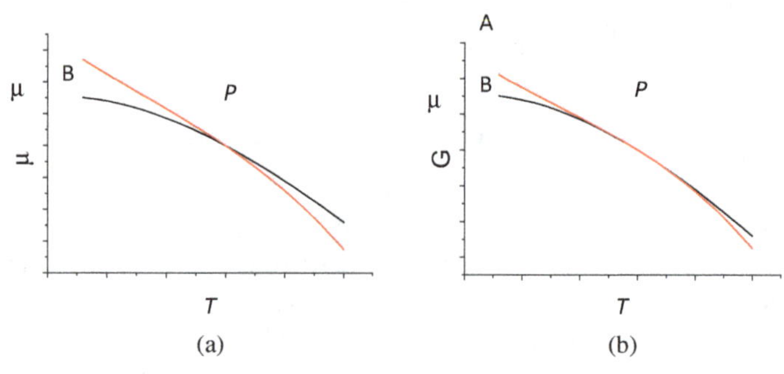

Figure 3. The red and black lines show the variation in the chemical potential for the phases A and B: (a) first-order phase transition at P; (b) continuous phase transition at P.

the first derivative of the chemical potential. Using the same logic, in a "**second-order transition**," the first derivatives of the chemical potential are continuous at the transformation temperature (Figure 3(b)), while the **second derivative** of the chemical potential, corresponding to quantities, such as **specific heat** and **isothermal compressibility**, exhibit a finite discontinuity. However, only a superconducting phase transition in a zero magnetic field satisfies this criterion, while in all the other phase transitions, including the ferromagnetic–paramagnetic and ferroelectric–paraelectric transitions and the liquid–vapor transition at the critical point, these properties diverge. Thus, one employs the terminology "**continuous phase transition**" for all these cases, where the term "continuous" signifies the continuity of the first derivatives across the transition temperature.

9. Van der Waals Equation of State

The van der Walls equation of state for a real gas was the first microscopic theory to explain all the essential features of the liquid–vapor phase transition, wherein the first-order phase boundary terminates at a **critical point**, and what is known as the "**law of corresponding states**" was also derived. Historically, this theory heralded a major victory for the atomistic theory of matter. It is easy to see that an ideal gas does not exhibit a phase transition at any temperature.

There are two main ingredients in the van der Walls model for a real gas. First, the molecules interact through a long-range attractive force, which leads to an additional compressive force so that

$$P_{eff} = P + \frac{a}{V^2} \tag{21}$$

where a is a measure of the long-range attractive interaction between the molecules and V is the volume of the system.

Second, the molecules have a finite size. So, one molecule cannot approach another to a distance closer than the diameter of the molecule due to short-range repulsive interaction. Thus, for the effective volume, V_{eff}, one has the expression,

$$V_{\mathrm{eff}} = V - b \tag{22}$$

where b is a measure of the excluded volume per particle. Thus, the van der Walls equation of state has the form

$$\left(P+\frac{a}{V^2}\right)(V-b)=nk_BT \tag{23}$$

This equation can be easily transformed into a cubic equation in volume V:

$$PV^3-(Pb+k_B\,T)V^2+aV-ab=0 \tag{24}$$

This is a cubic equation and thus has three solutions. It is instructive to plot P against V at different temperatures so that we have a family of isotherms.

Figure 4 shows a typical set of van der Walls isotherms for a real gas. It may be noted that the isotherms fall into two categories, namely those above a characteristic temperature T_c, called the "**critical temperature**" (marked "C" in the diagram), and those below this temperature.

For $T > T_C$, there are one real and two imaginary roots for this equation. Since only the real root is of physical significance, V is a single-valued function of P, and the isotherms in this regime closely resemble the isotherms of an ideal gas.

For $T < T_C$, there are three different real roots for this equation. If we draw a horizontal line corresponding to a particular value of pressure, it

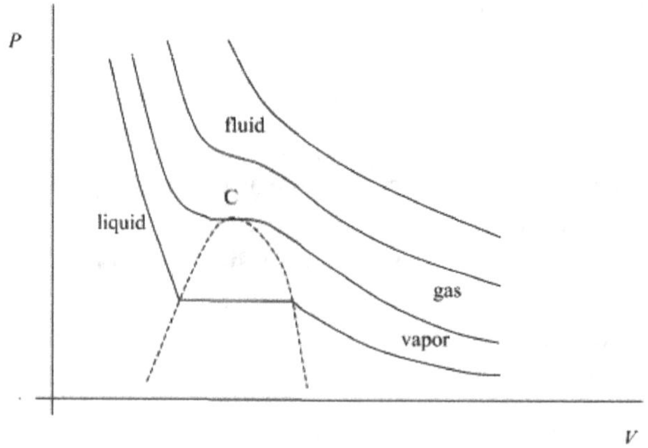

Figure 4. Van der Waals isotherms for a real gas.

will cut the isotherm at three different points, and the corresponding values of the volume are the three real roots of the equation.

The critical point C is a unique point in the phase diagram with coordinates P_C, V_C, and T_C, wherein all three roots merge into a single root (three-fold degenerate). Mathematically, the critical point is an inflection point satisfying the twin conditions $\frac{dP}{dV} = 0$ and $\frac{d^2P}{dV^2} = 0$. These conditions along with the equation of state leads to the relations

$$V_C = 3b \quad P_C = \frac{a}{27b^2} \quad \text{and} \quad k_B T_C = \frac{8a}{27b} \tag{25}$$

10. Critical Point Phenomenon

We now turn our attention to the most important topic, namely the physics close to the critical point. We first turn our attention to a response function called "**isothermal compressibility**," $\kappa_T = -\frac{1}{V}\left(\frac{\partial V}{\partial P}\right)$, which describes how the volume of the system responds to an external stimulus of pressure. We note that the critical point is defined by the condition $\left(\frac{dP}{dV}\right) = 0$ at $T = T_C$, and the Taylor expansion around this point leads to

$$\left(\frac{\partial P}{\partial V}\right) = -a\left(T - T_C\right) \tag{26}$$

Thus, the isothermal compressibility κ_T should diverge at the critical point, behaving as

$$\kappa_T \sim \frac{1}{T - T_C} \tag{27}$$

The divergence signifies that a small change in the applied pressure leads to a very large change in the volume of the system. This leads to large density fluctuations in the medium (and hence spatial variation in the refractive index), and light gets strongly scattered in all directions from such an inhomogeneous medium. This phenomenon is called "**critical opalescence**" and is a visibly spectacular demonstration of light scattering from the inhomogeneous medium, which turns opaque or milky white near the critical point. This type of power-law divergence is generally written as $\kappa_T \sim (T - T_C)^{-\gamma}$, where γ is called a "**critical exponent**." According to the van der Walls theory, $\gamma = 1$.

The next critical exponent is concerned with how volume changes with pressure along the critical isotherm. This is easy to decipher because the critical point has to obey the twin conditions $\frac{dP}{dV} = 0$ and $\frac{d^2P}{dV^2} = 0$. Thus, in the Taylor expansion, the lowest-order term is a cubic term, and hence,

$$P - P_C \sim (V - V_C)^3 \tag{28}$$

In general, this type of power law is expressed as

$$P - P_C \sim (V - V_C)^\delta$$

where δ is another critical exponent. According to the van der Walls theory, $\delta = 3$.

The most important critical exponent is concerned with the change in the magnitude of the volume discontinuity, $(V_{gas} - V_{liquid})$ as we approach the critical point. We have already noted that the volume discontinuity, which is characteristic of a first-order transition, continuously decreases and becomes zero at the critical point. Using the law of corresponding states and the Taylor expansion around T_C, it can be shown that

$$(V_{gas} - V_{liquid}) \sim (T_C - T)^{1/2} \tag{29}$$

In general, this power law is expressed as

$$(V_{gas} - V_{liquid}) \sim (T_C - T)^\beta$$

where β is the appropriate critical exponent.

11. Weiss Theory of Paramagnetic–Ferromagnetic Phase Transition

Pierre Weiss, in the year 1907, developed a remarkable phenomenological theory of ferromagnetism to describe the appearance of "spontaneous magnetization" in materials such as iron, nickel, and cobalt, at temperatures below a characteristic temperature called the "Curie temperature." These metals are paramagnetic at high temperatures and undergo a temperature-induced phase transition from a paramagnetic to a ferromagnetic state when cooled below the Curie temperature. It is to be appreciated that this theory predates the development of quantum theory and the understanding of the origin of magnetism. Weiss modeled these materials as consisting of tiny "atomic magnets," each with a magnetic moment of μ, and it is well known that its energy in the presence of a magnetic field $\boldsymbol{H}$ is $-\boldsymbol{\mu}.\boldsymbol{H}$. For the sake of simplicity, we consider that this tiny magnet can assume only two orientations, namely either parallel or antiparallel, with respect to the magnetic field. The energy is lower for the arrangement wherein the moment is parallel to the magnetic field and higher for the antiparallel arrangement. Furthermore, while the magnetic field tends to align the magnetic moment along its direction, an increase in temperature would disrupt this alignment. We now consider an aggregate of N atomic moments distributed far apart so that each of these moments effectively feel the presence of the external field only and there is no interaction between the moments. The concentration $N(\uparrow)$ of the moments parallel to $\boldsymbol{H}$, according to Boltzmann statistics, is given by the expression

$$N(\uparrow) = N\left[\frac{e^{\mu H/k_B T}}{e^{\mu H/k_B T} + e^{-\mu H/k_B T}}\right] \tag{30}$$

Similarly, the concentration $N(\downarrow)$ of the moments antiparallel to $\boldsymbol{H}$ is given by the expression

$$N(\downarrow) = N\left[\frac{e^{-\mu H/k_B T}}{e^{\mu H/k_B T} + e^{-\mu H/k_B T}}\right] \tag{31}$$

The magnetization $\boldsymbol{M}$ developed in the direction of the field is given by

$$\boldsymbol{M} = [N(\uparrow) - N(\downarrow)]\boldsymbol{\mu} \tag{32}$$

Thus, one arrives at the following expression for $\boldsymbol{M}$:

$$\boldsymbol{M} = N\mu\left[\frac{e^{\mu H/k_B T} - e^{-\mu H/k_B T}}{e^{\mu H/k_B T} + e^{-\mu H/k_B T}}\right] = N\mu\tanh\left(\frac{\mu H}{k_B T}\right) \tag{33}$$

We can define a dimensionless quantity, $m = \frac{M}{N\mu}$, known as "reduced magnetization." The equation that describes the paramagnetic state has the form

$$m = \tanh(x) \quad \text{where} \quad x = \frac{\mu H}{k_B T} \tag{34}$$

The plot of tanh(x) against x is shown in Figure 5.

It may be noted that tanh(x) is an odd function of x, i.e., tanh($-x$) = $-$tanh(x).

Furthermore, for small values of x,

$$\tanh(x) \approx x - \frac{x^3}{3} + \cdots \tag{35}$$

For $H \sim 10^4$ G and at room temperature, $x = \mu H/k_B T = 2 \times 10^{-3}$. x is small for small fields. So, we have the following relations for m and the magnetic susceptibility χ: $m = \frac{\mu H}{k_B T}$ and $\chi = \frac{C}{T}$. This 1/T dependence of the susceptibility is the well-known **Curie's law** and is generally obeyed in many materials with a dilute concentration of magnetic species.

For high fields or low temperatures, $x >> 1$, and m approaches the limiting value of ±1, corresponding to near-perfect parallel alignment of the atomic magnets in the direction of the magnetic field.

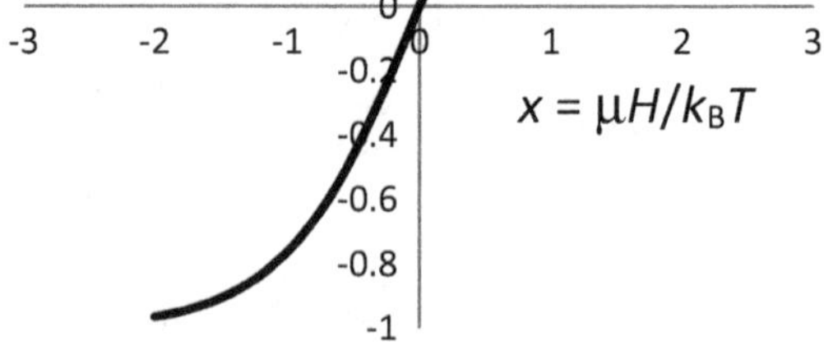

Figure 5. Plot of tanh(x) against x.

The important feature of this "**ideal paramagnet**" is the smooth behavior of magnetization with the magnetic field or with temperature. Weiss noted the close similarity of this behavior with the smooth behavior of an "**ideal gas**" in the P–T plane. In our discussion on the van der Walls equation of state, we noted that the inclusion of intermolecular attraction modifies the pressure of the gas and, under appropriate conditions, leads to a vapor–liquid phase transition. Similarly, Weiss postulated that a given atomic magnet experiences not only the applied magnetic field but also the "average" or "mean" magnetic field produced by the other atomic magnets in the system. It is natural to approximate this mean field $\boldsymbol{H}_{\text{mean}}$ to be proportional to the magnetization $\boldsymbol{M}$ of the system. Thus, one can write for $\boldsymbol{H}_{\text{mean}}$ the expression

$$\boldsymbol{H}_{\text{mean}} = \lambda \boldsymbol{M} \tag{36}$$

where λ is a constant, independent of temperature.

Weiss made another important simplification to address the effect of fluctuating magnetic fields produced by neighboring atomic moments by just taking their time-averaged values. Thus, one can add up this field to the external field $\boldsymbol{H}$ so that the effective field acting on an atomic magnet is given by

$$\boldsymbol{H}_{\text{eff}} = \boldsymbol{H} + \lambda \boldsymbol{M} \tag{37}$$

The magnetization $\boldsymbol{M}$ for a system of "***interacting***" magnetic moments has the form

$$M = N\mu \tanh\left[\frac{\mu(H + \lambda M)}{k_B T}\right] \tag{38}$$

It is convenient to recast this equation in terms of the "reduced magnetization m:

$$m = \tanh\left[\frac{\mu H}{k_B T} + \left(\frac{N\mu^2\lambda}{k_B T}\right)m\right] \tag{39}$$

One can define a "characteristic temperature" T_C for the system through the relation

$$T_C = \frac{N\mu^2\lambda}{k_B} \tag{40}$$

The physical significance of T_C would become clearer in a later analysis.

The magnetic equation of state, according to Weiss' mean field description, has the form

$$m = \tanh\left[\frac{\mu H}{k_B T} + \left(\frac{T_C}{T}\right)m\right] \tag{41}$$

We also define a dimensionless quantity, $\theta = \frac{T}{T_C}$, known as "reduced temperature." This relation bears close analogy with the van der Walls equation of state, and we will explore this aspect later in some detail.

The equation of state for the most important case of $H = 0$ assumes the rather simple form

$$m = \tanh\left[\frac{m}{\theta}\right] \tag{42}$$

Since m appears on both sides of the equation, the roots of this equation cannot be obtained in an explicit way. This is a transcendental equation, and an analytical solution is not possible. However, we can solve this equation in a "self- consistent" way by looking for the intersection of the curve $y = m$ with the curve $y = \tanh(m/\theta)$. Figure 6(a) depicts a typical plot for $T > T_C$, where $T = 1.25T_C$. The two curves intersect only at $m = 0$, and this is the solution. Thus, at high temperatures, the average magnetization of the system is zero, and thus, the system is in a **paramagnetic state.**

Figure 6(b) shows a similar situation with $m = 0$ prevailing until one reaches the temperature $T = T_C$, where the slopes of the two curves are identical at the origin. This characteristic temperature is the **Curie temperature** of the magnetic system.

Figure 6(c) presents typical data for $T < T_C$, where $T = 0.5T_C$. Now, there are three solutions: $m = 0.957$, $m = -0.957$, and $m = 0$. It would suffice here to mention that the solution $m = 0$ is unstable in this temperature regime as it corresponds to a higher-energy state as compared to the other two solutions, and hence, it is not considered. This solution is analogous to the middle unstable solution encountered in the van der Walls equation

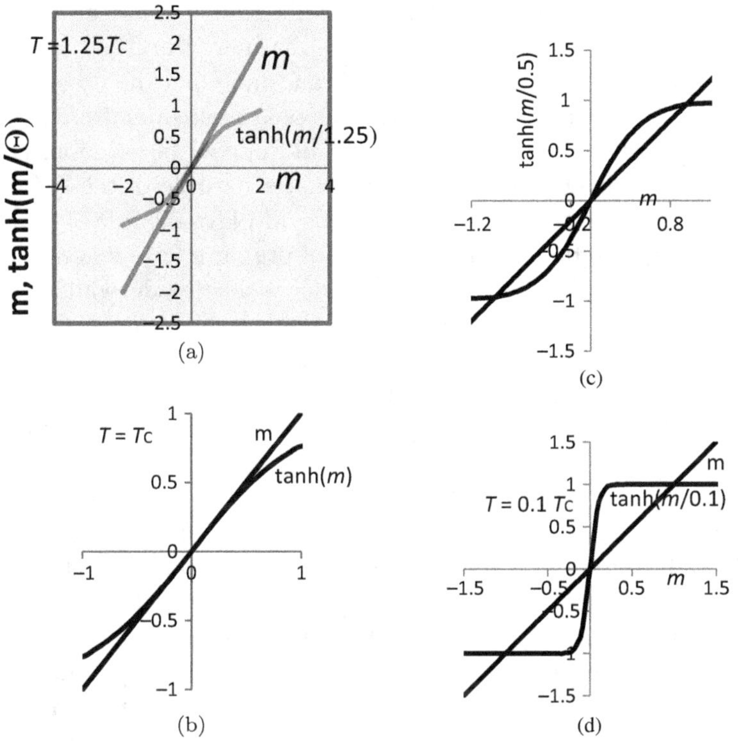

Figure 6. (a) $T = 1.25T_C$, (b) $T = T_C$, (c) $T = 0.5\ T_C$, and (d) $T = 0.1T_C$.

of state. **The two symmetrical solutions at $m = \pm m_0$ signify that the system develops spontaneous magnetization even in the absence of an external magnetic field.**

Figure 6(d) shows the data for $T \ll T_C$, where $T = 0.1T_C$. The two solutions $m \cong \pm 1$ signify **saturation magnetization**.

The physical picture that emerges from the analysis of Weiss' mean field theory of magnetism is that at temperatures higher than T_C, thermal disorder dominates the tendency for atomic moments to align parallel to each other, and all orientations of the magnetic moment are equally probable, resulting in the paramagnetic state. As the temperature is lowered below T_C, order continuously develops, and the magnitude of the magnetization increases progressively from 0 to reach the saturation value of either +1 or −1. The state with positive m is called the "**ferromagnetic up**" state, while the state with negative ***m*** is called the "**ferromagnetic down**" state. Both of these states have the same free energy, and as the system is cooled to T_C, the system will choose at random either the up state or the down state. It is instructive to classify this phase transition from a thermodynamic point of view discussed in detail in an earlier section. We note here that for a magnetic system, $-HM$ takes the role of the PV term in the expression for Gibbs free energy. Thus, M plays the role of V, and the external magnetic field H plays the role of pressure P. In this context, the magnetization M (or equivalently m), which is the first derivative of Gibbs free energy with respect to the magnetic field, is **continuous** at the transition temperature T_C. Furthermore, the magnetic part of the entropy is also **continuous** at T_C, implying that there is no latent heat involved in this transition. Based on the Ehrenfest scheme of classification, the **paramagnetic–ferromagnetic** transition in a zero magnetic field is a second-order or, more appropriately, a **continuous** phase transition. Figure 7 shows all the salient features of this phase transition in the m–T plane. It is to be noted that the "ferro up" and "ferro down" states indicated in Figure 7 are two distinct phases and bear a close analogy with the "liquid" and "vapor" phases encountered in the van der Walls equation of state. The coexistence curve clearly shows that the volume discontinuity $V_{Gas} - V_{Liquid}$ (which is indicative of a first-order transition) continuously decreases with temperature and becomes zero at the critical point.

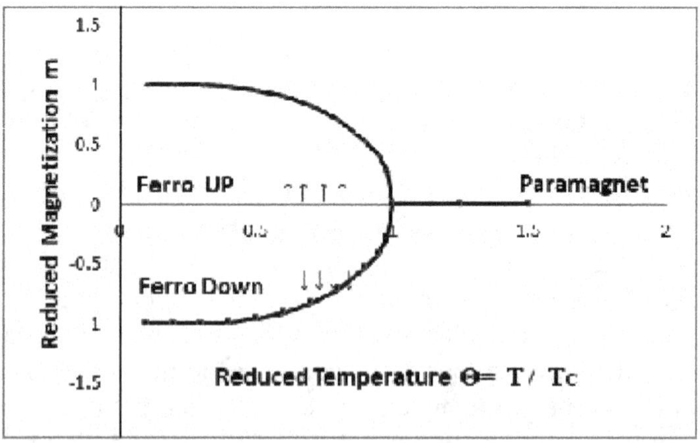

Figure 7. Magnetic phase diagram of a typical ferromagnet in the m–Θ plane. Note that the paramagnetic – "ferro up" or "ferro down" state at $T = T_C$ is a continuous phase transition. The "ferro up" and "ferro down" states are symmetry related.

Now, we explore the magnitude of the "critical magnetic field" $\boldsymbol{H}_C$, which is analogous to the "critical pressure" $\boldsymbol{P}_C$ associated with the liquid–vapor phase boundary. The first-order phase boundary in the P–T plane terminates at a critical point defined by P_C and T_C (Figure 2), and the slope of this curve, which is always positive, is given by the Clausius–Clapeyron relation. Since H plays the role of P and magnetization M is the analogue of volume V, the Clausius–Clapeyron equation for the magnetic system takes the form

$$\frac{\mathrm{d}H}{\mathrm{d}T} = \left[\frac{S\uparrow - S\downarrow}{M\uparrow - M\downarrow}\right] \tag{43}$$

At any temperature $T < T_C$, the "ferro down" state is realized by the reversal of magnetic moments in the "ferro up" state, and the entropies of these phases are identical, leading to $S\uparrow - S\downarrow = 0$. However, $M\uparrow - M\downarrow$ corresponding to a discontinuous change in magnetization is always finite so that we have the relation

$$\frac{\mathrm{d}H}{\mathrm{d}T} = 0 \tag{44}$$

Thus, one gets the important result that $H_C = 0$. The unique magnetic phase diagram is given in Figure 13. This phase diagram pertains to the case where only two orientations, namely either parallel or antiparallel arrangement, for the atomic moments is allowed. The thick horizontal line is at zero magnetic field and at a temperature regime lower than the critical temperature T_C. The magnetization is positive above this line and negative below this line. Further magnetization exhibits a jump by changing sign at zero magnetic field, signifying that it is a first-order phase transition. The magnitude of the jump in magnetization continuously decreases with temperature and reaches the value of zero at the critical point. **The point of disappearance of the jump in magnetization at $\mathbf{T_C}$ signifies the changeover to a continuous phase transition.** This is analogous to the first-order liquid–gas phase boundary terminating at a critical point.

The analogy between the fluid and magnet systems would be complete when we evaluate the critical exponents of the magnetic system. The variation in m with temperature just below T_C can be discussed by noting that m is close to zero at T_C, and thus,

$$m = \tanh\left[\frac{m}{\theta}\right] \approx \frac{m}{\theta} - \frac{1}{3}\left[\frac{m^3}{\theta^3}\right] \tag{45}$$

On simplification, one gets the important result

$$m = \sqrt{3\left(\frac{T_C - T}{T_C}\right)} \quad \text{or} \quad m \sim \left[(T_C - T)\right]^{1/2} \tag{46}$$

This "critical exponent" $\beta = \frac{1}{2}$ is identical with the one encountered while discussing the behavior of the analogous relation

$$(V_{gas} - V_{liquid}) \sim (T_C - T)^{1/2}$$

We now consider the behavior of the isothermal magnetic susceptibility, $\chi_T = \frac{\partial M}{\partial H}$, close to the critical temperature T_C. This requires an analysis of the Weiss equation of state with the addition of an external magnetic field H:

$$m = \tanh\left[\frac{\mu H}{k_B T} + \left(\frac{T_C}{T}\right)m\right]$$

Like the reduced magnetization m and reduced temperature, one can define the reduced magnetic field h such that

$$\frac{\mu H}{k_B T} = \frac{h}{\Theta} \tag{47}$$

H is a dimensionless quantity representing the external magnetic field in appropriate energy units. The equation of state can be recast in the form

$$m = \tanh\left[\frac{h}{\Theta} + \frac{m}{\Theta}\right] \tag{48}$$

Near the critical point, both m and h are small, leading to the relation

$$m \cong \left[\frac{h}{\Theta} + \frac{m}{\Theta}\right] \quad m \cong h\left[\frac{T_C}{T - T_C}\right] \quad \chi_T = \frac{\partial m}{\partial h} = \left[\frac{T_C}{T - T_C}\right] \tag{49}$$

This is the well-known Curie–Weiss law for the isothermal magnetic susceptibility, which exhibits divergence at the critical point:

$$\chi_T \sim (T - T_C)^{-\gamma}$$

and the critical exponent $\gamma = 1$ and is identical to the result from the analogous quantity, isothermal compressibility, encountered in the liquid–gas system.

The critical isotherm is the curve in the H–M plane or, equivalently, the h–m plane corresponding to $T = T_C$ ($\Theta = 1$). Its shape is described by the critical exponent δ in the expression $h \sim m^\delta$. It is straightforward to see that for $\Theta = 1$, one gets the result $h \sim m^3$. The critical exponent, $\delta = 3$, is identical with the analogous result $P - P_C \sim (V - V_C)^3$ for the fluid system.

12. Experimental Techniques for Phase Transitions Studies

There are numerous techniques employed to study phase transitions. However, differential thermal analysis (DTA) and differential scanning calorimeter (DSC) are the two most important thermal techniques employed to obtain the transformation temperature and enthalpy change accompanying the phase transition. In what follows, we discuss the underlying principles involved in these techniques and the related instrumentation, with some typical results. The AC specific heat technique (see Chapter 10) has been discussed, with special reference to the ferromagnetic–paramagnetic transition in metallic samples, such as nickel. A high-resolution AC resistivity measurement technique for studying the paramagnetic–ferromagnetic transition in nickel is discussed in the following. A simple technique involving capacitance measurement across the temperature-induced ferroelectric–paraelectric phase transition in a modified barium titanate is discussed in a later section. This technique also allows us to measure the dielectric susceptibility near the Curie point and establish the region of validity of the Curie–Weiss law. In the field of materials science, there is a unique phase transition involving a change from martensite to austenite structure with remarkable properties, such as shape memory effect and superelastic behavior. This weak first-order phase transition occurring near room temperature can be conveniently probed by employing what is generally known as the "delta technique" for resistivity measurement.

12.1. *Differential thermal analysis*

This technique, commonly known as DTA, is extensively used to determine the characteristic temperatures, such as the melting point, boiling point, sublimation point, and solid–solid transformation temperature, associated with first-order phase transitions and obtain a reasonable estimate of the enthalpy change or the latent heat involved in the process. The operating principle of this technique is based on the fact that during phase transition, temperature remains constant, and the heat supplied manifests as a discontinuous change in enthalpy, also called latent heat. Thus, if the sample and a reference material (which is devoid of any phase transition in the temperature region of interest) are subjected to a linear rate of heating, ensuring a good thermal contact between them, the temperature differential $\Delta T = T_{\text{sample}} - T_{\text{reference}}$ would be ideally zero except in the region of phase transition. Figure 8(a) shows a comparison of the temperature versus time plots for the sample and reference materials subjected to a programmed rate of linear heating. It is clear from the plot that the temperature of the sample, T_{sample}, lags that of the reference, $T_{\text{reference}}$, at the phase transition, leading to an "endothermic" peak in the ΔT versus T plot shown in Figure 8(b). As an example, this could be the melting transition of a pure sample. In the cooling cycle, this corresponds to solidification transition, which is "exothermic." In the region away from the phase transition, ΔT versus T is nearly flat and is referred to as the "baseline." The design of the DTA apparatus is primarily directed at achieving a nearly flat baseline in the temperature regime of interest. Note the constancy of the temperature of the sample during the phase change.

12.1.1. Experimental setup

In the DTA technique, we record a thermal event in the sample by comparing it with a reference material which is devoid of any physical or chemical changes in the temperature range of interest. Figure 9 presents a block diagram of a home-built DTA system. In the conventional experimental arrangement, a single block of a metal with symmetrical cavities for the sample and the reference is heated uniformly in the constant temperature zone of a tubular furnace. The block acts as a heat sink, ensuring that the sample and reference are heated at the same rate so that there is no temperature difference between them in the absence of a thermal event. Furthermore, the block also exchanges heat with the sample so that

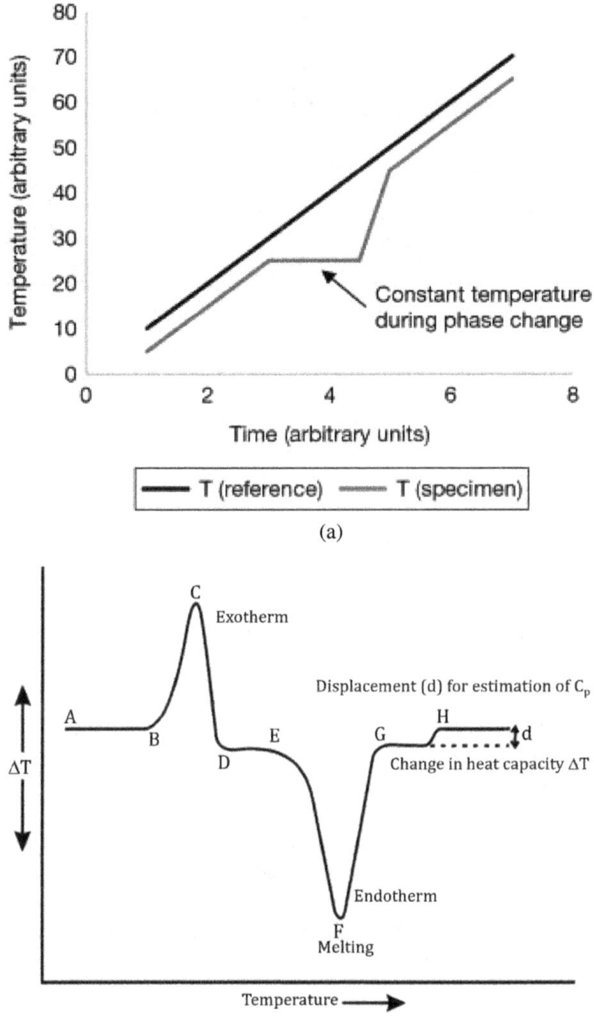

Figure 8. (a) Linear rate of heating of the sample compared with the reference. (b) Plot of $\Delta T = T_{sample} - T_{reference}$ vs. T. If ΔT is negative, the phase transition is "endothermic." If ΔT is positive, the phase transition is "exothermic." A shift in the baseline signifies a change in heat capacity. The displacement can be used to estimate the change in C_p.

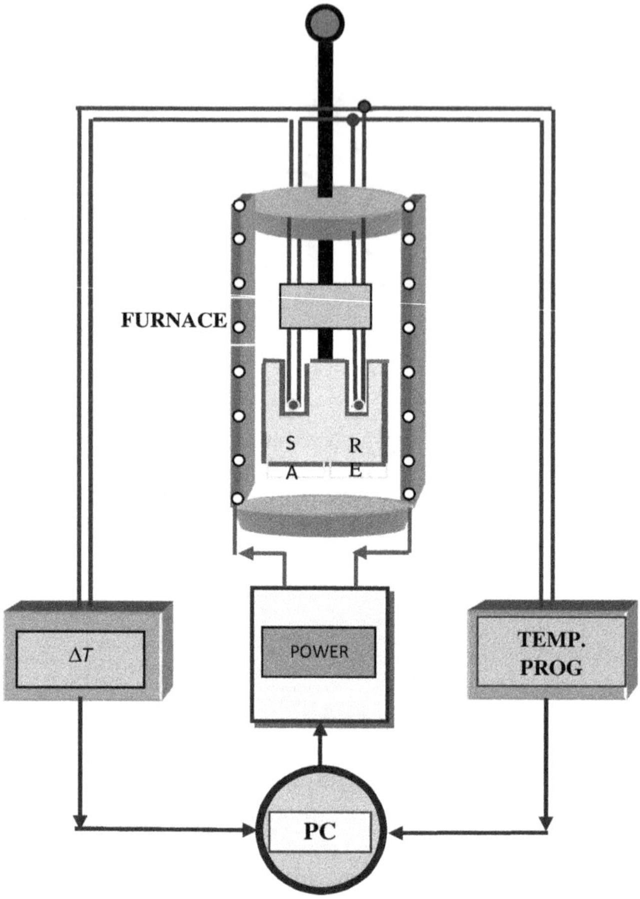

Figure 9. A block diagram of DTA. Apparatus

the sample temperature re-equilibrates with the furnace temperature after a thermal event in the sample. However, the temperature of the sample must not re-equilibrate too rapidly because the DTA signal (the differential temperature) will then be very small. It is thus recommended to have a sample container with low thermal conductivity so as to get an adequate DTA signal. It follows that the temperature change accompanying a transition is directly proportional to the enthalpy change of the transition. An absolute quantitative measurement of the heat of transition requires a mathematical treatment of the heat flow between the sample and the block. A practical method of evaluating the change in enthalpy or latent heat is to calibrate the DTA instrument with a material whose change in enthalpy is known from other studies and compare the peak areas.

In the conventional DTA technique, the thermocouples are housed in the sample and reference cavities of a block of aluminum for the temperature range of 25°C–500°C. For temperatures up to 800°C, the block is made of stainless steel. This sample block is located in the constant temperature zone of a tubular furnace. For temperatures up to 800°C, the conventional type-K (chromel–alumel) thermocouple is employed.

The DTA system consists of four subsystems, namely the furnace, temperature programmer, power amplifier, and data acquisition system, apart from the sample holder discussed earlier:

Furnace: A compact furnace built using nonmagnetic materials with provisions to house the DTA sample holder is an important subsystem. Nichrome wire of 32 SWG is wound around an alumina tube with dimensions of 27 mm OD and 23 mm ID and a length of 200 mm. The number of turns per inch of the nichrome wire wound on the alumina tube is 18. The resistance of the heating coil thus wound is around 10 Ω. The nominal current-carrying capacity of this wire is 1 A. Brackets made of Syndanio material, which is a good thermal insulator, are fitted at either end of the alumina tube to thermally isolate it from an outer enclosure made of powder-coated mild steel. The nichrome wire heater is covered with K-wool along its entire length for thermal insulation. The efficiency of this heater arrangement is around 10°C/W. The furnace is powered by a 100 W power amplifier. With these specifications, the operating temperature range for the sample holder would be from 30°C to 700°C.

Power amplifier: This subsystem powers the furnace designed for high-temperature DTA studies on materials. The PID signal from the D/A channel of the DAS card, which is an analog signal, forms the input of the DC power amplifier. The output voltage signal from the amplifier has the required current capability to power the furnace. The output rating of the DC power amplifier is 30 V and 5 A. The maximum power output of 150 W is optimum for attaining temperatures of around 800°C and for achieving linear rates of heating of up to 20°C/min.

Precision temperature programmer: The precision temperature programmer is the heart of the DTA system. The thermocouple input from the chromel–alumel thermocouple in thermal contact with the reference material (alumina) is routed through a differential input filter followed by a high-quality instrumentation amplifier configured for a gain of 200. Since temperature measurements are carried out using a single-junction thermocouple, reference-point compensation is required, and an IC temperature transducer provides the necessary cold junction compensation (CJC). The amplified thermocouple voltage with CJC is linearized using a proprietary circuit that employs a fourth-order polynomial curve fitting for the thermo-emf versus temperature data. The fitting accuracy is ±0.5°C for the temperature range of 25°C–800°C. The analog output of the linearized temperature (1 mV corresponding to 1°C) is available for further processing in a data acquisition system (DAS). The PCI DAS add-on card employed here is a high-resolution 16-bit A/D converter with a front-end 8 channel (differential) or 16-channel (single-ended) multiplexer operating at a maximum speed of 60 kHz. The gain of the preamplifier is optimized prior to A/D conversion to enhance the accuracy of measurement. The card also has 12-bit D/A converters, which are useful for close-loop control of parameters, such as temperature. This card with menu-driven software developed in a LabView environment provides the linkage between the PC and the hardware described earlier.

The analog voltage signal corresponding to the temperature of the reference material (alumina) is digitized in a 16-bit A/D converter channel of DAS and compared with the "set point" of the proportional–integral–derivative (PID) algorithm. The optimized PID digital signal is transformed into an analog voltage signal using the 16-bit D/A converter of the DAS and forms the input to the DC power amplifier. This ensures close-loop control of temperature, and one can achieve stability of temperature to within ±0.5°C of the set point. If the set point is incremented at a specified rate, the controlled temperature would follow suit, and one can achieve a predetermined rate of heating.

The significant features of the software are: (a) selectable rate of heating from 1°C/min to 20°C/min, (b) real-time digital filtering of the data for improved signal-to-noise ratio, and (c) graphic display of the ΔT versus T signal for real-time observations of thermal events occurring in the sample.

The data from the DTA run can be stored in a file for further analysis. Software can also be developed for the numerical computation of the peak areas to estimate enthalpy changes.

12.1.2. Typical experimental data

The performance of the system is demonstrated by some typical data on KNO_3 plotted in Figure 10. The DTA signals observed near 129°C and 340°C correspond to structural and melting transitions, respectively. An endothermic peak near 129°C corresponds to a crystallographic phase transition, while the one near 340°C corresponds to the melting transition.

12.2. *Differential scanning calorimeter*

DSC is the most extensively used thermal analysis technique in materials research to accurately determine transformation temperatures, such as the melting point, the heat of transition accompanying a first-order phase transition (latent heat), and the change in specific heat in a glass transition. This technique is also extensively used in checking the purity of a material.

Basically, there are two variants of this technique, known as "heat flux DSC" and "power-compensated DSC." We describe here the basic physical principles involved in the design of these systems.

12.2.1. Heat flux DSC

This technique is essentially a sequel to the DTA technique described earlier. Figure 11 presents a schematic diagram of the cell arrangement.

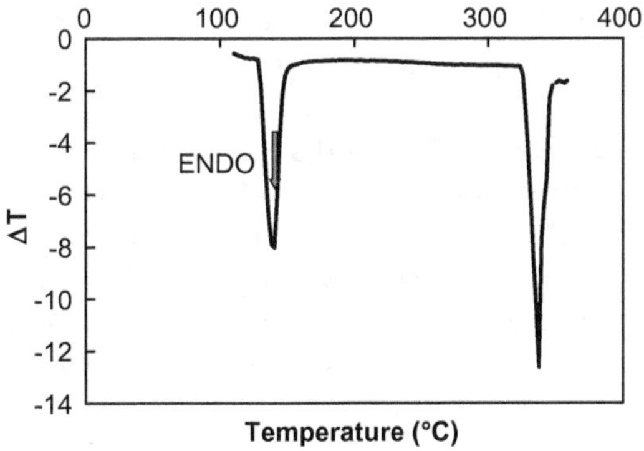

Figure 10. Structural and melting transition of KNO_3 studied through the DTA technique.

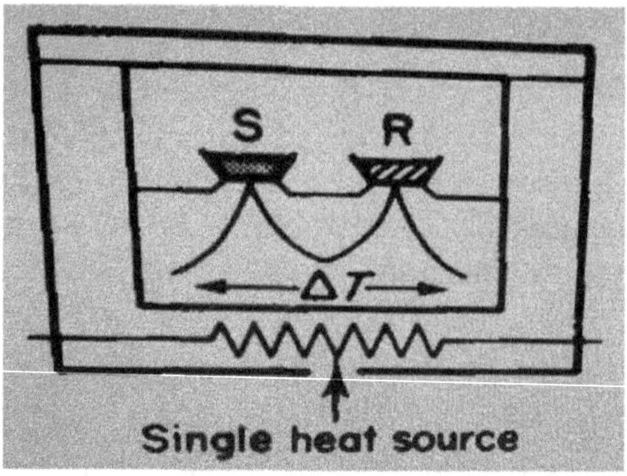

Figure 11. A schematic diagram of the cell arrangement for heat flux DSC.

The heating block, generally made of silver, is in close proximity to a single heater. A chromel–alumel thermocouple is attached to the heating block and forms the input to a closed-loop PID temperature programmer. The major change here is an arrangement to provide a very low resistance path for heat flow between the sample and the reference. A circular disk, made of constantan alloy with two elevated platforms for housing the holders for the sample and reference, facilitates heat flow between them. Another disk, made of chromel alloy, is inserted below the sample and reference holders. Wires attached to the chromel disks ensure that the temperature difference between the sample and the reference, ΔT, is conveniently measured by this chromel–constantan thermocouple in the differential configuration. An alumel wire spot-welded to the chromel disk below the sample holder ensures the measurement of the sample temperature T. It is to be emphasized that the magnitude of ΔT here is much smaller than in the DTA technique when there is no thermal event occurring in the sample. The constantan disk ensures a near-isothermal environment for the sample and reference. Unlike in DTA, where an inert material such as alumina is used as a reference, here an empty holder is used as a reference. A blank temperature run is carried out, keeping an empty holder in both the sample and reference platforms, and the data are stored in the DAS. These data values are subtracted from the actual data from the sample. This ensures a nearly flat baseline in the temperature region, devoid of a phase transition. The ΔT signal, although of lower magnitude, is processed in a high-quality pre-amplifier and digitized in a high-resolution A/D converter, which employs algorithms for improved signal-to-noise ratio. This processed ΔT signal is converted into a heat flux signal through calibration procedures involving a standard material with a known enthalpy change.

12.2.2. Power-compensated DSC

A significant advance took place in the field of thermal analysis with the development of the power-compensated DSC. Figure 12 depicts a block diagram of the system. There are two independent and identical compact heaters to separately house the sample and reference holders. Beneath the sample and reference holders, there is a distributed temperature sensor comprising over a hundred thin-film thermocouples configured for achieving high sensitivity. In some designs, thin-film platinum resistance sensors are used.

The central design principle of this technique is that, through closed-loop control, the temperature of the sample is made identical to the temperature of the reference. Thus, when a thermal event, such as an endothermic phase transition, takes place, additional power has to be supplied to the sample furnace relative to the reference furnace to maintain the same temperature in both the sample and reference holders. The difference in power applied to the sample and reference holders directly gives the enthalpy change or the "latent heat." Thus, this novel design provides a direct measurement of heat flow change or enthalpy change in the sample. The low thermal mass of the furnaces also makes it possible to achieve linear heating rates over a wide range, from 0.1°C/min to

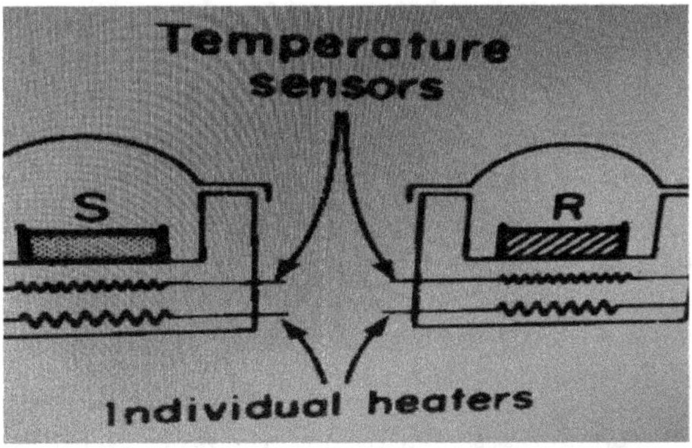

Figure 12. A schematic diagram of the setup for power-compensated DSC.

100°C/min. The electronic measurement system is designed to measure the difference in power supplied to heaters in the sample and reference holders. It may be noted that the baseline here has exceptional stability, and enthalpy changes as low as 1 μW can be measured.

12.3. *Ferromagnetic–paramagnetic transition in nickel studied through high-resolution resistivity technique*

Nickel is a 3d transition metal and has been extensively studied for its magnetic and electronic transport properties. Nickel is a prototype ferromagnetic metal with a Curie temperature of around 357°C. Although the DSC technique can provide an accurate value for the Curie temperature, the detailed behavior of the specific heat near T_C cannot be deciphered with this technique. We now discuss another useful technique, namely the high-resolution AC resistivity technique, to track the temperature-driven ferromagnetic–paramagnetic phase transition and also probe the detailed behavior of the specific heat near T_C.

12.3.1. Experimental setup

The system is based on the well-known four-probe technique for resistivity measurement, along with a chromel–alumel thermocouple probe spot-welded to the same sample for temperature measurement. In order to circumvent the effects of Seebeck voltages, which necessarily exist due to temperature gradients in the sample, an AC resistivity technique is employed. This technique has been described in Chapter 13 on the electrical conductivity of metals and semiconductors.

A typical plot of the resistivity data spanning the ferromagnetic–paramagnetic transition is given in Figure 13. It has been shown theoretically that the temperature derivative from resistivity is closely related to the magnetic contribution to the specific heat. A plot of dR/dT versus temperature extracted from the above data is given in Figure 13(b). This curve bears close resemblance to the λ-type specific heat anomaly observed in a continuous phase transition.

12.4. *Ferroelectric–paraelectric transition in modified barium titanate*

Ferroelectricity is a property of certain nonconducting crystals, or dielectrics, that exhibit spontaneous electric polarization (separation of the centers of positive and negative electric charges, making one side of the

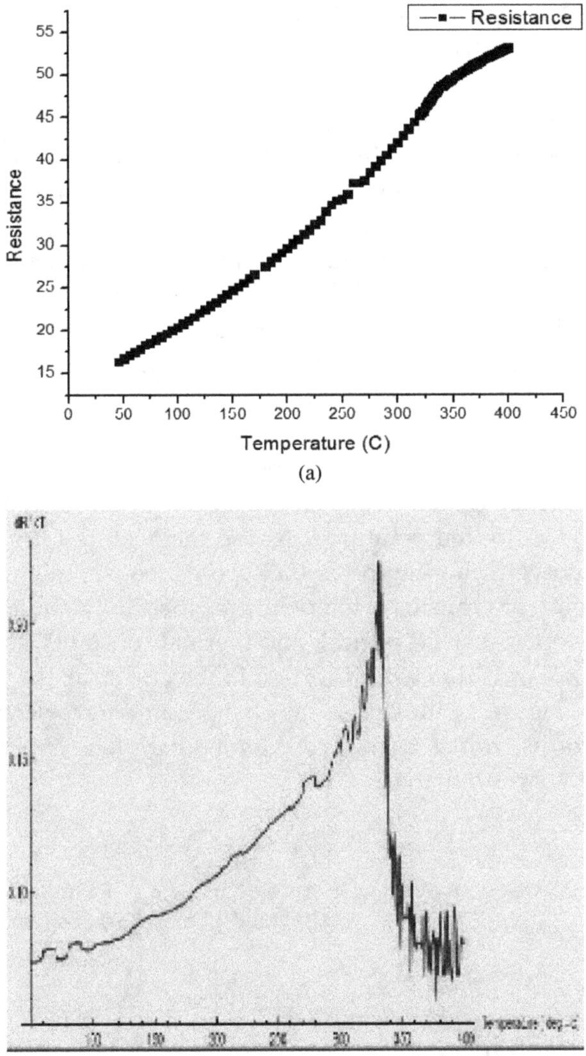

Figure 13. **(a) Ferromagnetic–paramagnetic transition in Ni tracked through resistivity measurement. (b) Plot of calculated d*R*/d*T* against temperature.**

crystal positive and the opposite side negative) that can be reversed in direction by the application of an appropriate electric field. Ferroelectricity is named in analogy with ferromagnetism, which occurs in such materials as iron. Iron atoms, being tiny magnets, spontaneously align themselves in clusters called ferromagnetic domains, which, in turn, can be oriented predominantly in a given direction by the application of an external magnetic field.

Ferroelectric materials, such as barium titanate ($BaTiO_3$) and Rochelle salt, are composed of crystals in which the structural units are tiny electrical dipoles. In each unit, the centers of the positive and negative charges are slightly separated. In some crystals, these electric dipoles spontaneously line up in clusters called domains, and in ferroelectric crystals, the domains can be oriented predominantly in one direction by a strong external electric field. Reversing the external field reverses the predominant orientation of the ferroelectric domains. However, there is a time lag between switching to a new direction and a change in the external electric field. This lag of electric polarization behind the applied electric field is ferroelectric hysteresis, named in analogy with ferromagnetic hysteresis.

Ferroelectricity ceases in a material above a characteristic temperature, called its Curie temperature, because the heat agitates the dipoles sufficiently to overcome the forces that spontaneously align them. Near the Curie point or transition temperature, thermodynamic properties, including dielectric, elastic, optical, and thermal constants, show anomalous behavior. This is due to a distortion in the crystal as the phase structure changes. Figure 14 illustrates the changes in a ferroelectric material which transforms from a ferroelectric tetragonal phase to a paraelectric cubic phase in barium titanate.

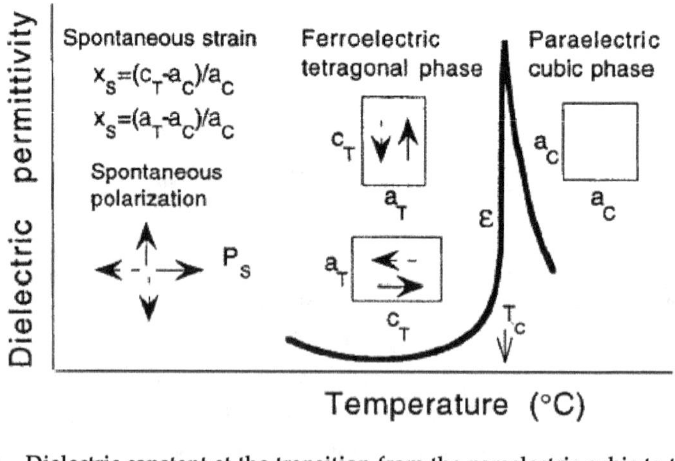

Figure 14. Dielectric constant at the transition from the paraelectric cubic to the ferroelectric tetragonal phase.

The permittivity curve represents data measured on a $BaTiO_3$ ceramic. The arrows show possible directions of spontaneous polarization (in two dimensions). The unit cell is represented by a square in the cubic phase and a rectangle in the tetragonal phase.

As ferroelectricity ceases above the Curie temperature, materials lose their intrinsic polarization in the paraelectric phase. The electrical susceptibility, χ, in the paraelectric phase follows the well-known Curie–Weiss law:

$$\chi = A/(T - T_C)$$

where A is a constant. This is the same form of expression as the Curie–Weiss law of magnetic susceptibility.

A convenient method of determining electrical susceptibility is to measure the capacitance of a parallel-plate capacitor containing the ferroelectric substance as a dielectric. The capacitance, C, is related to susceptibility through the expression

$$C = C_0(1 + \chi) \tag{50}$$

Here, C_O is the capacitance without a dielectric and is determined from the relation

$$C_O = \varepsilon_0 A/d \tag{51}$$

where $\varepsilon_0 = 8.85 \times 10^{-12}$ F/m, A is the area of the specimen, and d is its thickness.

12.4.1. Experimental arrangement

To measure the transition temperature of a ferroelectric sample under investigation, the sample is inserted in a simple planar heater arrangement consisting of a nichrome heater sandwiched between thermally insulating Syndanio blocks. The nichrome heater is powered by a 0–30 V, 0–3 A DC power supply. The capacitance of the specimen is measured through an LCR meter operating at 1 kHz. A chromel–alumel thermocouple is placed in proximity to the sample for temperature measurement. An electronic unit, incorporating the linearization circuitry for type-K thermocouple and CJC, facilitates the direct measurement of the temperature of the sample with a resolution of 0.1°C.

12.4.2. Ferro–para transition in modified barium titanate

Using the experimental arrangement described above, the Curie temperature, T_C, of modified $BaTiO_3$ can be directly determined. This polycrystalline form of $BaTiO_3$ is widely used as a dielectric in high-permittivity commercial capacitors. In its pure single-crystal form, $BaTiO_3$ has a T_C of ~120°C. Above T_C, the paraelectric phase has a cubic structure. Below T_C, it is ferroelectric with a tetragonal structure. The Curie temperature of $BaTiO_3$ decreases with the substitution of small amounts of ions, such as Ca^{2+} and Sr^{2+}, for the Ba^{2+} ions.

Figure 15 presents a typical set of data for the capacitance of a parallel-plate condenser incorporating the modified $BaTiO_3$ as the dielectric material. In the current experiment, $C_0 \sim 1$ *pfd* so that $C/C_0 \sim \chi$. Thus, Figure 15(a) is essentially a plot of the dielectric susceptibility χ plotted against temperature. The Curie temperature for the modified $BaTiO_3$ is around 80°C. In the paraelectric region above 80°C, it is expected that the susceptibility obeys the Curie–Weiss law. Accordingly, a plot of $1/\chi$ against $(T - T_C)$ should be a straight line. Figure 15(b) depicts this graph for the temperature regime above the Curie temperature. It is important to note that the data fits a linear plot very well for the region $(T - T_C) >$ 20°C and deviates markedly in the region close to the Curie temperature. The mean field theory, which forms the basis of the Curie–Weiss law, is validated experimentally in the temperature regime away from T_C. The law breaks down close to T_C due to the neglect of large fluctuations in the order parameter, i.e., electric polarization, in this region.

12.5. *Martensite–austenite phase transition in a shape memory alloy*

The shape memory effect in a near-equi-atomic alloy of nickel–titanium was discovered accidentally in 1959 by a team led by William J. Buehler of the U.S. Naval Ordnance Laboratory. This alloy, commercially known as Nitinol (Nickel Titanium Naval Ordinance Laboratory), exhibits two remarkable effects, namely, "shape memory," wherein the alloy can "memorize" a predetermined shape set at a high temperature and return to this shape under certain temperature conditions, and "pseudo-elasticity," where large strains of the order of 8%–10% can be recovered in

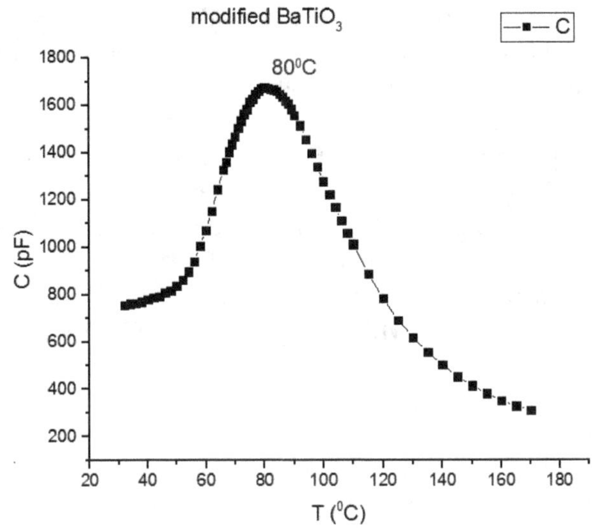

(a)

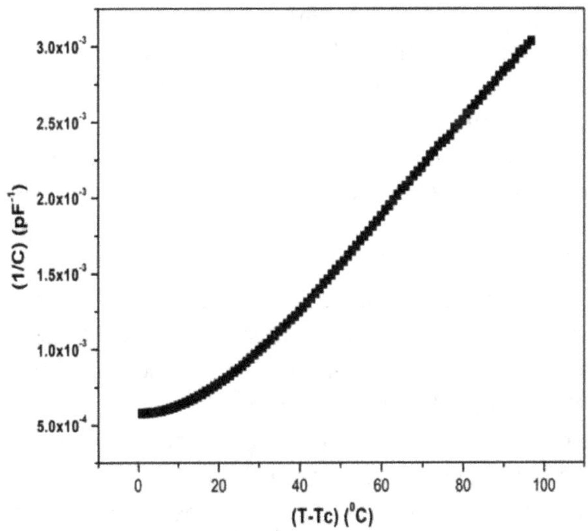

(b)

Figure 15. (a) Plot of capacitance against temperature for a parallel-plate capacitor with modified $BaTiO_3$ as dielectric. (b) Plot of $1/C$ or $1/\chi$ against $(T - T_C)$. Note the breakdown of linear behavior in the region $(T - T_C) < 20°C$.

applications. These unique features have made Nitinol a remarkable engineering material with applications in such diverse fields as cardiovascular surgery, orthodontics, solid-state heat engines, aerospace, and the toy industry.

The serendipitous discovery of the "memory metal" took place at the U.S. Naval Ordnance Laboratory when Buehler noticed a remarkable acoustic damping change of a Ni–Ti ingot with a temperature change near room temperature. This unusual event unfolded when an assistant of Buehler was transporting several melted Ni–Ti bars from an arc furnace to a table. One of the bars which had cooled to near room temperature accidentally fell on the concrete floor and made a dull "thud" sound, while a bar at a higher temperature made a characteristic "metallic sound" on being dropped to the floor. This quick test by Buehler to determine the damping capacity of the Ni–Ti alloy has now come to be known as pseudo-elasticity. Intuitively, Buehler reasoned that the startling change in the acoustic damping must be related to a major atomic structural change, which is related only to minor temperature variation. Further metallurgical studies, such as micro-hardness and microstructure tests, led Buehler and his group to the significant conclusion that in this alloy, major atomic movements occur in a rather low temperature regime near room temperature.

The revelation of the unique shape memory effect in Nitinol came a little later. In the early 1960s, Buehler prepared a long, thin Nitinol strip for use in demonstrations of the material's unique damping properties. The strip was bent into short folds longitudinally, forming a sort of metallic accordion. The strip could be repeatedly compressed and stretched (as an accordion) at room temperature. In a review meeting, this strip was passed around the conference table, and everyone flexed the strip repeatedly. One of the technical directors, who was a pipe smoker, accidentally heated the compressed strip, which, to everyone's amazement, transformed at once into the original longitudinal strip. The mechanical memory discovery, while not made in Buehler's metallurgical laboratory, was the missing piece of the puzzle of the earlier mentioned acoustic damping and other unique changes during temperature variation. This serendipitous discovery became the ultimate payoff for Nitinol. It is now well recognized that the phase change from the low-temperature martensite to the high-temperature austenite occurring near room temperature is responsible for the remarkable shape memory and pseudo-elastic behavior of Nitinol.

The martensite–austenite transformation involves cooperative movement of atoms, unlike nucleation and growth mechanisms commonly encountered in first-order phase transitions. It may be noted that a cooperative movement of atoms through shear transformation can bring about the cubic (high-temperature austenite phase) to monoclinic (low-temperature martensite phase) structural transition. This type of diffusion-less transformation is also known as "military transformation" and is characterized by a low enthalpy change. Furthermore, the martensitic phase is "twinned" so that the overall shape is retained. The volume change accompanying the martensite–austenite first-order phase transition is less than 1%, which emphasizes the subtle nature of this phase transformation. The process of "detwinning" on application of stress in the martensitic phase accounts for the large recoverable strain in this system.

Another unique feature of this transition is that the phase change occurs over a temperature range so that during the heating cycle, one can define the austenite start (A_S) and austenite finish (A_F) temperatures. The martensitic start (M_S) and martensitic finish (M_F) temperatures characterize the reverse transition during the cooling cycle.

12.5.1. Experimental

Electrical resistivity measurements provide a convenient probe to track the first-order martensitic phase transformation in a shape memory alloy. The experimental arrangement essentially consists of a Nitinol wire with spot-welded current and voltage probes. A chromel–alumel thermocouple is also spot-welded at the center for measuring the temperature of the sample.

This sample holder is sandwiched between two nichrome heaters which are connected in parallel. A 10 V, 1.5 A power supply is adequate to heat the sample to around 150°C. A constant current of 10 mA is passed through the current leads from a DC constant-current source. The emf developed across the voltage leads, which is proportional to the resistance of the sample, is amplified in a high-quality DC amplifier (typically with a gain of 100), and its output is read in a digital panel meter. The thermo-emf from the chromel–alumel thermocouple, after CJC, is again amplified in a DC amplifier, followed by linearization circuitry, whose output directly measures the temperature of the sample and is read through the digital panel meter. A typical plot of the resistivity versus temperature data is given in Figure 16.

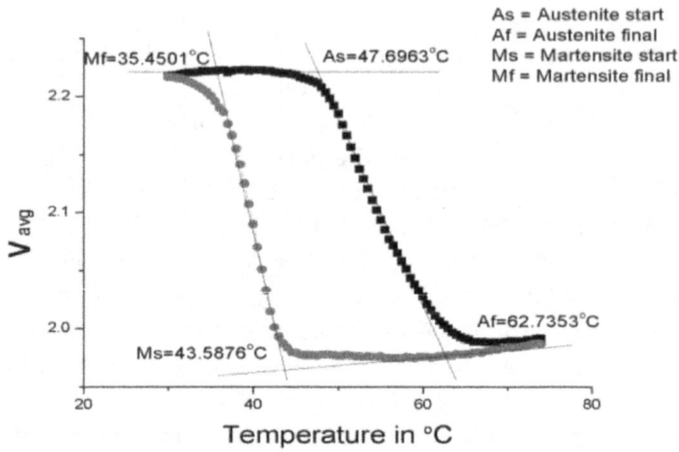

Figure 16. **Martensite–austenite phase transition in nitinol on heating and reverse transformation during cooling cycle. The austenite start (A_s), austenite finish (A_f), martensite start (M_s), and martensite finish (M_f) temperatures can be evaluated using the construction shown.**

References

1. Gitterman, M. and Halpern, V. H. (2004). *Phase Transitions*. World Scientific, Singapore.
2. Stanley, H. E. (1971). *Introduction to Phase transitions and Critical Phenomena*. Oxford University Press, Oxford.

INDEX

Absolute Seebeck coefficient, 321
Temperature variation of, 324
Absolute Seebeck coefficients of constantan and copper, 338
Absolute Seebeck coefficients of n-type and p-type Bismuth telluride, 342
Absorption coefficient of Electromagnetic wave, 357
Absorption spectrometer for IR, UV and Visible, 431
Double beam arrangement, 433
Single beam arrangement, 431
AC resistivity setup, 314
phase sensitive detection, 314
Acceptor state in p-type semiconductor, 304
Advantages of scanning probe microscopes, 158
Allowed energy levels, 292, 299
Amplitude reflection and transmission coefficients, 98
Analysis of Ellipsometric Data, 118
Cauchy dispersion formula, 119
Mean Square Error analysis, 119
Angular correlation in Positron annihilation spectroscopy, 187
Schematic diagram, 189
Applications of Electron Microscopes, 142
Atomic force microscope (AFM), 150
Contact AFM, 153
Intermittent contact AFM, 154
Non-contact AFM, 154
Surface roughness of clean glass plate, 157
Variation with distance of the force between surface and tip, 152
Auger electrons, 137
Auger process, 80

Baker-Jarvis, J., 368
Balanda, M., 382
Band theory of solids, 291
Berman, R., 259
B-H curve in magnetic materials, 374
Coercivity, 375
Hysteresis loop, 375
Retentivity, 374
Saturation of B, 374
Binding energy of an electron, 71
Influence of chemical environment, 73
Influence of valence state, 73
Bloch theorem, 292

Bloch wave function, 293
Bloch wave, 277
Bloch, Felix, 290
Brewster's angle of incidence, 97
Brock, J.C.F., 264

Calibration of the Ellipsometer, 116
Carrier concentration in intrinsic Si, 302
Chemical potential, 287, 474
    for an intrinsic semiconductor, 301
    Variation with temperature of, 478
Clausius-Mosotti relation, 353
Coaxial cylindrical capacitor, 358
Cold junction compensation, 323
Cole-Cole plot for n-Octanol, 363
Collimation of X-rays, 48
Conduction band, 77
Crystal momentum, 293, 294
Curie temperature, 372, 486
Curie's law for magnetic susceptibility, 488

Debye temperature, 291
Defects in solids, 161
    vacancies and interstitials, 161
Density of states, 77, 286
Deposition of Thin films, 23
    Chemical Vapour Deposition (CVD), 28
    Dip coating, 24
    Spin coating, 26
    Pulsed Laser deposition, 33
    Sputtering, 34
Dielectric constant, 349
    Cole-Cole plot, 362
    electronic contribution to, 353
    Free space method with VNA to measure, 367
    Frequency dependence of, 353
    imaginary part, 355
    Impedance Analyzer for measurement of, 358
    LCR meter for measurement of, 358
    of n-Octanol, 362
    of Teflon, 362
    Open ended probe technique with VNA to measure, 366
    orientational contribution to, 353
    real part, 355
    Resonator technique for measuring, 368
    S-parameters with VNA to measure, 364
    Transmission line technique with VNA to measure, 365
    Vector Network Work Analyzer (VNA) for measurement of, 357
    vibrational contribution to, 353
Dielectric Properties, 347
Dielectric susceptibility, 349
Differential Scanning Calorimeter (DSC), 501
    Heat flux DSC, 501
    Power compensated DSC, 503
    Schematic arrangement of setup in Power Compensated DSC, 503
    Schematic diagram of the cell arrangement in Heat flux DSC, 502
Differential Thermal Analysis (DTA), 496
    Block diagram of DTA apparatus, 498
    Experimental setup, 496
    Structural and melting transition of KNO_3, 501
Dipole moment, 348
Dislocations, edge and screw, 123
    Slip plane, 124

Dispersion curve of E vs. k, 292
Distinction between, 296
    metals, insulators and semiconductors, 296
Donor state in n-type semiconductor, 304
Doppler broadening line shape, 180
    variation with Positron energy, 186
Doppler broadening spectrometry in Positron annihilation, 178
    Measurement of, 179
Drift velocity, 279, 280, 288
Drude formula, 280
Drude, Paul, 278
Dynamic methods to measure elastic constants, 203
    Pulse echo method, 203
    Resonance method, 203

Eddy current, 376
Effective mass, 292, 298, 299
Effective mass, 292, 298, 299
    Negative, 292, 299
    Positive, 299
Ehrenfest, Paul, 482
Elastic constants, 199
    Block diagram of the sing around technique to measure, 204
    Pulse superposition method, 205
    Sound velocity, 200
    Vibrating Reed setup, 210
Elastic Properties, 195
Electrical conductivity, 277
    Carrier concentration, 285
    Drude Theory, 278, 281
    Sommerfeld model, 281
Electrical Polarization, 351
    relaxation time, 354
    surface contribution, 353
Electrochemical potential, 328
Electron Diffraction, 129
Electron gas model, 321
Electron magnetic moment, 397
Electron Microscope, 123
    Principle of operation, 128
Electron Paramagnetic Resonance (EPR), 415
    Anisotropy in g value, 421
    Applications of EPR, 425
    Electron-nuclear double resonance (ENDOR), 423
    ENDOR instrument, 424
    Hyperfine interaction, 418
    Hyperfine lines in the EPR spectrum of Benzene, 420
    Pulsed EPR spectrometer, 418
    Spectrometer, 416
    Spin-orbit interaction, 420
Electron scattering by impurities, 290
Electron scattering by phonons, 290
Electron Spectroscopy for Chemical Analysis (ESCA), 71
Electrons, 302
Ellipsometry, 93
    Advantages of, 110
    Basic principles of, 107
    Experimental techniques, 109
    for Thin film analysis, 93
    Null, 110
    PCSA Configuration, 111
    Photometric, 110
    PSA Configuration, 113
    Quantities measured in, 105
    Setup for measurements, 111
    Spectroscopic, 110
Energy band diagram, 333
    n-type semiconductor, 333
    p-type semiconductor, 333
Energy band gap, 292,294
    physical origin, 294
Energy band, 75
Energy dispersive X ray analysis (EDAX), 142

Enthalpy, 481
Entropy, 477
EPR Spectroscopy, 397
ESCA Spectra, 84
    Core level shifts in, 84
    Fixing the Fermi level, 87
    measuring Density of states, 87
    Plasmon peaks in metals, 89
ESCA spectrometer, 81
    Energy analyzer, 83
Ewald's sphere, 52
Exchange interaction, 372
Experimental Techniques for Phase Transition studies, 495
    A.C. resistivity technique, 495
    A.C. specific heat technique, 495
    Capacitance technique, 495
    Differential Scanning Calorimetry (DSC), 495
    Differential Thermal Analysis (DTA), 495
Experimental arrangement to study the para- and ferro-electric transition in modiffied $BaTiO_3$, 507
Extrinsic semiconductor, 297

Fermi energy, 84, 284, 285
Fermi surface, 288, 289
Fermi velocity, 287, 289
Fermi wave vector, 285
Fermi-Dirac distribution function, 286
Fermi-Dirac statistics, 78, 286, 320
Ferrites, 376
Forbidden energy levels, 292
Fourier Transform NMR, 406
Free electron gas, 281
Fresnel's equations for Reflection and Transmission, 94
    for p- polarized state, 96
    for s- polarized state, 96

Gamboa, F, 210
Gibbs Free Energy, 474

Hall coefficient for a doped semiconductor, 303
Hall effect in doped semiconductors, 315
Hall effect setup, 316
Hall voltage, 315, 317
    as a function of magnetic field in n-type Si, 316
High resolution resistivity technique, 504
    Ferromagnetic to paramagnetic transition in Nickel, 504
Hole, 299, 302
Hooke's Law, 199
Hydrothermal synthesis of CrO_2, 13

Ideal resistivity, 290
Image of the surface of Si(100) taken with Non-contact AFM, 156
Infrared (IR) Spectroscopy, 427
    Fourier Transform IR instrument, 433
    Rotational frequencies in the Far Infrared, 430
    Rotational modes of a molecule, 428
    Sample of IR transmission spectrum, 435
    Vibrational modes of a molecule, 429
Initial permeability in a ferro- or ferri-magnetic material, 376
Intensity reflection and transmission coefficients, 98
Intrinsic semiconductor, 297

Jahn-Teller distortion, 460
Jaleel, V.A., 13

Kinetic energy of the photoelectron, 79
Kraus-Rehberg, Reinhard, 162
K_α, K_β X-rays, 48

Landau, Lev, Davidovich, 481
Larmor precession frequency, 399
Lehloo, A.F.D., 467
Levitskaya, Tsylya, M., 359
Liedke, Maciej Oskar, 162
Limitations of X-ray diffraction, 62
Lorentz, H.A., 278
Low Energy Positron Beam Spectrometry, 183

Magnetic force microscopy, 157
Magnetic materials, 372
- anti-ferromagnetic, 372
- B-H Curve tracer, 389
- Diamagnetic, 372
- Domain walls, 375
- Ferrimagnetic, 373
- Ferromagnetic, 372
- Hard, 375
- magnetic permeability, 376
- Paramagnetic, 372
- Soft, 375

Magnetic moment, 372, 385
- Vibrating Sample Magnetometer (VSM) to measure, 383

Magnetic Resonance - Principle, 398
Magnetic susceptibility, 371
- AC susceptibility technique to measure, 380
- Curie Law, 372
- Faraday balance to measure, 378
- Gouy balance to measure, 376
- methods of measuring, 376
- SQUID magnetometer to measure, 386
- Torsion balance to measure, 379

Mahmood, S.H., 467
Main components of an Ellipsometer, 100
Maxwell-Boltzmann Statistics, 278
Maynard, J, 212
Mean free path, 280
Microstructure of the material, 126
Mobility, 302
- for electrons and holes in Si, 303
- for electrons and holes in InSb, 303

Molecular Beam Epitaxy (MBE), 40
Monochromatization of X-rays, 48
Mossbauer Spectroscopy, 447
- Application to Mixed valence states, 461
- Co^{57} source, 452
- Doppler shift, 451
- Emitter source, 451
- Experimental setup, 452
- Isomer shift, 454
- Nuclear hyperfine splitting in BCC Fe, 466
- Nuclear Quadrupole splitting in Fe^{57}, 457
- Nuclear Quadrupole splitting, 455
- Recoil-less photon, 449
- Resonant absorption, 448
- Shift in energy levels, 453
- Splitting due to Hyperfine interaction, 465

Mossbauer Spectrum of $FeCN_5(NO)^{2-}$, 459
Mossbauer Spectrum of $FeSO_4$-$7H_2O$, 460

Neel temperature of an antiferro- or ferri- magnetic material, 374, 376
Negative Seebeck coefficient, 321
- n-type semiconductor, 327

Neutral and inversion temperature of Iron-copper thermocouple, 339
Neutron Diffraction, 63
  Advantages of, 65
  Coherent and incoherent scattering in, 63
Neutron Powder Diffraction, 47
Neutron powder diffractometer, 68
NMR Spectroscopy, 397
Nuclear magnetic moment, 398
Nuclear Magnetic Resonance (NMR), 400
  Chemical shift, 402
  Integrated intensity, 404
  Multiplicity of peaks, 405
  Proton resonance, 402
  Pulsed NMR to measure relaxation time, 408
  Relaxation Phenomenon, 407
  Relaxation time, 408
  Schematic for pulsed NMR measurement, 411
  setup for, 401
  Spin echo, 413
  Spin-lattice relaxation time, 409
  Spin-spin relaxation time, 410
Null Ellipsometry Method, 114
Number of states in a band, 296

Parallel plate capacitor, 358
Peltier coefficient, 323
Peltier coolers, 334
Peltier effect, 323
  setup for, 323
Peltier, Jean Charles Athanase, 323
Phase Diagram, 478
  Boiling Line, 479
  Clausius-Clapeyron Relation, 479
  Coexistence curve, 479
  Critical point, 479
  Magnetic phase diagram in Weiss theory, 492
  Melting line, 478
  Sublimation line, 479
  Symmetry aspects, 481
  Triple point, 479
Phase equilibrium system, 477
  Single component system, 477
Phase equilibrium, 477
  Boiling point, 478
  Melting temperature, 478
Phase Transitions, 473
  Classification of, 482
  Continuous, 483
  Critical exponent, 485
  Critical opalescence, 485
  Critical Point Phenomenon, 485
  Ferroelectric to paraelectric, 473
  Ferromagnetic to paramagnetic, 473
  First order, 482
  Latent Heat, 481
  Order-disorder, 473
  Superconducting, 474
  Superfluidity in liquid Helium, 474
  Weiss Theory of paramagnetic to ferromagnetic, 486
Phonon-phonon scattering, 254
  normal process, 255
  Umklapp process, 255
Photoelectric effect, 78
Photoelectron spectrum, 80
Photometric Ellipsometry Method, 115
Physical basis for Seebeck coefficient, 327
Plasma assisted CVD, 29
Polarizability of a molecule, 353
Positive Seebeck coefficient, 321
  p-type semiconductor, 327
Positron Annihilation spectroscopy, 161
  Angular correlation, 163

- Doppler frequency shift, 163
- Positron density distribution, 165
- Positron lifetime, 163
- Schematic diagram, 164

Positron Annihilation techniques, 168
Positron lifetime measurements, 169
- Schematic diagram, 170
- Select examples of defect studies, 172

Positron lifetime vs. vacancy cluster size, 167
Positron sources, 168
Precursor method to prepare Barium Titanate, 8
Preparation of Nanomaterials, 13
- Bio-assisted methods, 20
- Flash spray pyrolysis, 14
- High energy ball milling, 17
- Laser pyrolysis, 16
- Microemulsion technique, 17
- Polyol process, 18

Preparation of Solid state materials, 3
- Combustion synthesis, 11
- High pressure synthesis, 12
- Precursor method, 8
- Sol-gel method, 9
- Solid state Reaction Technique, 4

Principle of Powder Diffraction, 51
Propagation of Electromagnetic waves in a medium, 356
Proton NMR signal in Lysozyme, 407
- Fourier transform of signal, 407
- Time domain signal, 407

Proton Resonance spectrum of methyl-terbutyl ether, 404

Ravindran, T.R., 264
Refractive index, 356
Relative Seebeck coefficient, 320
Relaxation time, 280, 332
Residual resistivity, 290
Resistivity measurement, 306
- AC technique, 314
- as a function of temperature, 313
- collinear four probe technique, 307
- in bulk samples, 306
- in thin films and pellets, 308
- in wires or thin strips, 310
- van der Pauw technique, 311

Resolution in a microscope, 127
Rotating Analyzer Ellipsometry, 115
Rotating Polarizer Ellipsometry, 115

Scanning Electron Microscopy (SEM), 134
- Schematic diagram, 136

Scanning Tunneling Microscope (STM), 146
- Constant current mode, 148
- Constant height mode, 148
- Reconstruction of Si(111) surface, 149

Schematic diagram of an optical and an Electron microscope, 127
Schematic diagram of Quasi-adiabatic calorimeter, 225
Schematic layout of a Scanning probe microscope, 145
Schrodinger wave equation, 282
Seebeck coefficient in Metals, 331
Seebeck coefficient, 327
- Differential method of measuring, 342
- Integral method of measuring, 334
- Landauer-Datta-Lundstrom formalism, 329

Seebeck effect, 319
Seebeck emf analyzer, 335
Seebeck, Thomas Johann, 319
Semiconductor
- Doping, 297

Shape Memory Alloy, 508
  Experimental setup to study phase transition in, 511
  Martensite to Austenite Phase transition, 508
Shear modulus, 201
Some examples of SEM images, 141
Some examples of TEM images, 134
Sommerfeld, Arnold, 281, 283, 287
Sound wave, 201
  Longitudinal, 201
  Transverse, 201
Specific Heat, 219
  1-τ relaxation method to measure, 228
  2-τ relaxation method to measure, 229
  A.C. Calorimetry to measure, 230
  at constant pressure Cp, 219
  at constant volume C_V, 219
  behaviour at the λ transition in liquid helium, 221
  Debye temperature, 220
  Debye theory, 219
  electronic contribution to, 221
  of YBCO, 226
  phonon contribution to, 221
  Procedure for measuring, 221
  Quasi-adiabatic calorimetry, 224
  Relaxation method to measure, 227
Sputter Deposition System, 37
  Diode, 37
  Magnetron sputtering, 39
  RF sputtering, 39
Sputtering Yield, 36
Sputtering, 34
  Rate of, 35
Stacking Fault, 125
State of Polarization SoP, 94
  Circularly polarized light, 94
  Elliptically polarized light, 94
  p- Polarized state, 96
  Plane polarized light, 94
  s- Polarized state, 96
  Unpolarized light, 94
Static methods to measure elastic constants, 202
Sternberg, Ben,K., 359
Strain tensor, 195, 196
Sundar, C.S., 162
Surface Probe Techniques, 145
Symmetry leading to splitting of atomic energy levels, 457

Techniques to measure thermal expansion, 237
  Interferometric method, 237
  Linear Voltage Differential Transformer method, 244
  Three terminal capacitance technique, 241
  X-ray technique, 248
Temperature variation of absolute and relative Seebeck coefficients of Chromel and Alumel, 342
Temperature variation of Specific Heat of Nickel, 233
Thermal conductivity, 253
  electron contribution, 257
  of copper ETP, 258
  of LiF, 259
  of YBCO, 265
  phonon contribution, 257
  Setup for good conductor, 261
  Setup for poor conductor, 263
  Steady state Method to measure, 259
Thermal diffusivity, 253
  Laser Flash method, 271
  Non-steady state methods to measure, 270
  Thermal wave method, 270

Thermal evaporation, 30
  Distribution of film thickness, 31
thermal expansion, 235, 237
  coefficient of linear expansion, 235, 236
  coefficient of volume expansion, 235, 236
Thermocouple, 320
  Chromel-Alumel, 335
  Inversion temperature, 335
  Iron-Copper and Iron-silver, 335
  Neutral temperature, 335
  single junction, 323
  two- junction, 320
Thermodynamic equilibrium, 473
Thermodynamics, 324
  applied to Seebeck and Peltier effects, 324
Thermoelectric power generators, 334
Thermoelectricity, 319
Thomson coefficient, 326
Thomson effect and Kelvin relations, 326
Thomson, William, 326
Three phonon scattering process, 255
Transmission Electron Microscopy (TEM), 130
  imaging and diffraction modes, 132
Tunneling phenomenon, 147

Universal Testing Machine to measure stress-strain curve, 197

Valence band, 78
van der Pauw, 311
Van der Waals equation of state, 483
Variation of magnetic susceptibility with temperature, 374
Vinothini, V, 3
Visible, UV Spectroscopy, 427
  Absorption and Fluorescence spectra of Anthracene, 441
  Absorption spectroscopy, 437
  Fluorescence and Phosphorescence, 439
  Raman Spectrometer, 443
  Raman Spectroscopy, 441
  Raman spectrum of Carbon-di-sulphide, 444
  Schematic of a Fluorescence spectrometer, 441

Wang,Q., 29
Weiss, Pierre-Ernest, 486

X-ray absorption edge, 49
X-ray Detectors, 50
X-ray Powder Diffraction pattern, 53
  Factors affecting, 53
  Indexing of, 55
X-ray Powder Diffraction, 47
  Applications, 58
  Data bank, 58
X-ray Powder Diffractometer, 51
  Sample preparation, 51

Young's Modulus, 201

www.ingramcontent.com/pod-product-compliance
Lightning Source LLC
LaVergne TN
LVHW010325220926
848484LV00004BB/842